OLIGONUCLEOTIDE ARRAY SEQUENCE ANALYSIS

OLIGONUCLEOTIDE ARRAY SEQUENCE ANALYSIS

MARTINA K. MORETTI

AND

LORENZO J. RIZZO

EDITORS

Nova Science Publishers, Inc.

New York

LIBRARY OF CONGRESS CATALOGING-IN-PUBLICATION DATA

Oligonucleotide array sequence analysis / Martina K. Moretti and Lorenzo J. Rizzo (editors).
p. ; cm.
Includes bibliographical references and index.
ISBN 978-1-60456-542-3 (hardcover)
1. DNA microarrays. 2. Oligonucleotides. 3. Nucleotide sequence. I. Moretti, Martina K. II. Rizzo, Lorenzo J.
[DNLM: 1. Oligonucleotide Array Sequence Analysis. QU 450 O46 2008]
QP624.5.D726O45 2008
572.8'636--dc22

2008013483

Published by Nova Science Publishers, Inc. ✢ New York

CONTENTS

PREFACE

Oligonucleotides are short sequences of nucleotides (RNA or DNA), typically with twenty or fewer bases. Automated synthesizers allow the synthesis of oligonucleotides up to 160 to 200 bases. The length of a synthesized base is usually denoted by 'mer' (from 'Greek' *meros* "part"). For example, a fragment of 25 bases would be called a 25-mer. Oligonucleotides are often used as probes for detecting complementary DNA or RNA because they bind readily to their complements. Examples of procedures that use oligonucleotides are DNA microarrays, Southern blots, fluorescent *in situ* hybridization (FISH), and the synthesis of artificial genes. This book brings together leading research in the field from around the globe.

Haplotypes of multiple genes have been reported their association with risk of adverse skin reactions after radiotherapy in breast cancer patients (*Int. J. Radiat. Oncol. Biol.* 2007, 69, 685-693). In this first Expert Commentary, a visible genotyping array device for a total of eight haplotype-tagSNPs in two genes (*PTTG1* and *CD44*) among above has been developed using previously established methodology (*Anal. Sci.* 2006, 22, 1537-1545). The device uses allele-specific extension reaction of covalently immobilized oligonucleotide primer to discriminate nucleotide at SNP site. Each primer oligonucleotide has 3'-end locked acid nucleotide modification to enhance its allelic specificity. Biotin-dUTPs provided in the reaction mixture are selectively incorporated only to the extenting oligonucleotides, enabling their use as tags for following streptavidin-conjugated alkaline phosphatase-mediated colored precipitation of substrates onto the surface of the array. Resulting colored spots on the device can be observed by the naked eye and conveniently recorded by a digital camera that is commercially available at reasonable cost. Allelic discrimination of all immobilized oligonucleotides was validated using 45 reference individuals previously genotyped by MassARRAY system. Reliability of genotyping by this device was quantitatively assessed by calculating Silhouette score, and resulting scores of all SNPs were proven to be beyond the cut-off value of 0.65. The eight haplotype-tagSNPs can be simultaneously analyzed in a single reaction mixture within a couple of hours, and multiple samples up to four can be analyzed simultaneously on a single device whose surface area is divided by a water-proof multiwell seal. This visible haplotype-tagSNP typing array device has intended benefits in convenient usage at reasonable cost, and thus it would provide easy access for bedside clinical diagnosis of radiation sensitivity.

Microarray hybridization specificity is one of the main measures of microarray result quality. In the second Expert Commentary, the authors consider oligonucleotide microarray hybridization specificity by first characterizing it at four different levels of hybridization between the oligo-probe and a target molecule (i.e., a single oligo-probe, a single cell, a cell-set and a microarray platform); the effect of low specificity on the obtained results will be suggested for each of these specificity levels, and across them. Second, specificity effectors will be presented and their effect on hybridization specificity will be summarized. Third, several example studies will be presented: these studies have either modelled or quantified the specificity effectors, or used data filtration based on specificity effectors, for increased specificity of oligonucleotide microarray hybridization results. Due to the complexity of microarray hybridization specificity, an ultimate solution to increasing hybridization specificity, taking into account its different effectors and levels, is not yet available. Rather, further developments are needed to accurately define and quantify microarray hybridization specificity.

The rapid development of microarray technology has led to important progress in many biomedical disciplines. In just over a decade, the technology has evolved from first small-scale prints to commercial and custom platforms of truly genomic scale. Though numerous types of microarray platforms exist, cDNA and oligonucleotide-based DNA microarrays are considered the platforms of choice for the range of applications. However, the success of every study depends upon a careful selection of a platform appropriate for a designed experiment. In Chapter 1, various technological and analytical aspects of microarray technology are discussed; some advantages and drawbacks of individual platform types are compared.

The spotlight for developing clinically relevant biomarkers has recently focused brightly on gene expression profiling in peripheral blood mononuclear cells (PBMCs). Since PBMCs circulate throughout the body via the blood stream, their gene expression can be altered by disease states even when the disease is localized to a less accessible compartment within the human host. PBMC gene expression may therefore represent a disease barometer that can be used to assess human health, and inquiries into this phenomenon have given rise to a burgeoning field known as *hemogenomics*. The clinical accessibility of blood by venipuncture, and the ease with which PBMCs can be isolated, add to the attractiveness of this tissue source for monitoring human health. Most studies of PBMC gene expression to date have identified genes expressed differentially between healthy and diseased patients, permitting the construction of potentially diagnostic gene classifiers. In addition, apparently prognostic gene classifiers have been developed using samples taken prior to the development of a particular disease state or the initiation of a therapeutic intervention. Gene classifiers have been developed for a variety of ailments such as: autoimmune diseases, cancer, chronic fatigue syndrome, hypertension, infectious diseases, neuropathy, post-traumatic stress disorder, toxic chemical exposures, and transplant tolerance. Chapter 2 will review multiple aspects of PBMC gene expression profiling including: PBMC isolation, storage considerations, sample preparation, gene expression analysis, and gene classifier construction and evaluation. A review of previous PBMC gene expression studies is presented to illustrate the current *status quo* in this particular field of biomarker development. Finally, the future of PBMC gene expression profiling is discussed with specific emphasis on its impact for medical practices and clinical care.

Finding relevant patterns in time-series gene expression data has been studied for quite a few years. Genes with similar expression profiles are expected to be functionally related or co-regulated, while in some cell cycles, like yeast and serum, it is quite important to group the genes based on different phases during the regulation process. Clustering gene expression data, given in terms of time series, is different from the general clustering problem, because exchanging two or more time points delivers quite different results, while doing this could not be biologically meaningful at all. As a result of this, traditional clustering methods, more specifically, conventional similarity measures, are not suitable in this context, and one has to resort to specialized methods. As the number of genes profiled in an experiment is large, clustering based on the similarity of their temporal profiles is a daunting task. Various clustering methods for time-series gene expression data have been proposed. While some of the proposed methods are more effective than others in some particular contexts, so far, none of them have been identified as *the best* technique for all scenarios. Clustering approaches for analyzing microarray time-series data are discussed and some key features of these methods are highlighted. Special challenges, advantages and drawbacks of the methods are discussed and an experimental comparison of some of these methods is also given in Chapter 3.

As explained in Chapter 4, much interest has been generated over the past few years aimed at microbial discovery using state-of–the-art technologies. Microbial communities of uncultured microorganisms are no longer an obstacle for the advance or understanding of 99% of microorganisms. Metagenomics, or culture-independent genomic analysis, has joined DNA microarray technology in recent years to discover many bacterial gene functions and identification in the environment. Biosensors and molecular subtyping as well as phage recombinant probe technologies continue to advance our knowledge in detecting and classifying bacterial isolates. These molecular tools are mainly dependent on DNA or protein sequence analysis. Our abilities to use these tools are helping us to detect and trace bacterial populations found in nature.

Genomic tiling-path microarray has become an economical and prompt way to efficiently investigate the functional DNA elements in many forms, such as coding/non-coding transcripts, transcription factor binding sequences, DNA content variation, DNA methylation, histone methylation, acetylation and other numerous essential epigenetic events that contribute to gene transcription regulation and chromatin organization. With the increasing of tiling array application in many fields of biology, various methods related to tiling array processing and analysis have been developed and evolved from simple assessments to more scientific model-based mathematical algorithms; however, a unified and regularized procedure must be established. Chapter 5 provides a general review of the application and statistical methodology involved in tiling array analysis from the following aspects:

Firstly, from an initial report of human ENCODE project, the importance of the application of tiling array in post-genomics era will be stressed. Secondly, some basic classifications of representative tiling array platforms with different definitions and several algorithms for selecting an optimal tile path of probes to cover a reference genome during tiling array design will be introduced. The third part relates to the methods and technical issues in preprocessing of tiling array raw data including standard within- and between-slide normalizations, spatial smoothing normalizations of unevenly distributed probe signals across array surface, background correction methods and some ChIP-chip, array-CGH specific normalizations methods. The fourth part will be focused on how to statistically infer the interrogated genomic regions with significantly enriched tiling array signals. A number of well-recognized statistical models and algorithms will be introduced, such as hidden Markov model (HMM), Wilcoxon signed-rank test, Model-based Analysis of Tiling-arrays (MAT) and dynamic programming algorithm, etc. In the last part, a brief introduction of the tiling array analysis software and online resources will be given.

Alcoholic and nonalcoholic liver diseases are increasingly recognized conditions that may progress to end-stage liver disease. Non alcoholic liver disease (NAFLD) and alcoholic liver disease (ALD) share similar pathological characteristics despite the etiological difference between two diseases. In both diseases, the first manifestation of injury is the accumulation of fat within hepatocytes (steatosis), followed by the development of necrosis and inflammation that leads to cirrhosis. Although NAFLD is a metabolic disease associated with obesity and insulin resistance, it is impossible to distinguish alcoholic from nonalcoholic steatohepatitis on histological grounds. This suggests that common pathogenic mechanisms are involved in both conditions. In addition to alcohol consumption, genetic factors, dietary fatty acids, gender and viral infection potentiate the severity of alcoholic liver injury. A high fat diet in rat and mice models reproduces many of the features of NAFLD found in humans. Female rats fed with fish oil and dextrose developed fatty liver, necrosis and inflammation. In an intragastric infusion rat model of ALD, saturated fatty acids (medium chain triglycerides) are protective, whereas unsaturated fatty acids (palm oil and fish oil) promote ALD. The rats fed with palm oil and ethanol developed only fatty liver, whereas rats fed with fish oil and ethanol developed fatty liver, severe necrosis and inflammation. In ALD and NAFLD, female rats developed more severe liver injury than males. In humans, significant gender differences in susceptibility to ALD are observed. The proposed molecular mechanism through which polyunsaturated fatty acid and ethanol develops pathogenesis includes enhanced lipid peroxidation-induced oxidative stress, mitochondrial dysfunction, endotoxin, and proinflammatory cytokines. To understand the effect of dietary fatty acids and gender on the development of more severe ALD and NAFLD in female rats, the authors performed affymetrix microarray based expression profiling of genes. A large number of genes showed significant changes in female livers compared to males. The authors employed EASE and Ingenuity computational pathway analysis tools to analyze the microarray data. Ingenuity computational pathway analysis tool identifies specific regulatory networks of genes operative in promoting liver injury. These networks allowed identification of genes involved in different signaling pathways for a better understanding of the role of cytokines in ALD in response to different types of fatty acids, the authors used inflammatory cytokines pathway specific superarray. Many cytokine genes underwent significant changes in the liver of rats fed with palm oil plus ethanol, and fish oil and ethanol as compared to animals fed with medium chain triglycerides and ethanol, Genes listed in Chapter 6 that significantly increased or decreased (expression two-fold or more) were assigned to pathological functional groups and reviewed in relevance to established hypotheses of NAFLD and ALD mechanisms of action in female liver injury.

The availability of information on genome sequences has prompted researchers to clarify the molecular pathogenesis and molecular mechanisms underlying diseases, predict the toxicity of drugs, and realize personalized therapy immediately. The available information can help in achieving these ends; however, the process that would be involved for doing this is not very simple. Under these circumstances, DNA microarray, one of the technologies that employ this information, has become a well-established and useful tool for life scientists, with improvement in its sensitivity and probe design, etc. since the post-genome era. Nevertheless, researchers would be interested in seeking options for obtaining comprehensive gene expression data for their studies through a simple and inexpensive process. Further, even though such comprehensive data are acquired, the difficulties involved in the processing and interpretation of the large quantities of the data remain unresolved. The author and colleagues have therefore developed a life science information-orchestrated platform termed KeyMolnet that focuses on "molecular relations." This system consists of content databases collecting life science information, and a search software that can generate molecular networks based on the contents. When importing experimental and/or clinical data such as DNA microarray data into the system, molecular networks can be generated based on these data. Since the various kinds of life science information collected in the KeyMolnet, such as that on biomolecules, established pathways, diseases and drugs, are mutually related through key molecules, researchers could obtain working hypotheses, suggestions, and new insights from the generated molecular networks based on their data. In Chapter 7, the author demonstrates new approaches toward analyzing and profiling DNA microarray data for medical, pharmaceutical, and biological applications in practice by using molecular networks reflecting molecular mechanisms.

Generalized ANOVA, including fixed and mixed ANOVA, are widely used statistical models in analyzing microarray experiments from various platforms. The flexibility of this approach allows investigators to simultaneously assess the contributions of multiple factors and their interactions to gene expression variation. It could account for all sources of variation in the complex experiment and provide easy interpretation of results. In the representative oligonucleotide platform Affymetrix GeneChips, each gene is represented by one or more probe sets (termed sibling probe sets) and each probe set is represented by multiple probes. The complicated many-to-one representation poses critical challenges to data analysts. While summary methods that provide an "average" measure of multiple probes to be the probe set expression score are useful, the mixed model ANOVA analysis using probe level expression data greatly increases sample size and hence increases power in identification of differentially expressed genes. Chapter 8 surveys some representative works in probe-level analysis of Affymetrix oligonucleotide data. For genes that are represented by multiple probe sets (sibling probe sets), differentially expressed genes between conditions could be identified using ANOVA by treating the sibling probe sets as a factor along with other interesting factors in the designed experiment. Expression levels of sibling probe sets that are proportional to each other across different conditions are more likely to be replicated measures of the same target gene. The decision on whether to consolidate sibling probe sets was made through statistical testing of the interaction effect in the ANOVA model. In the authors' recent work of applying ANOVA for sibling probe set analysis, results show marked improvement of gene expression estimates and identification of more biologically relevant genes. The authors demonstrate this advantage by analyzing three publicly available datasets and validate it using Gene Ontology (GO) terms enrichment analysis.

The complete sequencing and annotation of the *Arabidopsis thaliana* genome represented a major step in biological research. This knowledge enabled gene prediction, assignment of functional categories and gave an opportunity to characterize global gene expression patterns at the transcriptome level at different developmental stages and under various physiological and stress conditions. For the discovery of partial or whole regulatory or functional networks, the development of high-throughput technologies was inevitable with genome-wide transcriptomic studies providing an essential input. DNA microarray technologies are among the most frequently used methods for parallel global analysis of gene expression. Microarray technologies now belong to the standard functional genomics toolbox and have undergone massive development leading to increased genome coverage, accuracy and reliability. Whole genome microarrays developed by Affymetrix in collaboration with Syngenta represented the first standard in genome-wide transcriptomic studies in plants. Whole genome Affymetrix ATH1 GeneChips covers about 76% of the *Arabidopsis thaliana* genome. Moreover, the introduction of the Minimum Information About Microarray experiments (MIAME) has increased the value and reproducibility of microarray experiments and has become a standard in documentation of array experiments and in the creation of databases of comparable transcriptomic experiments. The number of experiments exploiting microarray technology has markedly increased in recent years. Not surprisingly, there are potential difficulties in the orientation in available data sets. Microarray expression data are deposited on servers, many of which are publicly accessible. Currently, these databases store several thousands of individual datasets and some of these offer on-line tools for data normalization, filtering, statistical testing and pattern discovery. In parallel with the rapid accumulation of transcriptomic data, on-line analysis tools are being introduced to simplify their use. Global statistical data analysis methods contribute to the development of overall concepts about gene expression patterns and to query and compose working hypotheses. More recently, these applications are being supplemented with more specialized products offering visualization and specific data mining tools. In Chapter 9, an overview of available online databases and web-based applications using microarray data is presented together with the information about their structure and elementary principles.

As presented in Chapter 10, cancer is a disease resulting from the breakdown of several checkpoints and tumor-suppressing mechanisms. In cancer research, the development of new technologies have produced genomic tools indispensable for understanding how gene products are regulated in normal and diseased conditions on a global genome scale; one of these technologies are the DNA arrays. Although the most common use of DNA arrays is gene expression profiling and mutation detection, scientists have successfully used them for multiple applications, including genotyping, re-sequencing, DNA copy number analysis and DNA-protein interactions.

This section will be dedicated to the use of a public sequence database that can be accessed, and the design of DNA oligonucleotide probes for oligoarrays derived from sequences of special interest. Single probe and stacking hybridization are explored as possible microarray designs. Then the use of thermodynamic models and *in silico* hybridization are explored in order to access the sensitivity and specificity of the oligoarray and how its design can be improved. However, many commercial and public applications do not consider that the hybridization between target DNAs and microarray probes is a chemical reaction which is influenced by several thermodynamic parameters. One of the most important of such parameters is the thermal stability of the nucleic acid duplexes, which are formed as a result of the hybridization. According to the reaction conditions, these duplexes can be perfectly or imperfectly paired, which is of critical importance when a diagnostic kit is developed in order to assess its sensitivity and specificity.

In silico hybridization can be used to avoid undesired hybridization events (such as multiple target-probe interactions and stable mismatched hybrids). Also, the strategy can facilitate (through use of different length probes) selection of a probe set that has a narrow duplex stability, allowing maximal specificity under a single hybridization condition.

Nonparametric methods for detecting differentially expressed (DE) genes in microarrays have attracted much attention recently. One of the most widely-used methods is the Significance Analysis of Microarrays (SAM) method. Empirical studies have shown that SAM can provide results highly consistent with biological findings. However, it has been recently reported that SAM tends to over-estimate the true false discovery rate (FDR). The cause of over-estimation was attributed to the over-dispersed null scores obtained by applying the SAM statistic to the permuted observations of DE genes. More recently, questions about SAM's choice of fudge factor have also been raised. It has been demonstrated in the literature that the fudge factor used in SAM does not generally have good power properties. In Chapter 11, the author proposes an improved method for detecting DE genes in microarrays. The authors' proposed method is based on a pair of recently-constructed test statistic and null statistic. Unlike the SAM statistic, the authors show that the new null statistic does not have the permutation-caused over-dispersion problem. The author further argues that, instead of using a global fudge factor, it is advantageous to use a local fudge factor that depends on the number of significant genes. The local fudge factor which minimizes the FDR is called the local optimal fudge factor. Contrary to the common understanding that the over-dispersed null scores are solely caused by the permutations, we demonstrate that the over-dispersion is also an intrinsic problem caused by the use of the fudge factor. In other words, the use of fudge factor will necessarily lead to over-dispersed null scores, and hence results in over-estimation of FDR. A two-stage FDR correction method is proposed to resolve this issue. Through extensive simulations, the author shows that the proposed method consistently outperforms SAM in the sense that it tends to provide lower FDR given that they have identified the same number of significant genes. The improvement of the author's proposed method is achieved by the use of more efficient test and null statistics, the local optimal fudge factor and a two-stage FDR correction. A real data example also shows that the proposed method provides more reasonable results and better control of FDR than SAM and a symmetric cutoff method based on the SAM statistic.

Information on protein expression levels can be used to gain insights about putative physiological gene function, for identifying target proteins that show differential expression in tumor versus normal cells and for formulating hypotheses regarding possible side effects of target inhibition.

However, obtaining systematic information at protein level is not trivial and currently standardized high throughput protein methods are not available. Gene expression levels, analyzed with microarray (MA) technology producing genome-wide information at the mRNA level, can partially compensate for the lack of information at protein level. Indeed, microarray-based transcription profiling is a consolidated methodology commonly used in areas such as pharmacogenomics, diagnostics and drug target identification.

Several high-density oligonucleotide array platforms are currently commercially available, the best known and most-widely used being Affymetrix. Affymetrix offers the largest panel of gene expression microarrays, which are available for a number of different organisms. Besides classical expression arrays, Human Gene/Exon 1.0 ST is a new emerging technology, suited for identifying both differential gene and exon expression allowing to infer alternative splicing events. It has recently been applied to the discovery of potentially interesting new targets for diseases such as cancer.

In Chapter 12, the authors review tools and methods for data analysis of both microarray transcription profiling and alternative splicing analysis.

An approach for multiplex qualitative and quantitative microarray-based PCR analysis has been proposed. The characteristics of PCR executed on a gel-based oligonucleotide microarray with immobilized forward primers and a single common reverse primer in solution were investigated for several DNA targets. One-stage multiplex on-chip PCR was studied for simultaneous amplification of herpes simplex viruses types 1 and 2, cytomegalovirus DNA, and bacteriophage lambda DNA as an internal control. Additionally, the joint analysis of increased number of targets (with addition of *Chlamydia trachomatis*, *Mycoplasma hominis*, and *Ureaplasma urealyticum* DNA) was done in a two-stage version of assay: the first stage was in-tube PCR with target-specific primers, while the reverse ones contained 5'-adapter region; the second stage was on-chip amplification with immobilized target-specific forward primers and adapter as common reverse primer in solution. The possible application of one-stage reaction for human cDNA analysis was additionally demonstrated with utilization of a common poly-T-containing primer in solution. SYBR green I and Cy-5 labeled dUTP were used for real-time and end-point detection of specific PCR products. The efficiencies of both one-stage and two-stage reactions were shown to be strongly dependent on magnesium and primers concentrations. Quantitative PCR in both versions was studied with 10-fold serial dilutions of phage lambda DNA. The method enabled detection of six DNA copies per reaction for both versions of assay. The quantitative interval for one-stage reaction covered eight orders of concentration. The revealed significant effect of gel pad size on microarray PCR effectiveness is discussed in Chapter 13.

Historically, the majority of microarray work has been restricted to well-defined model organisms. This was primarily due to the limited availability of genomic or transcriptomic sequence data and the then high cost involved in developing microarrays. However, recent technological developments have greatly enhanced the speed of generating the underpinning sequence data for non-model species, and have opened up more cost-effective approaches for microarray production to make them far more affordable for researchers at the lower end of the budget range. These developments have been seized upon by the environmental genomics community within the UK. The creation of a network of closely integrated facilities for sequencing, microarray printing and bioinformatics has opened the gateway for the study of environmentally relevant organisms. In Chapter 14, the authors describe the infrastructure for microarray development within the UK, and the diverse applications for which they are currently being used.

In biological sequence research, the positional weight matrix (PWM) is often used for motif signal detection. A set of experimentally verified oligonucleotides known to be functional subsequences, which can be bound by a transcription factor (TF), as translational initiation sites or pre-mRNA splicing sites, are collected and aligned. The frequency of each nucleotide A, C, G, or T at each column of the alignment is calculated in the matrix. Once a PWM is constructed, it can be used to search from a nucleotide sequence for the subsequences that possibly perform the same function. The match between a subsequence and a PWM is usually described by a score function, which measures the closeness of the subsequence to the PWM as compared with the given background. However, selection of threshold scores that legitimately qualify a functional subsequence has been a great challenge. Many laboratories have attempted to tackle this problem, but there is no significant breakthrough so far. In Chapter 15, the authors discuss the characteristics of a PWM and factors that affect motif predictions and propose a new score function that is tied into information content and statistical expectation of a PWM. The authors also apply this score function in the PWMs from public databases and compare it favorably with the broadly-used Match method.

The use of oligonucleotide microarrays to obtain gene expression measurements is a complex process with many potential sources of variability. Leaving apart those variations due to the processes or treatments under study, variability may arise from the sample preparation and hybridization or even from the analysis system itself. Subsequent changes in the hybridization values may obscure those changes due to the biological processes under study and thus lead the researchers to attain erroneous conclusions about gene expression phenomena. It becomes evident that any gene expression analysis should be preceded by a quality control of the hybridization data ensuring that following analyses are developed with appropriate and comparable data. To deal with this necessity, a number of QC methods have been developed, both to be applied before or after hybridizing the RNA to the microarray. These methods have been shown to differ in the obtained results and in the source of variability explored, making it difficult to choose the most appropriate QC measurement. The present contribution will provide the reader with a detailed and up-to-date review of the most important quality control methods available for Affymetrix oligonucleotide microarrays, the most widely used microarray platform. Chapter 16 begins with a description of all recognized sources of variability in microarray analysis followed by a brief review of the RNA quality analyses. Then follows a detailed description and discussion of the quality controls of the hybridization data, including information on the available bioinformatic tools.

Highly multiplex PCR or RT-PCR protocols have remained an elusive goal of molecular diagnostics almost since the introduction of nucleic acid amplification technologies. Implementation of these assays has generally faced technical challenges on three fronts: loss of sensitivity due to competition for limited reagents by multiple targets in a single reaction; loss of specificity due to an exponentially increasing number of possible primer interaction artifacts; and difficulty in distinguishing among multiple different end products. A recently-developed methodology by Genaco (now a Qiagen company) and marketed under the 'QIAplex' name is a combination of technologies which look to overcome these difficulties. Assay parameters which limit issues of sensitivity and specificity shortcomings are coupled to liquid-phase hybridization oligonucleotide bead array detection on the Luminex platform. The approach lends itself well to mid-to high-multiplicity (10 - 50 targets or more) PCR and RT-PCR assays. Along with new capabilities, this form of assay brings its own specific challenges, particularly in selecting and implementing appropriate control systems. Early studies on the clinical application of this method have been highly successful. Chapter 17 describes the underlying principles and technologies of the method, aspects of its clinical implementation in one context, ways to address the issue of appropriate controls, and future potential of the method. Two similar commercialized systems sharing the bead array hybridization approach are also briefly described.

As explained in Chapter 18, microarray is an exciting and promising technology for sequencing analysis, high-throughput mutation detection, SNP genotyping and gene expression profiling. Through the simultaneous analysis of thousands of genes, microarray technology is characterized by large-scale, highly-automatic and sensitive detection of genetic data at both qualitative and quantitative levels.

The most recent clinical application of microarray technology is aimed at DNA genotyping. A hot medical demand is the high-throughput scanning of cancer related DNA molecules, such as endogenous cytoplasmatic mitochondrial DNA (mtDNA) and exogenous

DNA (i.e, viral genomes). Both mtDNA and viral genomes are characterized by a small size and a high exon/intron ratio, suggesting a perfect match with the microarray requirements.

Mitochondria are cytoplasmatic organelles that play a vital role in energy metabolism, producing the ATP required for cellular functions. Despite a poor genotype/phenotype correlation, complex human disorders are increasingly being linked to mtDNA mutations that effect mitochondrial function. Mutations in mtDNA can be considered as useful biomarkers associated with several diseases, such as cancer, aging and metabolic disease. Mutations throughout the mitochondrial genome are recurrent events in primary tumor tissues and in corresponding non-invasively collected body fluids. Thus, mtDNA mutation analysis may provide a molecular tool for the early detection and prognosis of cancer.

Other than mtDNA, exogenous DNAs have been reported to be correlated to cancer onset. Hepatocellular carcinoma is one of the most common cancers throughout the world. Several viral oncogenes, such as the Hepatitis C Virus (HCV) core protein, are related to hepatocarcinogenesis. HCV is a serious, worldwide public health problem affecting an estimated 150–200 million individuals. HCV is a remarkably successful pathogen, establishing persistent infection in more than two-thirds of those who contract it. Its success is related to its abilities to blunt innate antiviral pathways and to evade adaptive immune responses. Up to 20% of the infected patients develop cirrhosis or liver cancer. There is currently no effective vaccine. Although antiviral therapies have improved considerably in recent years, they are associated with significant toxicity and are effective in only about one-half of those treated. Developing a specific microarray for HCV detection could have a direct effect on diagnostic applications.

Thus, the availability of such in-silico technology and the identification of potential cancer biomarkers in small genomes, such as mtDNA and HCV, could enable predictive cancer detection at early stages, improving patient management and treatment choices.

As presented in Chapter 19, the past decade has seen the initiation, development and subsequent adoption of DNA analysis in forensic science. Universally accepted standard procedures and databases are in place to organize the analysis of DNA in a manageable fashion. The potential to improve and enhance these current resources exists, and as technology continues to develop, the boundaries of our current capabilities are constantly being challenged. Improvements and enhancements are needed; a backlog of forensic DNA samples is present throughout the United States, and efforts are underway to alleviate the strain associated with the delays in sample processing. In addition to the sample backlog, situations exist in which current DNA typing methodologies are sub-optimal, such as those examining extensively degraded material or those seeking better resolution of the Y-chromosome. Genetic information is continually being made available, based on the sequencing and haplotyping of the human genome. This information, along with new technology, has identified other genetic polymorphisms with the potential to better address some of the current problems and solve the key issues involved in the field of DNA analysis in forensic science. Technologies employing higher throughput than the standard methods employed are better suited to capitalize on the limitless potential garnered from the sequence information and population data now available. Microarray analyses are one such format positioned to exploit this potential.

EXPERT COMMENTARIES

In: Oligonucleotide Array Sequence Analysis
Editors: M.K. Moretti and L.J. Rizzo, pp. 3-14

ISBN: 978-1-60456-542-3

VISIBLE HAPLOTYPE-TAGSNP TYPING ARRAY DEVICE FOR HUMAN RADIATION SENSITIVITY-ASSOCIATED GENES

Yuichi Michikawa**, Tomo Suga, Atsuko Ishikawa, Yoshimi Ohtsuka, Mayumi Iwakawa and Takashi Imai

RadGenomics Project, Research Center for Charged Particle Therapy, National Institute of Radiological Sciences, Chiba, Japan

ABSTRACT

Haplotypes of multiple genes have been reported their association with risk of adverse skin reactions after radiotherapy in breast cancer patients (*Int. J. Radiat. Oncol. Biol.* 2007, 69, 685-693). In this study, a visible genotyping array device for a total of eight haplotype-tagSNPs in two genes (*PTTG1* and *CD44*) among above has been developed using previously established methodology (*Anal. Sci.* 2006, 22, 1537-1545). The device uses allele-specific extension reaction of covalently immobilized oligonucleotide primer to discriminate nucleotide at SNP site. Each primer oligonucleotide has 3'-end locked acid nucleotide modification to enhance its allelic specificity. Biotin-dUTPs provided in the reaction mixture are selectively incorporated only to the extenting oligonucleotides, enabling their use as tags for following streptavidin-conjugated alkaline phosphatase-mediated colored precipitation of substrates onto the surface of the array. Resulting colored spots on the device can be observed by the naked eye and conveniently recorded by a digital camera that is commercially available at reasonable cost. Allelic discrimination of all immobilized oligonucleotides was validated using 45 reference individuals previously genotyped by MassARRAY system. Reliability of genotyping by this device was quantitatively assessed by calculating Silhouette score, and resulting scores of all SNPs were proven to be beyond the cut-off value of 0.65. The eight haplotype-tagSNPs can be simultaneously analyzed in a single reaction mixture within a couple of hours, and multiple samples up to four can be analyzed simultaneously on a single device whose surface area is divided by a water-proof multiwell seal. This visible haplotype-tagSNP typing array device has intended benefits in convenient usage at reasonable cost, and thus it would provide easy access for bedside clinical diagnosis of radiation sensitivity.

* E-mail address: y_michi@nirs.go.jp. Tel: 011-81-43-206-3072. Fax: 011-81-43-206-6267. Correspondence to: Yuichi Michikawa, RadGenomics Project, Research Center for Charged Particle Therapy, National Institute of Radiological Sciences, Anagawa 4-9-1, Inage, Chiba, 263-8555, Japan

INTRODUCTION

It has been increasingly recognized that preclinical prediction of radiation sensitivity of normal tissues surrounding tumor of individual cancer patients is essential to perform effective radiation therapy [1-5]. However, determinants of variations in severity of damage to normal tissues after radiation therapy have not yet been fully understood [6-7]. Genetic background of patients has been postulated to participate at certain degree to the above variations [8-11].

In the previous study, we identified haplotypes of single nucleotide polymorphism (SNP) markers in multiple genes that associated with the risk of early adverse skin reactions in breast cancer patients after radiotherapy [12]. A total of 399 breast cancer patients were enrolled in this study between 2001 and 2005. Statistical examination of a total of 999 SNPs on 137 genes of these patients revealed five risk-reducing haplotypes in genes such as *MAD2L2*, *PTTG1*, *RAD9A* and *LIG3*. A risk-increasing haplotype was found in *CD44* gene.

To validate clinical importance, it is necessary to perform secondary association study of these haplotype markers with larger number of patients. For this purpose, the MassARRAY SNP typing system [13] that we used to analyze large number of SNPs in the previous study is not an adequate choice of method. Since this system has been optimized for high throughput analysis in terms of SNP number, there are disadvantages in experimental processes, required operation time and costs for analysis of limited number SNPs. Thus, alternative method optimized for convenient analysis of the pre-selected limited number SNPs has been demanded. It should be desirable if the method is capable to be performed at hospitals without requiring specifically trained skills and additional experimental instruments.

We recently developed a new SNP typing device to meet the above demands [14]. It is a microarray using plastic base with specific chemical modifications on its surface to covalently immobilize aminoacylated oligonucleotides [15]. Genotype discrimination is achieved by allelic specificity-enhanced enzymatic extension of immobilized oligonucleotide primers with a locked nucleic acid (LNA) modification [16] at the 3'-end. Incorporation of multiple biotin-dUTP molecules during the allele-selective primer extension reaction, followed by binding of alkaline phosphatase-conjugated streptavidin allows visible detection of genotypes through precipitation of colored alkaline phosphatase substrates onto the surface of the plastic base. Thus, this device does not need expensive fluorescent scanner or mass spectrometer for detection, providing easy access by hospitals. Furthermore, overall processes can be easily performed within a couple of hours.

In the present study, this new SNP typing device has been successfully applied for the radiation sensitivity-associated haplotype-tagSNPs. A total of eight haplotype-tagSNPs in two genes (*PTTG1* and *CD44*) of the above genes were selected for the device. A water-proof seal was introduced to divide the surface area of the plastic base so that multiple numbers of patients up to four can be simultaneously analyzed by a single array slide. Silhouette score [17] was applied to quantitatively assess reliability of genotyping results by this device.

Materials and Methods

DNA Samples

Genomic DNAs from whole blood of 45 breast cancer patients were selected as references by their genotypes from the 399 patients enrolled in the previous study [12]. They were used for validation of reliability of the device. All the patients have provided written informed consent to participate in the previous study, which was approved by the Ethical Committee at the National Institute of Radiological Sciences and by those at collaborating institutions where they were treated radiation therapy.

PCR Amplification of Haplotype-tagSNP loci

Individual haplotype-tagSNP loci were respectively PCR amplified until saturation in distinct reaction wells. Ten microliters of reaction mixture contained 1 x LightCycler 480 SYBR Green I master mix (Roche Diagnostics, Basel, Switzerland), 10 ng of genomic DNA and 0.2 μM of each PCR primer pair (PCRF/PCRR, listed in Table 1). The reaction mixtures were initially heated at 95°C for 5 min and then subjected to 50 rounds of 2-step temperature cycling (95°C for 10 s and 65°C for 20 s) using LightCycler 480 (Roche Diagnostics, Basel, Switzerland). Then, equal volume of the products from same individuals were mixed together and stored at -20°C until use.

Table 1. Oligonucleotide sequences

Gene	SNP ID	Type	Sequence
PTTG1	rs2910190	PCRF*	5'-ACGTTGGATGTATGTGTAAGGAAGGGTCTC
		PCRR*	5'-ACGTTGGATGTGGAGTTGAGAGAGAAACAG
		Massxt†	5'-CTAACAGTATGTAGAGACCT
		Opcxt1‡	5'-(C6 Amino)CAGTGCTAACAGTATGTAGAGACCTC(LNA)
		Opcxt2‡	5'-(C6 Amino)GAAACAGTGCTAACAGTATGTAGAGACCTT(LNA)
	rs3811999	PCRF	5'-ACGTTGGATGTAGATTCTATTATGTTGGGC
		PCRR	5'-ACGTTGGATGTCCCCCCCATAACGTGTATC
		Massxt	5'-AGCTACCATTCCTGCCT
		Opcxt1	5'-(C6 Amino)GTAAAAGTAGCTACCATTCCTGCCTC(LNA)
		Opcxt2	5'-(C6 Amino)GTAAAAGTAAAAGTAGCTACCATTCCTGCCTT(LNA)
	rs1862391	PCRF	5'-ACGTTGGATGGCTGCTGTTTGATCTACCTG
		PCRR	5'-ACGTTGGATGGGGAAACAACCAATGTTGGG
		Massxt	5'-AATGTTGGGTCTAAAGAATAC
		Opcxt1	5'-(C6 Amino)GGAAACAACCAATGTTGGGTCTAAAGAATACT(LNA)
		Opcxt2	5'-(C6 Amino)GAAACAACCAATGTTGGGTCTAAAGAATACG(LNA)
	rs2961951	PCRF	5'-ACGTTGGATGAAGCACTTCCACTGTACACC
		PCRR	5'-ACGTTGGATGTCCAGGAAAATGTTCTCTAG
		Massxt	5'-CTCTAGTTTTAGCTTGCAACTCA
		Opcxt1	5'-(C6 Amino)GGAAAATGTTCTCTAGTTTTAGCTTGCAACTCAT(LNA)
		Opcxt2	5'-(C6 Amino)GAAAATGTTCTCTAGTTTTAGCTTGCAACTCAG(LNA)

Table 1. Continued

Gene	SNP ID	Type	Sequence
CD44	rs187116	PCRF	5'-ACGTTGGATGGGGGATAAGAGAGTAACTGC
		PCRR	5'-ACGTTGGATGACTTCCCTTTCAGGCTCTTC
		Massxt	5'-GACCTAAAGCGAAGGTGC
		Opcxt1	5'-(C6 Amino)GTGACCTAAAGCGAAGGTGCC(LNA)
		Opcxt2	5'-(C6 Amino)GAGTGACCTAAAGCGAAGGTGCT(LNA)
	rs3794116	PCRF	5'-ACGTTGGATGTGGATGGCATAGATGTCAGG
		PCRR	5'-ACGTTGGATGACTTCCCTTTCAGGCTCTTC
		Massxt	5'-AGGCTCCAGTGCTCCCCA
		Opcxt1	5'-(C6 Amino)CACTCCCAAGCCCTCTGAAC(LNA)
		Opcxt2	5'-(C6 Amino)CCACTCCCAAGCCCTCTGAAT(LNA)
	rs3794107	PCRF	5'-ACGTTGGATGCGCAAGGCCATTGGAAATTC
		PCRR	5'-ACGTTGGATGCCTTCCTCAAACTATGGCTC
		Massxt	5'-AACTATGGCTCCAAGCAAGATAA
		Opcxt1	5'-(C6 Amino)CCTCAAACTATGGCTCCAAGCAAGATAAT(LNA)
		Opcxt2	5'-(C6 Amino)CCTCAAACTATGGCTCCAAGCAAGATAAA(LNA)
	rs8193	PCRF	5'-ACGTTGGATGATGGGTCCATTTTGCCCTTC
		PCRR	5'-ACGTTGGATGTAACTAGTACACCCCAACCC
		Massxt	5'-ACCCCCAACCTCAGTGGAAA
		Opcxt1	5'-(C6 Amino)CAACCCCCAACCTCAGTGGAAAA(LNA)
		Opcxt2	5'-(C6 Amino)AACCCCCAACCTCAGTGGAAAG(LNA)

*, Oligonucleotides designed for amplification of target sequence by PCR. †, Oligonucleotides designed for extension reaction of Massarray SNP typing. ‡, Oligonucleotides designed for allele-specific primer extension on the haplotype-tagSNP typing array.

Design of Allele-Discriminating Oligonucleotide Primers for Haplotype-tagSNPs

A total of eight SNPs reported as radiation sensitivity-associated haplotype-tagSNPs of genes *PTTG1* and *CD44* were used for this study. The allele-discriminating oligonucleotide primers were designed based on the upstream nucleotide sequence of target SNPs as described previously [14]. Briefly, Tm of the primers was calculated to be 60°C by adjusting number of overall nucleotide while the SNP nucleotide always located at the 3'-end. Each oligonucleotide was synthesized with a 3'-end LNA modification for enhancement of allelic discrimination and a 5'-end aminoacylation for covalent attachment to the plastic base. Sequences of the oligonucleotides were listed in Table 1.

Preparation of Haplotype-tagSNP Typing Arrays

The allele-discriminating oligonucleotides (Opcxt1 and Opcxt2 in Table 1) were prepared as 0.03 μM solution in 1 x S-Bio spot solution (Sumitomo Bakelite, Tokyo, Japan). Drops of approximately 50 nl were spotted onto S-Bio Primesurface plastic bases (75 x 25 x 1 mm,

Sumitomo Bakelite, Tokyo, Japan) by MassARRAY Nanodispenser (Sequenom, San Diego, CA). The spotted plastic bases were heated at 80°C for 1 h then a Multiwell Geneframe (19 x 10 mm x 5 wells/frame, ABgene House, Surrey, UK) was placed on each base. The bases were washed in 0.1% Tween-20 at room temperature for 1 min then soaked into 1 x S-Bio blocking solution (Sumitomo Bakelite, Tokyo, Japan) containing 0.1% Tween-20 at room temperature for 5 min. They were washed in 1 x TBS-T (10 mM Tris, pH 7.6, 150 mM NaCl and 0.1% Tween-20) at room temperature for 1 min, then in 0.1% Tween-20 at 80°C for 1 h to reduce amount of surface adhesive chemicals of the Multiwell Geneframe and finally in 0.1% Tween-20 at room temperature for 1 min. They were centrifuged at 100 x g for 1 min and dried at room temperature for 10 min. They were placed in a desiccator and stored at 4°C until use.

Allele-Specific Primer Extension on the Haplotype-tagSNP Typing Array

Fifty microliters of a reaction mixture containing 1 x Mg-free ThermoPol II reaction buffer (New England Biolabs, Beverly, MA), 2.5 units of HotStar Taq DNA polymerase, 10 μM biotin-modified dUTP (Fermentas, Hanover, MD), 10 μM each of normal nucleotide (dATP, dCTP and dGTP), 4mM MgCl2, 1 μl of PCR product mixture. The reaction mixtures were initially heated at 95oC for 15 min, cooled down for a couple of minutes and poured onto the wells on the array. The arrays were placed in prewarmed, humidified plastic boxes. The boxes were then sealed thoroughly with Saran wrap (AsahiKASEI Life and Living, Tokyo, Japan) and incubated at 65oC for 30 min. The arrays were washed in 1 x TBS-T for 1 min. Fifty microliters of alkaline phosphatase-conjugated streptavidin (Bio-Rad Laboratories, Hercules, CA) diluted 100 times in 1 x TBS-T was treated to individual wells of the arrays for 10 min at room temperature. Then the arrays were washed again in 1 x TBS-T. One hundred microliters of 5-bromo-4-chloro-3'-indolylphosphate (BCIP)/nitro-blue tetrazolium chloride (NBT) (Perkin Elmer Optoelectronics, Fremont, CA) with 0.1% Tween-20 were treated to the individual wells of the arrays and incubated at room temperature for 30 min. The arrays were washed in 0.1% Tween-20 for 1 min then centrifuged at 100 x g for 1 min, then dried at room temperature for 10 min.

Array Image Capture and Data Processing

The image of each well on the array was captured by Nikon D70 digital camera (Nikon, Melville, NY). The 16-bit RAW format RGB files were converted to 8-bit grayscale TIFF format files by Adobe Photoshop Version 6.0 software (Adobe Systems, San Jose, CA). The signal intensities of individual spots on the array were quantified by the Daredemo DNA Array Kaiseki software version 1.0 (Dynacom, Chiba, Japan). Silhouette scores for cluster assessment of each haplotype-tagSNP typing oligonucleotides was calculated by ClusterA software [17].

RESULTS

Experimental Setup

A total of four haplotype-tagSNPs (rs2910190, rs3811999, rs1862391 and rs2961951) in *PTTG1* gene and four haplotype-tagSNPs (rs187116, rs3794116, rs3794107 and rs8193) in *CD44* gene associated with clinical radiation sensitivity (12) were selected for this study. Among them, 5 SNPs (rs2910190:G/A, rs3811999:G/A, rs187116:G/A, rs3794116:G/A and rs8193:T/C) were transitions and 3 SNPs (rs1862391:A/C, rs2961951:A/C and rs3794107:A/T) were transversions. Their locations on corresponding genetic regions are shown in Figure 1. *PTTG1* haplotype-tagSNPs spanned 11.8 kb of chromosome 5 whereas *CD44* haplotype-tagSNPs spanned 85.5 kb of chromosome 11. Allele-discriminating oligonucleotides with 3'-end LNA for individual SNPs were designed based on previously established method [14] and listed in Table 1. For the purpose of validating allelic discrimination efficiency of the synthesized oligonucleotides, genomic DNAs derived from a total of forty five breast cancer patients were selected as references based on their genotypes investigated by MassARRAY method [13] in the previous study [12].

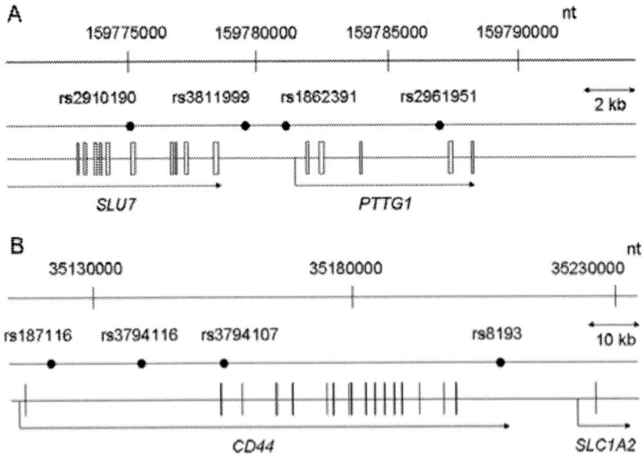

Figure 1. Map of haplotype-tagSNPs. A: *PTTG1* locus on chromosome 5. B: *CD44* locus on chromosome 11. Locations of SNPs were indicated by closed circles. Exons were indicated by open squares. Arrows indicate regions from transcription initiation site to the 3'-end of transcript.

Validation of Synthesized Oligonucleotides by Real-Time Quantitative PCR

Initially, allelic discrimination efficiency of synthesized oligonucleotides was investigated using the reference DNAs by real-time quantitative PCR (data not shown) as described previously [14]. Individual reference DNAs were amplified by allele 1-discriminating primer pairs (PCRF/Opcxt1) or allele 2-discriminating primer pairs (PCRF/Opcxt2), respectively. All of the primer pairs showed distinct amplification kinetics depends on genotype of template DNA, probing effectiveness of their allele specificity.

Allele-Specific Primer Extension on the Haplotype-tagSNP Typing Array

The allele-discriminating oligonucleotides were then immobilized on plastic bases and their allelic discrimination efficiencies were validated using the reference DNAs. Figure 2 shows the images of plastic arrays after reaction and an example of magnified spot images. All of the eight haplotype-tagSNPs of same individual were analyzed simultaneously in a single well of the Multiwell Geneframe. Conventional scatter plots of signal intensities measured for individual spots represented good separation of clusters made by data points (Figure 3). All alleles (8 alleles x 2 copies x 45 DNAs = 720 alleles) were correctly assigned to definite clusters. Format of the scatter plots of these data points were modified according to ref. 17 and shown in Figure 4. Two dimensional clusters in the conventional scatter plots were converted to one dimensional ones. Silhouette scores were then calculated to quantitatively assess the allelic discrimination efficiency of individual SNPs and shown in Table 2. The average Silhouette score was calculated to be 0.849 (±0.082) and all of SNPs on the device were proven to have higher Silhouette score than the recommended cut off score of 0.65 [17].

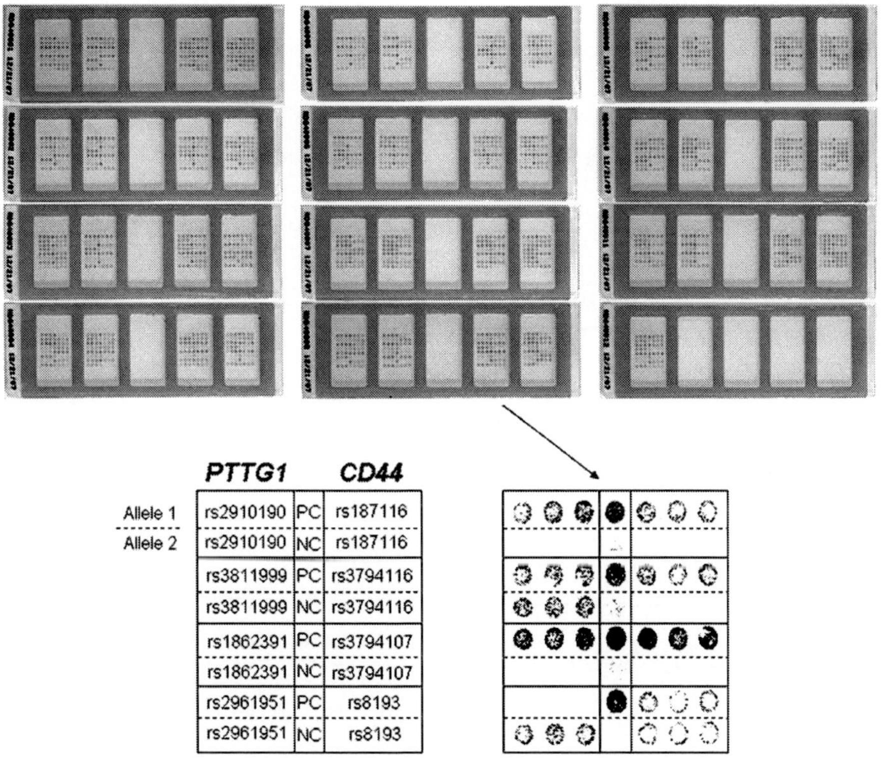

Figure 2. Simultaneous genotyping by allele-specific primer extension on the haplotype-tagSNP typing array. Digital camera-captured images of plastic arrays used for analysis of 45 patients and one example of enlarged spot image were shown. PC: positive control oligonucleotide (14). NC: Negative control oligonucleotide [14].

Table 2. Silhouette scores of individuals SNPs by allele-specific primer extension of the haplotype-tagSNP typing array

Gene	SNP	Type	Mean			Variances			F-Statistics	Silhouette score
			Allele1 homo	Heterozygote	Allele2 homo	Allele1 homo	Heterozygote	Allele2 homo		
PTTG1	rs2910190	G/A	0.326	0.507	0.686	3.784E-04	9.194E-05	1.733E-04	1760.058	0.895
PTTG1	rs3811999	G/A	0.328	0.502	0.668	6.118E-04	7.309E-05	3.293E-04	682.387	0.840
PTTG1	rs1862391	A/C	0.295	0.500	0.675	2.543E-04	2.421E-05	4.123E-04	1690.582	0.905
PTTG1	rs2961951	A/C	0.346	0.472	0.663	4.401E-04	1.067E-03	6.502E-04	374.753	0.667
CD44	rs187116	G/A	0.313	0.476	0.669	9.343E-04	3.120E-04	5.043E-04	680.115	0.802
CD44	rs3794116	G/A	0.324	0.515	0.735	7.522E-04	1.670E-05	1.157E-04	338.374	0.900
CD44	rs3794107	A/T	0.289	0.486	0.687	1.768E-04	1.662E-04	5.721E-04	1905.103	0.890
CD44	rs8193	T/C	0.273	0.490	0.693	4.704E-04	2.076E-04	1.940E-04	1785.150	0.891

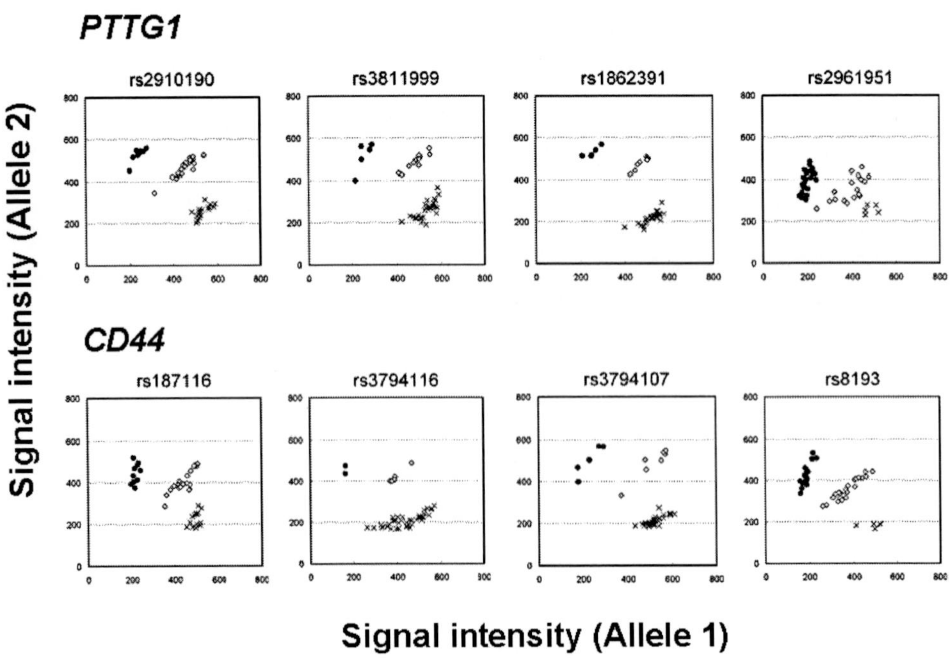

Figure 3. Conventional scatter plots of allele-specific primer extension on the haplotype-tagSNP typing array. Signal intensities of spots on the plastic array shown in Figure 2 were calculated and plotted. X-axis: Signal ntensity of allele-1 discriminating oligonucleotide. Y-axis: Signal intensity of allele-2 discriminating oligonucleotide. Crosses: Allele 1 homozygotes determined by MassARRAY. Open diamonds: Heterozygote determined by MassARRAY. Closed circles: Allele 2 homozygotes determined by MassARRAY.

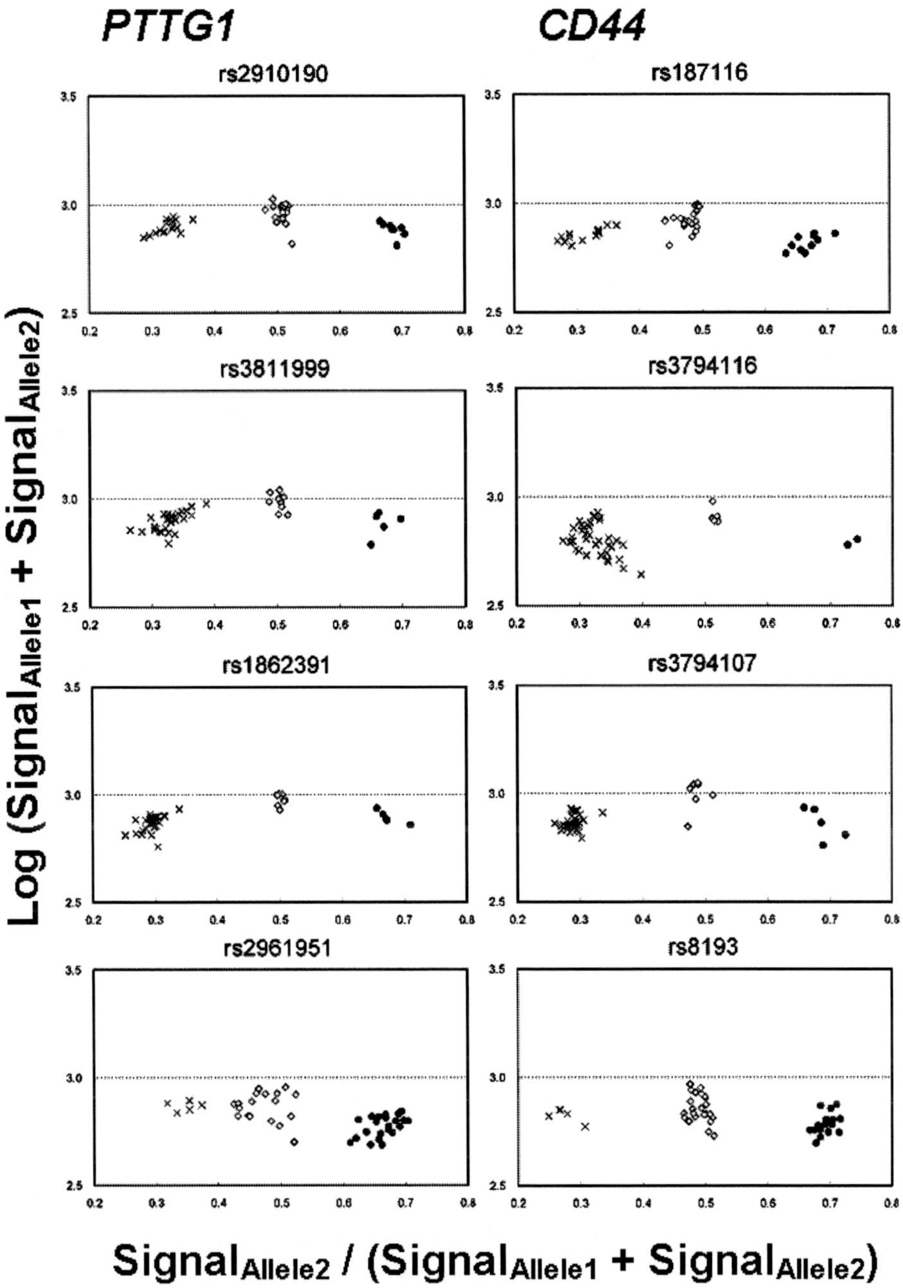

Figure 4. Modified scatter plots of allele-specific primer extension on the haplotype-tagSNP typing array. Signal intensity fraction [Signal intensity$_{Allele2}$ / (Signal intensity$_{Allele1}$ + Signal intensity$_{Allele2}$)] was plotted on the x-axis and the logarithm of the signal intensities of both alleles [Log (Signal intensity$_{Allele1}$ + Signal intensity$_{Allele2}$)] was potted on the y-axis. For Silhouette score calculation, the distance from each data point in a cluster to all other data points within the same cluster and to all data points in the closest cluster were determined by ClusterA software. Crosses: Allele 1 homozygotes determined by MassARRAY. Open diamonds: Heterozygotes determined by MassARRAY. Closed circles: Allele 2 homozygotes determined by MassARRAY.

Discussion

A plastic array device for simultaneous typing of radiation sensitivity-associated haplotype-tagSNPs on *PTTG1* and *CD44* has been successfully developed in this study. This device can be used with only conventional laboratory instruments such as a constant temperature incubator and a digital camera. Overall time required was less than 1h from applying template PCR products onto the device to getting the visible images. Up to four patients can be simultaneously analyzed by a single device and multiple numbers of devices can be conveniently handled in a single operation. These features and, especially, the visibility of results would provide easy access to most of hospitals.

The average Silhouette score for individual SNPs on the device was calculated to be 0.849 (±0.082). Lovmar *et al* reported that the Silhouette score>0.65 to be the cut off value to consider reliable discrimination, though the Silhouette score falling between 0.25-0.65 may be still accepted (17). The Silhouette score of the all SNPs were beyond 0.65 as shown in Table 2 thus typing results of this device should be considered reliable.

Benefits of this array device come from its convenient utilization at reasonable cost. Visible spots on the device are quite clear as shown in Figure 2 and enable easy and immediate observation of the results. Furthermore, easy operation and no requirement of specific experimental instruments would provide easy access for the radiotherapy hospitals where genotyping is not currently operating yet.

Radiation sensitivity prediction of cancer patients has been expected not only in wealthy advanced countries but also in developing countries to spare resources related to cure patients (18). In general, genetic approaches to improve human healthcare have been increasingly demanded in such not wealthy developing countries (19). It was idealized that DNA would be analyzed at local sites on demand rather than at remote sites after transporting of human materials. The array device system we have developed may participate as one of choices for the above purpose.

Acknowledgement

The authors acknowledge Ms. Masayo Terada for her assistance in manuscript preparation.

References

[1] Guirado D, Ruiz de Almodóvar JM. Prediction of normal tissue response and individualization of doses in radiotherapy. *Phys. Med. Biol.* 2003, 48, 3213-3223

[2] Mackay RI, Hendry JH. The modelled benefits of individualizing radiotherapy patients' dose using cellular radiosensitivity assays with inherent variability. *Radiother Oncol.* 1999, 50, 67-75

[3] Bentzen SM, Hendry JH. Variability in the radiosensitivity of normal cells and tissues. Report from a workshop organised by the European Society for Therapeutic Radiology and Oncology in Edinburgh, UK, 19 September 1998. *Int J Radiat Biol.* 1999, 75, 513-517

[4] Hill RP, Rodemann HP, Hendry JH, Roberts SA, Anscher MS. Normal tissue radiobiology: from the laboratory to the clinic. *Int J Radiat Oncol Biol Phys*. 2001, 49, 353-365

[5] West CM, McKay MJ, Hölscher T, Baumann M, Stratford IJ, Bristow RG, Iwakawa M, Imai T, Zingde SM, Anscher MS, Bourhis J, Begg AC, Haustermans K, Bentzen SM, Hendry JH. Molecular markers predicting radiotherapy response: report and recommendations from an International Atomic Energy Agency technical meeting. *Int J Radiat Oncol Biol Phys*. 2005, 62, 1264-1273

[6] Begg AC. Can the severity of normal tissue damage after radiation therapy be predicted? *PLoS Med*. 2006, 3, e440

[7] Dikomey E, Borgmann K, Peacock J, Jung H. Why recent studies relating normal tissue response to individual radiosensitivity might have failed and how new studies should be performed. *Int J Radiat Oncol Biol Phys*. 2003, 56, 1194-1200

[8] Iwata M, Iwakawa M, Noda S, Ohta T, Minfu Y, Kimura T, Shibuya H, Imai T. Correlation between single nucleotide polymorphisms and jejunal crypt cell apoptosis after whole body irradiation. *Int J Radiat Biol*. 2007, 83, 181-186

[9] Fernet M, Hall J. Genetic biomarkers of therapeutic radiation sensitivity. *DNA Repair*. 2004, 3, 1237-1243

[10] Andreassen CN, Alsner J, Overgaard M, Overgaard J. Prediction of normal tissue radiosensitivity from polymorphisms in candidate genes. *Radiother Oncol*. 2003, 69, 127-135

[11] Andreassen CN, Alsner J, Overgaard J. Does variability in normal tissue reactions after radiotherapy have a genetic basis--where and how to look for it? *Radiother Oncol*. 2002, 64, 131-140

[12] Suga T, Ishikawa A, Kohda M, Otsuka Y, Yamada S, Yamamoto N, Shibamoto Y, Ogawa Y, Nomura K, Sho K, Omura M, Sekiguchi K, Kikuchi Y, Michikawa Y, Noda S, Sagara M, Ohashi J, Yoshinaga S, Mizoe J, Tsujii H, Iwakawa M, Imai T. Haplotype-based analysis of genes associated with risk of adverse skin reactions after radiotherapy in breast cancer patients. *Int J Radiat Oncol Biol Phys*. 2007, 69, 685-693.

[13] Buetow KH, Edmonson M, MacDonald R, Clifford R, Yip P, Kelley J, Little DP, Strausberg R, Koester H, Cantor CR, Braun A. High-throughput development and characterization of a genomewide collection of gene-based single nucleotide polymorphism markers by chip-based matrix-assisted laser desorption/ionization time-of-flight mass spectrometry. *Proc. Natl. Acad. Sci. USA*. 2001, 98, 581-584

[14] Michikawa Y, Fujimoto K, Kinoshita K, Kawai S, Sugahara K, Suga T, Otsuka Y, Fujiwara K, Iwakawa M, Imai T. Reliable and fast allele-specific extension of 3'-LNA modified oligonucleotides covalently immobilized on a plastic base, combined with biotin-dUTP mediated optical detection. *Anal Sci*. 2006, 22, 1537-1545

[15] Kinoshita K, Fujimoto K, Yakabe T, Saito S, Hamaguchi Y, Kikuchi T, Nonaka K, Murata S, Masuda D, Takada W, Funaoka S, Arai S, Nakanishi H, Yokoyama K, Fujiwara K, Matsubara K. Multiple primer extension by DNA polymerase on a novel plastic DNA array coated with a biocompatible polymer. *Nucleic Acids Res*. 2007, 35, e3.

[16] Vester B, Wengel J. LNA (locked nucleic acid): high-affinity targeting of complementary RNA and DNA. *Biochemistry* 2004, 43, 13233-13241

[17] Lovmar L, Ahlford A, Jonsson M, Syvänen AC. Silhouette scores for assessment of SNP genotype clusters. *BMC Genomics*. 2005, 6, 35
[18] Bhadrasain V. Radiation therapy for the developing countries. *J Cancer Res Ther*. 2005, 1, 7-8
[19] Editorial of Nature Genetics. Developing genetics for developing countries. *Nat Genet*. 2007, 39, 1287

In: Oligonucleotide Array Sequence Analysis
Editors: M.K. Moretti and L.J. Rizzo, pp. 15-24
ISBN: 978-1-60456-542-3

SPECIFICITY OF OLIGONUCLEOTIDE MICROARRAY HYBRIDIZATION: EFFECTORS AND THEIR EFFECTS

Hinanit Koltai [a,*] ***and Carmiya Weingarten-Baror*** [b]

[a] Department of Ornamental Horticulture, ARO Volcani Center, Bet Dagan, Israel

[b] Department of Neurobiology, Weizmann Institute of Science, Rehovot, Israel

ABSTRACT

Microarray hybridization specificity is one of the main measures of microarray result quality. In the present mini-review, we will consider oligonucleotide microarray hybridization specificity by first characterizing it at four different levels of hybridization between the oligo-probe and a target molecule (i.e., a single oligo-probe, a single cell, a cell-set and a microarray platform); the effect of low specificity on the obtained results will be suggested for each of these specificity levels, and across them. Second, specificity effectors will be presented and their effect on hybridization specificity will be summarized. Third, several example studies will be presented: these studies have either modelled or quantified the specificity effectors, or used data filtration based on specificity effectors, for increased specificity of oligonucleotide microarray hybridization results. Due to the complexity of microarray hybridization specificity, an ultimate solution to increasing hybridization specificity, taking into account its different effectors and levels, is not yet available. Rather, further developments are needed to accurately define and quantify microarray hybridization specificity.

INTRODUCTION

Microarray hybridization is a powerful tool for profiling gene expression; the obtained results, however, are largely influenced by several main measures of microarray hybridization. These may be defined as accuracy ('the degree of conformity of the measured quantity to its actual [true] value'; [1]), sensitivity ('the concentration range within which accurate measurements can be made'; [1]), specificity ('the probe's ability to bind to a unique target sequence'; [1]) and reproducibility ('the degree to which repeat measurements of the same quantity will show the same or similar results'; [1]). Accuracy, sensitivity, and reproducibility are largely discussed by Dufva [2] and Draghici et al [1], and will not be

[*] E-mail address: hkoltai@agri.gov.il (Corresponding Author)

covered here. The present review will focus on the measure of oligonucleotide microarray specificity, as one of the key factors for obtaining valid results; microarray hybridization specificity is discussed for all microarray types in [3]. Nevertheless, the relations between the different measures will also be noted.

Microarrays consist of multiple probes. Short-oligomer microarrays (e.g. Affymetrix Genechips®, NimbleGen®, Agilent® and Febit®) may be either custom-designed or ready-made. Some of the array types are fabricated in situ, on the microarray slide (e.g., Affymetrix), whereas other types are spotted on the slide (e.g., Agilent). Oligomers on the microarrays (i.e., oligo-probes) are usually arranged as probe-sets for the detection of a single target sequence. Probe-sets may include probes that represent different segments of a target sequence (e.g., different exons of a unigene), or probe pairs of perfectly matched and mismatched probes: the perfectly matched probe is designed to be complementary to a target sequence; the mismatched probe is designed to be complementary to the target sequence except for a one-base mismatch (e.g., Affymetrix microarrays).

A key factor for specificity during microarray hybridization is the probe's ability to discriminate between different target molecules. Therefore, a probe with high specificity to its target molecule is expected to provide a signal that is influenced only by the presence of that molecule.

Although probes are designed to be complementary to target molecules according to the Watson-Crick rules of binding, a perfect match between a probe and its target molecule, in terms of sequence complementary, is not an assurance of specificity. This is due to the presence, during microarray hybridization, of thousands of target molecules, each of which is composed of hundreds or thousands of nucleotides. Hence, during hybridization of a microarray probe to a target molecule, a variable number of target molecules are hybridized to a microarray probe which is not strictly complementary to them, or *vice versa* (a variable number of microarray probes are hybridized to a target molecule that is not strictly complementary to them).

In the present review, we will consider oligonucleotide microarray hybridization specificity by characterizing the specificity of the hybridization process for targeting molecules at four different levels of hybridization (i.e., a single oligo-probe, a single cell, a cell-set and a microarray platform). For each of these levels, we will outline the effect of low specificity on the obtained results. Then, specificity effectors will be presented, and their effect on hybridization specificity will be summarized for the different levels of the hybridization and post-hybridization procedures. Next, we will present several example studies which have either modeled or quantified specificity effectors, or used data filtration based on modelling or quantification of specificity effectors, for increased specificity of oligonucleotide microarray hybridization results. Lastly, some future prospects will be discussed, with respect to the need for further developments aimed at improving microarray hybridization results.

Four Levels of Microarray Hybridization Specificity and the Effect of Low Specificity on the Obtained Results

Four levels of microarray hybridization specificity can be defined for oligonucleotide microarrays. These include the oligo-probe level (i.e., a single oligomer); the cell level (i.e., multiple and identical oligo-probes); the cell-set level (i.e., cells with perfect matches or mismatches to the target sequence, or cells that represent different segments of a target sequence), and the microarray platform level (i.e., multiple cells or cell-sets, depending on the oligo-array platform design) (Figure 1).

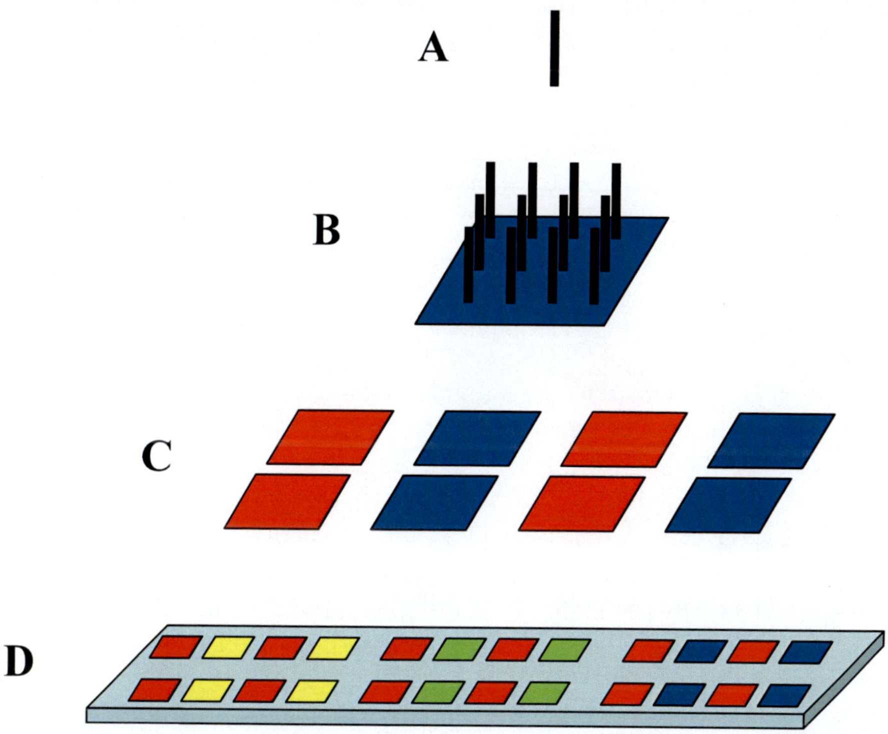

Figure 1. Schematic illustration of the defined units of an oligonucleotide array. (A) A single oligo-probe molecule; (B) a single cell composed of multiple oligo-probes; (C) a cell-set composed, for example, of matched (blue) and mismatched (red) cells; (D) a microarray platform composed of multiple cell-sets, each cell-set including matched (yellow, green or blue) and their mismatched (red) cells, and representing a particular target sequence.

The basic and most fundamental determinants lie within the level of hybridization of a single oligo-probe molecule and a single target molecule (Figure 1A). This hybridization may be perfect, partial, or non-existent. Since the hybridization signal is usually detected at the cell level (detailed below), partial or no hybridization between oligo-probes and target molecules might be detected at the cell level as a low or no signal, respectively.

The second level of specificity is that of the cell (Figure 1B), in which multiple oligo-probe molecules that compose one cell are hybridized to multiple target molecules. Here, the cell may exhibit perfect, partial or no hybridization with the target molecules. At the cell level, partial hybridization might be extensively introduced, due to 'cross-hybridization' of non-target molecules with some of the oligos present in the cell. Hence, a single cell may possess, simultaneously, oligo-probes that are perfectly hybridized to the target sequence, partially hybridized, by cross-hybridization, to non-target sequences, and not hybridized to the target molecule. This multiplicity of hybridization-specificity states making up the signal of a single cell may lead to a poor reflection of the amount of target molecule, which may be hard to trace due to the multiplicity of specificity states (e.g., both high- and low-specificity hybridizations non-separately composing a cell signal).

The third level of specificity is that of the cell-set (or, in Affymetrix terminology, 'probe-set'; Figure 1C), in which multiple cells represent the same reference sequence, but through different parts of its sequence (e.g., different exons of a gene). At this level, different cells of a cell-set may exhibit different forms of hybridization with the target molecule: perfect hybridization, partial hybridization or no hybridization. In some microarray designs, some of the cells are designed to perfectly match the target sequence, and hence exhibit perfect hybridization with the target molecule, and some to have a mismatch with the target sequence and hence, partial or no hybridization with the target molecule. Partial or no hybridization to target molecules may also be exhibited by cells with perfectly matched oligos. In this case, cross-hybridization may occur due to, for example, a perfectly matched cell, within a cell-set, that represents an evolutionarily conserved gene segment; this cell may hybridize with non-target molecules derived from various gene-family members.

Hence, at the cell-set level, some of the cells may be hybridized by perfect hybridization with the target molecules and some by partial hybridization to the target molecules. Since the determination of a target molecule's concentration at the cell-set level is a function of the cell-set's signal, results will be incorrect (i.e., lower than the true target molecule concentration; see, for example, [4]).

The fourth level of specificity is that of the microarray platform (Figure 1D), in which a variable number of cell-sets may exhibit different forms of hybridization with the target sequences: perfect hybridization, partial hybridization, and no hybridization may be present for a number of the cell-sets. Once cells are hybridized by cross-hybridization (e.g., target molecules of different genes hybridize to the same cell-set or target molecules of a particular gene hybridize to several different genes' cell-sets), competition between cell-sets may occur on the microarray platform: on the one hand, non-target molecules that are hybridized by cross-hybridization to a cell-set may be less available for their target cell-set; on the other, they may block the cross-hybridized cell-set probes for specific hybridization with target molecules. As a result, specious results for both involved cell-sets are expected. These different forms of hybridization may exist for a large number of different target molecules or cell-sets. At the microarray level, cumulative effects of partial hybridization, no hybridization and cross-hybridization across all levels of specificity may dramatically alter the genome-scale results obtained in an experiment.

The specificity of microarray hybridization, at its various levels is affected by several effectors; these can be found in all aspects of the microarray experiment. These effectors have been identified, modeled and quantified, and are detailed below.

EFFECTORS OF HYBRIDIZATION SPECIFICITY

Effectors of microarray hybridization specificity emerge from different categories, including sequence of the probe and target molecule, quantity of target and non-target molecules, and hybridization kinetics. These categories and effectors are summarized in Table 1.

Table 1. Specificity effectors and their effect on microarray hybridization specificity

Category	Specificity effectors	Effectors' effect on specificity	Reference
Sequences of the probe and target molecule	Sequence identity between probe and target molecule	Sequence identity underlies the matching between probe and target molecule	[7,8]
	The position of matching (or mismatching) between probe and target molecules	Probes with segments complementary to segments at either end of the target molecule have higher specificity than those that are complementary to segments within the central portion of the target molecule	[5,8]
	Probe length	Short probes are more specific than long probes	[5]
	Probe and target GC contents	Higher target GC content increases specificity	[5]
	Probe inter-molecular and probe-target molecule pairing	Increased thermodynamic stability of probe self-folding reduces specificity	[9]
	Neighboring bases in the probe and target molecule sequences	Neighboring bases in the probe and target molecule sequences affect hybridization specificity	[10]
	Contiguous base pair between probe and target molecule	The most contiguous base pairs between probe and target sequences (i.e., pairing segments) increase specificity	[8]
Quantity of target and non-target molecules	Sample complexity	Increase in sample complexity decreases specificity	[14]
Hybridization kinetics	Differences in hybridization kinetics between correct and incorrect probe-target molecule complexes	As a result of a mismatch to the probe, non-target molecules are less strongly bound, and as a result dissociate faster	[15, 16]
	Duration of hybridization	Increased hybridization time increases specificity	[17,18,20]
	Duration of post-hybridization washing	A short period of washing, which allows dissociation of many of the incorrect complexes and only a few of the correct complexes, results in higher specificity	[20]
	Mixing of hybridization solution	Mixing of hybridization solution during microarray hybridization leads to higher specificity of hybridization	[19]

Several studies, which have identified specificity effectors and modeled them, detected them empirically or filtered the data accordingly to increase hybridization specificity are detailed below; these studies are grouped together according to the category of the studied effecter.

Sequence of Probe and Target Molecule

The sequence complementary between probe and target molecule underlies the two molecules' hybridization. Hence, many of the effectors of specificity are related to the sequence of both the probe and target molecule. Effectors include the level of matching between probe and target molecule and the position of matching (or mismatching) [5-8], the length of the probe [5], the probe's G and C content [5], the primary molecular interactions the probe may form with itself [9], neighboring bases in the probe and target molecule sequences [10], and the length and composition of the contig formed during hybridization ([8, 11]; Table 1).

Several studies have used some of these effectors, in different combinations, for modeling and quantification, in order to better understand the measure of hybridization specificity.

One example is a study that modeled probe length and the region of match or mismatch along the target and probe molecules as specificity effectors. In their theoretical study, Jaramayan et al. [5] simulated an association between a single molecule attached to a hard surface and a single target molecule in solution, an association which may represent that occurring in microarray hybridization of a single probe and a single target molecule. Their model defined the effect of the examined effectors on hybridization specificity.

Another study modeled the specificity of effectors combining the primary molecular structures that may be formed during hybridization and the position of matching and mismatching between probe and target molecules [12] during the formation of RNA/DNA and RNA/RNA duplexes in solution. This was done for both perfectly matched and mismatched probes. Their model was able to quantify target concentration.

He et al. [8] examined several effectors: sequence identity, continuous stretch, and mismatch position, including the effect of neighboring bases along the molecule's sequence as well. Based on nearest-neighbor modeling, the binding free energy for probe and target molecule was calculated, using established thermodynamic parameters. Correlation of the effectors' values to the relative signal intensity obtained for a cell produced weighted parameters for each of the effectors; the weighted parameters were suggested for use in the rational design of probe pre-microarray printing.

Another step towards increasing hybridization specificity, following the modeling of effectors and quantification of target molecules, is filtration of the data, to include only those that are hybridized with increased specificity. The weighted parameters defined in He et al.'s [8] study might be used for the filtration of hybridization data post-microarray hybridization, to obtain microarray probes with increased specificity.

Another example is a study by Flikka et al. [13], who developed a tool for the detection of potential cross-hybridizations for DNA microarrays, based on the level of matching between the oligo-probe and target molecule, detected via homology searches: the tool compares probe sequences with an extensive sequence database representing a transcriptome of the target species, followed by filtration of the probes based on probe length and sequence-comparison results.

Quantity of Target and Non-target Molecules

The quantity of target and non-target molecules has been shown to affect hybridization specificity. Wick et al. [14] showed that an increase in sample complexity (i.e., an increase in the amount of non-target molecules) leads to an increase in signal resulting from incorrect complexes between probes and non-target molecules (i.e., cross-hybridization), and to a decrease in signal resulting from correct hybridization complexes. This relative reduction in signal for correct hybridization complexes and increase in signal for cross-hybridization may reflect a microarray-specificity-level effect of cross-hybridization between cells; perhaps target molecules are 'caught' by non-target cells, leading to a reduced signal from the target cells (Fig 1).

Hybridization Kinetics

The kinetics of correct and incorrect spot hybridization has been shown to differ: it was suggested that when there is a mismatch to the probe, non-target molecules are less strongly bound, and as a result dissociate faster [15, 16]. Similarly, Dai et al. [17] showed that hybridization kinetics may be substantially different for specific and non-specific binding. They performed a real-time study of microarray hybridization, on a microarray scale, using multiple oligo-probes with various mismatches to the target molecules. They showed that specific binding takes longer to reach hybridization equilibrium than non-specific binding.

The Dai et al. [17] study predicted increased specificity with increased hybridization time. This prediction was confirmed by Dorris et al. [18], who examined the signal obtained during microarray hybridization, as a fold change between test and sample, and found an increase in the accuracy of fold-change values with increased hybridization time. They suggested that this increase in accuracy may be due to increased specificity, as a result of the differences in specific and non-specific hybridization kinetics.

Finally, the effect of degree of mixing of the hybridization solution during hybridization on specificity was empirically examined. Schaupp et al. [19] compared mixing and non-mixing methods during hybridization: a higher specificity of hybridization was obtained when the hybridization solution was mixed during microarray hybridization; this was deciphered from the lower signal read from negative controls. Furthermore, both higher sensitivity (i.e., higher ratio of signal-to-background), and higher reproducibility were obtained by mixing during hybridization [19]. The degree of mixing needed for higher specificity of hybridization results was measured by Sorokin et al. [20]. Inducing flow pulsations to mix the buffer solution during oligonucleotide microchip hybridization significantly improved the discrimination between matched and mismatched duplexes.

To summarize, the various specificity effectors—probe and target molecule sequences, quantity of target molecule, and kinetics of hybridization between probe and target molecule, all affect specificity of the obtained microarray hybridization results. Nevertheless, to better illustrate the association between the specificity effectors and hybridization specificity, two issues need to be considered. One is the level of hybridization specificity: each of the specificity effectors may directly affect specificity on one or more specificity levels, and as a result, influence other levels of specificity as well. For example: the specificity effector of GC content of the oligo-probe sequence, although determined at the level of the oligo-probe, may influence the signal obtained at the cell level following hybridization. This signal may be different between cell-sets, as each cell may contain oligo-probes with a particular GC content which differs among cells. Hence, GC content may influence the cell-set specificity level as well.

The other issue that needs to be considered is the interconnection between the effectors themselves. For example, the effecter 'continuous stretch of base pairing between probe and target molecule' may not be easily separated from the effecter 'sequence matching between probe and target molecule' (e.g., [8]). Moreover, the effecter categories are also connected. For example, the quantity of target and non-target molecules (i.e., complexity of the hybridized samples) may not be easily separated from the sequence of probe and target molecules (i.e., mismatched base-paring and mismatch position) [14].

Conclusion

In this review, we provide a glimpse into the issue of microarray hybridization specificity and its complexity. This complexity stems from the different levels of hybridization specificity, the existence of various specificity effectors, and the hybridization and post-hybridization procedures used; each of the specificity effectors may directly affect specificity on one or more specificity levels, during hybridization or post-hybridization.

Moreover, other measures of microarray hybridization, namely accuracy, sensitivity and reproducibility [1], affect the obtained results. All of the measures are interconnected, and they both influence and are influenced by hybridization specificity. For example: in the case of low hybridization specificity, the signal is reflecting the presence of non-target molecules, in addition to that of target molecules. The presence of non-target molecules, in addition to target molecules, increases the complexity of the hybridization (e.g., different molecules have different kinetics of hybridization during a particular hybridization reaction). As a result, the accuracy of the results (i.e., 'the degree of conformity of the measured quantity to its actual [true] value') is lower. Once the accuracy is lower for a single hybridization reaction, it is likely that the reproducibility between hybridization reactions, each with reduced accuracy, will also be low.

Due to the complex nature of microarray hybridization, and despite the numerous studies (some examples of which were cited in this review) that, for correct quantification of target molecules, either modeled or empirically determined effectors of specificity, or filtered the data for improved specificity, further advances are needed [3]. These advances may precisely define and quantify specificity, accuracy, sensitivity and reproducibility of microarray hybridization at the different levels of hybridization. Subsequently, a full and comprehensive quantification and improved microarray hybridization results will become feasible.

REFERENCES

[1] Draghici, S., Khatri, P., Eklund, A.C. & Szallasi, Z. (2006) Reliability and reproducibility issues in DNA microarray measurements. *Trends Genet.*, **22**, 101-109.

[2] Dufva, M. (2005) Fabrication of high quality microarrays. *Biomol. Eng.*, **22**, 173-84.

[3] Koltai, H. & Weingarten-Baror,C. (2008) Specificity of DNA microarray hybridization: characterization, effectors and approaches for data correction. *Nucleic Acids Res.*,**36**, 2395-2405.

[4] Hammond, J.P., Broadley, M.R., Craigon, D.J., Higgins, J., Emmerson, Z.F., Townsend, H.J., White, P.J. & May, S.T. (2005) Using genomic DNA-based probe-selection to improve the sensitivity of high-density oligonucleotide arrays when applied to heterologous species. *Plant Methods*, **1**, 10.

[5] Jayaraman, A., Hall, C.K., & Genzer, J. (2006) Computer simulation study of molecular recognition in model DNA microarrays. *Biophys. J.*, **91**, 2227-2236.

[6] Bar-Or, C., Czosnek, H. & Koltai, H. (2007) Cross-species microarray hybridizations: a developing tool for studying species diversity. *Trends Genet.*, **23**, 200-207.

[7] Bar-Or, C., Bar-Eyal, M., Gal, T.Z., Kapulnik, Y., Czosnek, H. & Koltai, H. (2006) Derivation of species-specific hybridization-like knowledge out of cross-species hybridization results. *BMC Genomics*, **7**, 110.

[8] He, Z., Wu, L., Li, X., Fields, M.W. & Zhou, J. (2005) Empirical establishment of oligonucleotide probe design criteria. *Applied and Environmental Microbiology*, **71**, 3753-3760.

[9] Matveeva, O. V., Shabalina, S. A., Nemtsov, V. A., Tsodikov, A. D., Gesteland, R. F. & Atkins, J. F. (2003) Thermodynamic calculations and statistical correlations for oligo-probes design. *Nucleic Acids Res.*, **31**, 4211-4217.

[10] Dimitrov, R.A. & Zuker, M. (2004) Prediction of hybridization and melting for double-stranded nucleic acids. *Biophys. J.*, **87**, 215–226.

[11] Chen, Y.A., Chou, C.C., Lu, X., Slate, E.H., Peck, K., Xu, W., Voit, E.O. & Almeida, J.S. (2006) A multivariate prediction model for microarray cross-hybridization. *BMC Bioinformatics*, **7**, 101.

[12] Carlon, E. & Heim, T. (2006) Thermodynamics of RNA/DNA hybridization in high-density oligonucleotide microarrays. *Physica A*, **362**, 433-449.

[13] Flikka, K., Yadetie, F., Laegreid, A. & Jonassen, I. (2004) XHM: a system for detection of potential cross hybridizations in DNA microarrays, *BMC Bioinformatics*, **5,** 117

[14] Wick, L.M., Rouillard, J.M., Whittam, T.S., Gulari, E., Tiedje, J.M. & Hashsham, S.A. (2006) On-chip non-equilibrium dissociation curves and dissociation rate constants as methods to assess specificity of oligonucleotide probes, *Nucleic Acids Res.*, **34**, e26.

[15] Zhang, Y., Hammer, D. A. & Graves, D. J. (2005) Competitive hybridization kinetics reveals unexpected behavior patterns. *Biophys. J.*, **89**, 2950-2959.

[16] Wu, C., Carta, R. & Zhang, L. (2005) Sequence dependence of cross-hybridization on short oligo microarrays. *Nucleic Acids Res.*, **33**, e84.

[17] Dai, H., Meyer, M., Stepaniants, S., Ziman, M. & Stoughton, R. (2002) Use of hybridization kinetics for differentiating specific from non-specific binding to oligonucleotide microarrays. *Nucleic Acids Res.*, **30**, e86.

[18] Dorris, D.R., Nguen, A., Gieser, L., Lockner, R., Lublinsky, A., Patterson, M., Touma, E., Sendera T.J., Elghanian, R., & Mazumder, A. (2003) Oligodeoxyribonucleotide probe accessibility on a three-dimensional DNA microarray surface and the effect of hybridization time on the accuracy of expression ratios. *BMC Biotechnology*, **3**, 6.

[19] Schaupp, C.J., Jiang, G., Myers, T.G. & Wilson, M.A. (2005) Active mixing during hybridization improves the accuracy and reproducibility of microarray results. *Biotechniques*, **38**, 117-119.

[20] Sorokin, N. V., Chechetkin, V. R., Livshits, M. A., Pan'kov, S. V., Donnikov, M. Y., Gryadunov, D. A., Lapa, S. A. & Zasedatelev, A. S. (2005) Discrimination between perfect and mismatched duplexes with oligonucleotide gel microchips: role of thermodynamic and kinetic effects during hybridization. *J. Biomol. Struct. Dyn.*, **22**, 725-734.

Reviewed by an anonymous Reviewer as part of a peer review process.

RESEARCH AND REVIEW STUDIES

In: Oligonucleotide Array Sequence Analysis
Editors: M.K. Moretti and L.J. Rizzo, pp. 27-36

ISBN: 978-1-60456-542-3

Chapter 1

SELECTING A PROPER PLATFORM FOR MICROARRAY EXPERIMENT

Sergey V. Anisimov*

V.A. Almazov Federal Center for Heart, Blood & Endocrinology
Saint Petersburg 197341, Russia and
Cureline, Inc., Burlingame, CA 94010, USA

ABSTRACT

The rapid development of microarray technology has led to important progress in many biomedical disciplines. In just over a decade, the technology has evolved from first small-scale prints to commercial and custom platforms of truly genomic scale. Though numerous types of microarray platforms exist, cDNA and oligonucleotide-based DNA microarrays are considered, being the platforms of choice for the range of applications. However, the success of every study depends upon a careful selection of a platform appropriate for a designed experiment. In this short review, various technological and analytical aspects of microarray technology are discussed; some advantages and drawbacks of individual platform types are compared.

Keywords: DNA microarray, platform selection

The rapid development of microarray technology has led to important progress in many biomedical disciplines, facilitating unraveling basic mechanisms underlying many important biological processes, gene discovery, drug discovery, disease diagnostics, etc. [1,2]. An ability to identify novel prognostic and diagnostic markers, potential therapeutic targets and molecular markers of toxicogenomic response in fast and reliable manner by microarray means has undoubtedly transformed modern biology. This point could be vividly illustrated by a simple query to the PubMed database (a service of the U.S. National Library of Medicine and the National Institutes of Health), which lists 9 original microarray technology-related

* E-mail address: sergey-anisimov@cureline.com. Tel: +7 (812) 347 5546. FAX: +7 (812) 596 3896. Correspondence: Department of Hematology Research, V.A. Almazov Federal Center for Heart, Blood & Endocrinology, Akkuratova Str. 2, Saint Petersburg 197341, Russia

articles and 1 review in a whole period of 1993-1997 (the earliest reports by [3-5], 106 relevant papers in 1998-1999 and over 25000 papers since (Fig. 1). This expansion is in parallel with a volume of microarray platforms designed and utilized, as further illustrated with a number of that deposited in Gene Expression Omnibus (GEO, National Center for Biotechnology Information (NCBI); http://www.ncbi.nlm.nih.gov/geo/) database: data from over 3000 individual microarray platforms are present in this database alone (as of early 2008).

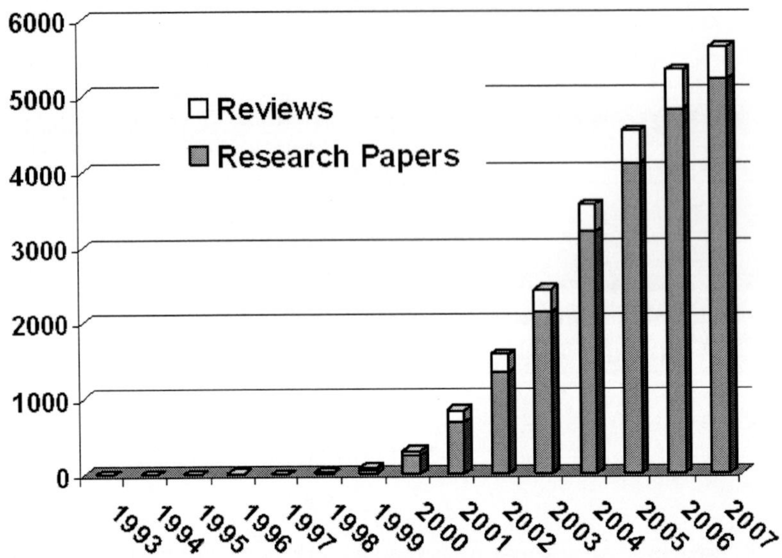

Figure 1. Microarray technology-relevant entries in PubMed: 1993-2007. Entries are sorted as original articles (nearly 22000 in total) and reviews (over 2500). Taken together, over the period of 1993-2007, >24000 published papers have relied on microarray technology. This expansion is in parallel with a volume of microarray platforms designed and utilized.

Although no established classification exists for microarray subtypes, it could be provisionally classified based on the object pursued (e.g., microarray-based expression analysis, mutation/single nucleotide polymorphism (SNP) analysis, comparative genomic hybridization), substrate type (e.g., whole tissue, DNA, RNA, antibody, protein and peptide microarrays), design (flat (2D) or volumetric (3D) microarrays), detection method (microelectronic, fluorescent and radioisotope detection-based microarrays), etc.. It was hard if possible to predict in the early years of microarray era what principle type of the microarray platform will be most suitable. Numerous types of the microarray platforms now considered 'unorthodox' were therefore tested, with a various degree of success. Even though some of those types (e.g., radio transponder-based, volumetric (3D) microarrays and some others) are still used in the limited niches of their specific applications, it was DNA microarray-based expression analysis that have gradually took over the largest share of the microarray technology-related applications.

DNA microarrays consist of two major categories, namely cDNA microarrays and oligonucleotide microarrays. The platforms belonging to the first group (printed on glass slides or on nylon filter membranes) rely on collections of bacterial clones containing inserts typically 500-2,000 nucleotides long, representing either fully sequenced genes of known function or partially sequenced cDNA derived from expressed sequence tags (ESTs). The latter corresponds to the messenger RNAs of genes representing either a certain degree of similarity to know genes, or to fully unknown, novel genes. Oligonucleotide microarrays rely on the synthesized molecules printed over the solid surfaces. Generally, microarray-bound oligonucleotide probes (matching targeted genes) could be 20-35 nucleotides (short oligonucleotide-based DNA microarrays), 35-45 nucleotides (intermediate) or 65-75 (long) nucleotides long. Each manufacturer seeks finding a better balance of economical (e.g., oligonucleotide synthesis expenses) and technical (e.g., hybridization-related parameters, including oligonucleotide design parameters like guanine/cytosine content, secondary structure, melting temperature, etc.) features when selecting the properties of oligonucleotides for the corresponding platforms. Furthermore, similar features affect a decision to place a large number of short oligonucleotide probe sets for each gene on the same platform, or rather use a single long oligonucleotide per gene. Thus, a wide assay of oligonucleotide-based DNA microarray platforms is available, and a scientist can select one most utterly satisfying the needs of the planned research experiment [6]. While longer oligonucleotides generally promise higher hybridization specificity and affinity (thus being more suitable for studying low-abundance genes), one may rather prefer using a higher number of shorter oligonucleotides spotted over the same microarray platform. Similarly, two major intrinsic factors affect the precision of gene expression data generated when utilizing cDNA microarray platforms. First, cDNA clones may have moderately different insert sizes, thus the signal intensities might to certain extent vary as a function of insert length. Second, cDNA clones can be derived from the alternative regions within the target gene, again affecting a quality of quantitative assessments of the hybridization data. Those factors contrast with the features of oligonucleotide microarrays, for which the probes are designed to be similar in regards to hybridization properties, - and therefore suggest cDNA microarray platforms might be more suitable for comparative studies than for gene expression profiling of individual samples ('snapshot gene expression profiling'). Interestingly, an emerging class of microarrays relying on extremely long oligonucleotides (synthesized molecules up to the thousands nucleotides long) might become a format of choice in the following decade.

Similarly, certain advantages and disadvantages are associated with available labeling techniques, with radioactive (radioisotope) and fluorescent labeling being most common among those. Although highly effective and most sensitive, radioactive labeling is relatively inexpensive; additionally, labeled membranes could be stripped and re-labeled, if necessary. At the same time, applying this type of labeling technique rightfully requires strict control at each step (storage and handling of both the radioactive isotope stock and labeled probes, liquid and solid waste disposal, etc.). All works related to the radioactive isotopes are subject to the governmental and institutional regulations in the context of safety risks, and in many cases require specific certification and licensing, designation of the dedicated work zones, etc. Radioactive labeling is also considered not being suitable for high-density DNA arrays, due to the risk of high radioactive signal 'leaking' from the neighboring spots, causing cross-contamination of individual spot-derived signal intensity value records (Fig. 2). At the same time, fluorescent labeling allows performing simultaneous two-color hybridizations (e.g., for the experiment vs. control) and its application is not associated with safety risks of the scale mentioned above. In comparison to radioactive isotopes routinely applied for microarray probe labeling, fluorescent dyes (including commonly used Cy3 and Cy5 cyanine dyes) have longer shelf life. An excess of the labeled probe could be frozen for re-hybridization, which could take place many weeks following the labeling procedure itself. At the same time, those dyes are highly light-sensitive, which requires a good practice of laboratory procedures performed. Moreover, cyanine dyes are oxidation-sensitive (Cy5 in particular).

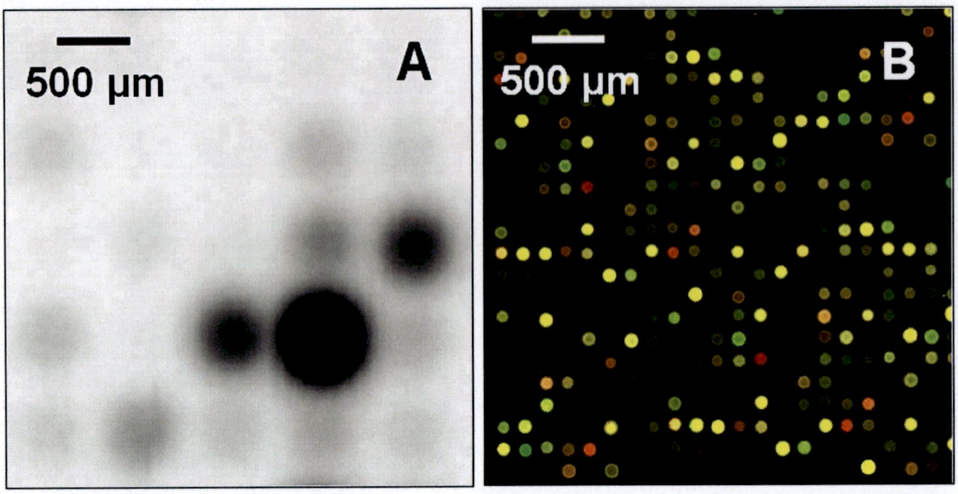

Figure 2. (A), Raw signal generated by cDNA microarray spotted over nylon membrane by GMS417 arrayer (Genetic Microsystems, USA). Distance between spot centers, 665 μm; spot diameter, 300 μm. (B) Raw signal generated by oligonucleotide microarray spotted over glass slides by MicroGrid II 600R arrayer (Harvard BioRobotics, USA). Distance between spot centers, 140 μm; average spot diameter, 90–110 μm. Scale bars, 500 μm (0.5 mm).

Over the course of just over a decade, numerous technological advances have led to the remarkable evolution of DNA microarrays [7-9]. While in the early years of microarray platform development those have carried hundreds of spots of DNA material [7,10-12], after just a few years large-scale DNA microarray platforms could carry some tens of thousands of these [13-16]. By the moment, a number of available DNA microarray platforms carry over hundred thousand spots of DNA material, thus reaching truly genomic scale. Still, specialized microarray platforms of small-to-intermediate scale are most valuable in many fields of study, allowing a concentrated (and thus less expensive) analysis of a limited number of pre-selected gene (or protein, etc.) targets: e.g., belonging to one or a few biological pathways [17]. Notably, a progress in microarray platform manufacturing (exemplified by the possibility to print a substrate material over the platform in the precise manner and with even higher density), closely follows the development of detection tools (high-resolution, high-throughput scanners). Clearly, individual properties of the materials utilized can affect platform selection: for example, cDNA microarrays printed over nylon membranes have significantly longer shelf life than oligonucleotide microarrays printed over the glass slides, and generally do not require controlled storage conditions.

Importantly, many areas of research could not directly benefit from the progress of microarray technology itself, due to the amount of biological material essential for analysis. It was therefore essential to refine techniques allowing highly accurate amplification of the starting material, for downstream microarray application [18,19]. Importantly, such techniques should match an existing range of available microarray platform types [20-23]. In its extreme, RNA material purified from a single cell is now enough to perform large-scale microarray analysis [24,25]. It is generally believed, however, that any amplification performed on pre-hybridization step introduces a degree of bias to the quantitative accuracy of microarray data generated. Novel technological developments (e.g., laser capture microdissection (LCM)) allow reliable and high-throughput selection of the material for the downstream analysis. The researcher must therefore balance between a desire to have cell population as pure and uniform as possible, and a requirement to have a larger volume of starting material, as a mean to decrease amplification-introduced bias. Numerous further measures aim improving quantitative accuracy of microarray-generated data. Among those, repeated hybridizations of microarray platforms with independently purified and labeled probes (so-called 'biological replicates') and/or selfsame probes ('technical replicates') are utilized extensively [26-31] (see also [32]). As already mentioned above, multiplication of the number of spots carrying the same cDNA clone/oligonucleotide within the same platform could serve as an alternative, or rather an addition to this approach [17,33]. A quantity of the spots carrying equal DNA material (being duplicated, triplicated, quadruplicated, etc.) is then limited by the technical properties of the particular microarray platform. This combined approach effectively prevents data loss by individual small-scale hybridization artifacts, in particular. It also allows more accurate estimation of the variability present in an experiment, increases the probability of detecting genes undergoing significant fold changes in expression, and substantially decreased the probability of observing fold changes due to chance rather than true differential expression [34]. 'Pooling' of the samples assessed by microarray technology represents an alternative to purifying those by LCM or any other mean. Though this approach is criticized by many reviewers, it can effectively eliminate a need for amplification of starting material and furthermore reduce the effect of biological variation in the individual samples [35-37] (see also extensive reviews in [38,39]).

As described in this short review, many DNA microarray platforms are currently available commercially or via custom design, providing a multiplicity of choices. Due to the diversity of available DNA microarray platforms, no single established protocol or reagent could be applied to every DNA microarray platform. Moreover, technological means employed for data acquisition and analysis (i.e., scanner model, analytical software available, etc.) vary in different laboratories, which obviously also affects platform selection (Fig. 3). Therefore, variations and adaptations of basic protocols are most common on every stage of microarray experiments (namely, sample labeling, pre-hybridization, hybridization and post-hybridization washes) [40]. Taken together, variations in the performance between individual laboratories, practiced protocols, and analytical algorithms applied to the same array platform are no less meaningful than variations between the platforms [41-43]. At the same time, many of those factors could be successfully compensated for, using appropriate manufacturing, handling and data processing techniques [44].

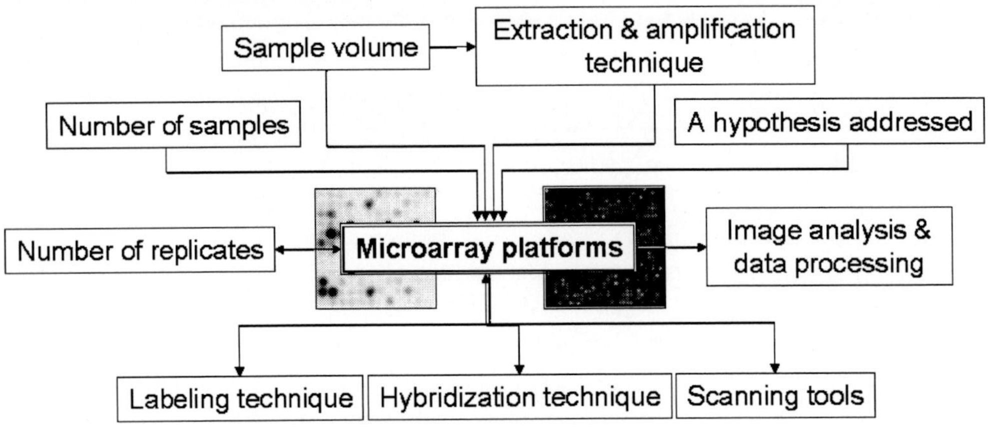

Figure 3. Some factors affecting microarray platform selection. Note that in the majority of cases the availability of the tools (e.g., access to the particular scanner model, license for radioactive isotope studies, etc.) affects the selection.

Finally, it is generally accepted that no standard way of microarray data analysis could address all the diversity of available microarray platforms and individual experiments performed [45]. Analytical algorhythm utilized should therefore be adjusted in each individual case according to the scale of DNA microarray platform, presence and scale of the intrinsic target multiplication (i.e., duplications of the target genes spotted within the same platform), biological replicates, total quantity of samples compared, etc. Numerous options for data analysis could be considered in each individual case [46-50]. Furthermore, blind design of microarray experiment could be a valuable factor [28,51,52]. It should be also stressed that no matter what platform has been selected for a particular microarray application, microarray-derived data should always be validated using an alternative approach (e.g., a conventional RT-PCR, real-time PCR, in-situ hybridization technology, etc.).

References

[1] Pusztai, L. Chips to bedside: incorporation of microarray data into clinical practice. (2006) *Clin. Cancer Res.,* **12**(24), 7209-7214.

[2] Trevino, V.; Falciani, F. and Barrera-Saldaña, H.A. DNA microarrays: a powerful genomic tool for biomedical and clinical research. (2007) *Mol. Med.*, **13**(9-10), 527-541.

[3] Fodor, S.P.; Rava, R.P.; Huang, X.C.; Pease, A.C.; Holmes, C.P. and Adams, C.L. Multiplexed biochemical assays with biological chips. (1993) *Nature*, **364**(6437), 555–556.

[4] Guo, Z.; Guilfoyle, R.A.; Thiel, A.J.; Wang, R. and Smith, L.M. Direct fluorescence analysis of genetic polymorphisms by hybridization with oligonucleotide arrays on glass supports. (1994) *Nucleic Acids Res*., **22**(24), 5456-5465.

[5] Pease, A.C.; Solas, D.; Sullivan, E.J.; Cronin, M.T.; Holmes, C.P. and Fodor, SP. Light-generated oligonucleotide arrays for rapid DNA sequence analysis. (1994) *Proc. Natl. Acad. Sci. USA,* **91**(11), 5022-5026.

[6] Walker, S.J.; Wang, Y.; Grant, K.A.; Chan, F. and Hellmann, G.M. Long versus short oligonucleotide microarrays for the study of gene expression in nonhuman primates. (2006) *J. Neurosci. Methods*, **152**(1-2), 179-189.

[7] Schena, M.; Shalon, D.; Davis, R.W. and Brown, P.O. Quantitative monitoring of gene expression patterns with a complementary DNA microarray. (1995) *Science*, **270**(5235), 467-470.

[8] Lipshutz, R.J.; Fodor, S.P.; Gingeras, T.R. and Lockhart, D.J. High density synthetic oligonucleotide arrays. (1999) *Nat. Genet.*, **21**(1 Suppl), 20-24.

[9] Müller, U.R. Evolution of high-sensitivity microarray technology. (2003) *JALA*, **8**(6), 96-100.

[10] Schena, M.; Shalon, D.; Heller, R.; Chai, A.; Brown, P.O. and Davis, R.W. Parallel human genome analysis: microarray-based expression monitoring of 1000 genes. (1996) *Proc. Natl. Acad. Sci. USA*, **93**(20), 10614-10619.

[11] DeRisi, J.; Penland, L.; Brown, P.O.; Bittner, M.L.; Meltzer, P.S.; Ray, M.; Chen, Y.; Su, Y.A. and Trent, J.M. Use of a cDNA microarray to analyse gene expression patterns in human cancer. (1996) *Nat. Genet.*, **14**(4), 457-460.

[12] Shalon, D.; Smith, S.J. and Brown, P.O. A DNA microarray system for analyzing complex DNA samples using two-color fluorescent probe hybridization. (1996) *Genome Res.*, **6**(7), 639-645.

[13] Kargul, G.J.; Dudekula, D.B.; Qian, Y.; Lim, M.K.; Jaradat, S.A.; Tanaka, T.S.; Carter, M.G. and Ko, M.S. Verification and initial annotation of the NIA mouse 15K cDNA clone set. (2001) *Nat. Genet.*, **28**(1), 17-18.

[14] Truckenmiller, M.E.; Vawter, M.P.; Cheadle, C.; Coggiano, M.; Donovan, D.M.; Freed, W.J. and Becker, K.G. Gene expression profile in early stage of retinoic acid-induced differentiation of human SH-SY5Y neuroblastoma cells. (2001) *Restor. Neurol, Neurosci.*, **18**(2-3), 67-80.

[15] VanBuren, V.; Piao, Y.; Dudekula, D.B.; Qian, Y.; Carter, M.G.; Martin, P.R.; Stagg, C.A.; Bassey, U.C.; Aiba, K.; Hamatani, T.; Kargul, G.J.; Luo, A.G.; Kelso, J.; Hide, W. and Ko, M.S. Assembly, verification, and initial annotation of the NIA mouse 7.4K cDNA clone set. (2002) *Genome Res.*, **12**(12), 1999-2003.

[16] Carter, M.G.; Hamatani, T.; Sharov, A.A.; Carmack, C.E.; Qian, Y.; Aiba, K.; Ko, N.T.; Dudekula, D.B.; Brzoska, P.M.; Hwang, S.S. and Ko, M.S. In situ-synthesized novel microarray optimized for mouse stem cell and early developmental expression profiling. (2003) *Genome Res.*, **13**(5), 1011-1021.

[17] Anisimov, S.V.; Christophersen, N.S.; Correia, A.S.; Li, J.Y. and Brundin P. "NeuroStem Chip": a novel highly specialized tool to study neural differentiation pathways in human stem cells. (2007) *BMC Genomics*, **8**, 46.

[18] Wilson, H.L. and O'Neill, H.C. Preparation of label from small cell numbers for microarray screening. (2004) *OMICS*, **8**(3), 190-196.

[19] Viale, A.; Li, J.; Tiesman, J.; Hester, S.; Massimi, A.; Griffin, C.; Grills, G.; Khitrov, G.; Lilley, K.; Knudtson, K.; Ward, B.; Kornacker, K.; Chu, C.Y.; Auer, H. and Brooks, A.I. Big results from small samples: evaluation of amplification protocols for gene expression profiling. (2007) *J. Biomol. Tech.*, **18**(3), 150-161.

[20] Nagy, Z.B.; Kelemen, J.Z.; Fehér, L.Z.; Zvara, A.; Juhász, K. and Puskás, L.G. Real-time polymerase chain reaction-based exponential sample amplification for microarray gene expression profiling. (2005) *Anal. Biochem.*, **337**(1), 76-83.

[21] Pettersson, E.; Lindskog, M.; Lundeberg, J. and Ahmadian A. Tri-nucleotide threading for parallel amplification of minute amounts of genomic DNA. (2006) *Nucleic Acids Res*., **34**(6), e49.

[22] Laurell, C.; Wirta, V.; Nilsson, P. and Lundeberg J. Comparative analysis of a 3' end tag PCR and a linear RNA amplification approach for microarray analysis. (2007) *J. Biotechnol*., **127**(4), 638-646.

[23] Hu, G.; Yang, Q.; Cui, X.; Yue, G.; Azaro, M.A.; Wang, H.Y. and Li, H. A highly sensitive and specific system for large-scale gene expression profiling. (2008) BMC Genomics, **9**, 9.

[24] Brandt, S.; Kloska, S.; Altmann, T. and Kehr J. Using array hybridization to monitor gene expression at the single cell level. (2002) *J. Exp. Bot*., **53**(379), 2315-2323.

[25] Saitou, M.; Yabuta, Y. and Kurimoto, K. Single-cell cDNA high-density oligonucleotide microarray analysis: detection of individual cell types and properties in complex biological processes. (2008) *Reprod. Biomed. Online*. **16**(1), 26-40.

[26] Baggerly, K.A.; Coombes, K.R.; Hess, K.R.; Stivers, D.N.; Abruzzo, L.V. and Zhang, W. Identifying differentially expressed genes in cDNA microarray experiments. (2001) *J. Comput. Biol*., **8**(6), 639-659.

[27] Andreadis, J.D.; Mann, T.T.; Russell, A.C.; Stenger, D.A. and Pancrazio, J.J. Identification of differential gene expression profiles in rat cortical cells exposed to the neuroactive agents trimethylolpropane phosphate and bicuculline. (2001) *Biosens. Bioelectron*., **16**(7-8), 593-601.

[28] Anisimov, S.V.; Boheler, K.R.; Khavinson, V.Kh. and Anisimov, V.N. Studies of the effects of Vilon and Epithalon on gene expression in mouse heart using DNA-microarray technology. (2002) *Bull. Exp. Biol. Med*., **133**(3), 293-299.

[29] Anisimov, S.V.; Boheler, K.R. and Anisimov, V.N. Microarray technology in studying the effect of melatonin on gene expression in the mouse heart. (2002) *Dokl. Biol. Sci.,* **383**, 90-95.

[30] Anisimov, S.V.; Khavinson, V.Kh. and Anisimov, V.N. Elucidation of the effect of brain cortex tetrapeptide Cortagen on gene expression in mouse heart by microarray. (2004) *Neuro Endocrinol. Lett*., **25**(1-2), 87-93.

[31] Anisimov, S.V.; Khavinson, V.Kh. and Anisimov, V.N. Effect of melatonin and tetrapeptide on gene expression in mouse brain. (2004) *Bull. Exp. Biol. Med*., **138**(5), 504-509.

[32] Wei, C.; Li, J. and Bumgarner, R.E. Sample size for detecting differentially expressed genes in microarray experiments. (2004) *BMC Genomics*, **5**(1), 87.

[33] Lee, M.L.; Kuo, F.C.; Whitmore, G.A. and Sklar, J. Importance of replication in microarray gene expression studies: statistical methods and evidence from repetitive cDNA hybridizations. (2000) *Proc. Natl. Acad. Sci. USA,* **97**(18), 9834-9839.

[34] Black, M.A. and Doerge, R.W. Calculation of the minimum number of replicate spots required for detection of significant gene expression fold change in microarray experiments. (2002) *Bioinformatics*, **18**(12), 1609-1616.

[35] Warrington, J.A.; Nair, A.; Mahadevappa, M. and Tsyganskaya, M. Comparison of human adult and fetal expression and identification of 535 housekeeping/maintenance genes. (2000) *Physiol. Genomics*, **2**(3), 143-147.

[36] Walker, D.G.; Lue, L.F. and Beach, T.G. Gene expression profiling of amyloid beta peptide-stimulated human post-mortem brain microglia. (2001) *Neurobiol. Aging*, **22**(6), 957-966.

[37] Agrawal, D.; Chen, T.; Irby, R.; Quackenbush, J.; Chambers, A.F.; Szabo, M.; Cantor, A.; Coppola, D. and Yeatman, T.J. Osteopontin identified as lead marker of colon cancer progression, using pooled sample expression profiling. (2002) *J. Natl. Cancer Inst.*, **94**(7), 513-521.

[38] Kendziorski, C.M.; Zhang, Y.; Lan, H. and Attie, A.D. The efficiency of pooling mRNA in microarray experiments. (2003) *Biostatistics*, **4**(3), 465-477.

[39] Zhang, W.; Carriquiry, A.; Nettleton, D. and Dekkers, J.C. Pooling mRNA in microarray experiments and its effect on power. (2007) *Bioinformatics*, **23**(10), 1217-1224.

[40] Yauk, C.L.; Berndt, M.L.; Williams, A. and Douglas, G.R. Comprehensive comparison of six microarray technologies. (2004) *Nucleic Acids Res.*, **32**(15), e124.

[41] Kuo, W.P.; Jenssen, T.K.; Butte, A.J.; Ohno-Machado, L. and Kohane, I.S. Analysis of matched mRNA measurements from two different microarray technologies. (2002) *Bioinformatics*, **18**(3), 405-412.

[42] Irizarry, R.A.; Warren, D.; Spencer, F.; Kim, I.F.; Biswal, S.; Frank, B.C.; Gabrielson, E.; Garcia, J.G.; Geoghegan, J.; Germino, G.; Griffin, C.; Hilmer, S.C.; Hoffman, E.; Jedlicka, A.E.; Kawasaki, E.; Martínez-Murillo, F.; Morsberger, L.; Lee, H.; Petersen, D.; Quackenbush, J.; Scott, A.; Wilson, M.; Yang, Y.; Ye, SQ. and Yu, W. Multiple-laboratory comparison of microarray platforms. (2005) *Nat. Methods*, **2**(5), 345-350.

[43] Johnson, D.S.; Li, W.; Gordon, D.B.; Bhattacharjee, A.; Curry, B.; Ghosh, J.; Brizuela, L.; Carroll, J.S.; Brown, M.; Flicek, P.; Koch, C.M.; Dunham, I.; Bieda, M.; Xu, X.; Farnham, P.J.; Kapranov, P.; Nix, D.A.; Gingeras, T.R.; Zhang, X.; Holster, H.; Jiang, N.; Green, R.D.; Song, J.S.; McCuine, S.A.; Anton, E.; Nguyen, L.; Trinklein, N.D.; Ye, Z.; Ching, K.; Hawkins, D.; Ren, B.; Scacheri, P.C.; Rozowsky, J.; Karpikov, A.; Euskirchen, G.; Weissman, S.; Gerstein, M.; Snyder, M.; Yang, A.; Moqtaderi, Z.; Hirsch, H.; Shulha, H.P.; Fu, Y.; Weng, Z.; Struhl, K.; Myers, R.M.; Lieb, J.D. and Liu, X.S. Systematic evaluation of variability in ChIP-chip experiments using predefined DNA targets. (2008) *Genome Res.*, **18**(3), 393-403.

[44] Wilkes, T.; Laux, H. and Foy, C.A. Microarray data quality - review of current developments. (2007) *OMICS*, **11**(1), 1-13.

[45] Butte, A. The use and analysis of microarray data. (2002) *Nat. Rev. Drug Discov.*, **1**(12), 951-960.

[46] Quackenbush, J. Computational analysis of microarray data. (2001) *Nat. Rev. Genet.*, **2**(6), 418-427.

[47] Churchill, G.A. Fundamentals of experimental design for cDNA microarrays. (2002) *Nat. Genet.*, **32**(Suppl), 490-495.

[48] Cheadle, C.; Vawter, M.P.; Freed, W.J. and Becker, K.G. Analysis of microarray data using Z score transformation. (2003) *J. Mol. Diagn.*, **5**(2), 73-81.

[49] Leung, Y.F. and Cavalieri, D. Fundamentals of cDNA microarray data analysis. (2003) *Trends Genet.*, **19**(11), 649-659.

[50] Hackl, H.; Sanchez Cabo, F.S.; Sturn, A.; Wolkenhauer, O. and Trajanoski, Z. Analysis of DNA microarray data. (2004) *Curr. Top. Med. Chem.*, **4**(13), 1357-1370.

[51] Cirelli, C.; Shaw, P.J.; Rechtschaffen, A. and Tononi, G. No evidence of brain cell degeneration after long-term sleep deprivation in rats. (1999) *Brain Res.,* **840**(1-2), 184-193.

[52] Hakak, Y.; Walker, J.R.; Li, C.; Wong, W.H.; Davis, K.L.; Buxbaum, J.D.; Haroutunian, V. and Fienberg, A.A. Genome-wide expression analysis reveals dysregulation of myelination-related genes in chronic schizophrenia. (2001) *Proc. Natl. Acad. Sci. USA,* **98**(8), 4746-4751.

In: Oligonucleotide Array Sequence Analysis
Editors: M.K. Moretti and L.J. Rizzo, pp. 37-82
ISBN: 978-1-60456-542-3

Chapter 2

THE CLINICAL RELEVANCE OF GENE EXPRESSION PROFILES IN PERIPHERAL BLOOD MONONUCLEAR CELLS

Christopher H. Woelk[1,*] and Michael E. Burczynski[2]
[1] Department of Pathology, University of California San Diego
La Jolla, CA, 92093, USA
[2] Biomarker Laboratory, Wyeth Research, 500 Arcola Road
Collegeville, PA, 19246, USA

ABSTRACT

The spotlight for developing clinically relevant biomarkers has recently focused brightly on gene expression profiling in peripheral blood mononuclear cells (PBMCs). Since PBMCs circulate throughout the body via the blood stream, their gene expression can be altered by disease states even when the disease is localized to a less accessible compartment within the human host. PBMC gene expression may therefore represent a disease barometer that can be used to assess human health, and inquiries into this phenomenon have given rise to a burgeoning field known as *hemogenomics*. The clinical accessibility of blood by venipuncture, and the ease with which PBMCs can be isolated, add to the attractiveness of this tissue source for monitoring human health. Most studies of PBMC gene expression to date have identified genes expressed differentially between healthy and diseased patients, permitting the construction of potentially diagnostic gene classifiers. In addition, apparently prognostic gene classifiers have been developed using samples taken prior to the development of a particular disease state or the initiation of a therapeutic intervention. Gene classifiers have been developed for a variety of ailments such as: autoimmune diseases, cancer, chronic fatigue syndrome, hypertension, infectious diseases, neuropathy, post-traumatic stress disorder, toxic chemical exposures, and transplant tolerance. This chapter will review multiple aspects of PBMC gene expression profiling including: PBMC isolation, storage considerations, sample preparation, gene expression analysis, and gene classifier construction and evaluation. A review of previous PBMC gene expression studies is presented to illustrate

[*] E-mail address: cwoelk@ucsd.edu. Tel: 858-552-8585 extn 7193. Fax: 858-552-7445. Christopher H. Woelk, University of San Diego California, Department of Pathology, Stein Clinical Research Building, Rm. 326, 9500 Gilman Drive # 0679, La Jolla, CA 92093-0679

the current *status quo* in this particular field of biomarker development. Finally, the future of PBMC gene expression profiling is discussed with specific emphasis on its impact for medical practices and clinical care.

Keywords: peripheral blood mononuclear cells, gene expression, microarray, pharmacogenomics

INTRODUCTION

The tissue compartment of the peripheral blood has been a rich source of biomarkers for decades, if not centuries. The ability of health-care providers and physicians to routinely and non-invasively sample blood from patients has enabled the rapid assessment of human pathophysiology (e.g., elevated cholesterol and tri-glycerides in metabolic syndromes), disease status (e.g., detection of circulating tumor cells in certain cancers, or alterations in $CD4^+$ T cell counts in HIV infected individuals), drug monitoring (e.g., quantitating the levels of administered therapeutics), immune responses (e.g., quantifying antibody titers after exposure to antigens), and a host of other measurable phenomena. The majority of these assays use peripheral blood samples to measure exogenous drug concentrations (often by high performance liquid chromatography [HPLC] or liquid chromatography/mass spectrometry [LC/MS]), endogenous metabolite levels (also often by HPLC or LC/MS), cell surface antigens (by flow cytometry) or proteins/peptides in serum or plasma (by enzyme linked immunosorbent assays [ELISA]).

In the last decade the sequencing of the human genome and the proliferation of microarray technologies, along with the availability of key reagents, have enabled the assessment of the human transcriptome in various tissues. A growing number of scientific investigators are applying microarray profiling technologies to the circulating cells of the peripheral blood in a variety of paradigms. By evaluating the transcriptomes of peripheral blood mononuclear cells (PBMCs), researchers hope to understand the roles and responses of these cells in pathogenesis, immunity, disease status, drug exposure and other medically relevant phenomena.

Isolated PBMCs are predominantly comprised of T-cells, B-cells and monocytes, which are transcriptionally active and known to be involved in a number of disease processes. They can be either protective or destructive depending on the disease, and their transcriptional responses may result from the presence of disease, the influence of a therapeutic intervention, or both. One of the prevailing hypotheses underlying the field of hemogenomics is that the transcriptional patterns observed in PBMCs will be correlated with specific diseases or the status of a disease sufficiently to enable better prognosis, diagnosis or drug monitoring *in vivo*. An ever-growing array of diseases and drug treatment paradigms has been investigated by this technology. This review summarizes the technical challenges involved in the application of transcriptional profiling in PBMCs, describes the various approaches for identifying and validating transcriptional signatures from a variety of settings, and reviews the literature published to date in the field of hemogenomics with respect to expression profiles measured in PBMCs.

Isolation of PBMCs

Different approaches are employed to isolate PBMCs and extract RNA for analysis. Purification of PBMC is most commonly achieved using either Ficoll density gradients [1], or cell-purification tubes (e.g., CPT tubes, Becton-Dickinson, Baltimore MD). Based on differential sedimentation of cell types through density gradients, these methods enrich the proportions of monocytes and lymphocytes relative to the number of granulocytes, red blood cells and platelets originally present in whole blood. Because these isolation methods do not stabilize the cellular transcriptome, the preferred method of choice is the immediate processing of isolated PBMCs to minimize the variability associated with the storage of whole blood.

Storage of PBMCs

Despite the recommendation to immediately isolate PBMCs from whole blood after collection, it is recognized that various limitations at the site of collection may mean that whole blood samples will sit under certain storage conditions until being processed. As with any biomarker assay, the impact of such pre-analytical variables should always be considered. A number of published reports have determined the effects of temperature, time of storage and other pre-analytical variables on PBMC gene expression profiles [2-8]. Fluorescence-activated cell sorting performed at Affymetrix showed that overnight incubation of whole blood prior to isolation of PBMCs by Ficoll did not significantly impact the proportion of T lymphocytes, B lymphocytes or natural killer cells, but caused a slight elevation in monocytes [2]. In this study, transcripts corresponding to 150 probe sets on the U133A GeneChip showed significant differences in expression using a 2-fold difference and a Bonferonni-corrected 95% significance cutoff after overnight storage.

Several other published studies have further characterized the impact of overnight incubation on PBMC gene expression profiles as measured by quantitative real-time RT-PCR (qRT-PCR) [3] or microarray analysis [4-8]. Debey et al. [6] annotated transcripts identified in their analyses as differentially expressed following overnight incubation and showed that pathways related to hypoxia and stress responses were elevated, while metabolism, transcription regulation, cell cycle, motility, apoptosis and immune responses appeared down-regulated. The main conclusions of this study on the impact of pre-analytical variables on the stability of PBMC transcriptional profiles strongly suggests that PBMCs should be isolated from whole blood: 1) as soon as possible, and 2) under tightly controlled and consistent conditions especially with respect to temperature of whole blood storage, PBMC isolation methods employed, and RNA extraction technology used. Appropriately controlling for these factors limits their impact and enables meaningful comparisons of transcriptomes to be made between samples and across studies.

PREPARATION OF PBMCS FOR GENE EXPRESSION PROFILING

Although the optimal recommendation is to isolate RNA as soon as possible from PBMCs purified from whole blood collection, Cobb et al. [9] demonstrated that PBMCs purified following centrifugation may be safely frozen prior to RNA processing. The gene expression profiles from PBMCs isolated in this manner were reported to be extremely concordant with PBMCs isolated and processed immediately. This approach better facilitates large multi-center studies because any site capable of collecting blood, centrifuging PBMC samples and subsequently storing frozen samples can ship these samples to other centralized laboratories for the more specialized purpose of RNA isolation and gene expression analysis.

RNA can be isolated from PBMCs by any number of commercialized methods and it is beyond the scope of this review to compare and contrast each of the methods available. In general, approximately one microgram of RNA is typically purified from one million PBMCs, which is the number typically found in one milliliter of blood. As the tubes most often used for whole blood collection contain between 4 to 8 mL of whole blood, a good estimate of total RNA yield from a standard blood collection is 5-10 micrograms. RNA can be quantified using absorption of light at 260 and 280 nm whereby an A260/280 ratio between 1.8 and 2.0 is indicative of high quality RNA. RNA integrity numbers (RIN) can be generated using the Agilent Bioanaylzer and should be 7.0 or higher. Samples with these characteristics are expected to yield reliable transcriptional profiles for analysis.

Precautions should be taken to remove genomic DNA (gDNA) from RNA preparations because gDNA can artificially inflate observed expression levels. On-column DNAse treatment is strongly recommended. One excellent, low-cost method for determining the extent of gDNA contamination in samples prior to global transcriptional profiling is to analyze all RNA samples in a single PCR assay, which analyzes a single exon in both the presence and the absence of a reverse transcription (RT) step. Samples with insignificant gDNA contamination will exhibit high levels of expression for the target RNA after RT, but will give low or undetectable levels in the absence of an RT step. In contrast, samples with significant gDNA contamination will exhibit high levels of expression even in the absence of the RT step. A good general rule is that samples exhibiting more than 10 cycles difference (by real-time PCR) between minus-RT and plus-RT controls in the above assay can be considered free of gDNA contamination.

A large number of cDNA preparation/labeling methodologies for global transcriptional profiling of RNA isolated from PBMCs are available and the selection of the appropriate method often depends on the platform selected for probing gene expression (see the next section). As long as the requisite characteristics of input RNA and the fidelity of the cDNA conversion methodology are established, then highly reproducible results can be obtained. The next section explores the various methods available for assessing the transcriptome using the RNA extracted from PBMC samples that have been processed for gene expression profiling.

PLATFORMS FOR EVALUATING PBMC GENE EXPRESSION

Several comprehensive review articles [10-15] and books [16, 17] provide complete descriptions of microarray gene expression analysis with respect to manufacturing, sample preparation, hybridization, scanning and normalization. A brief overview of the types of microarray platforms that have been used for PBMC gene expression follows. As shown in Figure 1, microarrays with probes constructed from short-oligonucleotides (e.g., Affymetrix GeneChips), long-oligonucleotides (e.g., Illumina BeadChips) or cDNAs (e.g., Research Genetics GeneFilters, Sigma Genosys Panorama Arrays and other custom microarrays) have been used to analyze gene expression in PBMCs. In addition, qRT-PCR has been used to assess the expression levels of large numbers of candidate genes in PBMCs [18].

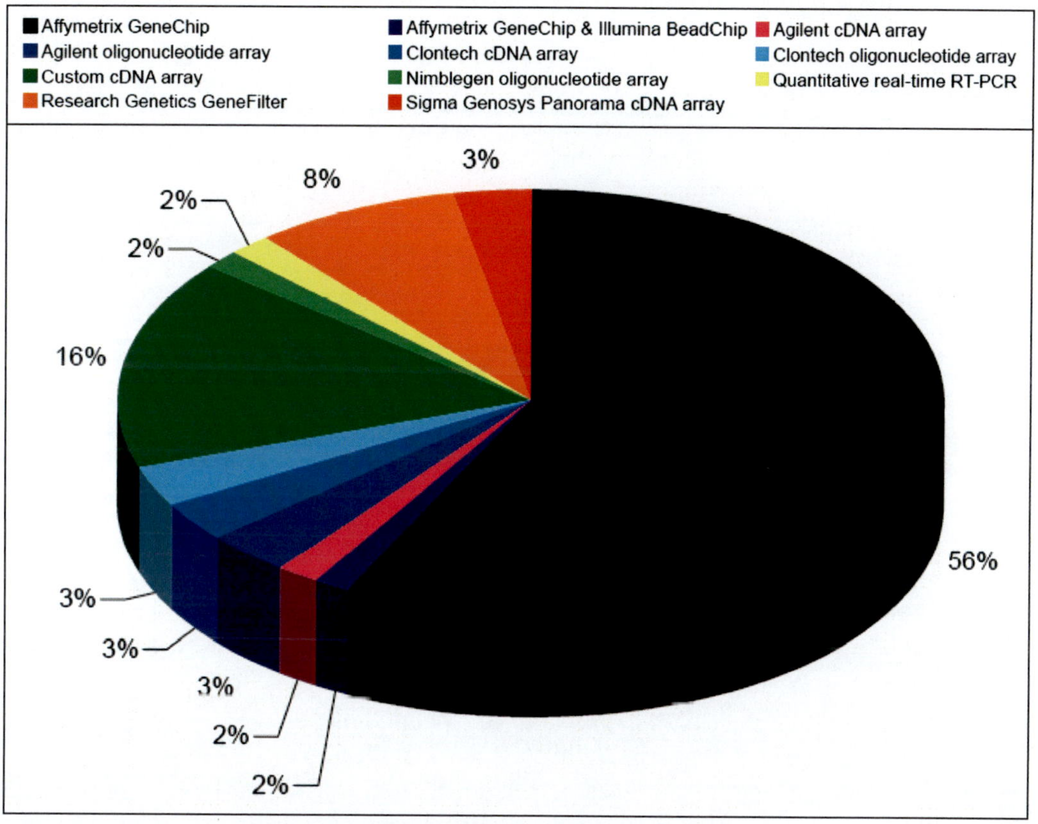

Figure 1. Platforms used for PBMC gene expression profiling.

Determining which platform is the best for analyzing gene expression in PBMCs is very much influenced by the resources available to the individual researcher. A number of platforms are capable of interrogating gene expression for the entire human genome, including the Affymetrix HG-U133 Plus 2.0 GeneChip (38,500 genes), the Agilent Whole Genome Oligo Microarray (>41,000 genes), the Applied Biosystems (ABI) Human Genome Survey Microarray V2.0 (29,098 genes), the GE CodeLink Human Whole Genome Bioarray (approx. 45,000 genes), the Illumina Human-6 v2 BeadChip (46,114 genes), and the Nimblegen Whole Genome Microarray (37,364 genes). Reassuringly, different microarray platforms yield similar results as demonstrated by studies comparing gene expression data generated by Affymetrix GeneChips and Illumina BeadChips [6, 12]. Debey and colleagues [6] used principal component analysis (PCA) to analyze the gene expression data generated using both the Affymetrix U133A GeneChip and the Illumina HumanRef-8 BeadChip for each of 14 samples, and found no distinction between the two platforms. More than half of PBMC gene expression studies performed to date have employed Affymetrix GeneChips (Figure 1). Thus, if normalization of gene expression data with previously completed studies is required, then the use of Affymetrix GeneChips may be advantageous. However, when PBMC source material is limited as in a multipurpose clinical trial or in an archived sample library, the Illumina platform provides an advantage in that only 5-500 ng of total RNA is required compared to the 1-5 μg needed for Affymetrix GeneChips [6]. Affymetrix further provides a Human Exon 1.0 ST GeneChip, which targets genes at the exon-level as opposed to the gene-level. This array may be used if the intent is to identify differentially expressed transcripts of the same gene between two clinical conditions (i.e., disease vs. control) in an effort to reveal novel biomarkers [19, 20]. Ultimately, irrespective of the platform selected for PBMC gene expression profiling, it is essential to confirm that a sufficient number of samples are incorporated into the analysis in order to identify differentially expressed genes. A number of tools exist for such power calculations including the *samr* module in R [21] and an online power calculator (http://www.scai.fhg.de/special/bio/howmanyarrays/ calculator.html) developed by Zien and coworkers [22].

Gene Classifier Construction and Validation

PBMC gene expression studies that yield potentially useful diagnostic or prognostic classifiers (Table 1) generally include four steps: (1) Class discovery, (2) Class comparison, (3) Class prediction, and (4) Gene classifier validation [23, 24]. Multiple methods and applications for performing these steps have been thoroughly documented and include: GenePattern [25], BRB ArrayTools [26], CLEAVER [27], and R/Bioconductor [28]. A brief description of each step follows so that the text in Table 1 may be better interpreted.

Following filtering to select significantly modulated genes, unsupervised methods are often used in class discovery to ensure that disease samples cluster distinctly from control samples on the basis of their gene expression. One popular method of clustering is average linkage hierarchical clustering, which clusters genes and samples in expression space as defined by a distance metric, and it can be performed with Cluster and TreeView software [29]. Such unsupervised methods may reveal unknown biologically relevant taxonomic structure, and the validity of new disease classes represented by distinct clusters of patient samples can be assessed by class comparison [30]. Class comparison is used to identify differentially expressed genes and ranks each gene according to the difference in expression level between the two classes (i.e. disease vs. control) and the consistency of this expression difference across the multiple samples. For example, Statistical Analysis of Microarrays (SAM) is a supervised method, which calculates a rankable score for each gene using a modified *t*-test statistic with sample-label permutations to evaluate statistical significance [31].

Table 1. Summary of PBMC gene expression studies that resulted in the construction of diagnostic or prognostic classifiers.[1]

Disease	Arrays	Study Design	Study Summary	Platform	Validation[2]	Classifier performance[3]	qRT-PCR	GEO	Ref.
Detection of exposure									
Cigarette exposure	102	Peripheral blood leukocyte (PBL) samples were taken from 41 smokers and 44 nonsmokers and divided into a training (32 smokers and 33 nonsmokers) and a test (TEST1: 10 smokers and 10 nonsmokers) data set. PBL samples taken from 17 patients during a follow up visit were used to construct a second test data set (TEST2: 10 smokers and 7 non-smokers). Therefore, TEST1 and TEST2 data sets were derived from the same individuals at two different time points.	The training data set identified 861 genes differentially expressed between smokers and nonsmokers. On the basis of lowest error rate, a 36-gene diagnostic classifier was constructed from these differentially expressed genes capable of correctly classifying 19/20 and 16/17 samples in the TEST1 and TEST2 data sets, respectively.	Agilent custom oligonucleotide array (Hu25K, 21,000 genes)	Test data	*95% for TEST1, and 94% for TEST2*	No	No	[80]
Diagnosis of disease stage									
Acute ischemic stroke	57	PBMC samples were isolated from 30 ischemic stroke patients and 30 elderly healthy controls. Samples were divided into a training (19 stroke and 19 controls) and a test (9 stroke and 10 controls) data set for gene expression analysis.	Bonferonni corrected *t*-tests identified 190 genes differentially expressed between acute ischemic stroke patients and healthy controls in the training data set. Differentially expressed genes were predominantly upregulated and were involved in activation and differentiation of white blood cells consistent with an	Affymetrix HG-U133A GeneChip (≈13,000 genes)	Test data	*79%*	Yes	No	[127]

Table 1. Continued

Disease	Arrays	Study Design	Study Summary	Platform	Validation[2]	Classifier performance[3]	qRT-PCR	GEO	Ref.
			inflammatory response. Prediction Analysis of Microarrays (PAM) was used to construct a diagnostic classifier of 22 genes that was able to correctly classify 7/9 stroke patients and 8/10 controls in the test data set.						
Acute myeloid leukemia (AML)	293	For improved diagnosis of AML leading to better disease prognosis, bone marrow aspirates or PBMC samples were isolated from 285 AML patients. CD43 cells were taken from 3 healthy controls and bone marrow aspirates from 5 healthy controls. Specimens from AML patients were divided into a training (n=189) and a test (n=96) data set.	Initially, all 285 AML samples were used in an unsupervised analysis to identify 16 subgroups of AML with specific molecular signatures. After dividing the samples into training and test data, supervised methods were used to construct diagnostic gene classifiers of varying size capable of assigning samples to an AML disease subgroup with high accuracy. This study demonstrates that cases of AML with known cytogenic abnormalities and new clusters of AML with characteristic gene-expression signatures can be identified with the use of a single assay.	Affymetrix HG-U133A GeneChip (≈13,000 genes)	Test data	100% (for AML subgroups with cytogenic abnormalities vs. other abnormalities)	Yes	GSE1159	[110]

Table 1. Continued

Disease	Arrays	Study Design	Study Summary	Platform	Validation[2]	Classifier performance[3]	qRT-PCR	GEO	Ref.
Auto-immune diseases	70	In order to determine whether gene expression profiles are shared between autoimmune disease patients and their first-degree relatives, PBMCs were isolated from 19 systemic lupus erythematosus (SLE), 9 rheumatoid arthritis (RA), 17 early RA (ERA), 5 insulin dependent diabetes mellitus (IDDM) and 4 multiple sclerosis (MS) patients, 8 unaffected family members and 8 healthy controls. Samples were divided into matched training and test data sets of 27 autoimmune disorders, 4 unaffected family members and 4 healthy controls.	Statistical Analysis of Microarrays (SAM) and Principle Component Analysis (PCA) were both able to significantly segregate autoimmune disease patients, unaffected family members and healthy controls into discrete groups on the basis of their gene expression profiles. PCA suggested that the unaffected family member samples were more like the control samples than they were like the autoimmune samples. SAM analysis of the training data set identified a 99-gene diagnostic classifier that could accurately classify 27/27 autoimmune patient samples and 7/8 control and unaffected family member samples.	Research Genetics GeneFilter 211 (4132 genes)	Test data	*97%*	No	GSE3447	[111]

Table 1. Continued

Disease	Arrays	Study Design	Study Summary	Platform	Validation[2]	Classifier performance[3]	qRT-PCR	GEO	Ref.
Auto-immune diseases	43	PBMCs were taken for gene expression analysis from 11 RA, 18 SLE, 5 IDDM, and 4 multiple sclerosis (MS) patients, as well as from 9 control patients both before and after influenza vaccination.	Initially the authors wished to define a normal immune response and compared gene expression in 7 control patients before and after immunization. Immunized patients could be distinguished from those with autoimmune disease on the basis of gene expression clustering but the autoimmune diseases did not segregate from each other. Differentially expressed genes were then identified between preimmune controls and autoimmune patients and a 35-gene diagnostic classifier was constructed from differentially expressed genes between SLE and healthy controls, which was able to correctly classify 6/6 additional SLE and 5/5 additional RA samples.	Research Genetics GeneFilter 211 (4132 genes)	Test data	100%	No	No	[128]

Table 1. Continued

Disease	Arrays	Study Design	Study Summary	Platform	Validation[2]	Classifier performance[3]	qRT-PCR	GEO	Ref.
Cardiac allograft rejection	285	To develop better non-invasive assays for rejection surveillance, 285 PBMC samples were taken from 98 cardiac allograft patients. In order to correlate PBMC gene expression with rejection status, endomyocardial biopsies (EMB) were also taken from these patients and assigned an International Society for Heart and Lung Transplantation (ISHLT) biopsy score (0 = quiescent, ≥3A = rejection).	SAM analysis identified 97 candidate genes from microarray expression data, which were combined with 155 genes from the literature involved in transplant rejection, to yield a total of 252 candidate genes. The expression of these 252 genes was than assessed by qRT-PCR in a training data set of 145 samples (36 rejection and 109 quiescent) from 86 patients. Linear discriminant analysis (LDA) was used to construct an 11-gene diagnostic classifier, which was capable of classifying the rejection status in an independent test data set of 31 rejection and 32 quiescent samples with only 60% accuracy. However, when samples taken >6 months or >12 months post-transplant were considered the classification accuracy increased to 77% in both cases.	Clontech custom oligonucleotide array (7370 leukocyte-focused genes)	Test data	*60%*	Yes	GSE2445	[129]

Table 1. Continued

Disease	Arrays	Study Design	Study Summary	Platform	Validation[2]	Classifier performance[3]	qRT-PCR	GEO	Ref.
HIV-1	87	PBMCs were taken from 22 drug-naive HIV-seropositive and 7 seronegative persons to constitute a training data set, and from 53 HIV-seropositive and 5 seronegative persons to constitute a test data set, in order to determine the biomarkers that can separate individuals on the basis of their HIV-1 serostatus.	Supervised learning algorithms identified a 10-gene classifier from the training data set capable of accurately determining HIV-1 serostatus in the test data set. In addition, 8 HIV-seropositive persons whose $CD4^+$ T cell counts increased after 27 months were compared to 8 HIV-seropositive patients whose counts decreased to derive a 6-gene classifier capable of segregating these two groups with 93% accuracy as assessed by LOOCV.	Affymetrix HG-Focus GeneChip (8,400 genes)	Test data	93%	No	GDS1449	[118]
Infectious diseases	137	PBMC were used in 144 microarray analyses performed on 137 samples taken from 131 patients who were infected with influenza A (36), *Staphylococcus aureus* (50), *Streptococcus pneumoniae* (16), *Escherichia coli* (29) or were healthy controls (7). In order to discriminate viral from bacterial infections a training data set of 11 influenza A, 6 *E. coli* and 6 *S. pneumoniae* samples was constructed. Gene classifier perfor-mance was then validated using 2 test data sets: (1) 7 influenza A, 23 *E. coli* and 7 *S. pneumoniae* samples, or (2) 7 influenza A and 31 *S. aureus* samples.	A Mann-Whitney test identified 854 differentially expressed genes between the influenza and bacterial samples in the training data set. A kNN algorithm was used to construct a 35-gene classifier that exhibited a high accuracy as assessed by LOOCV (91%) or by using test data sets (95 and 87%). Gene classifiers were also constructed that could distinguish between *E. coli* and *S. aureus*, or *E. coli* and *S pneumoniae* infection. This study showed that PBMC transcriptional patterns can be used to discriminate different microbial pathogens even when they manifest similar clinical symptoms.	Affymetrix HG-U133A GeneChip (13,000 genes) and HG-U133 Plus 2.0 GeneChip (38,500) and Illumina Human-6 BeadChip (46,114 genes)	Test data	95% for Test data 1 and 87% for Test data 2	Yes	GSE6269	[24]

Table 1. Continued

Disease	Arrays	Study Design	Study Summary	Platform	Validation[2]	Classifier performance[3]	qRT-PCR	GEO	Ref.
Inflamma-tory bowel disease (IBD)	127	In order to develop diagnostic gene classifiers capable of discriminating different causes of IBD, PBMCs were isolated from 59 Crohn's disease (CD) patients, 26 ulcerative colitis (UC) patients, and 42 healthy individuals.	Analysis of covariance of all samples indicated that there were 220 differentially expressed transcripts between CD patients and healthy controls, and 120 differentially expressed transcripts between UC patients and healthy controls, with only 45 transcripts in common between the two causes of IBD. Focusing on IBD samples, supervised class prediction using a training data set of 44 CD and 20 UC samples identified a 12-gene diagnostic classifier capable of correctly assigning all 15 CD and 6 UC patients in an independent test data set.	Affymetrix HG-U133A GeneChip (≈13,000 genes)	Test data	100%	Yes	GDS1615	[112]
Juvenile rheuma-toid arthritis (JRA)	37	PBMCs were isolated from juvenile patients with pauciarticular (5) or polyarticular rheumatoid arthritis (15), spondyloarthropathy (6) and healthy controls (11).	*T*-tests identified 342 differentially expressed genes between polyarticular patients and healthy controls which were able to correctly identify all controls and 13/15 polyarticular samples.	Affymetrix HG-U95Av2 GeneChip (≈10,000 genes)	LOOCV	*92%*	Yes	GDS711	[130]

Table 1. Continued

Disease	Arrays	Study Design	Study Summary	Platform	Validation[2]	Classifier performance[3]	qRT-PCR	GEO	Ref.
Multiple sclerosis (MS)	44	PBMCs were isolated from 26 MS patients and 18 healthy controls of whom the majority were women (n=36, 81.8%). Fourteen MS patients were exhibiting acute relapse and the remaining 12 were undergoing remission at the time of sampling.	A variety of scores (*t*-test, threshold number of misclassifications and the Info score) were used to identify 1019 differentially expressed genes between MS patients and healthy controls that were mainly involved in T-cell activation and expansion, inflammation, and apoptosis. These 1,019 genes were used in a diagnostic gene classifier that was able to classify all healthy control samples correctly and 24/26 MS patient samples. Finally, 721 differentially expressed genes were identified between MS patients in acute relapse and those undergoing remission but the classification accuracy of these genes was not assessed.	Affymetrix HG-U95Av2 GeneChip (≈10,000 genes)	10-fold	*95%*	Yes	No	[131]
Multiple sclerosis (MS)	46	PBMCs were isolated from 27 MS patients and 19 healthy controls and samples were segregated into training (17 MS and 9 controls) and test (10 MS and 10 controls) data sets.	When all the samples were used 112 differentially expressed genes were identified between MS patients and healthy controls. The training data set was used to construct a number of 2-gene diagnostic classifiers of which *HSPAIL* and *HIF2* correctly classified 10/10 MS patients and 6/10 controls, and *CKS2* and *PAFAH1B1* correctly classified 9/10 MS patients and 7/10 controls, in the independent test data set. Thus both classifiers have 80% classification accuracy.	Custom cDNA arrays (6500 or 7500 probes)	Test Data	80%	No	No	[132]

Table 1. Continued

Disease	Arrays	Study Design	Study Summary	Platform	Validation[2]	Classifier performance[3]	qRT-PCR	GEO	Ref.
Multiple sclerosis (MS) and systemic lupus erythematosus (SLE)	36	PBMCs were isolated from 13 relapse-remitting MS patients, 5 SLE patients and 18-age matched healthy controls. The MS and control samples were originally analyzed by Achiron et al. [131].	Three statistical tests (*t*-test, TNoM, and the Info score) were used to identify 541 differentially expressed genes between the autoimmune disorders (MS and SLE) and healthy controls that were related to immune surveillance, immune response, inflammation and proliferation. Three diagnostic gene classifiers were constructed of 163, 19, and 343 genes that were capable of distinguishing autoimmune diseases, MS alone and SLE alone from healthy controls, respectively, with 0 classification error.	Affymetrix HG-U95Av2 GeneChip (≈10,000 genes)	LOOCV	100%	Yes	No	[133]
Post - traumatic stress disorder (PTSD)	33	PBMCs were isolated from 24 trauma survivors in the emergency room (ER, n=15) and 4 months later (M4, n=18) who were subsequently diagnosed with PTSD (8), intermediate PTSD (10) or no PTSD (6).	574 differentially expressed genes were identified using 5 control and 6 PTSD samples taken in the ER. 408 differentially expressed genes were identified using 4 control and 5 PTSD samples taken 4 months later. These differentially expressed genes were used in conjunction with a naïve Bayesian classifier to correctly classify 9/11 ER samples and 8/9 M4 samples. In addition, potential biomarkers that may be capable of predicting continued PTSD were identified.	Affymetrix HG-U95A GeneChip (≈10,000 genes)	LOOCV	*82% for ER samples, and 89% for M4 samples*	No	GDS1020	[134]

Table 1. Continued

Disease	Arrays	Study Design	Study Summary	Platform	Validation[2]	Classifier performance[3]	qRT-PCR	GEO	Ref.
Pulmonary arterial hyper-tension (PAH)	20	PBMCs were taken from 14 drug-treated PAH patients and 6 healthy controls. Two thirds of the samples were used for a training data set and the remainder for a test data set.	A two sample *t*-test was used to construct a 106-gene classifier that was capable of distinguishing PAH patients from healthy controls with 100% accuracy in the test data set.	Affymetrix HuGeneFL GeneChip (5,600 genes)	Test data	100%	Yes	GDS504	[135]
Renal allograft tolerance	66	PBMC samples were isolated for microarray analysis from 58 transplant recipients who were immunosuppressive (IS) drug-free and graft tolerant (TOL, n=11), had stable graft function on minimal immunosupression (MIS, n=10), were treated with multiple IS drugs and had stable kidney graft function for >5 years (STA, 12), or suffered from acute (AR, n= 14) or chronic (CR, n=11) rejection, as well as from 8 healthy subjects (N). A training data set of 5 TOL, 11 CR and 8 N samples was constructed and the remaining samples were used as independent test data.	PAM three-class prediction applied across TOL, CR and N samples in the training data (n=24) identified a 49-gene diagnostic classifier capable of correctly classifying all CR and N controls and 10 of 11 TOL patients. This diagnostic gene classifier was further reduced to 33 genes, which correctly classified 5 of 6 TOL samples and 6 of 6 CR samples, by qRT-PCR in the independent test data. These diagnostic gene classifiers were used in further analyses to identify CR and TOL profiles in samples from MIS, STA and AR patients.	Custom cDNA array (12,400 genes)	Test data	*92%*	Yes	No	[136]

Table 1. Continued

Disease	Arrays	Study Design	Study Summary	Platform	Validation[2]	Classifier performance[3]	qRT-PCR	GEO	Ref.
Renal cell carcinoma (RCC)	65	PBMCs isolated from 45 RCC patients and 20 healthy volunteers were subdivided into a training (32 RCC and 14 controls) and a test (13 RCC and 6 controls) data set.	Initially, all samples were used to identify 184 genes differentially expressed (> 2-fold) using a Student's *t*-test between RCC patients and healthy controls. Samples in the training data set were analyzed using a correlation metric to construct an 8-gene diagnostic classifier that was capable of correctly identifying 13/13 RCC patients and 6/6 controls in the independent test data.	Affymetrix HG-U95A GeneChip (≈10,000 genes)	Test data	100%	No	No	[113]
Rheuma-toid arthritis (RA)	50	PBMCs were isolated from 29 patients with RA and 21 control subjects.	Student's *t*-test identified 81 genes differentially expressed between RA patients and controls, which were related to several different inflammatory and signaling pathways. Logistic regression (LR) analysis identified a 4-gene diagnostic gene classifier (*QPCT*, *S100A12*, *IL1RN* and *GAB2*) capable of discriminating RA patients and healthy controls with good accuracy.	Affymetrix HG-U95Av2 GeneChip (≈10,000 genes)	LR	80%	No	No	[137]
Rheuma-toid arthritis (RA)	53	To identify biomarkers indicative or early rheumatoid arthritis (ERA), PBMCs were isolated from 8 patients with established RA, 11 patients with ERA, 14 patients with SLE, 11 patients with a clinical diagnosis of allergic disease or asthma (ALL), and 9 control patients.	Using stringent criteria, a student's *t*-test identified 53 gene differentially expressed between RA and ERA patients of which the majority were downregulated and associated with immune and inflammatory functions. Using 8 RA and 8 ERA patients, LDA was used to construct an 8-gene ERA signature capable of correctly classifying all ERA patients, every ALL patient and all control patients but only 8/14 SLE patients in an independent test data set. Therefore a subset of SLE patients appears to share the ERA signature.	Research Genetics GeneFilter 211 (4132 genes)	Test data	*86%*	Yes	No	[138]

Table 1. Continued

Disease	Arrays	Study Design	Study Summary	Platform	Validation[2]	Classifier performance[3]	qRT-PCR	GEO	Ref.
Systemic lupus erythema-tosus (SLE)	90	To identify biomarkers of SLE, PBMCs were isolated from 48 SLE patients and 42 healthy controls.	A non-paired student's *t*-test identified 161 genes differentially expressed between SLE patients and healthy controls including a number of interferon stimulated genes (ISG) that were dysregulated in half of the SLE patients. Therefore, the expression of 14 ISGs forming an interferon (IFN) gene expression signature was correlated with the 11 criteria used to diagnose SLE. SLE patients were divided into an IFN-high (n=24) and IFN-low (n=24) group based on the expression of genes in the IFN signature and patients in the former group were shown to have a significantly higher number of SLE criteria than those in the latter. For example, 15/24 and 18/24 patients in the IFN-high group fulfilled criteria for kidney/CNS complications and hematologic involvement, respectively, compared to only 5/24 in the IFN-low group.	Affymetrix HG-U95A GeneChip (≈10,000 genes)	IFN signature	*71% for kidney/CNS complications, and 77% for hematologic involvement*	No	No	[139]
Systemic lupus erythema-tosus (SLE)	26	PBMCs were isolated from 26 SLE patients of whom 12 had active and 14 had inactive disease as determined using the SELENA-SLEDAI score.	PAM analysis using the "nearest shrunken centroids" approach identified a 29-gene diagnostic classifier that could discriminate active from inactive SLE patients with 70% accuracy as assessed by 10-fold cross-validation.	Sigma Genosys Panorama Cytokine Gene Array (375 genes)	10-fold	70%	No	No	[140]

Table 1. Continued

Disease	Arrays	Study Design	Study Summary	Platform	Validation[2]	Classifier perform-ance[3]	qRT-PCR	GEO	Ref.
Systemic onset juvenile idiopathic arthritis (SoJIA)	123	In order to establish a diagnosis for SoJIA patients during the systemic phase of disease to avoid lengthy hospitalizations and initiate effective therapy, PBMC samples were taken from pediatric patients with SoJIA (n=58) or systemic lupus erythematosus (SLE, n=38), 6 patients with PAPA syndrome, and 21 healthy patients. Gene expression analysis of these 123 samples was performed with both the HG-U133A and B (Set) GeneChips resulting in a total of 246 microarray runs. This gene expression data was analyzed in conjunction with a subset of the data extracted from the study performed by Ramilo et al. 2007 [24]. This subset consisted of gene expression data from PBMCs of 31 *S. aureus*, 15 *S. pneumoniae*, 30 *E. coli* and 18 Influenza A infected patients.	Initially, 8 SoJIA and 8 healthy samples were used to construct a 50-gene diagnostic classifier capable of identifying SoJIA patients with 100% accuracy using independent test data also consisting of 8 SoJIA patients and 8 healthy controls. However, this 50-gene diagnostic classifier was unable to consistently distinguish SoJIA patients from those with inflammation resulting from other conditions (pathogen infection, SLE and PAPA syndrome). Therefore a new 12-gene diagnostic classifier was constructed using stricter criteria from 16 SoJIA and 10 healthy patient samples and by removing genes that were also differentially expressed between each of the other disease conditions and their matched controls. This 12-gene diagnostic classifier was able to distinguish 6 of 7 SoJIA patient samples when combined in an independent test data set with non-SoJIA samples: 22 SLE, 6 PAPA, 15 *S. aureus*, 5 *S. pneumoniae*, 20 *E. coli*, 8 Influenza A and 10 healthy samples, and 2 misdiagnosed SoJIA cases.	Affymetrix HG-U133 Set GeneChips (≈33,000 genes)	Test data	100% for SoJIA vs. healthy controls, and 93% accuracy for SoJIA vs. non-SoJIA.	Yes	GSE8650	[86]

Table 1. Continued

Disease	Arrays	Study Design	Study Summary	Platform	Validation[2]	Classifier performance[3]	qRT-PCR	GEO	Ref.
Prognosis of disease outcome									
Cuta-neous t Cell lymphoma (CTCL)	68	To develop biomarkers capable of identifying patients most likely to develop progressive disease, PBMC samples were taken from 48 patients during the leukemic phase of CTCL and 20 healthy controls. The 48 CTCL patients were divided into 12 short-term (ST), 11 intermediate (MT) and 25 long-term (LT) survivors.	A Student's *t*-test identified 385 genes differentially expressed between 17 CTCL patients exhibiting high Sezary cell counts (>60%) and 9 Th2-skewed healthy controls. Penalized discriminant analysis (PDA) was used to construct a 10-gene prognostic classifier capable of discriminating ST from LT survivors with 100% accuracy.	Custom cDNA array (4,500 genes)	LOOCV	100%	Yes	No	[91]
HIV-1	24	PBMCs were extracted from 3 healthy subjects and 21 HIV-positive Ugandans who were classified as CDC clinical Category A (n=10), B (n=4), or C (n=7). The clinical status of 14 subjects was reassessed almost 2 years later, 5 of whom had died in the interim.	Unsupervised methods grouped HIV-infected patients into 3 clusters based on their gene expression profiles which correlated perfectly with CDC clinical diagnosis expect for two cases: a Category C subject which clustered with Category A patients in Cluster I and a Category B subject that clustered with Category C patients in Cluster III. Follow up clinical evaluation 2 years later showed that this category C patient survived and thus belonged in the cluster I, while the category B subject died and thus belonged in cluster III. A 40-gene prognostic classifier was constructed capable of perfectly predicting whether patients belonged to Cluster I (Category A) or Cluster III (Category C).	Custom cDNA array (731 immune-related genes)	LOOCV	100%	No	No	[141]

Table 1. Continued

Disease	Arrays	Study Design	Study Summary	Platform	Validation[2]	Classifier performance[3]	qRT-PCR	GEO	Ref.
Myco-bacterium tubercu-losis	36	Altogether PBMC samples were isolated from 37 Tuberculosis (TB) patients, 22 *M. tuberculosis*-infected healthy donors and 15 non-infected healthy controls for microarray and qRT-PCR analysis. Thirty-six two-channel microarray experiments were performed in order compare PBMC gene expression between TB patients and *M. tuberculosis*-infected healthy donors via a reference RNA pool from non-infected controls.	Using microarray generated expression data, a *t*-test identified 95 differentially expressed genes between TB patients and *M. tuberculosis*-infected healthy donors. A 9-gene subset of these differentially expressed genes were analyzed by qRT-PCR in a training data set of 23 Tuberculosis patients and 27 healthy controls that were either *M. tuberculosis* infected (n=17) or non-infected (n=10). LDA identified a 3 gene diagnostic classifier (RAB33A, CD64 and LTF) capable of correctly discriminating 9/10 Tuberculosis patients and 8/10 healthy controls in an independent test data set.	Agilent custom oligonucleotide array (8,033 genes)	Test Data	85%	Yes	GSE6112	[142]

Table 1. Continued

Disease	Arrays	Study Design	Study Summary	Platform	Validation[2]	Classifier performance[3]	qRT-PCR	GEO	Ref.
Pharmacogenomic assessment									
Asthma and gluco-corticoids (GCs)	296	In order to predict which asthma sufferers are resistant to GCs, 106 patients were initially identified as glucocorticoid-sensitive (GC-S, n=64) or resistant (GC-R, n=42). Following cessation of inhaled corticosteroids for at least 24 hours, PBMCs were collected at baseline (n=106), following IFNB/TNFA treatment *in vitro* in the presence or absence of GC treatment (n=87), as well as from 16 healthy controls. Baseline and *in vitro* treated samples were divided into matched training and test data sets.	Initially, 5,011 differentially expressed genes between GC-S and GC-R samples in the training data set could only predict response to GCs in the test data with 59% accuracy. An 11-gene prognostic classifier was constructed by combining the baseline training and test data sets but this only correctly classified samples in 60% of cases as assessed by LOOCV. In order to construct a better prognostic gene classifier, gene expression was analyzed in an additional 87 samples treated *in vitro* with cytokines IFNB and TNFA in the absence or presence of GC. These 87 samples were divided into training and test data and the analysis of the training data revealed that 1,334 genes were modulated by cytokine treatment of which 923 were reversed by GC treatment in the GC-S group. A 15-gene diagnostic classifier was constructed from these 923 genes, which was able to classify samples with 84% accuracy in the test data, and a single gene - *NFKB* - had a prognostic accuracy of 81%. qRT-PCR analysis of 11 of these genes in an additional 40 GC-S and 39 GC-R samples resulted in a classification accuracy of 84%.	Affymetrix HG-U95A GeneChip (≈10,000 genes)	Test data	84%	Yes	No	[143]

Table 1. Continued

Disease	Arrays	Study Design	Study Summary	Platform	Validation[2]	Classifier performance[3]	qRT-PCR	GEO	Ref.
Chronic hepatitis C virus (HCV) and IFN-α2b with or without ribavirin	40	PBMCs were isolated from 13 chronic HCV patients before treatment (n=13), 2 weeks after treatment initiation (n=12) and 6 months after the cessation of treatment (n=9), as well as from 6 healthy controls. These patients were later classified as complete-(CR, n=4), biomedical-(BR, n=4) or non-(NR, n=5) responders.	Support vector machine methods identified 48 genes, which comprised a diagnostic classifier capable of discriminating chronic HCV patients from healthy subjects with 100% accuracy. To identify genes predictive of a complete IFN response, PBMC gene expression was assessed in the 8 patients undergoing IFN-α2b monotherapy. Protective adaptive resonance theory (PART) was used to eliminate nonspecific genes or genes with errors so that a fuzzy neural network combined with a SWEEP operator method could be used to construct a 3-gene prognostic classifier capable of predicting response to IFN therapy with 91% accuracy. In addition, 3-gene prognostic classifiers capable of predicting therapeutic efficacy with 87% accuracy were also constructed by comparing gene expression before and 2 weeks after the initiation of therapy. Finally, it was noted that the predictors for IFN monotherapy were not always relevant for IFN and ribavirin combination therapy.	Custom cDNA array (1305 genes)	LOOCV	100% for discriminating CH-C patients from healthy controls, and 91% for predicting a good response to IFN therapy.	Yes	No	[144]

Table 1. Continued

Disease	Arrays	Study Design	Study Summary	Platform	Validation[2]	Classifier performance[3]	qRT-PCR	GEO	Ref.
Chronic hepatitis C virus (HCV) and HIV-1 coinfection treated with peg-IFN-α2b and ribavirin	36	PBMCs were collected from 29 patients with chronic HCV infection and stable HIV disease prior to therapy and from 17 of these patients following 48 weeks of treatment. Peg-IFN-α2b treated coinfected patients were identified as non-responders (NRs, n=10), sustained virological responders (SVRs, n=9) or end of treatment responders (ETRs, n=10) who had a relapse of HCV viral load following the cessation of treatment.	Analysis of treatment effects, response group effects, and treatment by response group interaction using a mixed-model analysis of variance identified 1352 significantly modulated genes. PAM using samples taken prior to treatment identified 79 genes that constitute a prognostic gene classifier capable of predicting the correct end of treatment outcome for all NR samples, 8/10 ETR samples, and 7/9 SVR samples. Post-therapy samples were also analyzed to determine which responders would relapse after treatment (ETR, n=9) or sustain HCV inhibition (SVR, n=8). One hundred and five genes were selected for a prognostic classifier capable of predicting all ETR patients and 7/8 SVR patients. In general, gene expression profiling suggested that immune activation before anti-HCV therapy was associated with non-response.	Affymetrix HG-U133A GeneChip (≈13,000 genes)	10-fold	86% for prediction of treatment outcome at 48 weeks, and 94% for prediction of progression to ETR or SVR following treatment cessation.	Yes	No	[98]

Table 1. Continued

Disease	Arrays	Study Design	Study Summary	Platform	Validation[2]	Classifier performance[3]	qRT-PCR	GEO	Ref.
Colorectal cancer and SU5416.	23	PBMC samples harvested from cancer patients before and after 6 weeks of treatment with either the standard care for chemotherapy (n=12) or the standard care plus SU5416 (n=11).	A *t*-test identified over 100 genes differentially expressed between the SU5416-treated arm and the control arm. Discriminant analysis was used to identify a 3-gene (*CD24*, *LTF*, and *LCN2*) diagnostic classifier which had 84% accuracy for assigning samples to the appropriate treatment arm as assessed by qRT-PCR analysis of an additional 67 samples (35 SU5416 and 32 controls).	Affymetrix HG-U95A GeneChip (≈10,000 genes)	Test data	84%	Yes	No	[145]
Multiple sclerosis and IFN-β.	336	Serial PBMCs samples were taken from 52 multiple sclerosis patients and the gene expression of 70 genes assessed in those samples taken prior to therapy for correlation with good (n=33) and poor (n=19) response. PBMC samples were split into training (75% of samples) and test (25% of samples) data sets with the same relative proportion of each class (63% good and 37% poor responders).	Nine high scoring gene triplets capable of predicting response outcome were identified using an LDA-based Bayesian interference system of which the triplet with the highest accuracy consisted of *Caspase 2*, *Caspase 10* and *FLIP*.	qRT-PCR	Test data	86%	Yes	No	[18]

Table 1. Continued

Disease	Arrays	Study Design	Study Summary	Platform	Validation[2]	Classifier performance[3]	qRT-PCR	GEO	Ref.
Renal cell carcinoma (RCC) and CCI-779 (Torisel)	45	To determine if PBMC gene expression profiles could be used to predict clinical outcome, PBMC samples from 45 patients with advanced RCC were taken prior to CCI-779 therapy and classified using clinical parameters to have a favorable (n=6), intermediate (n=17), or poor outcome (n=22). PBMC samples were assigned to training (70% of samples) and test (25% of samples) data sets. Time to progression (TTP), defined as the time until the first measurement of progressive disease with <106 days representing a poor outcome, and time to death (TTD) were recorded following the initiation of therapy.	Unsupervised analysis indicated that gene expression profiles grouped patients on the basis of their survival outcomes. Cox regression analysis identified a 30-gene prognostic classifier capable of accurately predicting TTP in 11/13 samples in the independent test data set. Similar analysis was used to construct a 20-gene prognostic classifier capable of accurately predicting TTD in 10/14 samples in the independent test data set.	Affymetrix HG-U95A GeneChip (≈10,000 genes)	Test data	85% for TTP, and 71% for TTD.	No	GDS1021	[146]

[1] *Arrays*, refers to the number of samples used for gene expression analysis except for the study of Baranzini et al. [18] where it refers to the number of samples analyzed by qRT-PCR; *qRT-PCR*, indicates whether the gene expression of select transcripts was confirmed by quantitative real-time RT-PCR; *GEO*, indicates whether gene expression data has been deposited at the Gene Expression Omnibus hosted by the NCBI (http://www.ncbi.nlm.nih.gov/geo/); *Ref.*, refers to reference.

[2] The validation column indicates the method used to validate the gene classifier: *Test data*, independent test data set; *LOOCV*, leave one out cross validation; *10-fold*, ten fold cross validation; *Logistic regression (LR)* and *IFN signature*, it should be stressed that classification schemes using these procedures do not represent independent assessment of the classification rate as obtained by procedures with test data or LOOCV.

[3] Classifier performance is reported as percent accuracy of classification and was calculated from data presented in the corresponding publication when italicized.

It is common at this stage to identify the biological processes that are different between the two classes under study by way of ontology or pathway analysis for which a number of software programs are now available: GenMapp [32], PathwayAssist [33], ArrayXPath [34], OntoExpress [35] and the Database for Annotation, Visualization and Integrated Discovery (DAVID) [36]. Next, class prediction can be used to construct a gene classifier whereby the top ranked differentially expressed genes are assessed for their ability to discriminate patient classes. For example, another supervised method known as Prediction Analysis of Microarrays (PAM) [37] uses the "nearest shrunken centroids" method to identify the subset of genes that best characterizes a predefined class and then uses this subset to predict the class of new samples [37]. Ideally, for gene classifier validation, samples will be divided into a training data set for gene classifier construction and a test data set for class prediction so that an unbiased estimate of the classification error rate can be obtained [23]. However, when samples are limited, it is common to use leave-one-out cross validation (LOOCV) or 10-fold cross validation [38] in order to assess classification accuracy (Table 1).

SELECTED PBMC GENE EXPRESSION PROFILING STUDIES

Thirty-one studies of PBMC gene expression resulting in the generation of a validated gene classifier of potential clinical value have been reviewed in Table 1. In addition, a further 32 PBMC gene expression studies that identified differentially expressed genes but did not construct a gene classifier have been considered for the generation of summary statistics presented in the figures [39-70]. For the main part, only gene expression studies analyzing PBMCs have been considered in this review, but it should be noted that similar analyses have been conducted with other blood fractions including whole blood, peripheral blood leukocytes (PBLs) and monocytes [71-79]. When considering all 62 studies, autoimmune (36%) and infectious (24%) disease studies represent the majority of PBMC gene expression research to date (Figure 2). Publication of PBMC gene expression studies peaked in 2004 and 2005 (Figure 3), and for each year there appears to be a balance of studies between those that only identified differentially expressed genes and those that went on to construct gene classifiers. Now that the statistical methods for constructing gene classifiers have become more firmly established in the microarray field, as detailed in the previous section, it is expected that most future PBMC gene expression profiling studies will attempt to identify gene classifiers capable of answering clinical questions of interest. The PBMC gene expression studies reviewed in Table 1 have been divided into four categories: (1) Detection of exposure, (2) Diagnosis of disease stage, (3) Prognosis of disease outcome, and (4) Pharmacogenomic assessment. The first two categories contain studies that generated diagnostic gene classifiers, the third category deals with studies that constructed prognostic gene classifiers, and the last category contains studies that generate both types of classifier. Exemplary papers are reviewed for each of these categories.

Detection of exposure. Lampe and colleagues [80] investigated whether PBL gene expression could be used as a sensor to detect environmental exposures in human subjects. Although this study did not strictly analyze gene expression in PBMCs, but rather PBLs, it is reviewed here because it provides one of the better examples of gene expression modulation in blood following toxic exposure. PBL samples were obtained from 41 self-identified smokers (16 women and 25 men) and 44 self-identified nonsmokers (21 women and 23 men) for gene expression analysis. In addition, smoking status was verified in these patients by assessing the concentration in the plasma of cotinine, the primary metabolite of nicotine [81].

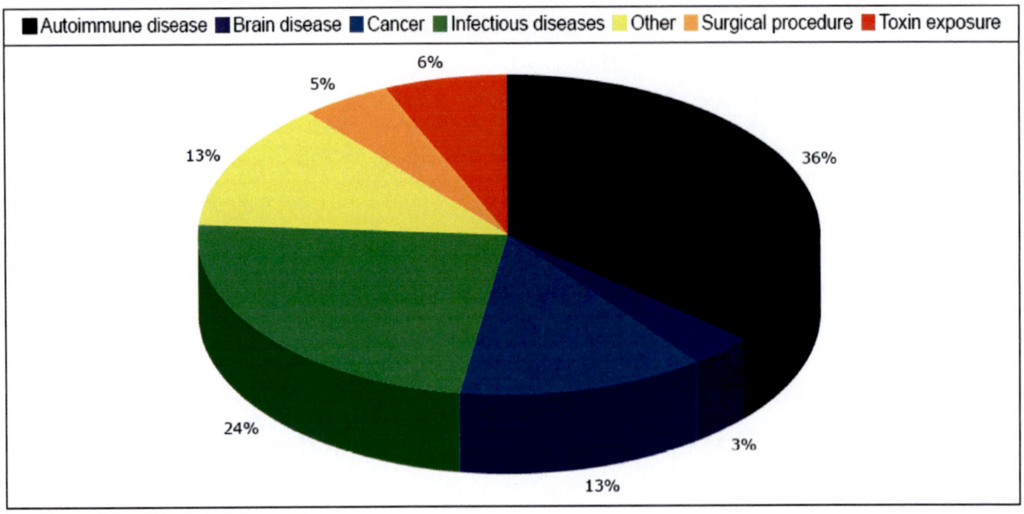

Figure 2. Diseases studied using PBMC gene expression profiling.

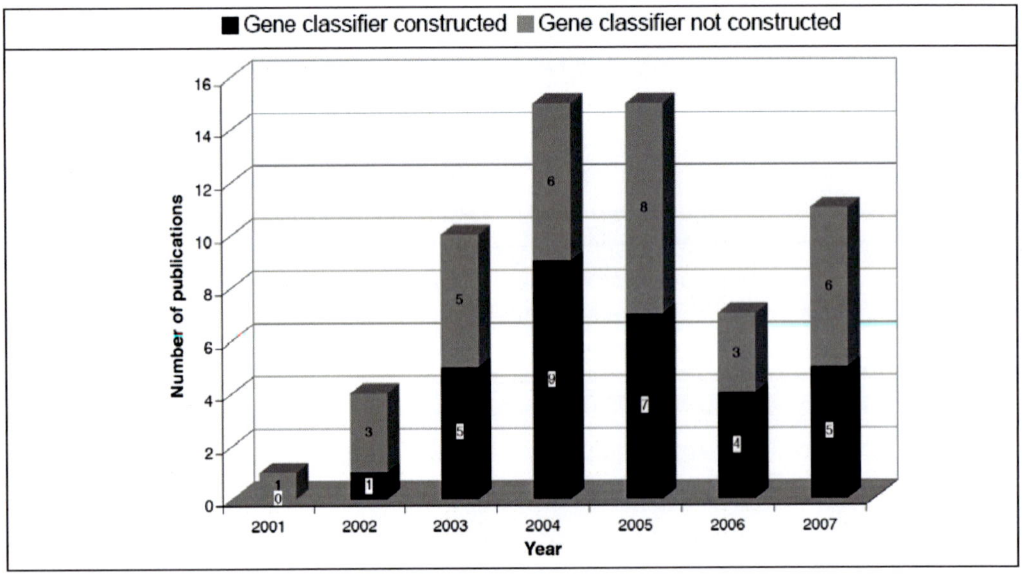

Figure 3. PBMC gene expression studies completed between 2001 and 2007.

RNA isolated from PBLs was hybridized to the Hu25K microarray (Agilent Technologies, Santa Clara, CA, USA), which is a long-oligonucleotide (60 mer) competitive hybridization (two-dye) platform capable of detecting the expression levels for 21,000 genes. Gene expression data from 32 smokers and 33 nonsmokers were selected for a training data set and data from 10 of the smokers and 10 of the nonsmokers as a test data set (TEST 1). An additional test data set (TEST 2) was constructed from 17 samples collected on a follow up visit from 10 smokers and 7 nonsmokers in TEST 1. Analysis of the training data set identified 861 genes that were differentially expressed between smokers and nonsmokers. Pearson correlation coefficients and permutation tests were used to rank genes for their correlation with plasma cotinine concentrations and 36 top candidate genes were selected on the basis of the lowest error rate. This 36-gene diagnostic classifier was used to predict the class of samples in the test data sets and only misclassified a single sample in both data sets corresponding to the same individual, a smoker misclassified as a non-smoker. Therefore, the expression levels of these 36 genes had a high accuracy for predicting whether a sample came from a smoker or a non-smoker (Table 1). Genes associated with immune function and inflammation, pathophysiology of smoking-induced injury, and response to polycyclic aromatic hydrocarbons present in cigarette smoke (such as *IL1B* [82, 83] and *CYP1B1* [84]) were among those in the 36-gene classifier. This study suggests that gene expression patterns in blood cells can be used to identify complex environmental exposure in humans.

Additional studies aiming to diagnose toxin exposure using PBMC gene expression profiling have also been performed. These studies identified differentially expressed genes between healthy controls and benzene exposed shoe factory workers [56], subjects intravenously exposed to endotoxin [64], and dioxin exposed patients 20 years after an industrial accident in Seveso, Italy [66].

Diagnosis of disease stage. The clinical diagnosis of systemic onset juvenile idiopathic arthritis (SoJIA) requires the presence of arthritis and the exclusion of an alternative diagnosis of infection or malignancy, which normally requires extensive diagnostic testing and hospitalization [85]. In order to establish a specific early diagnosis for SoJIA, Allantaz and colleagues [86] isolated PBMCs from 58 SoJIA patients and two other diseases with similar clinical features, systemic lupus erythmatosus (SLE, n=38) and Pyogenic arthritis, pyoderma gangrenosum and acne (PAPA) syndrome (n=6), as well as from 21 healthy control patients. RNA extracted from these samples was hybridized to the Affymetrix HG-U133 Set GeneChip, which consists of two separate GeneChips, HG-U133A and B, capable of probing the expression levels of approximately 33,000 genes between them. Initially, class comparison methods were used to identify the 50 most differentially expressed transcripts in a training data set of 8 SoJIA and 8 control patients. Class prediction was then performed using the K-Nearest Neighbors (*kNN*) method and showed that this 50-gene diagnostic classifier was capable of correctly classifying 100% of the patient samples in a matched test data set (8 SoJIA and 8 control patients). However, when combining gene expression data generated in this study from SLE and PAPA syndrome patients samples with those from a previous infectious disease study conducted by Ramillo et al. [24] of patients infected with *Staphylococcus aureus*, *Streptococcus pneumoniae*, *Escherichia coli* and Influenza A virus, the 50 genes in this diagnostic classifier were not shown to be SoJIA specific since they were also dysregulated by these other inflammatory conditions. Subsequently, further gene expression analysis identified 88 differentially expressed genes between SoJIA (n=16) and control patients (n=10) of which 12 were not differentially expressed between the other disease groups compared to their respective controls. This 12-gene diagnostic classifier was validated in an independent test data set (9 SoJIA, 22 SLE, 6 PAPA syndrome, 15 *S. aureus*, 5 *S. pneumoniae*, 20 *E. coli*, 8 Influenza A, and 10 control patients) and was capable of identifying 2 misdiagnosed SoJIA patients and correctly classifying 6 out of the 7 remaining SoJIA patients from all of the other non-SoJIA inflammatory disorders. This study suggests that it should be possible to develop clinically relevant classifiers capable of determining the specific cause of disease even when samples are obtained from a heterogeneous afflicted population exhibiting overlapping symptoms associated with distinct diseases.

Prognosis of disease outcome. Cutaneous T cell lymphoma (CTCL) is the most common T cell lymphoma but the etiology of this disease remains unclear [87, 88]. Sezary Syndrome (SS) is a leukemic and erythrodermic variant of CTCL characterized by the presence of Sezary cells, which are circulating lymphocytes with atypical cerebriform nuclei, in the skin, lymph nodes and peripheral blood. Fifty percent of patients with advanced SS disease do not respond to therapies, such as extracorporeal photopheresis or interferon(IFN)-α treatment, and more than 25% of those who initially respond eventually relapse and progress to fatal disease [89, 90]. In order to identify clinical markers capable of identifying those CTCL patients at risk of progressive disease, Kari and colleagues [91] isolated PBMCs from 48 leukemic phase CTCL patients and 20 healthy controls. Circulating Sezary cell counts were high (>60%) in 17 CTCL patients and low in 27 (5-53%) patients. CTCL patients were also characterized as short-term (0-6 months, n=12), intermediate (12-40 months, n=11) or long-term survivors (>40 months, n=25). Initially, 385 differentially expressed genes were identified between 18 samples from 17 patients with high Sezary cell count and 12 healthy control samples using a trio of custom cDNA arrays (nylon filters HA-01, -02 and -03) capable of probing 4,500 unique genes. To control for the Th2 phenotype exhibited in $CD4^+$ T cells from patients with a high tumor burden, PBMCs from healthy controls were skewed towards the Th2 phenotype by treatments with IL-4 and anti-IL-2 for 4 days. Supervised analysis using penalized discriminant analysis (PDA) was used to generate a 4-gene diagnostic classifier (i.e. *STAT4*, *TOB1*, *CD26* and *TRAIL*) that was capable of classifying CTCL and control samples with 100% accuracy as assessed by LOOCV. Gene expression data generated using the HA-03 array (2,032) were then stratified by survival data and PDA analysis identified a 10-gene prognostic classifier capable of predicting whether a CTCL patient would survive less than 6 months (short-term survivor) or survive for longer than 40 months (long-term survivor) with 100% accuracy, also determined by LOOCV. Since Sezary cell counts varied widely within both short- and long-term CTCL patient pools, they do not appear to be an optimal index of the severity of disease. This study highlights the potential prognostic capabilities of classifiers generated from PBMC gene expression data even when patients may die from various mechanisms of pathogenesis, as is the case with CTCL.

Pharmacogenomic assessment. Up to 30% of HIV-infected individuals in the United States may be coinfected with hepatitis C virus (HCV) [92]. Combination therapy with pegylated(peg)-IFN and ribavirin is used to treat patients with HCV infection but it is associated with significant toxicities. Efficacy of treatment varies with genotype and only moderate numbers (≈50%) of genotype I infected patients can be classified as sustained virologic responders (SVRs) [93, 94]. Coinfection with HIV increases the risk of relapse in HCV viral load following combination therapy among those patients classified as end-of-treatment responders (ETRs) [95-97]. In order to predict those patients unlikely to be SVRs with combination therapy and who could thus be spared from treatment-associated toxicities, Lempicki and colleagues [98] isolated PBMCs from 29 patients with chronic HCV infection and stable HIV disease prior to treatment with peg-IFN-α2b and ribavirin. Following treatment, 10 coinfected patients were classified as non-responders, 9 as SVRs, and 10 as ETRs who thus exhibited a subsequent relapse of HCV viral load. Gene expression analysis of the RNA extracted from these PBMC samples revealed that 1352 genes showed significant variation in expression across the 3 patient response groups. Ontology (functional category enrichment) analysis using DAVID [36] indicated that non-responders exhibited increased expression of interferon stimulated genes (ISGs) compared to the other response groups. This led the authors to suggest that patients showing an endogenously activated type I IFN response prior to therapy are resistant to the beneficial antiviral effects of IFN-α administration. Class prediction using PAM identified a 79-gene prognostic classifier, which correctly predicted the treatment outcome for all non-responders, 7 out of 9 SVRs and 8 out 10 ETRs, resulting in a classification accuracy of 86% as assessed by 10-fold cross validation. Similar analysis using PBMC samples taken at the end of treatment (48 weeks), were used to construct a 105-gene prognostic classifier capable of correctly predicting all of the ETRs and 7 out of 8 SVRs, giving a classification rate of 94%. Although this study did not strictly control for HCV genotype, it exemplifies the potential utility of gene expression classifiers for predicting therapy outcomes in patients treated with multiple drugs and infected with multiple pathogens.

Future Directions and Conclusions

Mechanistic implications of PBMC gene expression profiling. Gene expression studies are often conducted in the target tissue of interest in order to elucidate the underlying mechanisms of disease. It should be stressed that the differentially expressed genes and the components of gene classifiers identified by PBMC gene expression profiling should not be over-interpreted with respect to their mechanistic contribution to disease states. In fact, a member of a gene expression classifier is often designated as a "probably valid biomarker" by the US Food and Drug Administration in order to distinguish it from a "known valid biomarker" for which the precise mechanistic role has been established and accepted as a marker of disease activity [38, 99]. According to the NIH Biomarkers Definition Working group, a biomarker is any characteristic that is objectively measured and evaluated as an indicator of normal biological processes, pathogen infections, or pharmacologic responses to a therapeutic intervention [100]. It could be argued that the mechanistic interpretation of PBMC gene expression attains greater validity the closer the disease is to the blood compartment. If longitudinal sampling exists from a patient cohort with a disease affecting blood cells, then those genes with temporally stable expression may represent good candidates for further mechanistic study. Conversely, Glatt and colleagues [79] identified differentially expressed genes in the dorsolateral prefrontal complex (DLPFC) and in peripheral blood cells (PBCs) between schizophrenic patients and healthy controls, and noted that only two genes exhibited similar modulation in gene expression between these two compartments. Therefore, PBMC gene expression may be mechanistically linked to diseases that directly affect blood cells (e.g., HIV, autoimmune disorders, and leukemias and lymphomas), but probably represent secondary responses in diseases that affect distant compartments (e.g., renal cell carcinoma and psychiatric disorders).

Additional genomic technologies. Gene expression data generated from PBMC samples represents just one type of data that may be used to better understand the biological processes underlying complex disease and for the construction of clinically useful classifiers. Other genomic technologies can detect the expression of microRNAs (miRNAs), the presence of serum proteins and their activation states, as well as the occurrence of single-nucleotide polymorphisms (SNPs) and epigenetic changes in patient DNA [101]. It is possible that diagnostic and prognostic classifiers of greater accuracy will be obtained through multivariate analysis of the data generated when the same subset of PBMC samples are subjected to multiple different genomics technologies.

Due to cost restrictions, when a PBMC sample resource from a patient cohort becomes available, a decision must often be made whether to perform gene expression profiling or SNP genotyping. These two approaches, however, should be viewed as entirely complementary [102, 103]. For example, when SNP genotyping studies identify SNPs that correlate with a disease phenotype, a better understanding of pathogenesis may be revealed by investigating the expression of those genes that are in genomic proximity to the relevant SNPs. Conversely, once a gene classifier has been constructed from expression data, a better understanding of why the genes in the classifier are differentially expressed could be gleaned by investigating SNPs in regulatory and coding regions. Kristensen and colleagues [102] performed SNP genotyping using blood samples taken from 50 unrelated patients with breast cancer for which tumor gene expression data had been previously generated [104-106]. A total of 583 SNPs were analyzed that were associated with 203 candidate genes, of which 115 genes contained regulatory SNPs associated with gene expression in *trans* (i.e., associated with the expression of another gene), and 43 of these genes harbored SNPs associated with gene regulation in both *cis* (i.e., associated with expression of their own gene) and *trans*. Therefore, gene expression data were used to identify putative regulatory SNPs and thus shed further light on the underlying biology of breast cancer.

Both gene expression profiling and SNP genotyping provide the potential to identify prognostic biomarkers for disease [103]. Each technology has pros and cons. For example, a person's SNP genotype remains stable over time and therefore a sample may be taken at any time during a patient's life [107]. In contrast, a sample prior to drug treatment or disease onset is needed for gene expression profiling in order to develop prognostic biomarkers. Furthermore, SNP genotyping requires patient DNA, which is a more stable molecule than the RNA required for gene expression analysis [108]. Therefore, it is possible that SNP genotyping may be more successful with archived samples for which RNA preservation was not optimized. The sample selected for SNP genotyping can be collected from any tissue in the body, whereas the sample taken for gene expression profiling has historically been collected from the site of disease (e.g., a tumor) and more recently from the PBMC compartment as reviewed in this chapter. Although genome wide SNP genotyping arrays are now available (Affymetrix and Illumina), they do not contain all the SNPs relevant to the human population since human genetic diversity has not been fully mapped. Therefore, when the aim is to build a classifier, missing values are more common in SNP data than in gene expression data and it is possible that some of the features needed for highly accurate prediction will be missing [103]. Finally, many more samples are required to attain statistical significance with SNP genotype data compared to gene expression data, and this must be factored into study design [109].

Clinical Implications. The ultimate goal of PBMC gene expression studies should be the development of clinically useful classifiers that can diagnose disease states, predict disease outcomes and guide treatment options. The studies reviewed in this chapter (**Table 1**) can be characterized as developmental studies and are analogous to phase II clinical trials [38]. A number of these studies validated classifier performance using an independent subset of the data and this is referred to as "internal" validation, in contrast to "external" validation, which uses a completely independent cohort of patients or samples from a prospectively planned study. A number of studies used a sufficient number of samples to construct classifiers that subsequently attained high classification accuracy with internal validation [24, 80, 86, 110-113]. These classifiers could now be evaluated for their ability to improve standard clinical diagnostic or prognostic factors. This could initially be done by comparing the sensitivity, specificity, accuracy and predictive values of gene classifiers to those currently used in clinical practice [114]. Specifically, the predictive power of the multivariable model that contains the gene classifier and pre-existing prognostic factors could be compared to the predictive power of the model that lacks the gene classifier [114-116]. For example, the contribution of a prognostic gene classifier used to predict survival outcome to the standard clinical prognostic test can be computed from the test data using a log-rank test or receiver operating curve (ROC) analysis [117].

The diagnostic gene classifier constructed by Ockenhouse et al. [118] to discriminate HIV seropositive from seronegative individuals provides a good example of a gene classifier that does not improve the prognostic value of the current clinical test. Training and test data sets were used for gene classifier construction and validation, respectively, resulting in a sensitivity of 92%, a specificity of 100%, and an overall accuracy of prediction of 93%. The current standard for detecting HIV seropositivity uses a repeatedly reactive ELISA followed by a confirmatory Western blot [119]. For ELISA alone, the sensitivity and specificity is 99.7% and 98.5%, respectively [120], which when coupled with Western blot analysis results in a false positive rate of 1 in 250,000 (95% CI, 1 in 173 000 to 1 in 379 000) [121]. Therefore, it is unlikely that a diagnostic gene classifier will supersede or add to the accuracy of the current clinical test. However, there are many other aspects of HIV pathogenesis and drug treatment for which diagnostic and prognostic gene classifiers may be of benefit.

Once the classifier has been shown to enhance current clinical assays, the reproducibility of classifier performance should be externally validated using a prospectively planned phase III study [114, Simon, 2005 #1246]. Such studies attempt to utilize the gene classifier in a manner reflective of how it might be adopted for use in a broad clinical practice. External validation does not require samples to be analyzed by microarrays capable of interrogating the whole genome since constituent genes of the classifiers can be formulated into smaller qRT-PCR assays (e.g., TaqMan Low Density Arrays, ABI, Foster City CA). Such external validation is essential since internal validation can be biased by: (1) tissue handling and assay reagent differences between cases and controls, (2) the selection of patient cohort for primary analysis, and (3) the use of archived samples which may not reflect the patient population as it stands today [38].

Encouragingly, gene expression classifiers are becoming commercially available as exemplified by 3 gene expression tests for evaluating recurrence risk in early stage breast cancer: Oncotype DX™ (Genomic Health Inc., Redwood City, CA) The Breast Cancer Gene Expression Ratio (Quest Diagnostics, Madison, HJ), and MammaPrint® (Agendia, Netherlands). Only MammaPrint® has been cleared as an "*in vitro* diagnostic multivariate index assay" (IVDMIA) by the FDA, although the other two gene expression classifiers seem to have been as rigorously assessed for their ability to improve the diagnosis of early stage breast cancer patients for the risk of recurrence [122]. Currently, both Oncotype DX™ and MammaPrint® are involved in further large, high-powered, prospective, randomized, and controlled clinical trials [101, 123]. Based on these advances in the field of oncology, it is likely that gene expression classifiers emerging from appropriately planned PBMC gene expression studies with internal and external validation, will become commercially available for clinical use in various therapeutic areas in the coming years.

Conclusion. Now that concerns about the reliability and reproducibility of microarray gene expression data have been satisfactorily addressed [6, 12, 101, 124-126] the field of hemogenomics is blossoming. This is apparent from the large number of PBMC gene expression studies reviewed here (Figure 3), which provide a solid foundation upon which to build. Many of these developmental studies (Table 1) have already identified promising diagnostic and prognostic gene classifiers that must be further investigated in order to assess their potential clinical impact [24, 80, 86, 110-113]. Encouragingly, gene classifiers are not only effective for binary comparisons (i.e., control vs. disease) but are also capable of distinguishing a specific disease in a population of patients exhibiting multiple disease phenotypes [55, 86]. Gene expression analysis of RNA from whole blood that was stabilized in PAXgene tubes and subjected to globin mRNA reduction has been used to generate gene expression classifiers with comparable accuracy to those developed with PBMC samples [6]. Since whole blood is even more readily accessible than PBMCs and the transcriptome can be stabilized almost immediately after draw utilizing the PAXgene methodology, this may become the specimen of choice for future gene expression studies. Finally, prognostic gene classifiers show the greatest promise since these classifiers may be used to predict disease outcomes and to spare patients from unnecessary toxicities by identifying the subset of patients most likely to respond to a particular drug therapy [38].

Author's Disclosure of Potential Conflicts of Interest

The authors indicated no potential conflicts of interest.

Acknowledgements

CHW wrote this review with the support of the Genomics Core at the UCSD Center for AIDS Research (2P30 AI36214) and further grant funding from the NIH (AI065242). The authors wish to thank Drs. David Butler, Miguel Goicoechea, Steffney Rought and Mr. Jean Lozach for useful comments and suggestions.

References

[1] English, D. and Andersen, B.R., (1974) Single-step separation of red blood cells. Granulocytes and mononuclear leukocytes on discontinuous density gradients of Ficoll-Hypaque. *J Immunol Methods*, **5**(3): 249-252.

[2] Affymetrix technical note, (2003) An analysis of blood processing methods to prepare samples for GeneChip® expression profiling. [cited 2007 January 3rd]; Available from: http://www.affymetrix.com/support/technical/technotes/blood_technote.pdf.

[3] Tanner, M.A., Berk, L.S., Felten, D.L., Blidy, A.D., Bit, S.L., and Ruff, D.W., (2002) Substantial changes in gene expression level due to the storage temperature and storage duration of human whole blood. *Clin Lab Haematol*, **24**(6): 337-341.

[4] Baechler, E.C., Batliwalla, F.M., Karypis, G., Gaffney, P.M., Moser, K., Ortmann, W.A., Espe, K.J., Balasubramanian, S., Hughes, K.M., Chan, J.P., Begovich, A., Chang, S.Y., Gregersen, P.K., and Behrens, T.W., (2004) Expression levels for many genes in human peripheral blood cells are highly sensitive to ex vivo incubation. *Genes Immun*, **5**(5): 347-353.

[5] Feezor, R.J., Baker, H.V., Mindrinos, M., Hayden, D., Tannahill, C.L., Brownstein, B.H., Fay, A., MacMillan, S., Laramie, J., Xiao, W., Moldawer, L.L., Cobb, J.P., Laudanski, K., Miller-Graziano, C.L., Maier, R.V., Schoenfeld, D., Davis, R.W., and Tompkins, R.G., (2004) Whole blood and leukocyte RNA isolation for gene expression analyses. *Physiol Genomics*, **19**(3): 247-254.

[6] Debey, S., Schoenbeck, U., Hellmich, M., Gathof, B.S., Pillai, R., Zander, T., and Schultze, J.L., (2004) Comparison of different isolation techniques prior gene expression profiling of blood derived cells: impact on physiological responses, on overall expression and the role of different cell types. *Pharmacogenomics J*, **4**(3): 193-207.

[7] Breit, S., Nees, M., Schaefer, U., Pfoersich, M., Hagemeier, C., Muckenthaler, M., and Kulozik, A.E., (2004) Impact of pre-analytical handling on bone marrow mRNA gene expression. *Br J Haematol*, **126**(2): 231-243.

[8] Fan, H. and Hegde, P.S., (2005) The transcriptome in blood: challenges and solutions for robust expression profiling. *Curr Mol Med*, **5**(1): 3-10.

[9] Cobb, J.P., Mindrinos, M.N., Miller-Graziano, C., Calvano, S.E., Baker, H.V., Xiao, W., Laudanski, K., Brownstein, B.H., Elson, C.M., Hayden, D.L., Herndon, D.N., Lowry, S.F., Maier, R.V., Schoenfeld, D.A., Moldawer, L.L., Davis, R.W., Tompkins, R.G., Baker, H.V., Bankey, P., Billiar, T., Brownstein, B.H., Calvano, S.E., Camp, D., Chaudry, I., Cobb, J.P., Davis, R.W., Elson, C.M., Freeman, B., Gamelli, R., Gibran, N., Harbrecht, B., Hayden, D.L., Heagy, W., Heimbach, D., Herndon, D.N., Horton, J., Hunt, J., Laudanski, K., Lederer, J., Lowry, S.F., Maier, R.V., Mannick, J., McKinley, B., Miller-Graziano, C., Mindrinos, M.N., Minei, J., Moldawer, L.L., Moore, E., Moore, F., Munford, R., Nathens, A., O'Keefe, G., Purdue, G., Rahme, L., Remick, D., Sailors, M., Schoenfeld, D.A., Shapiro, M., Silver, G., Smith, R., Stephanopoulos, G., Stormo, G., Tompkins, R.G., Toner, M., Warren, S., West, M., Wolfe, S., Xiao, W., and Young, V., (2005) Application of genome-wide expression analysis to human health and disease. *Proc Natl Acad Sci U S A*, **102**(13): 4801-4806.

[10] Sasik, R., Calvo, E., and Corbeil, J., (2002) Statistical analysis of high-density oligonucleotide arrays: a multiplicative noise model. *Bioinformatics*, **18**(12): 1633-1640.

[11] Irizarry, R.A., Hobbs, B., Collin, F., Beazer-Barclay, Y.D., Antonellis, K.J., Scherf, U., and Speed, T.P., (2003) Exploration, normalization, and summaries of high density oligonucleotide array probe level data. *Biostatistics*, **4**(2): 249-264.

[12] Barnes, M., Freudenberg, J., Thompson, S., Aronow, B., and Pavlidis, P., (2005) Experimental comparison and cross-validation of the Affymetrix and Illumina gene expression analysis platforms. *Nucleic Acids Res*, **33**(18): 5914-5923.

[13] Elvidge, G., (2006) Microarray expression technology: from start to finish. *Pharmacogenomics*, **7**(1): 123-134.

[14] Olson, N.E., (2006) The microarray data analysis process: from raw data to biological significance. *NeuroRx*, **3**(3): 373-383.

[15] Ahmed, F.E., (2006) Microarray RNA transcriptional profiling: part I. Platforms, experimental design and standardization. *Expert Rev Mol Diagn*, **6**(4): 535-550.

[16] Parmigiani, G., Garrett, E.S., Irizarry, R.A., and Zeger, S.L., (2003) The Analysis of Gene Expression Data: Methods and Software. New York: Springer-Verlag.

[17] Schena, M., (2003) Microarray Analysis. New York: Wiley-Liss, John Wiley & Sons, Inc.

[18] Baranzini, S.E., Mousavi, P., Rio, J., Caillier, S.J., Stillman, A., Villoslada, P., Wyatt, M.M., Comabella, M., Greller, L.D., Somogyi, R., Montalban, X., and Oksenberg, J.R., (2005) Transcription-based prediction of response to IFNβ using supervised computational methods. *PLoS Biol*, **3**(1): e2.

[19] Gardina, P.J., Clark, T.A., Shimada, B., Staples, M.K., Yang, Q., Veitch, J., Schweitzer, A., Awad, T., Sugnet, C., Dee, S., Davies, C., Williams, A., and Turpaz, Y., (2006) Alternative splicing and differential gene expression in colon cancer detected by a whole genome exon array. *BMC Genomics*, **7**: 325.

[20] Kruzelock, R.P. and Short, W., (2007) Colorectal cancer therapeutics and the challenges of applied pharmacogenomics. *Curr Probl Cancer*, **31**(5): 315-366.

[21] Tibshirani, R., (2006) A simple method for assessing sample sizes in microarray experiments. *BMC Bioinformatics*, **7**: 106.

[22] Zien, A., Fluck, J., Zimmer, R., and Lengauer, T., (2003) Microarrays: how many do you need? *J Comput Biol*, **10**(3-4): 653-667.

[23] Simon, R., Radmacher, M.D., Dobbin, K., and McShane, L.M., (2003) Pitfalls in the use of DNA microarray data for diagnostic and prognostic classification. *J Natl Cancer Inst*, **95**(1): 14-18.

[24] Ramilo, O., Allman, W., Chung, W., Mejias, A., Ardura, M., Glaser, C., Wittkowski, K.M., Piqueras, B., Banchereau, J., Palucka, A.K., and Chaussabel, D., (2007) Gene expression patterns in blood leukocytes discriminate patients with acute infections. *Blood*, **109**(5): 2066-2077.

[25] Reich, M., Liefeld, T., Gould, J., Lerner, J., Tamayo, P., and Mesirov, J.P., (2006) GenePattern 2.0. *Nat Genet*, **38**(5): 500-501.

[26] Simon, R. and Lam, A., (2004) BRB ArrayTools user guide, version 3.2: Biometric Research Branch, National Cancer Institute.

[27] Raychaudhuri, S., Sutphin, P.D., Chang, J.T., and Altman, R.B., (2001) Basic microarray analysis: grouping and feature reduction. *Trends Biotechnol*, **19**(5):189-193.

[28] Gentleman, R., Carey, V., Huber, W., Irizarry, R.A., and Dudoit, S., (2005) Bioinformatics and Computational Biology Solutions Using R and Bioconductor. New York: Springer.

[29] Eisen, M.B., Spellman, P.T., Brown, P.O., and Botstein, D., (1998) Cluster analysis and display of genome-wide expression patterns. *Proc Natl Acad Sci USA*, **95**(25): 14863-14868.

[30] Slonim, D.K., Tamayo, P., Mesirov, J., Golub, T.R., and Lander, E.S., (2000). Class prediction and discovery using gene expression data. in *Proceedings of the fourth annual international conference on computational molecular biology (RECOMB)*. Tokyo: Universal Academy Press.

[31] Tusher, V.G., Tibshirani, R., and Chu, G., (2001) Significance analysis of microarrays applied to the ionizing radiation response. *Proc Natl Acad Sci U S A*, **98**(9): 5116-5121.

[32] Dahlquist, K.D., Salomonis, N., Vranizan, K., Lawlor, S.C., and Conklin, B.R., (2002) GenMAPP, a new tool for viewing and analyzing microarray data on biological pathways. *Nat. Genet.*, **31**(1): 19-20.

[33] Nikitin, A., Egorov, S., Daraselia, N., and Mazo, I., (2003) Pathway studio--the analysis and navigation of molecular networks. *Bioinformatics*, **19**(16): 2155-2157.

[34] Chung, H.J., Kim, M., Park, C.H., Kim, J., and Kim, J.H., (2004) ArrayXPath: mapping and visualizing microarray gene-expression data with integrated biological pathway resources using Scalable Vector Graphics. *Nucleic Acids Res*, **32**(Web Server issue): W460-464.

[35] Draghici, S., Khatri, P., Martins, R.P., Ostermeier, G.C., and Krawetz, S.A., (2003) Global functional profiling of gene expression. *Genomics*, **81**(2): 98-104.

[36] Dennis, G., Jr., Sherman, B.T., Hosack, D.A., Yang, J., Gao, W., Lane, H.C., and Lempicki, R.A., (2003) DAVID: Database for Annotation, Visualization, and Integrated Discovery. *Genome Biol*, **4**(5): P3.

[37] Tibshirani, R., Hastie, T., Narasimhan, B., and Chu, G., (2002) Diagnosis of multiple cancer types by shrunken centroids of gene expression. *Proc Natl Acad Sci U S A*, **99**(10): 6567-6572.

[38] Simon, R., (2005) Roadmap for developing and validating therapeutically relevant genomic classifiers. *J Clin Oncol*, **23**(29): 7332-7341.

[39] Ramanathan, M., Weinstock-Guttman, B., Nguyen, L.T., Badgett, D., Miller, C., Patrick, K., Brownscheidle, C., and Jacobs, L., (2001) In vivo gene expression revealed by cDNA arrays: the pattern in relapsing-remitting multiple sclerosis patients compared with normal subjects. *J Neuroimmunol*, **116**(2): 213-219.

[40] Gu, J., Marker-Hermann, E., Baeten, D., Tsai, W.C., Gladman, D., Xiong, M., Deister, H., Kuipers, J.G., Huang, F., Song, Y.W., Maksymowych, W., Kalsi, J., Bannai, M., Seta, N., Rihl, M., Crofford, L.J., Veys, E., De Keyser, F., and Yu, D.T., (2002) A 588-gene microarray analysis of the peripheral blood mononuclear cells of spondyloarthropathy patients. *Rheumatology (Oxford)*, **41**(7): 759-766.

[41] Rus, V., Atamas, S.P., Shustova, V., Luzina, I.G., Selaru, F., Magder, L.S., and Via, C.S., (2002) Expression of cytokine- and chemokine-related genes in peripheral blood mononuclear cells from lupus patients by cDNA array. *Clin Immunol*, **102**(3): 283-290.

[42] Vernon, S.D., Unger, E.R., Dimulescu, I.M., Rajeevan, M., and Reeves, W.C., (2002) Utility of the blood for gene expression profiling and biomarker discovery in chronic fatigue syndrome. *Dis Markers*, **18**(4): 193-199.

[43] Bennett, L., Palucka, A.K., Arce, E., Cantrell, V., Borvak, J., Banchereau, J., and Pascual, V., (2003) Interferon and granulopoiesis signatures in systemic lupus erythematosus blood. *J Exp Med*, **197**(6): 711-723.

[44] Ma, J. and Liew, C.C., (2003) Gene profiling identifies secreted protein transcripts from peripheral blood cells in coronary artery disease. *J Mol Cell Cardiol*, **35**(8): 993-998.

[45] Nowicki, M.O., Pawlowski, P., Fischer, T., Hess, G., Pawlowski, T., and Skorski, T., (2003) Chronic myelogenous leukemia molecular signature. *Oncogene*, **22**(25): 3952-3963.

[46] Sturzebecher, S., Wandinger, K.P., Rosenwald, A., Sathyamoorthy, M., Tzou, A., Mattar, P., Frank, J.A., Staudt, L., Martin, R., and McFarland, H.F., (2003) Expression profiling identifies responder and non-responder phenotypes to interferon-beta in multiple sclerosis. *Brain*, **126**(6): 1419-1429.

[47] Whistler, T., Unger, E.R., Nisenbaum, R., and Vernon, S.D., (2003) Integration of gene expression, clinical, and epidemiologic data to characterize Chronic Fatigue Syndrome. *J Transl Med*, **1**(1): 10.

[48] Achiron, A., Gurevich, M., Magalashvili, D., Kishner, I., Dolev, M., and Mandel, M., (2004) Understanding autoimmune mechanisms in multiple sclerosis using gene expression microarrays: treatment effect and cytokine-related pathways. *Clin Dev Immunol*, **11**(3-4): 299-305.

[49] Bovin, L.F., Rieneck, K., Workman, C., Nielsen, H., Sorensen, S.F., Skjodt, H., Florescu, A., Brunak, S., and Bendtzen, K., (2004) Blood cell gene expression profiling in rheumatoid arthritis. Discriminative genes and effect of rheumatoid factor. *Immunol Lett*, **93**(2-3): 217-226.

[50] Bullinger, L., Dohner, K., Bair, E., Frohling, S., Schlenk, R.F., Tibshirani, R., Dohner, H., and Pollack, J.R., (2004) Use of gene-expression profiling to identify prognostic subclasses in adult acute myeloid leukemia. *N Engl J Med*, **350**(16): 1605-1616.

[51] Feezor, R.J., Baker, H.V., Xiao, W., Lee, W.A., Huber, T.S., Mindrinos, M., Kim, R.A., Ruiz-Taylor, L., Moldawer, L.L., Davis, R.W., and Seeger, J.M., (2004) Genomic and proteomic determinants of outcome in patients undergoing thoracoabdominal aortic aneurysm repair. *J Immunol*, **172**(11): 7103-7109.

[52] Jison, M.L., Munson, P.J., Barb, J.J., Suffredini, A.F., Talwar, S., Logun, C., Raghavachari, N., Beigel, J.H., Shelhamer, J.H., Danner, R.L., and Gladwin, M.T., (2004) Blood mononuclear cell gene expression profiles characterize the oxidant, hemolytic, and inflammatory stress of sickle cell disease. *Blood*, **104**(1): 270-280.

[53] Mannick, E.E., Bonomolo, J.C., Horswell, R., Lentz, J.J., Serrano, M.S., Zapata-Velandia, A., Gastanaduy, M., Himel, J.L., Rose, S.L., Udall, J.N., Jr., Hornick, C.A., and Liu, Z., (2004) Gene expression in mononuclear cells from patients with inflammatory bowel disease. *Clin Immunol*, **112**(3): 247-257.

[54] Boni, J.P., Leister, C., Bender, G., Fitzpatrick, V., Twine, N., Stover, J., Dorner, A., Immermann, F., and Burczynski, M.E., (2005) Population pharmacokinetics of CCI-779: correlations to safety and pharmacogenomic responses in patients with advanced renal cancer. *Clin Pharmacol Ther*, **77**(1): 76-89.

[55] Chaussabel, D., Allman, W., Mejias, A., Chung, W., Bennett, L., Ramilo, O., Pascual, V., Palucka, A.K., and Banchereau, J., (2005) Analysis of significance patterns identifies ubiquitous and disease-specific gene-expression signatures in patient peripheral blood leukocytes. *Ann N Y Acad Sci*, **1062**: 146-154.

[56] Forrest, M.S., Lan, Q., Hubbard, A.E., Zhang, L., Vermeulen, R., Zhao, X., Li, G., Wu, Y.Y., Shen, M., Yin, S., Chanock, S.J., Rothman, N., and Smith, M.T., (2005) Discovery of novel biomarkers by microarray analysis of peripheral blood mononuclear cell gene expression in benzene-exposed workers. *Environ Health Perspect*, **113**(6): 801-807.

[57] Kaushik, N., Fear, D., Richards, S.C., McDermott, C.R., Nuwaysir, E.F., Kellam, P., Harrison, T.J., Wilkinson, R.J., Tyrrell, D.A., Holgate, S.T., and Kerr, J.R., (2005) Gene expression in peripheral blood mononuclear cells from patients with chronic fatigue syndrome. *J Clin Pathol*, **58**(8): 826-832.

[58] Maas, K., Chen, H., Shyr, Y., Olsen, N.J., and Aune, T., (2005) Shared gene expression profiles in individuals with autoimmune disease and unaffected first-degree relatives of individuals with autoimmune disease. *Hum Mol Genet*, **14**(10): 1305-1314.

[59] O'Toole, M., Janszen, D.B., Slonim, D.K., Reddy, P.S., Ellis, D.K., Legault, H.M., Hill, A.A., Whitley, M.Z., Mounts, W.M., Zuberek, K., Immermann, F.W., Black, R.S., and Dorner, A.J., (2005) Risk factors associated with beta-amyloid(1-42) immunotherapy in preimmunization gene expression patterns of blood cells. *Arch Neurol*, **62**(10): 1531-1536.

[60] Reghunathan, R., Jayapal, M., Hsu, L.Y., Chng, H.H., Tai, D., Leung, B.P., and Melendez, A.J., (2005) Expression profile of immune response genes in patients with Severe Acute Respiratory Syndrome. *BMC Immunol*, **6**: 2.

[61] Yu, S.Y., Hu, Y.W., Liu, X.Y., Xiong, W., Zhou, Z.T., and Yuan, Z.H., (2005) Gene expression profiles in peripheral blood mononuclear cells of SARS patients. *World J Gastroenterol*, **11**(32): 5037-5043.

[62] He, X.S., Ji, X., Hale, M.B., Cheung, R., Ahmed, A., Guo, Y., Nolan, G.P., Pfeffer, L.M., Wright, T.L., Risch, N., Tibshirani, R., and Greenberg, H.B., (2006) Global transcriptional response to interferon is a determinant of HCV treatment outcome and is modified by race. *Hepatology*, **44**(2): 352-359.

[63] Montano, M., Rarick, M., Sebastiani, P., Brinkmann, P., Russell, M., Navis, A., Wester, C., Thior, I., and Essex, M., (2006) Gene-expression profiling of HIV-1 infection and perinatal transmission in Botswana. *Genes Immun*. **7**(4): 298-309.

[64] Talwar, S., Munson, P.J., Barb, J., Fiuza, C., Cintron, A.P., Logun, C., Tropea, M., Khan, S., Reda, D., Shelhamer, J.H., Danner, R.L., and Suffredini, A.F., (2006) Gene expression profiles of peripheral blood leukocytes after endotoxin challenge in humans. *Physiol Genomics*, **25**(2): 203-215.

[65] Achiron, A., Feldman, A., Mandel, M., and Gurevich, M., (2007) Impaired expression of peripheral blood apoptotic-related gene transcripts in acute multiple sclerosis relapse. *Ann N Y Acad Sci*, **1107**: 155-167.

[66] McHale, C.M., Zhang, L., Hubbard, A.E., Zhao, X., Baccarelli, A., Pesatori, A.C., Smith, M.T., and Landi, M.T., (2007) Microarray analysis of gene expression in peripheral blood mononuclear cells from dioxin-exposed human subjects. *Toxicology*, **229**(1-2): 101-113.

[67] Scherer, C.A., Magness, C.L., Steiger, K.V., Poitinger, N.D., Caputo, C.M., Miner, D.G., Winokur, P.L., Klinzman, D., McKee, J., Pilar, C., Ward, P.A., Gillham, M.H., Haulman, N.J., Stapleton, J.T., and Iadonato, S.P., (2007) Distinct gene expression profiles in peripheral blood mononuclear cells from patients infected with vaccinia virus, yellow fever 17D virus, or upper respiratory infections. *Vaccine*, **25**(35): 6458-6473.

[68] Taylor, M.W., Tsukahara, T., Brodsky, L., Schaley, J., Sanda, C., Stephens, M.J., McClintick, J.N., Edenberg, H.J., Li, L., Tavis, J.E., Howell, C., and Belle, S.H., (2007) Changes in gene expression during pegylated interferon and ribavirin therapy of chronic hepatitis C virus distinguish responders from nonresponders to antiviral therapy. *J Virol*, **81**(7): 3391-3401.

[69] Wang, Y., Dennehy, P.H., Keyserling, H.L., Tang, K., Gentsch, J.R., Glass, R.I., and Jiang, B., (2007) Rotavirus infection alters peripheral T-cell homeostasis in children with acute diarrhea. *J Virol*, **81**(8): 3904-3912.

[70] Zilliox, M.J., Moss, W.J., and Griffin, D.E., (2007) Gene expression changes in peripheral blood mononuclear cells during measles virus infection. *Clin Vaccine Immunol*, **14**(7): 918-923.

[71] Kim, S.J., Dix, D.J., Thompson, K.E., Murrell, R.N., Schmid, J.E., Gallagher, J.E., and Rockett, J.C., (2007) Effects of storage, RNA extraction, genechip type, and donor sex on gene expression profiling of human whole blood. *Clin Chem*, **53**(6): 1038-1045.

[72] Timofeeva, A.V., Goryunova, L.E., Khaspekov, G.L., Kovalevskii, D.A., Scamrov, A.V., Bulkina, O.S., Karpov, Y.A., Talitskii, K.A., Buza, V.V., Britareva, V.V., and Beabealashvilli, R., (2006) Altered gene expression pattern in peripheral blood leukocytes from patients with arterial hypertension. *Ann N Y Acad Sci*, **1091**: 319-335.

[73] Jarvis, J.N., Dozmorov, I., Jiang, K., Frank, M.B., Szodoray, P., Alex, P., and Centola, M., (2004) Novel approaches to gene expression analysis of active polyarticular juvenile rheumatoid arthritis. *Arthritis Res Ther*, **6**(1): R15-R32.

[74] Han, G.M., Chen, S.L., Shen, N., Ye, S., Bao, C.D., and Gu, Y.Y., (2003) Analysis of gene expression profiles in human systemic lupus erythematosus using oligonucleotide microarray. *Genes Immun*, **4**(3): 177-186.

[75] Wang, Y., Barbacioru, C.C., Shiffman, D., Balasubramanian, S., Iakoubova, O., Tranquilli, M., Albornoz, G., Blake, J., Mehmet, N.N., Ngadimo, D., Poulter, K., Chan, F., Samaha, R.R., and Elefteriades, J.A., (2007) Gene expression signature in peripheral blood detects thoracic aortic aneurysm. *PLoS ONE*, **2**(10): e1050.

[76] McLean, L.A., Gathmann, I., Capdeville, R., Polymeropoulos, M.H., and Dressman, M., (2004) Pharmacogenomic analysis of cytogenetic response in chronic myeloid leukemia patients treated with imatinib. *Clin Cancer Res*, **10**(1): 155-165.

[77] Jost-Albrecht, K. and Hofstetter, W., (2006) Gene expression by human monocytes from peripheral blood in response to exposure to metals. *J Biomed Mater Res B Appl Biomater*, **76**(2): 449-455.

[78] Nikpour, M., Dempsey, A.A., Urowitz, M.B., Gladman, D.D., and Barnes, D.A., (2007) Association of a gene expression profile from whole blood with disease activity in systemic lupus erythematosus. *Ann Rheum Dis*: [Epub ahead of print].

[79] Glatt, S.J., Everall, I.P., Kremen, W.S., Corbeil, J., Sasik, R., Khanlou, N., Han, M., Liew, C.C., and Tsuang, M.T., (2005) Comparative gene expression analysis of blood and brain provides concurrent validation of SELENBP1 up-regulation in schizophrenia. *Proc Natl Acad Sci U S A*, **102**(43): 15533-15538.

[80] Lampe, J.W., Stepaniants, S.B., Mao, M., Radich, J.P., Dai, H., Linsley, P.S., Friend, S.H., and Potter, J.D., (2004) Signatures of environmental exposures using peripheral leukocyte gene expression: tobacco smoke. *Cancer Epidemiol Biomarkers Prev*, **13**(3): 445-453.

[81] International Agency for Research on Cancer (IARC), (1986) Tobacco smoking. *IARC monographs on the evaluation of the carcinogenic risk of chemicals to humans*. Vol. 38. Switzerland: IARC Sci. Publ.

[82] Kuschner, W.G., D'Alessandro, A., Wong, H., and Blanc, P.D., (1996) Dose-dependent cigarette smoking-related inflammatory responses in healthy adults. *Eur Respir J*, **9**(10): 1989-1994.

[83] Mikuniya, T., Nagai, S., Tsutsumi, T., Morita, K., Mio, T., Satake, N., and Izumi, T., (1999) Proinflammatory or regulatory cytokines released from BALF macrophages of healthy smokers. *Respiration*, **66**(5): 419-426.

[84] Piipari, R., Savela, K., Nurminen, T., Hukkanen, J., Raunio, H., Hakkola, J., Mantyla, T., Beaune, P., Edwards, R.J., Boobis, A.R., and Anttila, S., (2000) Expression of CYP1A1, CYP1B1 and CYP3A, and polycyclic aromatic hydrocarbon-DNA adduct formation in bronchoalveolar macrophages of smokers and non-smokers. *Int J Cancer*, **86**(5): 610-616.

[85] Cassidy, J.T., Levinson, J.E., Bass, J.C., Baum, J., Brewer, E.J., Jr., Fink, C.W., Hanson, V., Jacobs, J.C., Masi, A.T., Schaller, J.G., and et al., (1986) A study of classification criteria for a diagnosis of juvenile rheumatoid arthritis. *Arthritis Rheum*, **29**(2): 274-281.

[86] Allantaz, F., Chaussabel, D., Stichweh, D., Bennett, L., Allman, W., Mejias, A., Ardura, M., Chung, W., Wise, C., Palucka, K., Ramilo, O., Punaro, M., Banchereau, J., and Pascual, V., (2007) Blood leukocyte microarrays to diagnose systemic onset juvenile idiopathic arthritis and follow the response to IL-1 blockade. *J Exp Med*, **204**(9): 2131-2144.

[87] Li, G., Vowels, B.R., Benoit, B.M., Rook, A.H., and Lessin, S.R., (1996) Failure to detect human T-lymphotropic virus type-I proviral DNA in cell lines and tissues from patients with cutaneous T-cell lymphoma. *J Invest Dermatol*, **107**(3): 308-313.

[88] Kim, Y.H. and Hoppe, R.T., (1999) Mycosis fungoides and the Sezary syndrome. *Semin Oncol*, **26**(3): 276-289.

[89] Gottlieb, S.L., Wolfe, J.T., Fox, F.E., DeNardo, B.J., Macey, W.H., Bromley, P.G., Lessin, S.R., and Rook, A.H., (1996) Treatment of cutaneous T-cell lymphoma with extracorporeal photopheresis monotherapy and in combination with recombinant interferon alfa: a 10-year experience at a single institution. *J Am Acad Dermatol*, **35**(6): 946-957.

[90] Rook, A.H., Suchin, K.R., Kao, D.M., Yoo, E.K., Macey, W.H., DeNardo, B.J., Bromely, P.G., Geng, Y., Junkins-Hopkins, J.M., and Lessin, S.R., (1999) Photopheresis: clinical applications and mechanism of action. *J Investig Dermatol Symp Proc*, **4**(1): 85-90.

[91] Kari, L., Loboda, A., Nebozhyn, M., Rook, A.H., Vonderheid, E.C., Nichols, C., Virok, D., Chang, C., Horng, W.H., Johnston, J., Wysocka, M., Showe, M.K., and Showe, L.C., (2003) Classification and prediction of survival in patients with the leukemic phase of cutaneous T cell lymphoma. *J Exp Med*, **197**(11): 1477-1488.

[92] Sherman, K.E., Rouster, S.D., Chung, R.T., and Rajicic, N., (2002) Hepatitis C Virus prevalence among patients infected with Human Immunodeficiency Virus: a cross-sectional analysis of the US adult AIDS Clinical Trials Group. *Clin Infect Dis*, **34**(6): 831-837.

[93] Strader, D.B., Wright, T., Thomas, D.L., and Seeff, L.B., (2004) Diagnosis, management, and treatment of hepatitis C. *Hepatology*, **39**(4): 1147-1171.

[94] Fried, M.W., Shiffman, M.L., Reddy, K.R., Smith, C., Marinos, G., Goncales, F.L., Jr., Haussinger, D., Diago, M., Carosi, G., Dhumeaux, D., Craxi, A., Lin, A., Hoffman, J., and Yu, J., (2002) Peginterferon alfa-2a plus ribavirin for chronic hepatitis C virus infection. *N Engl J Med*, **347**(13): 975-982.

[95] Chung, R.T., Andersen, J., Volberding, P., Robbins, G.K., Liu, T., Sherman, K.E., Peters, M.G., Koziel, M.J., Bhan, A.K., Alston, B., Colquhoun, D., Nevin, T., Harb, G., and van der Horst, C., (2004) Peginterferon Alfa-2a plus ribavirin versus interferon alfa-2a plus ribavirin for chronic hepatitis C in HIV-coinfected persons. *N Engl J Med*, **351**(5): 451-459.

[96] Torriani, F.J., Rodriguez-Torres, M., Rockstroh, J.K., Lissen, E., Gonzalez-Garcia, J., Lazzarin, A., Carosi, G., Sasadeusz, J., Katlama, C., Montaner, J., Sette, H., Jr., Passe, S., De Pamphilis, J., Duff, F., Schrenk, U.M., and Dieterich, D.T., (2004) Peginterferon Alfa-2a plus ribavirin for chronic hepatitis C virus infection in HIV-infected patients. *N Engl J Med*, **351**(5): 438-450.

[97] Carrat, F., Bani-Sadr, F., Pol, S., Rosenthal, E., Lunel-Fabiani, F., Benzekri, A., Morand, P., Goujard, C., Pialoux, G., Piroth, L., Salmon-Ceron, D., Degott, C., Cacoub, P., and Perronne, C., (2004) Pegylated interferon alfa-2b vs standard interferon alfa-2b, plus ribavirin, for chronic hepatitis C in HIV-infected patients: a randomized controlled trial. *Jama*, **292**(23): 2839-2848.

[98] Lempicki, R.A., Polis, M.A., Yang, J., McLaughlin, M., Koratich, C., Huang, D.W., Fullmer, B., Wu, L., Rehm, C.A., Masur, H., Lane, H.C., Sherman, K.E., Fauci, A.S., and Kottilil, S., (2006) Gene Expression Profiles in Hepatitis C Virus (HCV) and HIV Coinfection: Class Prediction Analyses before Treatment Predict the Outcome of Anti-HCV Therapy among HIV-Coinfected Persons. *J Infect Dis*, **193**(8): 1172-1177.

[99] FDA, (2003) Draft guidance for industry: Pharmacogenomics data submission. Food and Drug Administration, Rockville, MD. Available from: http://www.fda.gov/cder/guidance/5900dft.pdf.

[100] Atkinson, A.J., Colburn, W.A., DeGruttola, V.G., DeMets, D.L., Downing, G.J., Hoth, D.F., Oates, J.A., Peck, C.C., Schooley, R.T., Spilker, B.A., Woodcock, J., and Zeger, S.L., (2001) Biomarkers and surrogate endpoints: Preferred definitions and conceptual framework. *Clin Pharmacol Ther* 69(**3**):89-95.

[101] Sotiriou, C. and Piccart, M.J., (2007) Taking gene-expression profiling to the clinic: when will molecular signatures become relevant to patient care? *Nat Rev Cancer*, **7**(7): 545-553.

[102] Kristensen, V.N., Edvardsen, H., Tsalenko, A., Nordgard, S.H., Sorlie, T., Sharan, R., Vailaya, A., Ben-Dor, A., Lonning, P.E., Lien, S., Omholt, S., Syvanen, A.C., Yakhini, Z., and Borresen-Dale, A.L., (2006) Genetic variation in putative regulatory loci controlling gene expression in breast cancer. *Proc Natl Acad Sci U S A*, **103**(20): 7735-7740.

[103] Waddell, M., Page, D., and Shaughnessy, J., (2005). Predicting cancer susceptibility from single-nucleotide polymorphism data: A case study in multiple myeloma. in *5th International Workshop on Data Mining and Bioinformatics (BIOKDD '05)*. Chicago, Illinois, USA: ACM.

[104] Perou, C.M., Sorlie, T., Eisen, M.B., van de Rijn, M., Jeffrey, S.S., Rees, C.A., Pollack, J.R., Ross, D.T., Johnsen, H., Akslen, L.A., Fluge, O., Pergamenschikov, A., Williams, C., Zhu, S.X., Lonning, P.E., Borresen-Dale, A.L., Brown, P.O., and Botstein, D., (2000) Molecular portraits of human breast tumours. *Nature*, **406**(6797): 747-752.

[105] Sorlie, T., Perou, C.M., Tibshirani, R., Aas, T., Geisler, S., Johnsen, H., Hastie, T., Eisen, M.B., van de Rijn, M., Jeffrey, S.S., Thorsen, T., Quist, H., Matese, J.C., Brown, P.O., Botstein, D., Eystein Lonning, P., and Borresen-Dale, A.L., (2001) Gene

expression patterns of breast carcinomas distinguish tumor subclasses with clinical implications. *Proc Natl Acad Sci U S A*, **98**(19): 10869-10874.

[106] Sorlie, T., Tibshirani, R., Parker, J., Hastie, T., Marron, J.S., Nobel, A., Deng, S., Johnsen, H., Pesich, R., Geisler, S., Demeter, J., Perou, C.M., Lonning, P.E., Brown, P.O., Borresen-Dale, A.L., and Botstein, D., (2003) Repeated observation of breast tumor subtypes in independent gene expression data sets. *Proc Natl Acad Sci U S A*, **100**(14): 8418-8423.

[107] Burgner, D., Rockett, K., Ackerman, H., Hull, J., Usen, S., Pinder, M., and Kwiatkowski, D.P., (2003) Haplotypic relationship between SNP and microsatellite markers at the NOS2A locus in two populations. *Genes Immun*, **4**(7): 506-514.

[108] Onken, M.D., Worley, L.A., Person, E., Char, D.H., Bowcock, A.M., and Harbour, J.W., (2007) Loss of heterozygosity of chromosome 3 detected with single nucleotide polymorphisms is superior to monosomy 3 for predicting metastasis in uveal melanoma. *Clin Cancer Res*, **13**(10): 2923-2927.

[109] Dove, A., (2005) The SNPs are down: genotyping for the rest of us. *Nature Methods*, **2**: 989-994.

[110] Valk, P.J., Verhaak, R.G., Beijen, M.A., Erpelinck, C.A., Barjesteh van Waalwijk van Doorn-Khosrovani, S., Boer, J.M., Beverloo, H.B., Moorhouse, M.J., van der Spek, P.J., Lowenberg, B., and Delwel, R., (2004) Prognostically useful gene-expression profiles in acute myeloid leukemia. *N Engl J Med*, **350**(16): 1617-1628.

[111] Liu, Z., Maas, K., and Aune, T.M., (2006) Identification of gene expression signatures in autoimmune disease without the influence of familial resemblance. *Hum Mol Genet*, **15**(3): 501-509.

[112] Burczynski, M.E., Peterson, R.L., Twine, N.C., Zuberek, K.A., Brodeur, B.J., Casciotti, L., Maganti, V., Reddy, P.S., Strahs, A., Immermann, F., Spinelli, W., Schwertschlag, U., Slager, A.M., Cotreau, M.M., and Dorner, A.J., (2006) Molecular classification of Crohn's disease and ulcerative colitis patients using transcriptional profiles in peripheral blood mononuclear cells. *J Mol Diagn*, **8**(1): 51-61.

[113] Twine, N.C., Stover, J.A., Marshall, B., Dukart, G., Hidalgo, M., Stadler, W., Logan, T., Dutcher, J., Hudes, G., Dorner, A.J., Slonim, D.K., Trepicchio, W.L., and Burczynski, M.E., (2003) Disease-associated Expression Profiles in Peripheral Blood Mononuclear Cells from Patients with Advanced Renal Cell Carcinoma. *Cancer Res*, **63**(18): 6069-6075.

[114] Dunkler, D., Michiels, S., and Schemper, M., (2007) Gene expression profiling: does it add predictive accuracy to clinical characteristics in cancer prognosis? *Eur J Cancer*, **43**(4): 745-751.

[115] Kattan, M.W., (2003) Judging new markers by their ability to improve predictive accuracy. *J Natl Cancer Inst*, **95**(9): 634-635.

[116] Kattan, M.W., (2004) Evaluating a new marker's predictive contribution. *Clin Cancer Res*, **10**(3): 822-824.

[117] Rosenwald, A., Wright, G., Chan, W.C., Connors, J.M., Campo, E., Fisher, R.I., Gascoyne, R.D., Muller-Hermelink, H.K., Smeland, E.B., Giltnane, J.M., Hurt, E.M., Zhao, H., Averett, L., Yang, L., Wilson, W.H., Jaffe, E.S., Simon, R., Klausner, R.D., Powell, J., Duffey, P.L., Longo, D.L., Greiner, T.C., Weisenburger, D.D., Sanger, W.G., Dave, B.J., Lynch, J.C., Vose, J., Armitage, J.O., Montserrat, E., Lopez-Guillermo, A., Grogan, T.M., Miller, T.P., LeBlanc, M., Ott, G., Kvaloy, S., Delabie, J.,

Holte, H., Krajci, P., Stokke, T., and Staudt, L.M., (2002) The use of molecular profiling to predict survival after chemotherapy for diffuse large-B-cell lymphoma. *N Engl J Med*, **346**(25): 1937-1947.

[118] Ockenhouse, C.F., Bernstein, W.B., Wang, Z., and Vahey, M.T., (2005) Functional genomic relationships in HIV-1 disease revealed by gene-expression profiling of primary human peripheral blood mononuclear cells. *J Infect Dis*, **191**(12): 2064-2074.

[119] Chou, R., Huffman, L.H., Fu, R., Smits, A.K., and Korthuis, P.T., (2005) Screening for HIV: a review of the evidence for the U.S. Preventive Services Task Force. *Ann Intern Med*, **143**(1): 55-73.

[120] United States Department of Health and Human Services, (1990) Update: serologic testing for HIV-1 antibody—United States, 1988 and 1989. *Morbidity and Mortality Weekly Report (MMWR)*, **39**(380-383).

[121] Kleinman, S., Busch, M.P., Hall, L., Thomson, R., Glynn, S., Gallahan, D., Ownby, H.E., and Williams, A.E., (1998) False-positive HIV-1 test results in a low-risk screening setting of voluntary blood donation. Retrovirus Epidemiology Donor Study. *Jama*, **280**(12): 1080-1085.

[122] Blue Cross and Blue Shield Association Technology Evaluation Center, (2007) Gene Expression Profiling of Breast Cancer to Select Women for Adjuvant Chemotherapy. [cited 2007 December 12th]; Available from: http://www.bcbs.com/betterknowledge/tec/press/gene-expression-profiling-of.html.

[123] Schubert, C.M., (2003) Microarray to be used as routine clinical screen. *Nat Med*, **9**(1): 9.

[124] Shi, L., Reid, L.H., Jones, W.D., Shippy, R., Warrington, J.A., Baker, S.C., Collins, P.J., de Longueville, F., Kawasaki, E.S., Lee, K.Y., Luo, Y., Sun, Y.A., Willey, J.C., Setterquist, R.A., Fischer, G.M., Tong, W., Dragan, Y.P., Dix, D.J., Frueh, F.W., Goodsaid, F.M., Herman, D., Jensen, R.V., Johnson, C.D., Lobenhofer, E.K., Puri, R.K., Schrf, U., Thierry-Mieg, J., Wang, C., Wilson, M., Wolber, P.K., Zhang, L., Amur, S., Bao, W., Barbacioru, C.C., Lucas, A.B., Bertholet, V., Boysen, C., Bromley, B., Brown, D., Brunner, A., Canales, R., Cao, X.M., Cebula, T.A., Chen, J.J., Cheng, J., Chu, T.M., Chudin, E., Corson, J., Corton, J.C., Croner, L.J., Davies, C., Davison, T.S., Delenstarr, G., Deng, X., Dorris, D., Eklund, A.C., Fan, X.H., Fang, H., Fulmer-Smentek, S., Fuscoe, J.C., Gallagher, K., Ge, W., Guo, L., Guo, X., Hager, J., Haje, P.K., Han, J., Han, T., Harbottle, H.C., Harris, S.C., Hatchwell, E., Hauser, C.A., Hester, S., Hong, H., Hurban, P., Jackson, S.A., Ji, H., Knight, C.R., Kuo, W.P., LeClerc, J.E., Levy, S., Li, Q.Z., Liu, C., Liu, Y., Lombardi, M.J., Ma, Y., Magnuson, S.R., Maqsodi, B., McDaniel, T., Mei, N., Myklebost, O., Ning, B., Novoradovskaya, N., Orr, M.S., Osborn, T.W., Papallo, A., Patterson, T.A., Perkins, R.G., Peters, E.H., Peterson, R., Philips, K.L., Pine, P.S., Pusztai, L., Qian, F., Ren, H., Rosen, M., Rosenzweig, B.A., Samaha, R.R., Schena, M., Schroth, G.P., Shchegrova, S., Smith, D.D., Staedtler, F., Su, Z., Sun, H., Szallasi, Z., Tezak, Z., Thierry-Mieg, D., Thompson, K.L., Tikhonova, I., Turpaz, Y., Vallanat, B., Van, C., Walker, S.J., Wang, S.J., Wang, Y., Wolfinger, R., Wong, A., Wu, J., Xiao, C., Xie, Q., Xu, J., Yang, W., Zhang, L., Zhong, S., Zong, Y. and Slikker, W., Jr., (2006) The MicroArray Quality Control (MAQC) project shows inter- and intraplatform reproducibility of gene expression measurements. *Nat Biotechnol*, **24**(9): 1151-1161.

[125] Brazma, A., Hingamp, P., Quackenbush, J., Sherlock, G., Spellman, P., Stoeckert, C., Aach, J., Ansorge, W., Ball, C.A., Causton, H.C., Gaasterland, T., Glenisson, P., Holstege, F.C., Kim, I.F., Markowitz, V., Matese, J.C., Parkinson, H., Robinson, A., Sarkans, U., Schulze-Kremer, S., Stewart, J., Taylor, R., Vilo, J., and Vingron, M., (2001) Minimum information about a microarray experiment (MIAME)-toward standards for microarray data. *Nat Genet*, **29**(4): 365-371.

[126] Guo, L., Lobenhofer, E.K., Wang, C., Shippy, R., Harris, S.C., Zhang, L., Mei, N., Chen, T., Herman, D., Goodsaid, F.M., Hurban, P., Phillips, K.L., Xu, J., Deng, X., Sun, Y.A., Tong, W., Dragan, Y.P., and Shi, L., (2006) Rat toxicogenomic study reveals analytical consistency across microarray platforms. *Nat Biotechnol*, **24**(9): 1162-1169.

[127] Moore, D.F., Li, H., Jeffries, N., Wright, V., Cooper, R.A., Jr., Elkahloun, A., Gelderman, M.P., Zudaire, E., Blevins, G., Yu, H., Goldin, E., and Baird, A.E., (2005) Using peripheral blood mononuclear cells to determine a gene expression profile of acute ischemic stroke: a pilot investigation. *Circulation*, **111**(2): 212-221.

[128] Maas, K., Chan, S., Parker, J., Slater, A., Moore, J., Olsen, N., and Aune, T.M., (2002) Cutting edge: molecular portrait of human autoimmune disease. *J Immunol*, **169**(1):5-9.

[129] Deng, M.C., Eisen, H.J., Mehra, M.R., Billingham, M., Marboe, C.C., Berry, G., Kobashigawa, J., Johnson, F.L., Starling, R.C., Murali, S., Pauly, D.F., Baron, H., Wohlgemuth, J.G., Woodward, R.N., Klingler, T.M., Walther, D., Lal, P.G., Rosenberg, S., and Hunt, S., (2006) Noninvasive discrimination of rejection in cardiac allograft recipients using gene expression profiling. *Am J Transplant*, **6**(1): 150-160.

[130] Barnes, M.G., Aronow, B.J., Luyrink, L.K., Moroldo, M.B., Pavlidis, P., Passo, M.H., Grom, A.A., Hirsch, R., Giannini, E.H., Colbert, R.A., Glass, D.N., and Thompson, S.D., (2004) Gene expression in juvenile arthritis and spondyloarthropathy: pro-angiogenic ELR+ chemokine genes relate to course of arthritis. *Rheumatology (Oxford)*, **43**(8): 973-979.

[131] Achiron, A., Gurevich, M., Friedman, N., Kaminski, N., and Mandel, M., (2004) Blood transcriptional signatures of multiple sclerosis: unique gene expression of disease activity. *Ann Neurol*, **55**(3): 410-417.

[132] Bomprezzi, R., Ringner, M., Kim, S., Bittner, M.L., Khan, J., Chen, Y., Elkahloun, A., Yu, A., Bielekova, B., Meltzer, P.S., Martin, R., McFarland, H.F., and Trent, J.M., (2003) Gene expression profile in multiple sclerosis patients and healthy controls: identifying pathways relevant to disease. *Hum Mol Genet*, **12**(17): 2191-2199.

[133] Mandel, M., Gurevich, M., Pauzner, R., Kaminski, N., and Achiron, A., (2004) Autoimmunity gene expression portrait: specific signature that intersects or differentiates between multiple sclerosis and systemic lupus erythematosus. *Clin Exp Immunol*, **138**(1): 164-170.

[134] Segman, R.H., Shefi, N., Goltser-Dubner, T., Friedman, N., Kaminski, N., and Shalev, A.Y., (2005) Peripheral blood mononuclear cell gene expression profiles identify emergent post-traumatic stress disorder among trauma survivors. *Mol Psychiatry*, **10**(5): 500-513, 425.

[135] Bull, T.M., Coldren, C.D., Moore, M., Sotto-Santiago, S.M., Pham, D.V., Nana-Sinkam, S.P., Voelkel, N.F., and Geraci, M.W., (2004) Gene microarray analysis of peripheral blood cells in pulmonary arterial hypertension. *Am J Respir Crit Care Med*, **170**(8): 911-919.

[136] Brouard, S., Mansfield, E., Braud, C., Li, L., Giral, M., Hsieh, S.C., Baeten, D., Zhang, M., Ashton-Chess, J., Braudeau, C., Hsieh, F., Dupont, A., Pallier, A., Moreau, A., Louis, S., Ruiz, C., Salvatierra, O., Soulillou, J.P., and Sarwal, M., (2007) Identification of a peripheral blood transcriptional biomarker panel associated with operational renal allograft tolerance. *Proc Natl Acad Sci U S A*, **104**(39): 15448-15453.

[137] Batliwalla, F.M., Baechler, E.C., Xiao, X., Li, W., Balasubramanian, S., Khalili, H., Damle, A., Ortmann, W.A., Perrone, A., Kantor, A.B., Gulko, P.S., Kern, M., Furie, R., Behrens, T.W., and Gregersen, P.K., (2005) Peripheral blood gene expression profiling in rheumatoid arthritis. *Genes Immun*, **6**(5): 388-397.

[138] Olsen, N., Sokka, T., Seehorn, C.L., Kraft, B., Maas, K., Moore, J., and Aune, T.M., (2004) A gene expression signature for recent onset rheumatoid arthritis in peripheral blood mononuclear cells. *Ann Rheum Dis*, **63**(11): 1387-1392.

[139] Baechler, E.C., Batliwalla, F.M., Karypis, G., Gaffney, P.M., Ortmann, W.A., Espe, K.J., Shark, K.B., Grande, W.J., Hughes, K.M., Kapur, V., Gregersen, P.K., and Behrens, T.W., (2003) Interferon-inducible gene expression signature in peripheral blood cells of patients with severe lupus. *Proc Natl Acad Sci U S A*, 100(**5**): 2610-2615.

[140] Rus, V., Chen, H., Zernetkina, V., Magder, L.S., Mathai, S., Hochberg, M.C., and Via, C.S., (2004) Gene expression profiling in peripheral blood mononuclear cells from lupus patients with active and inactive disease. *Clin Immunol*, **112**(3): 231-234.

[141] Motomura, K., Toyoda, N., Oishi, K., Sato, H., Nagai, S., Hashimoto, S., Tugume, S.B., Enzama, R., Mugewa, R., Mutuluuza, C.K., Mugyeyi, P., Nagatake, T., and Matsushima, K., (2004) Identification of a host gene subset related to disease prognosis of HIV-1 infected individuals. *Int Immunopharmacol*, **4**(14): 1829-1836.

[142] Jacobsen, M., Repsilber, D., Gutschmidt, A., Neher, A., Feldmann, K., Mollenkopf, H.J., Ziegler, A., and Kaufmann, S.H., (2007) Candidate biomarkers for discrimination between infection and disease caused by Mycobacterium tuberculosis. *J Mol Med*, **85**(6): 613-621.

[143] Hakonarson, H., Bjornsdottir, U.S., Halapi, E., Bradfield, J., Zink, F., Mouy, M., Helgadottir, H., Gudmundsdottir, A.S., Andrason, H., Adalsteinsdottir, A.E., Kristjansson, K., Birkisson, I., Arnason, T., Andresdottir, M., Gislason, D., Gislason, T., Gulcher, J.R., and Stefansson, K., (2005) Profiling of genes expressed in peripheral blood mononuclear cells predicts glucocorticoid sensitivity in asthma patients. *Proc Natl Acad Sci U S A*, **102**(41): 14789-14794.

[144] Tateno, M., Honda, M., Kawamura, T., Honda, H., and Kaneko, S., (2007) Expression profiling of peripheral-blood mononuclear cells from patients with chronic hepatitis C undergoing interferon therapy. *J Infect Dis*, **195**(2): 255-267.

[145] DePrimo, S.E., Wong, L.M., Khatry, D.B., Nicholas, S.L., Manning, W.C., Smolich, B.D., O'Farrell, A.M., and Cherrington, J.M., (2003) Expression profiling of blood samples from an SU5416 Phase III metastatic colorectal cancer clinical trial: a novel strategy for biomarker identification. *BMC Cancer*, **3**: 3.

[146] Burczynski, M.E., Twine, N.C., Dukart, G., Marshall, B., Hidalgo, M., Stadler, W.M., Logan, T., Dutcher, J., Hudes, G., Trepicchio, W.L., Strahs, A., Immermann, F., Slonim, D.K., and Dorner, A.J., (2005) Transcriptional profiles in peripheral blood mononuclear cells prognostic of clinical outcomes in patients with advanced renal cell carcinoma. *Clin Cancer Res*, **11**(3): 1181-1189.

In: Oligonucleotide Array Sequence Analysis ISBN 978-1-60456-542-3
Editors: M.K. Moretti and L.J. Rizzo, pp. 83-121

Chapter 3

CLUSTERING TIME-SERIES GENE EXPRESSION DATA

***Luis Rueda*[1*] *and Ataul Bari*[2†]**
[1]IEEE. Department of Computer Science and
Center for Biotechnology, University of Concepción.
Edmundo Larenas 215, Concepción, VIII Region, Chile
[2]School of Computer Science, University of Windsor.
401 Sunset Ave., Windsor, ON, N9B3P4, Canada

Abstract

Finding relevant patterns in time-series gene expression data has been studied for quite a few years. Genes with similar expression profiles are expected to be functionally related or co-regulated, while in some cell cycles, like yeast and serum, it is quite important to group the genes based on different phases during the regulation process. Clustering gene expression data, given in terms of time series, is different from the general clustering problem, because exchanging two or more time points delivers quite different results, while doing this could not be biologically meaningful at all. As a result of this, traditional clustering methods, more specifically, conventional similarity measures, are not suitable in this context, and one has to recur to specialized methods. As the number of genes profiled in an experiment is large, clustering based on the similarity of their temporal profiles is a daunting task. Various clustering methods for time-series gene expression data have been proposed. While some of the proposed methods are more effective than others in some particular contexts, so far, none of them have been identified as *the best* technique for all scenarios. Clustering approaches for analyzing microarray time-series data are discussed and some key features of these methods are highlighted. Special challenges, advantages and drawbacks of the methods are discussed and an experimental comparison of some of these methods is also given.

*E-mail address: lrueda@udec.cl. Phone: +56 41 2204305, fax: +56 41 2221770.
†E-mail address: bari1@uwindsor.ca.

1. INTRODUCTION

Clustering genes based on the similarity of their temporal expression profile is important for many biological studies, since genes with similar expression patterns are expected to be functionally related or co-regulated [10]. Clustering gene expression data given in terms of time series is different from the general clustering problem, because exchanging two time points delivers quite different results, while doing this could not be biologically meaningful. Many unsupervised methods for gene clustering based on the similarity (or dissimilarity) of their microarray temporal profiles have been proposed in the past few years [1, 2, 3, 5, 7, 10, 11, 16, 20, 23, 25, 27, 29, 30, 31, 33, 34]. One of the methods for clustering microarray time-series data is based on a hidden phase model (similar to a hidden Markov model) to define the parameters of a mixture of normal distributions in a Bayesian-like manner, which are estimated by using expectation maximization (EM) [3]. Clustering time-series data has also been studied using a Bayesian approach in [27], and a hidden Markov model (HMM) in [30]. A partitional clustering based on k-means and Euclidian distance has been studied in [34], and in [33], self-organizing maps (SOM) have been applied to visualize and interpret gene temporal expression profiles patterns. Other methods based on correlation measures have been proposed for clustering genes using microarray time-series data [6, 16]. The method proposed in [6] requires computing the mean expression levels of some candidate profiles using some pre-identified, arbitrarily selected profiles. In [16], a method for clustering microarray time series data employing a jack-knife correlation coefficient with or without using the seeded candidate profiles is proposed. Specifying expression levels for the candidate profiles in advance for these correlation-based procedures requires estimating each candidate profile, which is made using a small sample of arbitrarily selected genes. This makes it vulnerable to the possibility of missing important genes, since the resulting clusters depend upon the initially chosen template genes. A regression-based approach to address the challenges for clustering short time-series expression datasets has been proposed in [11], which is suitable for analyzing single or multiple microarray temporal data. The study in [20] focused on analyzing gene temporal expression profiles datasets that are non-uniformly sampled and can have missing values. Gene temporal expression profiles are represented as continuous curves using statistical spline estimation. In [7], clustering of gene expression temporal profiles are done by identifying homogeneous clusters of genes. This method has focused on the *shapes of the curves* instead of the *absolute levels of expression*. Fuzzy Clustering gene temporal profiles has been the focus of [23], where the similarity measure for the co-expressed genes is computed based on the rate of change of the expression levels across time. Another method is to select and cluster genes using the ideas of order-restricted inference, where estimation makes use of known inequalities among parameters [25]. In this method, at first, potential candidate profiles of interest are defined and expressed in terms of inequalities between the expected gene expression levels at various time points. For a given candidate profile, the estimated mean expression level of each gene is computed and the best fitting profile for a given gene is selected using the goodness-of-fit criterion and the bootstrap test procedure. In this approach, two genes expression profiles $\mathbf{x}_1$ and $\mathbf{x}_2$ fall into the same cluster if they show similar profiles in terms of direction of the changes of expression ratios (e.g. up-up-up-down-down), regardless how big/small is the change.

In [1], a minimum-square-error profile alignment approach to cluster microarray time series data was proposed. The idea is to pairwisely align two temporal profiles in such a way that the sum of square errors between two aligned vectors is minimized. The alignment procedure, however, does not consider the length of the interval between the time points at which individual measurements are taken. Profile alignment that takes into consideration the effect of unequal time interval was studied in [29] and an integral based distance function was proposed to hierarchically cluster gene temporal data.

A comprehensive review of the clustering approaches for analyzing microarray time-series data is presented. In the following sections, the key features of each method are highlighted, providing a methodological comparison, their main advantages and drawbacks, and performing an experimental comparison of some of these methods.

2. Clustering Temporal Expression Profiles

In this section, clustering methods for time-series expression data are discussed. In order to introduce these approaches, the clustering problem is formally stated. Consider a dataset $\mathcal{D} = \{\mathbf{x}_1, \mathbf{x}_2, ..., \mathbf{x}_n\}$, where $\mathbf{x}_i = [x_{i_1}, x_{i_2}, ..., x_{i_m}]^t$ is an m-dimensional feature vector that represents the expression level of gene i at m different time points, $\mathbf{t} = [t_1, t_2, ..., t_m]^t$. The aim is to partition $\mathcal{D}$ into k disjoint subsets $\mathcal{D}_1, \mathcal{D}_2, ..., \mathcal{D}_k$, where $\mathcal{D} = \mathcal{D}_1 \cup \mathcal{D}_2 \cup ... \cup \mathcal{D}_k$, and $\mathcal{D}_i \cap \mathcal{D}_j = \emptyset$, for $\forall i, j, i \neq j$, in such a way that a similarity (dissimilarity) cost function $\phi : \{0, 1\}^{n \times k} \rightarrow \Re$ is maximized (minimized).

Various clustering methods for time-series gene expression data have been proposed, including k-means clustering [34], self-organizing maps (SOM) applied to hematopoietic differentiation in four well studied models [33], a jack-knife correlation coefficient model [16], an order-restricted inference-based method [25], a statistical two-regression step approach [5], fuzzy clustering of time-course data based on the similarity of relative change of expression level and the corresponding temporal information of the profiles [23], combined spline smoothing and first derivative computation [7], a method for assigning genes to a pre-defined set of model profiles [11], and statistical spline estimation [20]. Other approaches based on Bayesian models [3, 27], and hidden Markov models (HMM) have been proposed as well [3, 30]. In [1, 29], a minimum-square-error profile alignment approach to cluster micro-array time-series data was proposed to pairwisely align two temporal profiles in such a way that the sum of square errors between two aligned gene expression vectors is minimized.

2.1. Traditional Clustering

Various methods for clustering expression time-series data have been proposed. Most of these approaches are either directly adopted, or somewhat modified from the standard clustering methodology available in the field of classification and pattern recognition. These methods, in general, take gene temporal expression profiles and apply some conventional metrics and algorithms for the purpose of clustering, like any general clustering approach, and do not perform any profile alignment before applying a similarity measure.

2.1.1. *k*-Means Based Methods

Tavazole *et al.* [34] applied a partitional clustering method based on k-means [14], to cluster gene expression temporal data. In k-means, a dataset is partitioned into k predefined clusters by iteratively reallocating cluster members such that the overall within-cluster dispersion is minimized. In [34], this partitional clustering approach was applied to the yeast dataset [4] that contains expression levels of 6,220 mRNA in synchronized Saccharomyces cerevisiae batch cultures, measured using oligonucleotide arrays at 15 time points, across two cell cycles. Based on the average expression level, the top 6,000 genes were selected. Among them, the most variable 3,000 ORFs, measured by using a variation metric based on the normalized dispersion in the expression level of each gene across the time points, were chosen for further analysis. The 3,000 genes expression data, transformed such that each gene variance is normalized across the 15 time points, constitute the transformed data matrix in the 15-dimensional *feature space*. The transformed *variance-normalized* datset of 3,000 ORFs was then partitioned, based on similarities in their expression patterns, into 30 clusters. Each cluster was assigned with 49–186 ORFs. For the similarity measure, the Euclidean distance was used to compute the dissimilarity between each pair of genes in the feature space. The application of k-means to this dataset revealed new sets of co-regulated genes and their putative *cis*-regulatory elements.

This approach does not require any prior knowledge about the structure or to make any assumptions about the dynamics of the expression profile. But the value of k needs to be known *a priori*. Also, the complete lack of structure in the clustering may result in inconsistent clusters, since the clustering may reinforce local groups [33].

2.1.2. Self-organizing Maps

Self-organizing maps (SOM) have been applied to visualize and interpret the patterns of gene temporal expression profiles [33]. The SOM is a type of mathematical cluster analysis and is especially suitable for large, high-dimensional datasets. This method consists of maintaining a set of nodes, a topology and a distance function (on the nodes). The algorithm iteratively maps the nodes into the feature space of the genes.

This approach was tested on yeast cell cycle data [4], and applied to hematopoietic differentiation in four other models, HL-60, U937, Jurkat, and NB4 cells, showing to be able to capture the patterns in an efficient manner. It was also able to uncover genes that were earlier not known to be regulated.

As in k-means, this approach does not require any prior knowledge about the structure or making any assumptions about the dynamics of the expression profiles. The SOM suites well with exploratory analysis of the data and to reveal relevant patterns in a dataset. But deciding upon the number of nodes is important, since if the number is small, the patterns can not be distinguished properly (due to a large within-cluster scatter). On the other hand, adding nodes exceeding a certain value (that produces distinct patterns) fails to produce any new patterns.

2.1.3. Bayesian Approaches

In [3], an approach for improving the clustering results of gene expression series using rough knowledge regarding the *general shapes* of the classes was proposed. Knowledge about the general shapes can be *elementary* or *incomplete-information* regarding the change of the mean expression level over time. The examples of such knowledge are classes with bell-curved shapes, and classes with high expression level at the beginning of the series.

In this approach, the information regarding the shape(s) of the class(es), when available, are directly integrated into the model so that classes with the desired profiles are favored. But if no such information is available, then classical clustering using a Bayesian approach is performed, which intuitively deals with the temporal nature of the data. Two types of models are involved in this approach, one is a *probabilistic mixture model* that describes and classifies the temporal expressions. For this model, some parameters are needed to be estimated for the clustering. The second model, referred to as *Hidden Phase Model (HPM)* – close to the *Hidden Markov Model (HMM)*, is used to express prior knowledge or features of the temporal data. The HPM works on the assumption that the genes of a component pass through *phases* or *biological states* over time. Therefore, it can be assumed that, for a given component, a range of consecutive time points correspond to the same biological state. Although these phases are hidden, still the evolution of the mean expression level of the component is affected by them. Two types of HPMs are proposed: (i) probabilistic HPM: Used when strong constraints can be expressed and is suitable for expressing knowledge about simple or well defined profiles, and (ii) non-probabilistic HPM: Suitable for more complex knowledge, or when no information about the profiles is available and only to represent these series data.

The effectiveness of the method to recover a particular class of genes, for which there was prior knowledge, was demonstrated using a dataset composed of a mixture of real and synthetic data. It was constructed by injecting some synthetic data into the original Fibroblast dataset, used in [19]. The experiments show that the method can recover the synthetic data with high accurecy. The other experiments were performed on real classes, such as the datasets presented in [19], to uncover classes that show a quick over expression at the beginning of the series, and on the datasets presented in [32] to uncover classes with sinusoidal shapes.

In [27], Ramoni et al. also used a Bayesian method for model-based clustering, where the models are autoregressive curves of fixed order. In that work, the gene expression dynamics were represented as autoregressive equations, and an agglomerative procedure was used to search for the most likely set of clusters out of the given temporal expression data. During clustering, the approach explicitly takes into account the dynamic nature of gene expression time series data and principled a way that identify the number of distinct clusters.

This method can be viewed as a specialized version of Bayesian Clustering by Dynamics (BCD), where the concept of similarity is defined in such a way that two time series are considered similar if they are generated by the same stochastic process. The proposed method models temporal gene-expression profiles by autoregressive equations that derive a posterior probability, and agglomeratively picks models having maximum posterior probabilities. This method is also able to identify the optimal number of clusters based on the well-known Akaike information criterion.

The method was tested on a dataset for temporal deployment of the transcriptional program underlying the response of human fibroblasts to serum [19] that recorded the changes of expression levels of 8,613 human genes over 24 h. using two-dye cDNA microarrays and selected 517 genes whose expression level changed significantly in response to serum stimulation. A number of experiments were conducted, on the log-transformed data, with four different autoregressive orders (0, 1, 2, and 3) and with different similarity measures including Euclidean distance, correlation, and lag-correlation (correlation between two series where one series lags the other) to account for the temporal dependency of the data. The statistical analysis shows that the Euclidean distance produces the best results by systematically returning the cluster models that have higher posterior probabilities than correlation-based distances.

A Hidden Markov Model (HMM) that accounts for horizontal dependencies in gene expression time-series data was proposed in [30]. The HMMs can be seen as probabilistic functions of a Markov chain [26], where each state of the chain produces emissions based on *emission probabilities* or *densities*. The approach proposed in [30] focuses on univariate emission probability densities. Usually, *model-based* clustering approaches are effective when *grammatical* or *structural* properties of the data can be explicitly expressed, and the HMM can be used by these types of approaches to partition gene expression time-series data into clusters. The process starts with an initial collection of HMMs that encompasses typical qualitative behavior and iteratively finds cluster models and assigns data points to these models in such a way that the joint likelihood of the clustering is maximized. The approach also provides a method for partially supervised learning, which allows to add groups of labeled data to the initial collection of clusters.

The method was tested on the datasets of yeast cell cycle [32] and fibroblast serum response data [19]. The effectiveness was evaluated by comparing the results with that of the autoregressive curve method [27] finding that this method, when combined with the *Viterbi-based* cluster decomposition, resulted in homogeneous and fine grained clusters, which helps resolving the multitude of distinct regulatory process classes in the data.

The method can cope effectively with unlabeled data, as well as with additional labeled data. The information contained in the labeled data is used to produce high quality clusters. This method is also robust with respect to noisy and frequently missing data.

2.1.4. Correlation-Based Clustering

The approach for clustering gene expression time series data proposed in [16] employed a new correlation-coefficient-based method for the similarity measure. Any clustering technique attempts to group genes with *similar* expression patterns (i.e., coexpressed genes). A similarity measure is a pairwise measure of coexpression that, ideally, assigns high scores to coexpressed ORFs and low scores to unrelated ORFs. Noting that sometimes related genes are not coexpressed, and some unrelated genes can have similar expression profiles, and further noting that conventional distance measures can sometimes fail to capture such similarities, a new similarity measure was proposed, referred to as *jack-knife* correlation, and which is based on a jackknifing procedure in computational statistics. Jack-knife correlation gives high scores to gene pairs that exhibit similar behavior throughout the time points and is robust to outliers. It can also reduce false positives, i.e., giving a high score to a pair of dissimilar profiles by the similarity measure. Heyer et al. [16] also proposed a clustering algorithm, referred to as *quality cluster algorithm (QT_Clust)*, for the gene expression temporal profiles, that iteratively groups the ORFs based on their similarity decided by the jack-knife correlation.

The method was applied to the yeast cell cycle dataset [4] and found to be able to group similar genes in the same cluster. The results were supported by the available biological knowledge, and can be extended to discover more information about gene functions and regulations.

2.1.5. Order-Restricted Inference

The order-restricted inference method was used by Peddada et al. in [25] for selecting and clustering genes expression profiles for time-series or dose-response data. This approach uses known inequalities among parameters and applies the ideas of order-restricted inference. The entire process is carried out in multiple steps. In the first step, potential candidate profiles of interest are defined (e.g., a candidate profile for monotonically increasing expression with four time points can be defined as *up-up-up-up*) and expressed in terms of inequalities between the expected gene expression levels at various time points. In the next step, the mean expression level of each gene is estimated by using the method proposed in [18]. Then, the *goodness-of-fit criterion* and the *bootstrap test procedure* is used to chose the best fitting profile for a given gene [24]. Finally, the gene is assigned to a profile and the process continues for each gene in the dataset under consideration. Using this procedure, two profiles are put in the same cluster only if all the inequalities between the expected expression levels at various time points are the same.

The effectiveness of the approach was demonstrated using log-transformed relative expression data from [21], where the MCF-7 breast cancer cell line was treated with 17β-estradiol or ethanol (vehicle control). After treatment, the samples were harvested at six different time points, 1, 4, 12, 24, 36 and 48 hours. A total of 8 hybridizations were performed at each time point and each array consisted of 1,900 genes. The method was tested for each gene, where the alternative hypothesis is the union of 10 profiles that include *monotone decreasing*, *monotone increasing*, *up down profiles* and *down-up profiles*. The results identified 11 new genes that display typical phenotypes for estrogen-treated MCF-7 cells. This method makes use of the ordering in a time-series study and can detect genes more sensitively using their temporal ordering and finding consistent patterns over time.

2.1.6. Regression-Based Clustering

Cones et al. [5] proposed a statistical procedure that identifies genes with different expression profiles across analytical groups in time-series experiments. It is a general regression-based approach suitable for analyzing single or multiple microarray temporal data. This method, referred to as *maSigPro* (microarray Significant Profiles), is a two-step regression strategy where the experimental groups are identified by dummy variables. In the first step, the procedure adjusts a global regression model with all the defined variables to identify differentially expressed genes. In the next step, statistically significant different profiles are found by applying a variable selection strategy that studies the differences between the groups. The model parameters are adjusted based on the data under study and the specific interests of the researcher.

To show the effectiveness of the approach, the *maSigPro* method was applied on a real-life dataset, as well as on a simulated data. The real-life dataset was obtained from a study on *toxicogenomics* [15] that recorded the effect of the *hepatotoxicant bromobenzene* in rats. In that study, three groups of rats were treated with *low, medium*, and *high* doses of *bromobenzene*, dissolved in corn oil; one group treated only with the drug administration vehicle (*corn oil*), while the other group was left *untreated*. Each individual RNA rat sample was processed and recorded at different measurement time-points, 6, 24 and 48 hrs., and 2,665 genes were made available for statistical analysis. The application of *maSigPro* to this dataset selected 155 genes showing statistically significant differences between high-dose and corn-oil reference groups. Various numbers of other genes were also selected having statistically significant differences between the experimental groups, and *maSigPro* was able to detect 104 new genes. The simulated dataset contained 600 genes with profile differences that should be classified into three expression patterns. The application of *maSigPro* to this dataset resulted in the detection of fewer false positives than the contemporary method.

The *maSigPro* method does not require multiple pairwise comparisons, although it is able to detect significant profile differences. It allows for unbalanced designs and heterogeneous sampling times. This approach can be used to find genes with significant temporal expression changes between experimental groups, as well as, to analyze the magnitude of these differences.

2.1.7. Clustering Short Time-Series Data

Considering the challenges inherent to short time-series expression datasets (datasets containing measurements for at most eight time points), Ernst *et al.* [11] proposed an algorithm that specifically addresses the issues for clustering such expression data. The algorithm first selects a set of potential distinct patterns that can be expected from the experiment and then assigns genes to the profile that best represents them among the pre-selected profiles. The work detailed on choosing such a set of profiles, and proposed a method for assigning genes to profiles and determining the significance of each profile. Significant profiles are retained for further analysis and can be combined to form clusters. The clustering approach makes use of the correlation coefficient to measure the similarity between expression profiles.

The method was tested on simulated and real-life biological datasets. The real-life data is taken from the work on immune response [13], which used human cDNA microarrays to study the gene expression program of gastric AGScells infected with various strains of Helicobacter pylori. The work of [11] focused on the response of the wildtype G27 strain, for which data were obtained from two replicates on the same biological sample at five time points (0, 0.5, 3, 6, and 12 hrs.). From the 24,192 array probes, 2,243 genes were selected for further analysis, based on the agreement between the two repeats and their change at any of the experiment time points. The data was analyzed against a set of 50 model profiles. The results, when compared with contemporary clustering methods based on k-means, demonstrate that this approach was able to identify a more coherent set of genes, by grouping together the temporal profiles of relevant functional categories. This method can work with a dataset with no repeats by leveraging the statistical power obtained from the large number of genes being profiled simultaneously.

2.1.8. Spline-Smoothing Approaches

In [20], Bar-Joseph et al. focused on the analysis of gene temporal expression profiles that can cope with the problem of missing values (unobserved time-points) and non-uniformly sampled data. In this work, gene temporal expression profiles are represented as continuous curves using statistical spline estimation. Each expression profile is modeled as a cubic spline, estimated from the observed data and the spline coefficients of genes are constrained in such a way that similar expression patterns fall into the same class, while allowing for the gene-specific parameters. The work also includes a model-based clustering algorithm and an alignment procedure. The clustering algorithm operates directly on the continuous representations of gene expression profiles, thus, permitting interpolation of missing values. The number of clusters is accepted as an input to the process and the algorithm iterates, using a modified Expectation Maximization (EM) algorithm, until it converges, at which point the class for each gene is discovered.

The alignment algorithm also uses the spline representation and is proposed to warp the time-scale of one realization into another. A set of genes, known (by some other means) to have the same temporal pattern of expression is chosen and a parameterized warping function is then selected. The algorithm attempts to produce an optimal alignment (that maximizes the similarity between the two sets of expression profiles) by adjusting the function parameters.

Experiments were conducted for unobserved data estimation, clustering and alignment on the cell-cycle time-series dataset [32], and the Fkh1/Fkh2 knockout experiment dataset [35]. The results demonstrated that this approach is able to reconstruct unobserved time-points with a 10-15% fewer error than other methods. Also, the clustering approach was effective with non-equidistant sampled time-series data and the continuous alignment algorithm, that allows to control the number of degrees of freedom of the warp, helping avoid over fitting and the difficulties faced by the discrete methods. While this method is suitable for the analysis of relatively long time series (10 time points or more), the suitability of the approach in shorter time-series experiments has not been demonstrated.

In [7], Déjean et al. have also attempted to obtain relevant clustering of gene expression temporal profiles by identifying homogeneous clusters of genes. This method focuses on the *shapes of the curves* and not on the *absolute levels of expression*. It combines spline smoothing and first derivative computation with hierarchical and partitional clustering. The rationale for not primarily depending on the absolute level of expression was that the measured intensities could be significantly influenced by the probes on the microarray, and a relevant clustering method should consider the data specificity and integrate the temporal aspects of the data. In contrast, curves shapes can provide meaningful information on coordinate gene regulations. The *derivatives* are proposed to be a mathematical tool to describe the specific information on the data. Smoothing the temporal profiles requires to obtain regular and differentiable functions. Therefore, the approach is based on the framework of *functional data analysis* [28] with focus on the first derivative of curves by means of a priori spline smoothing. The objectives are achieved in two steps. In the first step, spline smoothing is performed, which is followed by, in the second step, the clustering of the derivatives of the continuous smoothed curves, resulted from the smoothing, which in fact, represent the temporal variations of mRNA concentrations. Both hierarchical clustering and k-means were applied to obtain the clusters. Tuning the smoothing parameter in the first step is achieved by using a heuristic approach that takes into account both statistical and biological considerations.

The effectiveness of the method was demonstrated on a dataset of postprandial fasting experiment, conducted in mouse, where 200 genes were monitored with a dedicated microarray at 11 time points between 0 and 72 hours (0, 3, 6, 9, 12, 18, 24, 36, 48, 60, and 72 hrs.) of fasting. The results mostly agree with other works that have focused on the effects of fasting on the mouse liver.

2.1.9. Fuzzy Clustering Temporal Profiles

The work in [23] proposed a similarity measure for the co-expressed genes based on the expression level rate of change across time. The gene expression time-series were considered as piecewise linear functions and the similarity was calculated by measuring the difference of slopes between the functions. The variable-time intervals in a time-series experiment is taken into consideration and is viewed as a weight, where farther apart (in terms of time) expressions take smaller weights in the comparison. A clustering algorithm was also proposed, motivated by the advantages of fuzzy clustering, which is a partitioning-optimization technique for extracting biological insights and identifying distinct features of genetic function and regulation [12]. The algorithm is referred to as Fuzzy Short Time-Series (FSTS) clustering, and incorporates the distance measure proposed in this work in the *fuzzy-c-means* clustering scheme.

The approach was applied to simulated data sets and a subset of the transcriptional program of sporulation in budding yeast [6], showing that the method clustered genes in such a way that it agrees with the model profiles defined in [6]. It was further observed that the method performs better than the conventional approaches (fuzzy c-means, k-means, average linkage hierarchical clustering and random clustering) for unevenly distributed time points.

The proposed method benefited from the fuzzy approach that inherently accounts for noise in the data and allows genes to belong to more than one group (hence revealing individual genes distinctive features). It was designed to cope with short and unevenly sampled time-series data.

2.2. Profile Alignment Clustering

The aim of clustering gene expression temporal data is to group together profiles with similar patterns. Deciding upon the similarity often involves pairwise distance measures of co-expressions. Conventional distance measures include correlation, rank correlation, Euclidean distance, angle between vectors of observations, among others. However, clustering algorithms that apply a conventional distance function (e.g., the Euclidian distance, correlation coefficient) on a pair of profiles, can fail to reflect the temporal information embedded in the underlying expression profiles. Some alignment techniques resolve the issue effectively prior to applying the distance function. The basic idea and the effect of profile alignment when computing the distance between two profiles can be visualized in the following example.

Typical examples for the profile alignment are depicted in Figure 1 with different cases of temporal profiles for three pairs of genes. Figures 1(a), (b) and (c) show three pairs of genes prior to alignment. Using the Pearson correlation distance [25], genes in (b) are most likely to be clustered together, since they produce the largest value for the correlation coefficient among all three pairs of genes, which yields 0.9053. If the prime interest is to cluster genes according to the variation of their expression level at different time points, then, genes from (c) would be better candidates to be clustered together than the genes in (a) and (b). However, the value of the correlation coefficient between the pairs of genes in (c) is the minimum (0.8039) among all three pairs of genes. Figures 1(d), (e) and (f) show the pairs of genes after aligning the expression profiles from (a), (b) and (c), respectively. A careful visual inspection shows that the expression profiles in (f) are closer to each other compared to the expression profiles from the other two figures, (d) and (e).

2.2.1. Mean-Square-Error Profile Alignment

The idea of profile alignment based on mean-square-error, proposed in [1], is to consider two feature vectors, and produce two new vectors in such a way that the sum of square-error difference between the "aligned" expression ratio at each time point is minimized. This approach treats the weights of the intervals between the measurements of an experiment equally, irrespective to the actual size of the interval of the measurement. The profile alignment scheme works as follows. Let $\mathbf{x}_1, \mathbf{x}_2 \in \mathcal{D}$ be two feature vectors. The aim is to find two new vectors $\mathbf{x}_1$ and $\mathbf{x}_2' = \mathbf{x}_2 - a$ (e.g. to find a scalar a) such that:

$$f(a) = \|\mathbf{x}_2 - a - \mathbf{x}_1\|^2 = \sum_{i=1}^{m}(x_{2_i} - a - x_{1_i})^2, \tag{1}$$

is minimized. To minimize $f(a)$, the first order necessary condition is obtained as follows:

$$\begin{aligned} \frac{\partial f}{\partial a} &= \sum_{i=1}^{m}(x_{1_i} - x_{2_i}) + ma = 0 \\ &\Rightarrow \quad a = \frac{1}{m}\sum_{i=1}^{m}(x_{2_i} - x_{1_i}) \,. \end{aligned} \tag{2}$$

The necessary condition only confirms a critical point, which can represent a minimum, maximum or an inflexion point. To this, the second order sufficient condition is derived as follows:

$$\frac{\partial^2 f}{\partial^2 a} = \frac{\partial}{\partial a} 2\sum_{i=1}^{m}(x_{1_i} - x_{2_i} + a) = 2m > 0\,, \tag{3}$$

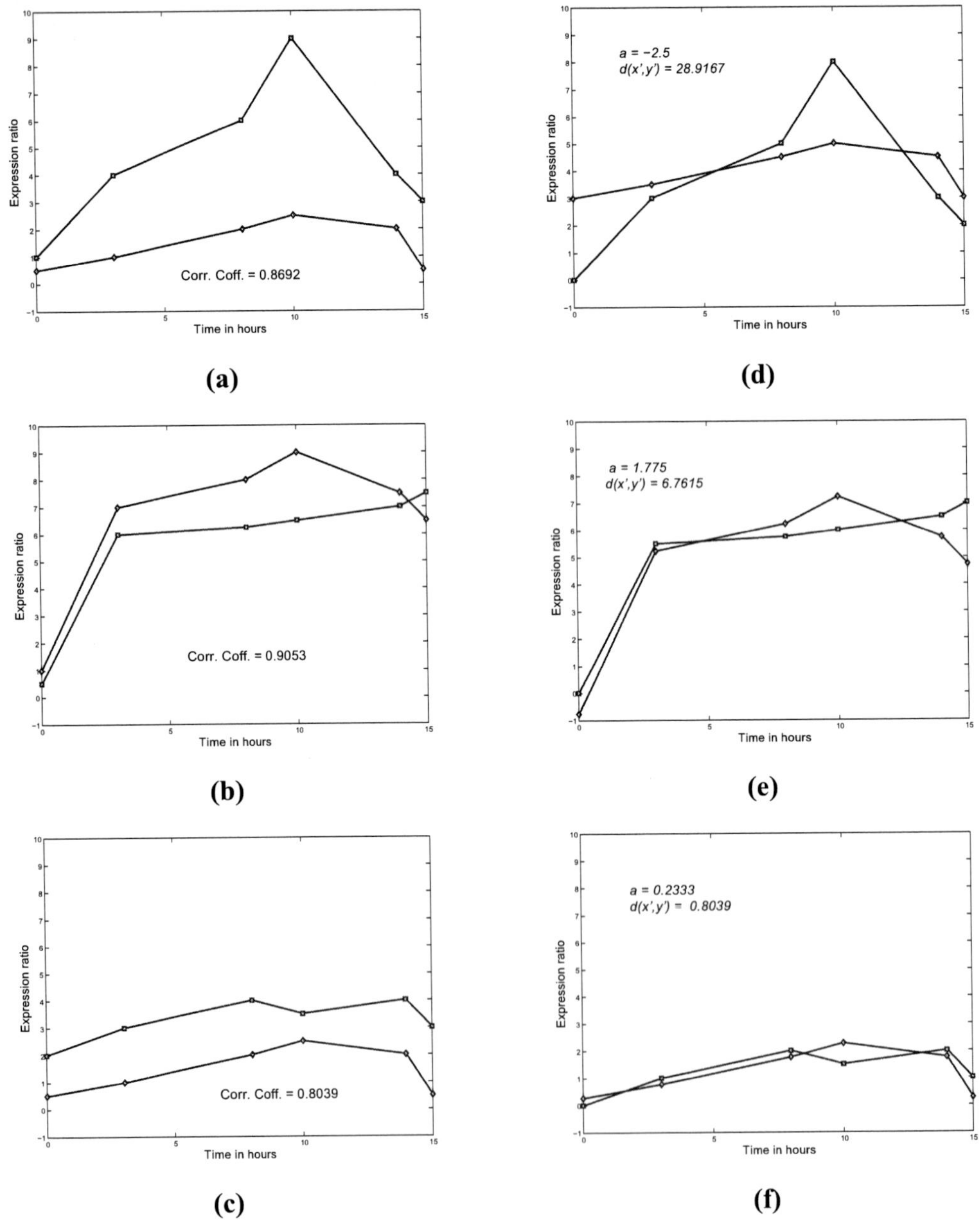

Figure 1. (a) Two genes that are likely to be clustered as in [25], although the difference among them in terms of rate of expression ratio changes between different time points is large; (b) Two genes with different profiles that are likely to be clustered together by correlation-based methods; (c) Two genes with similar profiles in terms of rate of expression ratio changes between different time points that may not be clustered together by the method proposed in [10] and the correlation-based methods; (d), (e) and (f) depict the result after aligning the two genes from (a), (b) and (c), respectively.

which is positive. This indicates that a minimum is reached at $a = \frac{1}{m}\sum_{i=1}^{m}(x_{2_i} - x_{1_i})$. Using (2) to obtain the value of a, i.e. the scalar that minimizes (1), two vectors are aligned by following the procedure given in Algorithm *Profile-Alignment*. The algorithm takes two feature vectors from the *original* space as input (which are two temporal gene expression data in this case) and outputs two feature vectors in the *transformed* space after aligning them in such a way that the sum of square errors is minimized. The algorithm works as follows: it first picks one vector, $\mathbf{x}_1$, from the input and produces the first aligned vector, $\mathbf{y}_1$. Then, it computes the value of the scalar a using (2) and produces the second aligned vector, $\mathbf{y}_2$, as $\mathbf{y}_2 \longleftarrow \mathbf{x}_2 - a$.

Algorithm 1 *Profile-Alignment*

Input: Two vectors, $\mathbf{x}_1 = [x_{1_1}, x_{1_2}, \ldots, x_{1_m}]$ and $\mathbf{x}_2 = [x_{2_1}, x_{2_2}, \ldots, x_{2_m}]$.
Output: Two vectors $\mathbf{y}_1$ and $\mathbf{y}_2$ such that the mean-square error is minimized.
begin
  $\mathbf{y}_1 \longleftarrow \mathbf{x}_1 - x_{1_1}$
  $a \longleftarrow \frac{1}{m}\sum_{i=1}^{m}(x_{2_i} - y_{1_i})$.
  $\mathbf{y}_2 \longleftarrow \mathbf{x}_2 - a$
end
return $\mathbf{y}_1, \mathbf{y}_2$

Using the *Profile-Alignment* algorithm and a conventional distance function, a new distance function is defined, namely $d_{MSE}(\mathbf{x}_1, \mathbf{x}_2)$, which aligns $\mathbf{x}_1$ and $\mathbf{x}_2$, and invokes any conventional metric. It is not difficult to show that the *Profile-Alignment* algorithm used in conjunction with any metric d is also a metric. This can be easily proved as follows. Let $d : \Re^d \times \Re^d \rightarrow \Re$ be a metric and let $d_{MSE}(\mathbf{x_1}, \mathbf{x_2})$ be the result of algorithm *Distance* with input $(\mathbf{x_1}, \mathbf{x_2})$, then:

1. $d_{MSE}(\mathbf{x}_1, \mathbf{x}_2) \geqslant 0$. It follows since the *Profile-Alignment* algorithm returns two new vectors, $\mathbf{y}_1 \in \Re^d$ and $\mathbf{y}_2 \in \Re^d$. Then we have $d(\mathbf{y}_1, \mathbf{y}_2) \geqslant 0$, since d is a metric.

2. $d_{MSE}(\mathbf{x}_1, \mathbf{x}_2) = d_{MSE}(\mathbf{x}_2, \mathbf{x}_1)$. Again, the *Profile-Alignment* algorithm returns two new vectors $\mathbf{y}_1 \in \Re^d$ and $\mathbf{y}_2 \in \Re^d$. Thus, since d is a metric, it implies that, $d(\mathbf{y}_1, \mathbf{y}_2) = d(\mathbf{y}_2, \mathbf{y}_1)$, and hence, $d_{MSE}(\mathbf{x}_1, \mathbf{x}_2) = d_{MSE}(\mathbf{x}_2, \mathbf{x}_1)$.

3. $d_{MSE}(\mathbf{x}_1, \mathbf{x}_3) \leqslant d_{MSE}(\mathbf{x}_1, \mathbf{x}_2) + d_{MSE}(\mathbf{x}_2, \mathbf{x}_3)$. From the *Profile-Alignment* algorithm, it is know that $\mathbf{y}_i = \mathbf{x}_i - a \in \Re^d$, for $i = 1, 2, 3$. Then:

$$\begin{aligned} d(\mathbf{y}_1, \mathbf{y}_3) &\leqslant d(\mathbf{y}_1, \mathbf{y}_2) + d(\mathbf{y}_2, \mathbf{y}_3) \\ &\Rightarrow d(\mathbf{x}_1 - a, \mathbf{x}_3 - a) \leqslant d(\mathbf{x}_1 - a, \mathbf{x}_2 - a) + d(\mathbf{x}_2 - a, \mathbf{x}_3 - a), \end{aligned}$$

   since d is a metric. Thus $d_{MSE}(\mathbf{x}_1, \mathbf{x}_3) \leqslant d_{MSE}(\mathbf{x}_1, \mathbf{x}_2) + d_{MSE}(\mathbf{x}_2, \mathbf{x}_3)$.

Note that once the *Profile-Alignment* algorithm is applied, any metric d can be used.

2.2.2. Area-Based Profile Alignment

The profile alignment approach proposed in [29] takes into consideration the length of the intervals, which could be unequal, between the time points at which the measurements

are taken in an experiment. As in [1], this approach also used the *Profile-Alignment* algorithm, but the alignment of the profiles are done using an area based distance. The idea of this scheme works as follows.
Let $\mathbf{t} = [t_1, t_2, ..., t_m]^t$ be the vector representing the time points at which the measurements are taken, and let $\mathbf{x} = [x_1, x_2, ..., x_m]^t$, and $\mathbf{y} = [y_1, y_2, ..., y_m]^t$ be two expression profiles, whose expression ratios were measured at time points given in $\mathbf{t}$, which are to be aligned. Again, the aim is to find a scalar a that minimizes the total area between the two profiles, e.g., between the lines that join the expression ratios. To do this, first, $\mathbf{x}$ is slid down and a new vector, $\mathbf{x}'$, is obtained as follows:

$$\mathbf{x}' = \left[x'_1, x'_2, ..., x'_m\right]^t \leftarrow \mathbf{x} - x_1 \,. \tag{4}$$

Now, let the straight line that joins points (t_{i-1}, x'_{i-1}) and (t_i, x'_i) be denoted by $x'_{i-1} + \frac{(x'_i - x'_{i-1})}{t_i - t_{i-1}} u$, and for points (t_{i-1}, y_{i-1}) and (t_i, y_i) be denoted by $y_{i-1} - a + \frac{(y_i - y_{i-1})}{t_i - t_{i-1}} u$. Since the aim is to minimize the area between $\mathbf{x}$ and $\mathbf{y}$ (aligned), for all $t_1, t_2, ..., t_m$, it is necessary to find a value of a that minimizes the sum of square errors between each pair of points in the continuous times, equivalent to the following sum of integrals:

$$f(a) = \sum_{i=2}^{m} \int_{t_{i-1}}^{t_i} \left[x'_{i-1} + a - y_{i-1} + \frac{(x'_i - x'_{i-1}) - (y_i - y_{i-1})}{t_i - t_{i-1}} u\right]^2 du \,, \tag{5}$$

by means of the first and second order conditions. Solving the integral and isolating a results in:

$$a = -\frac{\sum_{i=2}^{m} \left[\left(x'_{i-1} - y_{i-1}\right)(t_i - t_{i-1}) + \frac{(x'_i - x'_{i-1}) - (y_i - y_{i-1})}{t_i - t_{i-1}} \frac{(t_i - t_{i-1})^2}{2}\right]}{\sum_{i=2}^{m} (t_i - t_{i-1})} \tag{6}$$

Then, a new vector, $\mathbf{y}'$, is computed as follows:

$$\mathbf{y}' = \mathbf{y} - a \tag{7}$$

or equivalently in the way that Algorithm Profile Alignment produces. Letting $f_i = \frac{(x'_i - x'_{i-1}) - (y'_i - y'_{i-1})}{t_i - t_{i-1}}$, and by computing the integral distance between the two new vectors $\mathbf{x}'$ and $\mathbf{y}'$, $d(\mathbf{x}', \mathbf{y}')$, results in:

$$d\left(\mathbf{x}', \mathbf{y}'\right) = \sum_{i=2}^{m} \left[\left(x'_{i-1} - y'_{i-1}\right)^2 u + \left(x'_{i-1} - y'_{i-1}\right) f_i u^2 + \frac{f_i^2 u^3}{3}\Bigg|_{t_{i-1}}^{t_i}\right]$$

$$d\left(\mathbf{x}', \mathbf{y}'\right) = \sum_{i=2}^{m} \left(x'_{i-1} - y'_{i-1}\right)^2 (t_i - t_{i-1}) + \left(x'_{i-1} - y'_{i-1}\right) f_i (t_i - t_{i-1})^2 + f_i^2 \frac{(t_i - t_{i-1})^3}{3} \,. \tag{8}$$

It is not difficult to show that the second order condition is satisfied: $\frac{\partial^2 f}{\partial^2 a} > 0$. Using (6) to obtain the value of a, two profile vectors are aligned first, and then a distance function is computed in the usual manner. It can be shown that the alignment, used in conjunction with any metric d, is also a metric [1]. It may be noted that once the alignment is applied, any metric d can be used.

The idea is described in Figure 2. In Figure 2(a) two vectors are shown before alignment. Figure 2(b) shows the "aligned" vectors such that the area between the profiles is minimized, i.e. they were aligned in such a way that the total area covered by the triangle $\{u, v, z\}$ and the polygon $\{z, w, q, e, r, k, h, g, s\}$ is minimized.

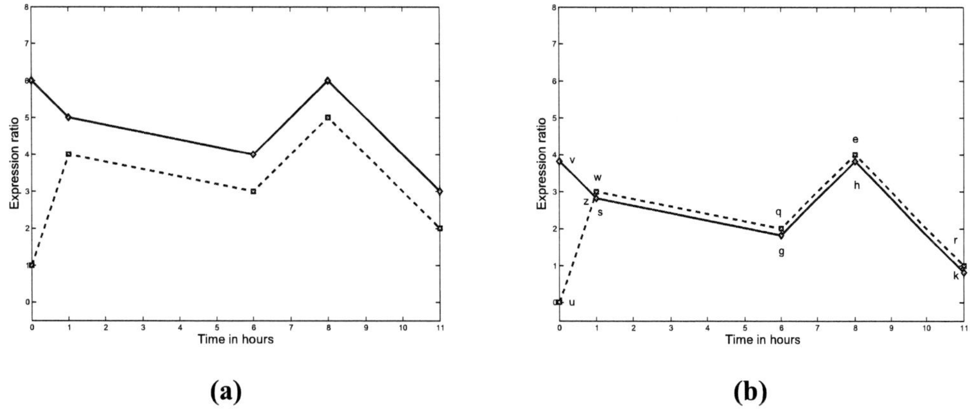

Figure 2. (a) Two unaligned profiles. (b) The two "aligned" profiles obtained after applying (4) - (7) such that the area between each pair of lines is minimized.

2.2.3. Clustering Methods for Profile Alignment

Clustering methods can generally be classified into two main categories, *partitional* and *hierarchical*. Many clustering approaches have been proposed in the past few years, including the methods proposed in [6][10][19][32]. Each of these methods has its own pros and cons depending upon handling noise levels in the measurements and the properties of the particular dataset being clustered, and hence, none of them can be regarded as *the best* method.

Hierarchical clustering is one of the most widely used approaches applied for the purpose of clustering microarray data. *Hierarchical Clustering* is divided into two main categories, *divisive methods*, and *agglomerative methods*. The basic idea of the hierarchical agglomerative clustering method is to successively fuse n objects into smaller groups, by applying the process of merging clusters based on their similarity, until all objects are joined together and form a single cluster. In this process, the data are not partitioned into clusters in a single step, but in a series of merging processes, which run from n clusters each containing a single object to a single cluster containing all n objects.

The various hierarchical agglomerative clustering methods differ in the decision rules (i.e., the way distances are computed among the clusters) used for merging two clusters. These are known as *linkage type*, and there are mainly four such linkage types that are commonly used in hierarchical agglomerative clustering [8]: *complete linkage* or *furthest neighbors*, *single linkage* or *nearest neighbors*, *centroid linkage* and *average linkage*. In our proposed method, we use *complete linkage* or *furthest neighbors* [8]. *Complete linkage* or *furthest neighbors* clustering computes the distance between the furthest pair of points for each pair of clusters and merges the pair of clusters that has the minimum distance among all such distances between the pair of clusters under consideration.

For the works of both [1] and [29], agglomerative clustering is used. The decision rule is based on the *furthest-neighbor* distance between two clusters [8], which is computed using $d_{MSE}(\mathbf{x}_1, \mathbf{x}_2)$, in [1], and area-based distance, in [29]. The *complete linkage* or *furthest neighbors* computes the distance between the furthest pair of points for each pair of clusters and merges the pair of clusters that has the minimum distance among all such distances between the pair of clusters under consideration. The generalized algorithm of *hierarchical agglomerative clustering* is slightly modified to obtain the desired number of clusters instead of a hierarchy of clusters, which is given in Algorithm *Agglomerative-Clustering*. The algorithm receives two parameters as input, a complete microarray temporal dataset, $\mathcal{D}$, and the desired number of clusters, k, and returns the dataset after partitioning it into k clusters. The decision rule is based on the *furthest-neighbor* distance between two clusters, which is computed using formula (8). The latter involves the alignment of each pair of profiles before applying a conventional distance function.

Algorithm 2 ***Agglomerative-Clustering***

Input: The dataset, $\mathcal{D} = \{\mathbf{x}_1, \mathbf{x}_2, ..., \mathbf{x}_n\}$, and k, the desired number of clusters.
Output: k disjoint subsets $\mathcal{D}_1, \mathcal{D}_2, ..., \mathcal{D}_k$.
begin
  Create n clusters, $\mathcal{D}_1, \mathcal{D}_2, ..., \mathcal{D}_n$, where $\mathcal{D}_i = \{\mathbf{x}_i\}$
  $\mathcal{D}_{currentClustersSet} \longleftarrow \{\mathcal{D}_1, \mathcal{D}_2, ..., \mathcal{D}_n\}$
  for $q \longleftarrow n$ **down to** k **do**
    For each pair of cluster $(\mathcal{D}_i, \mathcal{D}_j)$, find furthest neighbor $(\mathbf{x}_i, \mathbf{x}_j)$, using the profile alignment and then computing the distance between them.
    Select $(\mathcal{D}_k, \mathcal{D}_l)$ as the pair of clusters with the closest furthest neighbors.
    $\mathcal{D}_{mergedClusters} \longleftarrow \{\mathcal{D}_k \cup \mathcal{D}_l\}$
    $\mathcal{D}_{currentClustersSet} \longleftarrow \{\mathcal{D}_{currentClustersSet} \cup \mathcal{D}_{mergedClusters}\} \setminus \{\mathcal{D}_k, \mathcal{D}_l\}$
  end
end
return $\mathcal{D}_{currentClustersSet}$

The rationale behind choosing an agglomerative method over a partitional method is that, in order to measure the similarity between two profiles, they are needed to be aligned first. This requirement complicates partitional clustering methods, since they work with *all* genes simultaneously. Thus, in profile alignment, *multiple alignment* would be needed. However, the application of multiple alignment to the whole data is like creating a single cluster. Thus, the best option could be to partition the data into k clusters, multiple-aligning the profiles in a cluster, and after this, applying a criterion that maximizes the similarity among genes in the same cluster, while maximizing the dissimilarity among clusters. However, the latter problem is really complicated, since *clusters* are needed to be somehow aligned to measure the similarity. Then, it would be required to define the alignment between two (or more) clusters. Thus, the problem is quite intricate, and is open for a future work

2.2.4. Optimizing the Number of Clusters

Finding the optimal number of clusters is a well known open problem in clustering. For this, it is desirable to validate the number of clusters that is the best for a given dataset using a validity index. The most commonly used cluster validity indices are the Davies-Bouldin (DV) Index, the Dunn's index, the Calinski Harabasz (CH) index, and the $\mathcal{I}$-index [22]. For the validity purpose of the clustering, a variant of the $\mathcal{I}$-index [22] is used. The traditional $\mathcal{I}$-index is defined as follows:

$$\mathcal{I}(k) = \left(\frac{1}{k} \times \frac{E_1}{E_k} \times \mathcal{D}_k\right)^p, \tag{9}$$

where, $E_k = \sum_{i=1}^{k} \sum_{j=1}^{n} u_{ij} \|\mathbf{x}_j - \boldsymbol{\mu}_i\|$, $\mathcal{D}_k = max_{i,j=1}^{k} \|\boldsymbol{\mu}_i - \boldsymbol{\mu}_j\|$, n is the total number of samples in the dataset, $\{u_{ij}\}_{k \times n}$ is the partition (or membership) matrix for the data, $\boldsymbol{\mu}_i$ is the center of cluster $\mathcal{D}_i$ and k is the number of clusters. The best number of clusters k^* is chosen to maximize $\mathcal{I}(k)$.

Although this index has been found to work well in many cases, it has been argued, in [1] and [29], that it is not appropriate for the time series microarray data clustering due to a few reasons. A superficial analysis of $\mathcal{I}(k)$ suggests the following:

- $\mathcal{I}(k)$ is large when $\mathcal{D}_k$ is large. This occurs when the inter-cluster distances are large, i.e. the clusters are far away from each other.
- $\mathcal{I}(k)$ is large when $\frac{E_1}{E_k}$ is large. This occurs when E_k is small, or equivalently when the points are closer to the cluster center they belong to.
- $\mathcal{I}(k)$ is large when $\frac{1}{k}$ is large, which implies that k is small.

The third observation contradicts the aim of finding differentially expressed genes, which in many cases, needs to form clusters containing only one or two genes. Since this is penalized by the $\mathcal{I}$-index (also, in the case of the other indices), the following variant of the $\mathcal{I}$-index is proposed for Profile Alignment and Agglomerative Clustering (PAAC) methods [1] [29]:

$$\mathcal{I}_{PAAC}(k) = \left(\frac{1}{k}\right)^q \times \left(\frac{E_1}{E_k} \times \mathcal{D}_k\right)^p \tag{10}$$

The partition matrix $\{u_{ij}\}$ is defined as a membership function such that $u_{ij} = 1$, if $\mathbf{x}_j$ belongs to cluster $\mathcal{D}_i$, and zero otherwise. Again, the best number of clusters is the value of k that maximizes $\mathcal{I}_{PAAC}(k)$.

The implementation of index $\mathcal{I}_{PAAC}(k)$ is not straightforward. It includes the mean and scatter for each cluster, and these two are meant to include the profile alignment concept. For the mean, algorithm *Cluster-Mean*, given in Algorithm 3, is used. This algorithm takes a cluster of gene expression profiles and returns the mean expression profile of that cluster. The algorithm uses a greedy approach for computing the mean expression profile. It picks the first profile from the cluster and assigns it to be the current mean. Then, it picks the next gene profile from the cluster, applies the pairwise alignment algorithm with this profile to the current mean, and updates the current mean by taking the average of these two aligned profiles. The process continues until all the gene profiles are aligned and the current mean is updated correspondingly for each profile. The final value of the current mean is returned by the algorithm as the mean profile of the input cluster. Note that computing the mean in this way has the particularity that changing the order of the genes produces different means. Also, the mean obtained does not represent the *actual* mean; it is just an estimate. Another way of computing the mean would be, for example, by aligning *all* profiles in a cluster and then, compute the mean. However, this strategy for alignment includes multiple alignment, which remains an open problem.

Algorithm 3 ***Cluster-Mean***

Input: A cluster $\mathcal{D}_i$ with n_i samples $\mathcal{D}_i = [\mathbf{x}_{i_1}, \mathbf{x}_{i_2}, \ldots, \mathbf{x}_{in_i}]$.
Output: The mean of cluster $\mathcal{D}_i$, $\boldsymbol{\mu}_i$.

```
begin
|  μ_i ⟵ x_{i_1}
|  for j ⟵ 2 to n_i do
|  |  [y_1, y_2] ⟵ d_PAAC(μ_i, x_{i_j})
|  |  μ_i ⟵ ½(y_1 + y_2)
|  end
end
return μ_i
```

Once the *Cluster-Mean* is defined, it is used to compute the scatter of a cluster, using the algorithm *Within-Cluster-Scatter*, given in Algorithm 4. The algorithm takes a cluster of gene expression profile and its mean profile as inputs, and computes the sum of the distances between each gene expression profile and the cluster mean. Each expression profile and the mean profile are aligned before a distance function is applied.

Algorithm 4 ***Within-Cluster-Scatter***

Input: A cluster $\mathcal{D}_i$ with n_i samples, $\mathcal{D}_i = [\mathbf{x}_{i_1}, \mathbf{x}_{i_2}, \ldots, \mathbf{x}_{in_i}]$, and its mean, $\boldsymbol{\mu}_i$.
Output: The sum of the distances of each gene from the cluster mean, E_i.

```
begin
|  E_i ⟵ 0
|  for j ⟵ 1 to n_i do
|  |  E_i ⟵ E_i + d_PAAC(μ_i, x_{i_j})
|  end
end
return E_i
```

3. Experimental Comparison

In this section, experiments performed on well-known datasets are discussed. The clustering methods tested include the agglomerative clustering when used with the Pearson and the Spearman correlation to measure the similarity between two profiles. The other method that is tested is PAAC. For finding the best number of clusters, the index I_{PAAC} defined in (10) is used.

3.1. Datasets for the Experiments

To test the performance of the clustering methods proposed in [1] and [29] and clustering using correlation measures, three real-life datasets were used. Details of the datasets and their corresponding sources are briefly discussed here.

1. The serum dataset was obtained from the experimental data used in [19], and contains data on the transcriptional response of cell cycle-synchronized human fibroblasts to serum. These experiments have measured the expression levels of 8,613 human genes after a serum stimulation at twelve different time points, at 0 hr., 15 min., 30 min., 1 hr., 2 hrs., 3 hrs., 4 hrs., 8 hrs., 16 hrs., 20 hrs. and 24 hrs. From these 8,613 gene profiles, 517 profiles were separately analyzed, as their expression ratio has changed substantially at two or more time points. The experiments and analysis have focused on this dataset[1], which is the same group of the 517 genes used in [19].
2. The second is a dataset of DNA microarrays containing the changes in gene expression during sporulation of 6,118 yeast genes [2] [6]. The expression levels in the dataset were measured at seven different time points, at 0 hr., 0.5 hr., 2 hrs., 5 hrs., 7 hrs., 9 hrs. and 11.5 hrs.
3. Finally, in order to provide a biological significance of the results, the methods were applied to a dataset containing the changes in gene expression during the cell cycle of the budding yeast S. cerevisiae[3] [4]. The expression levels in the dataset were measured at seventeen different time points, from 0 min. to 160 min. with an interval of 10 min. The experiment monitored 6,220 transcripts for cell cycle-dependent periodicity, and 221 functionally characterized genes with periodic fluctuation were listed in Table 1 of [4].

3.2. Finding the Best Number of Clusters

It is well known that, in general, the best number of clusters for any real-life detaset is usually less than or equal to $\sqrt{n}$, where n is the number of samples in the dataset [22]. However, in time-series gene expression data, some useful clusters contain one or two differentially expressed genes. Thus, a range for potential numbers of clusters, which includes the values of k that lie between $\left\lceil\sqrt{\frac{1}{2}n}\right\rceil$ and $\left\lfloor\sqrt{\frac{3}{2}n}\right\rfloor$, is considered.

The index used for the computation is the $\mathcal{I}_{PAAC}$ as in (10), where the distance is computed by using $d_{MSE}(.,.)$. For the serum dataset the range for the number of clusters was set to $k = 16$ to 27. The values of the $\mathcal{I}_{PAAC}$ index were computed for each of these values of k. For each k, the $\mathcal{I}_{PAAC}$ index was computed using values of q from 0.3 to

[1]Expression data for this subset were obtained from the website: http://genome-www.stanford.edu/serum/.

[2]Expression data for this dataset were obtained from the website: http://cmgm.stanford.edu/pbrown/sporulation/additional/.

[3]Expression data for this dataset were obtained from the website: http://genome-www.stanford.edu/cellcycle/data/rawdata/individual.html.

Table 1. Values of the $\mathcal{I}_{PAAC}$ index for 517 genes, where k = 16 to 27 and q = 0.3 to 1.0

k	q=0.3	q=0.4	q=0.5	q=0.6	q=0.7	q=0.8	q=0.9	q=1.0
16	17344.58	13144.73	9961.84	7549.67	5721.58	4336.14	3286.18	2490.46
17	18791.87	14155.50	10663.02	8032.21	6050.49	4557.70	3433.21	2586.16
18	18685.75	13995.33	10482.28	7851.07	5880.32	4404.27	3298.73	2470.70
19	27333.00	20361.60	15168.29	11299.55	8417.55	6270.62	4671.27	3479.84
20	27169.35	20136.14	14923.59	11060.39	8197.23	6075.25	4502.58	3337.02
21	29057.98	21431.06	15805.99	11657.35	8597.61	6340.97	4676.64	3449.15
22	29577.07	21712.65	15939.35	11701.14	8589.86	6305.85	4629.15	3398.28
23	29554.55	21599.89	15786.24	11537.35	8432.05	6162.55	4503.89	3291.66
24	29493.69	21463.87	15620.21	11367.52	8272.65	6020.38	4381.29	3188.46
25	29981.57	21730.04	15749.49	11414.91	8273.29	5996.32	4346.01	3149.90
26	30081.89	21717.39	15678.71	11319.13	8171.76	5899.54	4259.13	3074.85
27	29959.90	21547.85	15497.71	11146.31	8016.69	5765.79	4146.89	2982.54

1.0, as shown in Table 1. Each row in the table corresponds to a value of k, each column corresponds to a value of q, and each cell shows the value of the index for the corresponding values of k and q. Of all the values of q, $q = 0.7$ is considered, based on the experimental results carried out. The value is selected to $q = 0.7$, since the intention was to favor a large number of clusters, as $q = 0.7 < 2.0 = p$. The value 0.7 has been found experimentally and confirmed with pre-clustered yeast data for which the method is found to provides a high classification accuracy. Also, note that any value of q between 0.4 and 0.8 would produce quite similar results to what was obtained. The table shows that when $q = 0.7$, the value of the index reaches to a maximum level when $k = 21$. Therefore, the value of $k = 21$ was taken as the optimal number of clusters[4] and plotted the profiles, as clustered using the PAAC method, and the agglomerative clustering with the Pearson and the Spearman correlations.

For the Yeast dataset, the values of the $\mathcal{I}_{PAAC}$ index was computed for $k = 55$ to 95, while keeping the range of values for q the same (from 0.3 to 1.0). The values for the $\mathcal{I}_{PAAC}$ index for $k = 55$ to 75 and $q = 0.3$ to 1.0 are shown in Table 2. Again, the value of $q = 0.7$ was selected, for which the value of the index reaches to a maximum level when[5] $k = 60$. Therefore, $k = 60$ was taken as the optimal number of clusters and plotted the profiles, as clustered using the PAAC method, and the Pearson and the Spearman correlations.

3.3. The Serum Dataset

The serum data was clustered into 21 clusters, using the Pearson correlation distance within the same hierarchical agglomerative clustering method, and the PAAC method. The 21 clusters of the serum dataset, using the PAAC method corresponding to $k = 21$, are plotted in Figure 3. The x-axis in each plot represents the time in hours and the y-axis

[4]As in the $\mathcal{I}$-index, the best number of clusters corresponds to the value in which the index is maximized.

[5]Although the table does not show the index values for $k > 75$, these values tend to decrease, and hence, confirming the validity of $k = 60$ as the optimal number of clusters.

Table 2. Values of the $\mathcal{I}_{PAAC}$ index for 6,118 genes, where k = 55 to 75 and q = 0.3 to 1.0

k	q=0.3	q=0.4	q=0.5	q=0.6	q=0.7	q=0.8	q=0.9	q=1.0
55	365.1896	244.6145	163.8498	109.7513	73.5146	49.2422	32.9838	22.0935
56	363.7068	243.1826	162.5975	108.7164	72.6902	48.6024	32.4966	21.728
57	362.8381	242.1728	161.6359	107.8823	72.005	48.0591	32.0766	21.4092
58	361.3878	240.7857	160.4308	106.892	71.22	47.4525	31.6167	21.0656
59	359.8542	239.3544	159.2048	105.8939	70.4346	46.849	31.1613	20.7267
60	382.1949	253.7873	168.5213	111.9024	**74.3061**	49.3412	32.7638	21.756
61	381.4872	252.8989	167.654	111.1427	73.6797	48.8444	32.3804	21.4659
62	381.258	252.3363	167.0093	110.5354	73.1581	48.4198	32.0467	21.2102
63	380.7769	251.615	166.2656	109.8673	72.5996	47.9734	31.7005	20.9475
64	379.4638	250.3527	165.1712	108.9724	71.8949	47.433	31.2941	20.6464
65	379.333	249.8787	164.6031	108.4293	71.4258	47.0505	30.9936	20.4165
66	381.8707	251.1667	165.1991	108.6559	71.4659	47.005	30.9165	20.3346
67	380.5694	249.9346	164.1417	107.7982	70.7952	46.494	30.5344	20.0531
68	381.0179	249.8587	163.8489	107.4466	70.4599	46.2052	30.2998	19.8696
69	379.7648	248.6737	162.8339	106.6252	69.8192	45.7183	29.9368	19.6029
70	379.032	247.8369	162.0527	105.9611	69.2846	45.303	29.6222	19.369
71	378.1795	246.929	161.2302	105.2738	68.7376	44.8816	29.3051	19.1345
72	377.0762	245.8645	160.3107	104.5272	68.1547	44.4389	28.9754	18.8928
73	378.0785	246.1783	160.2941	104.3723	67.9599	44.2507	28.813	18.761
74	389.7858	253.4561	164.8085	107.1659	69.684	45.3117	29.4637	19.1586
75	388.6166	252.357	163.8737	106.415	69.103	44.8736	29.1397	18.9225

represents the expression ratio. Each plot represents a cluster. The resulting 21 clusters found using the Pearson correlation are shown in Figure 4.

The comparison among the plots for PAAC and Pearson (figure 4) reveal the effectiveness of the method to discover new profiles. For example, PAAC left clusters 1 to 5 containing a single gene each (IDs 328692, 470934, 361247, 147050 and 310406, respectively). The Pearson correlation method, however, placed these genes in clusters 19, 2, 2, 11 and 19, respectively. By visual inspection of all the temporal expression profiles, it can be observed that these genes are *differentially expressed* and should be left alone in separate clusters, which is clearly done by PAAC. Also, PAAC produced four clusters containing only two profiles each, clusters 9 (IDs 356635 and 429460), 11 (IDs 26474 and 254436), 13 (IDs 280768 and 416842) and 16 (IDs 130476 and 130482). The Pearson correlation method clustered these genes as follows: 356635 and 429460 in cluster 16, 26474 and 254436 in cluster 21, 280768 and 416842 in cluster 2 and 130476 and 130482 in cluster 16. Although the Pearson correlation method placed each pair of genes in the same cluster, it also placed some other genes with them. By looking at the plots of the profiles of the clusters produced by the Pearson correlation method and comparing them to the plots of the clusters of the corresponding genes produced by PAAC, it is clear that these pairs of genes are differentially expressed.

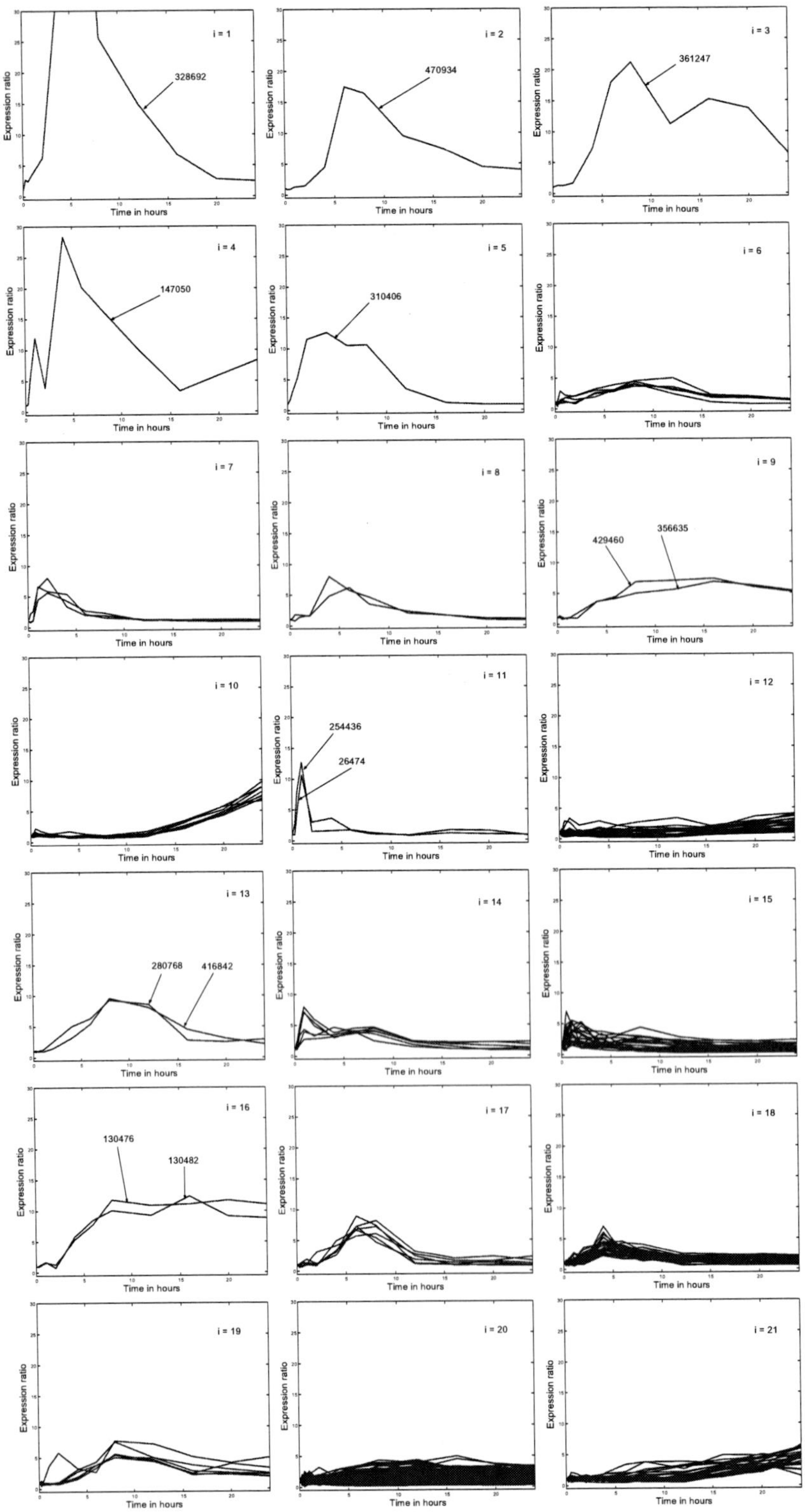

Figure 3. Different clusters obtained using PAAC on the 517 gene temporal expression profiles, where $k = 21$.

Also, to further analyze the performance of other methods, the same dataset was clustered using the Spearman rank correlation distance [17]. The 21 clusters obtained are shown in Figure 5. It can be observed that the Spearman correlation, though non-linear, is not able to identify and separate the differentially expressed genes properly. This situation is clearly observed in clusters 5, 6, 12, 18, and 21.

3.4. The Yeast Dataset

In the case of the yeast dataset, the value of $k = 60$ was taken as the optimal number of clusters. The profiles are plotted, as clustered using the PAAC method. The plots are shown in figures 6 and 7. Each figure shows 30 out of 60 clusters. The same dataset was also clustered into 60 clusters using the Pearson correlation distance. The plots of the resulting 60 clusters are shown in Figures 8 and 9. Again, each figure shows the plots of 30 out of 60 clusters.

Form the figures, it can be observed that although PAAC produced a number of clusters containing a single gene each, these genes are very differently expressed as compared to the other genes in the dataset. Therefore, it has been argued that these genes are better to be placed in individual clusters. For example, genes YER150w, YGL055W, YGR142W, YKL152C, YKL153W, YLR194C, YLR377C, YMR316W and YPR074C were placed in clusters with a single sample each by PAAC ($i = 34, 37, 40, 51, 52, 53, 56, 58,$ and 60, respectively). It is clear that all these expression profiles are completely different and hence they should be placed alone in a single cluster. The Pearson correlation method, on the other hand, placed these genes in clusters 45, 4, 1, 28, 15, 6, 38, 37 and 6, respectively ($i = 45, 4, 1, 28, 15, 6, 38, 37$ and 6 in figures 8 and 9). However, simply looking at these figures of the Pearson correlation distance, in terms of expression ratio changes, each of these genes has very different temporal profiles and can be easily distinguished from the rest of the genes placed in the same cluster by the Pearson correlation method. Therefore, PAAC produced clusters with better intra-cluster profile similarity.

3.5. Biological Significance

The PAAC method was finally applied to the 221 genes from the dataset of the budding yeast S. cerevisiae. The parameters were kept the same as those for serum dataset and for q=0.7, the best number of clusters $k = 28$ was obtained. The clusters obtained using PAAC are shown in Figure 10. The same dataset was also clustered using the Pearson correlation coefficient and the plots are shown in Figure 11. A comparison of the two figures shows that the PAAC separates the genes by profiles in a wise manner, while the Pearson correlation method is not able to capture all variations in the time series. This can be observed in various clusters that include genes with quite different profiles.

To show the biological significance of the results, the 221 genes are listed in Table 3, where for each gene, it is shown the cluster number PAAC assigns and the class (phase) that the gene is categorized as in Table 1 of [4]. An objective measure for comparing the two clusterings was taken by computing the overall classification accuracy, as the number of genes that PAAC *correctly* assigned to one of the phases. The *correct* class (phase) is the one that PAAC assigns the largest number of genes. This criterion is reasonable, as PAAC

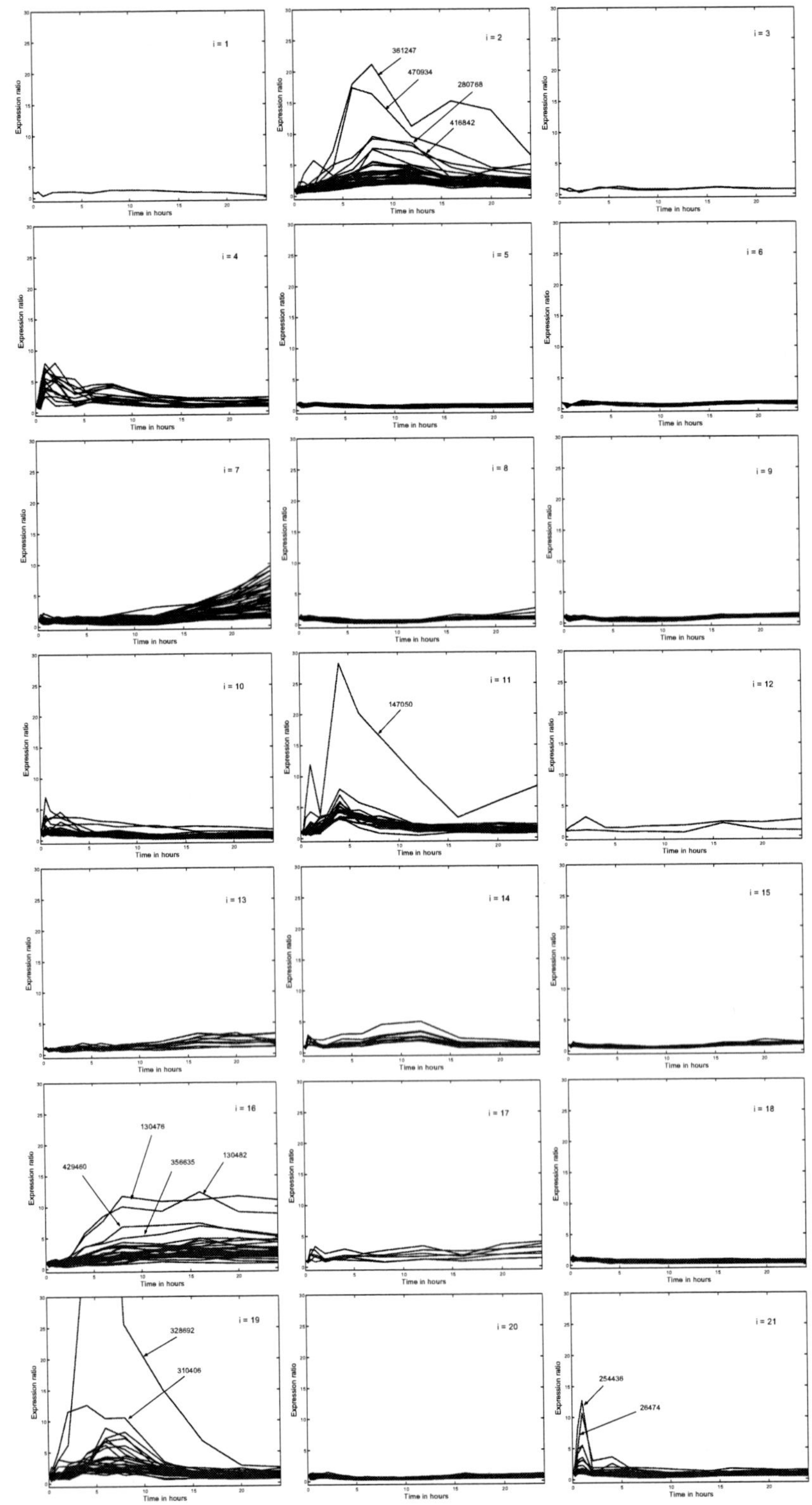

Figure 4. Different clusters obtained using the Pearson correlation distance on the 517 gene temporal expression profiles, where $k = 21$.

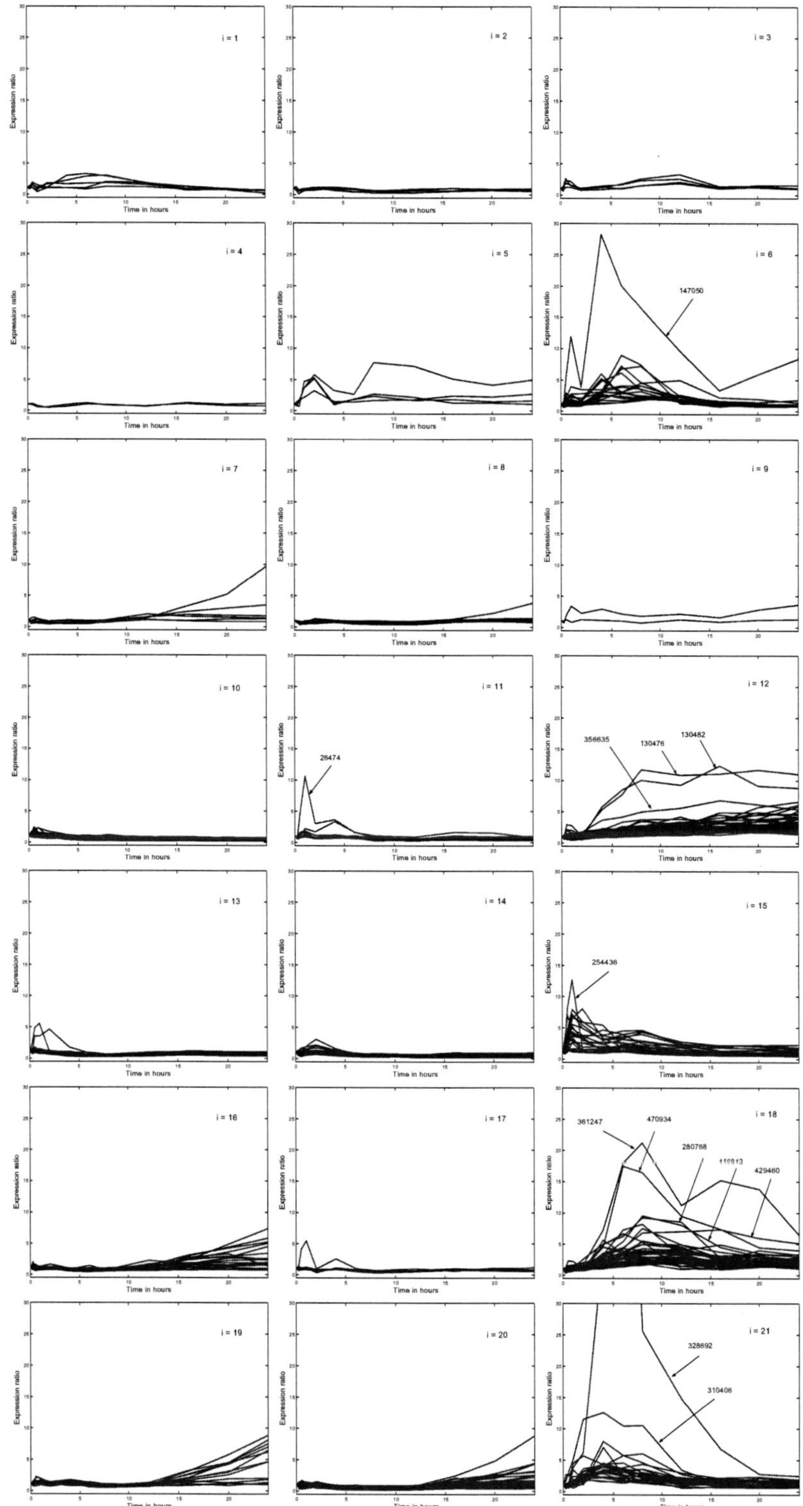

Figure 5. Different clusters obtained using the Spearman correlation distance on the 517 gene temporal expression profiles, where $k = 21$.

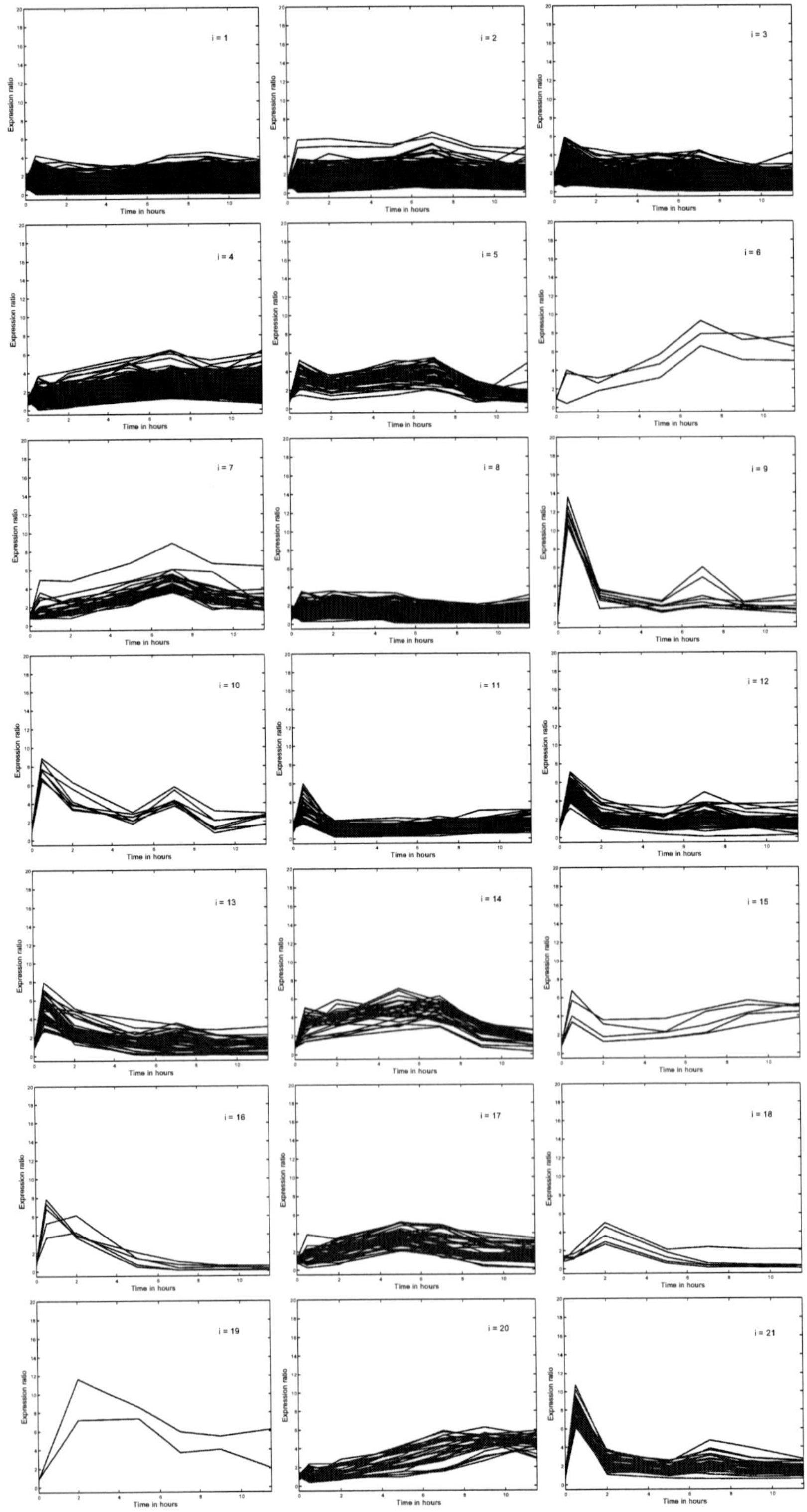

Figure 6. Continued on next page.

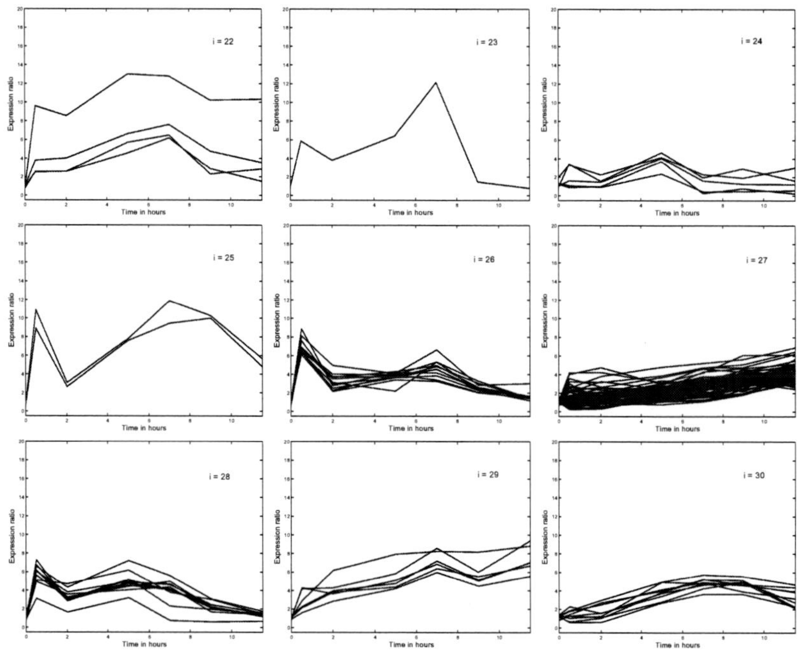

Figure 6. Clusters $i = 1$ to 30, obtained using the PAAC on the 6,118 gene temporal expression profiles of the yeast dataset.

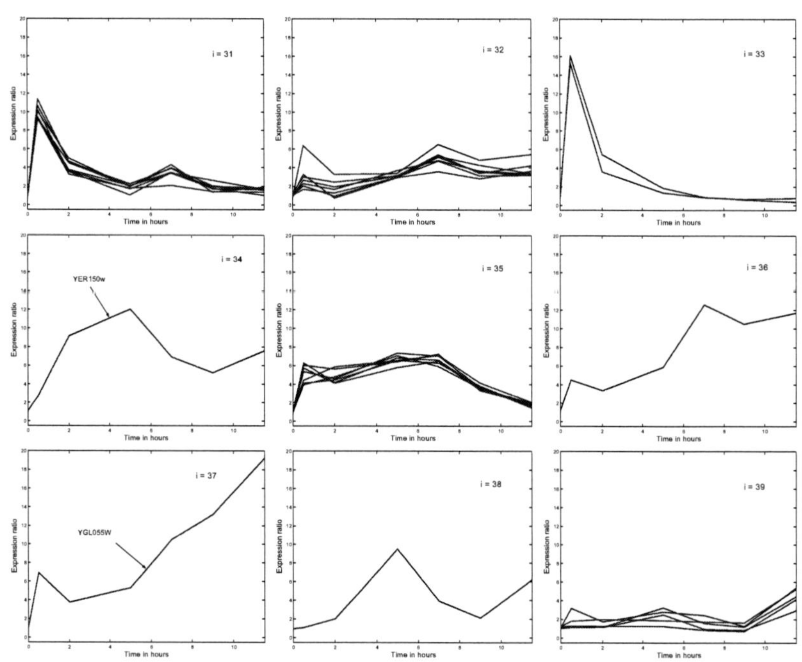

Figure 7. Continued on next page.

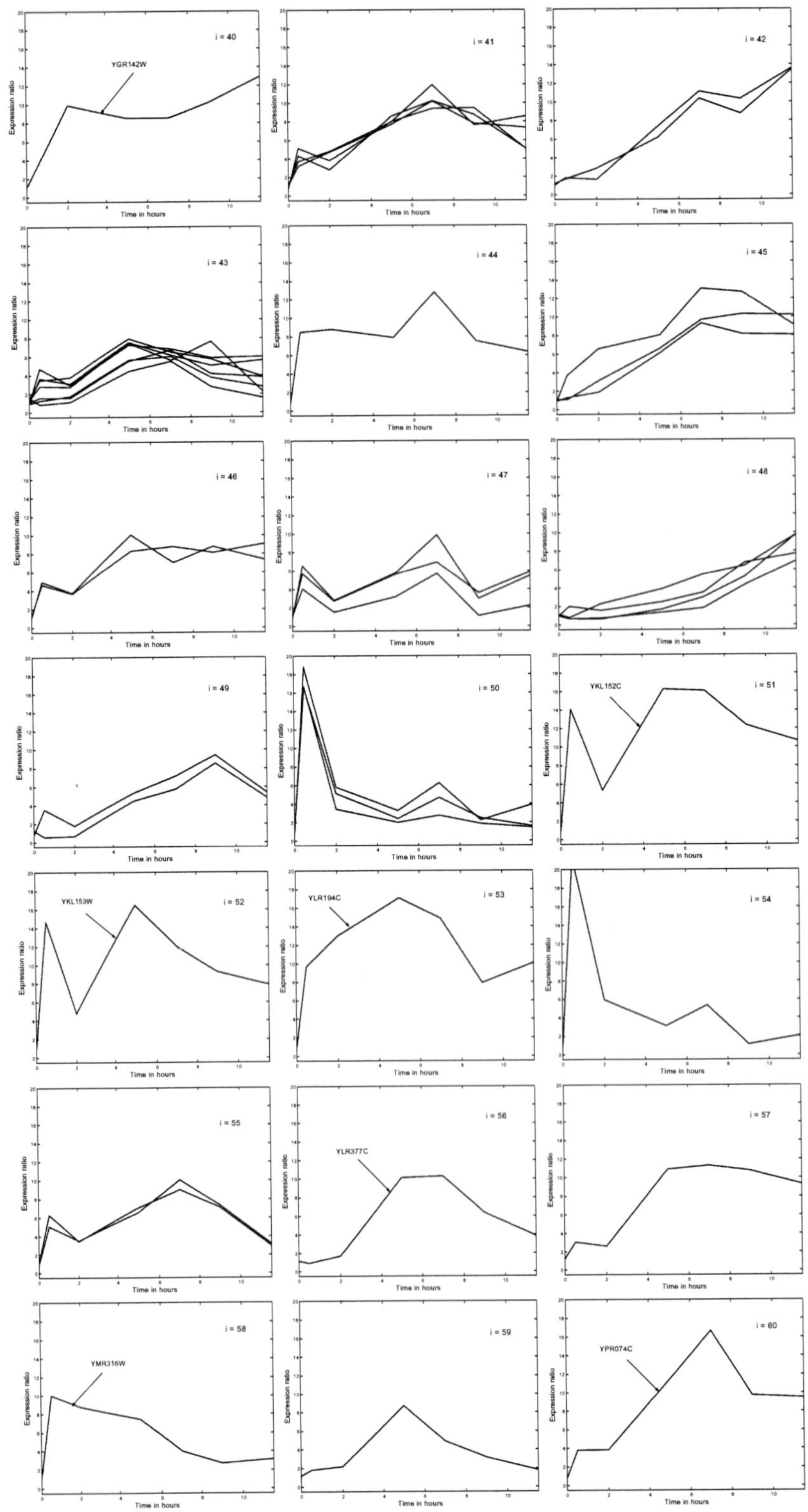

Figure 7. Clusters $i = 31$ to 60, obtained using the PAAC on the 6,118 gene temporal expression profiles of the yeast dataset.

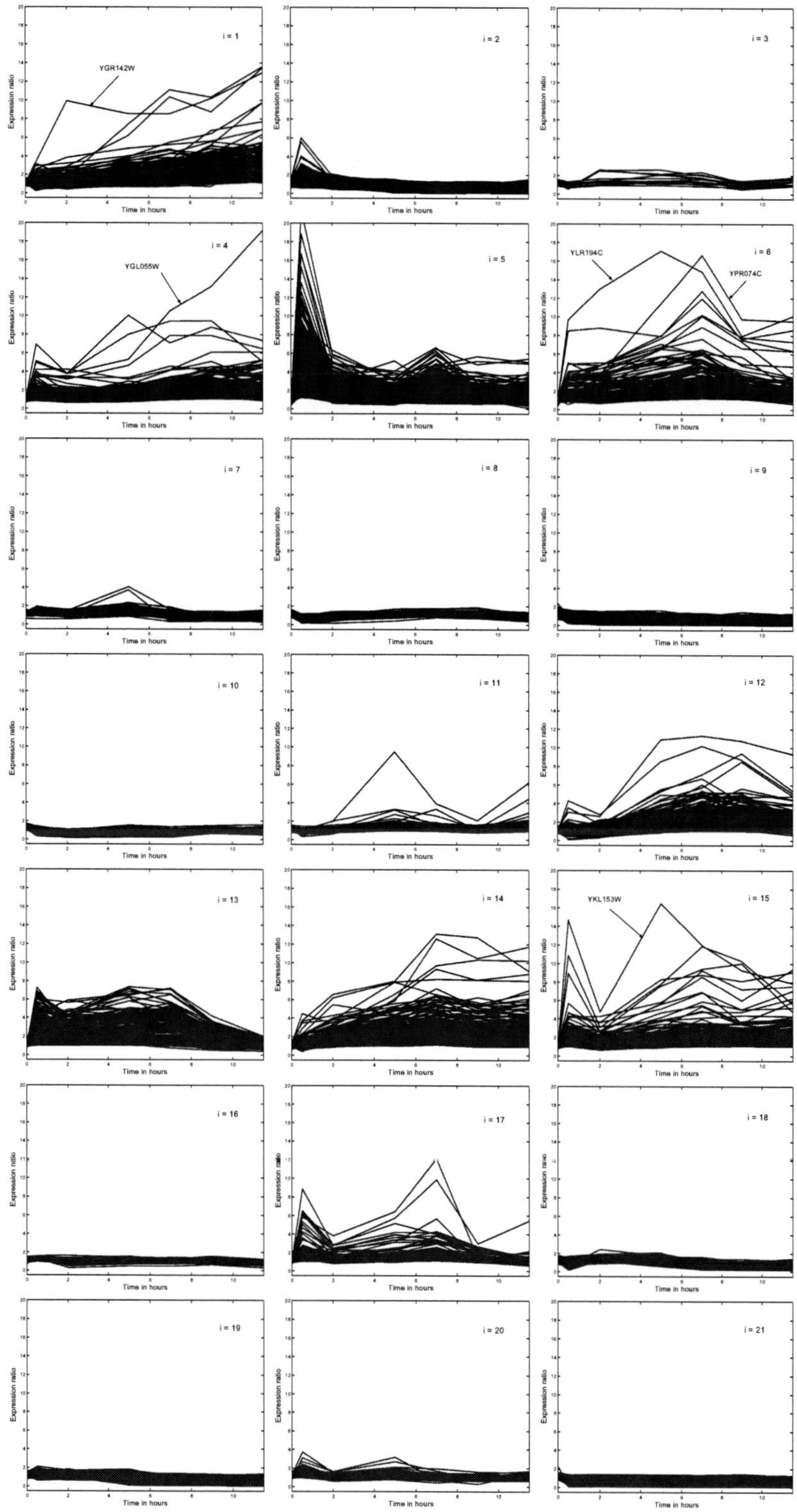

Figure 8. Continued on next page.

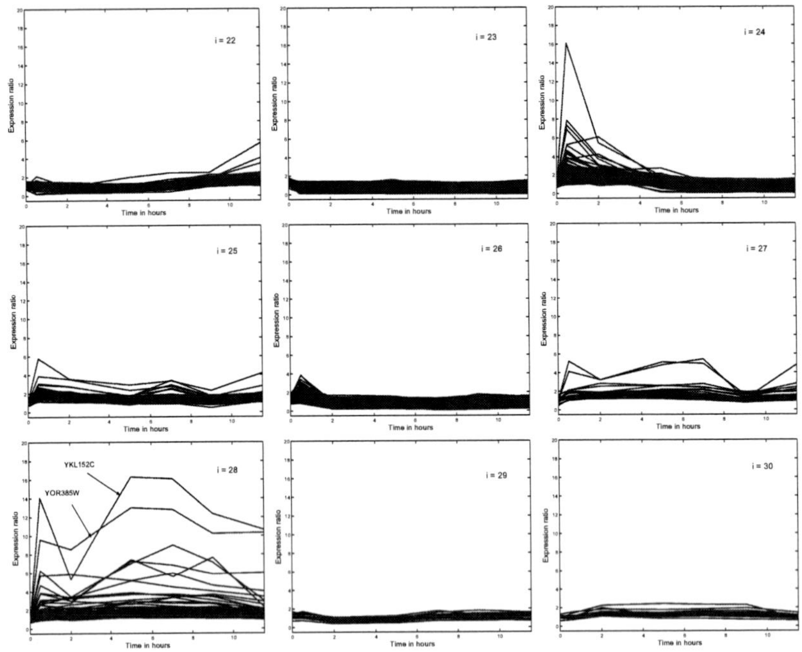

Figure 8. Clusters $i = 1$ to 30, obtained using the Pearson correlation distance on the 6,118 gene temporal expression profiles of the yeast dataset.

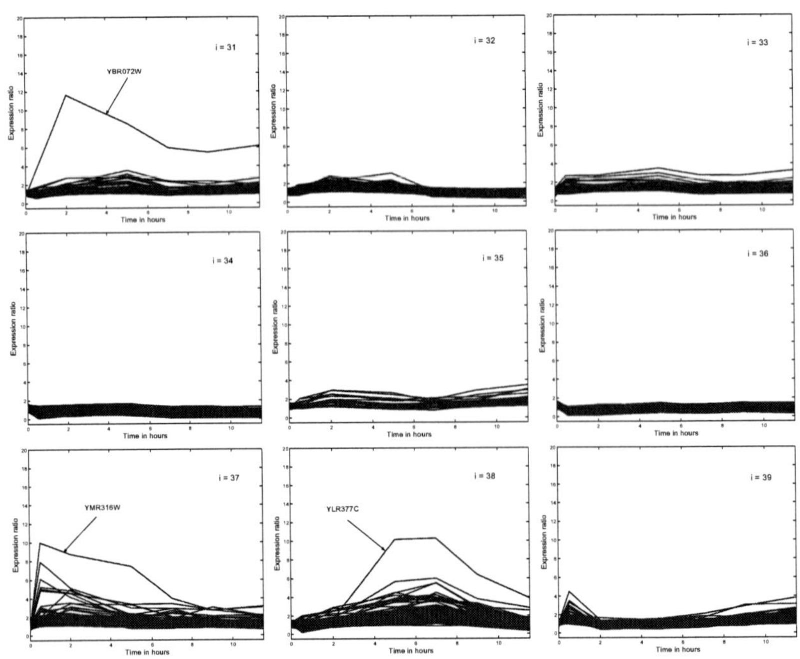

Figure 9. Continued on next page.

is an unsupervised learning approach that does not know the classes beforehand, and hence the aim is to "discover" new classes. The overall classification accuracy was computed as the average of the individual accuracies for each cluster, resulting in 83.47%. The accuracy

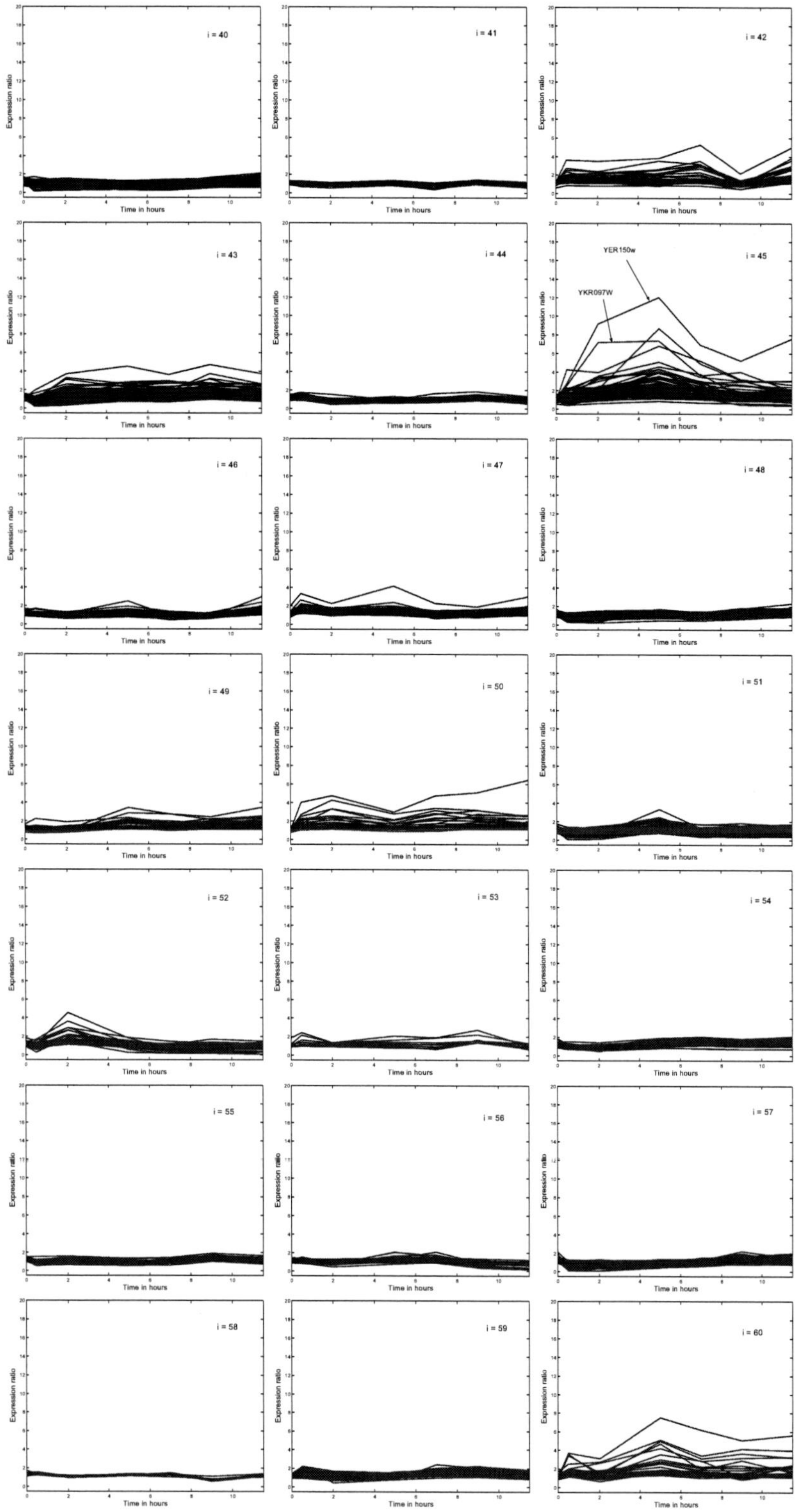

Figure 9. Clusters $i = 31$ to 60, obtained using the Pearson correlation distance on the 6,118 gene temporal expression profiles of the yeast dataset.

is very high considering the fact that PAAC is an *unsupervised* classification algorithm.

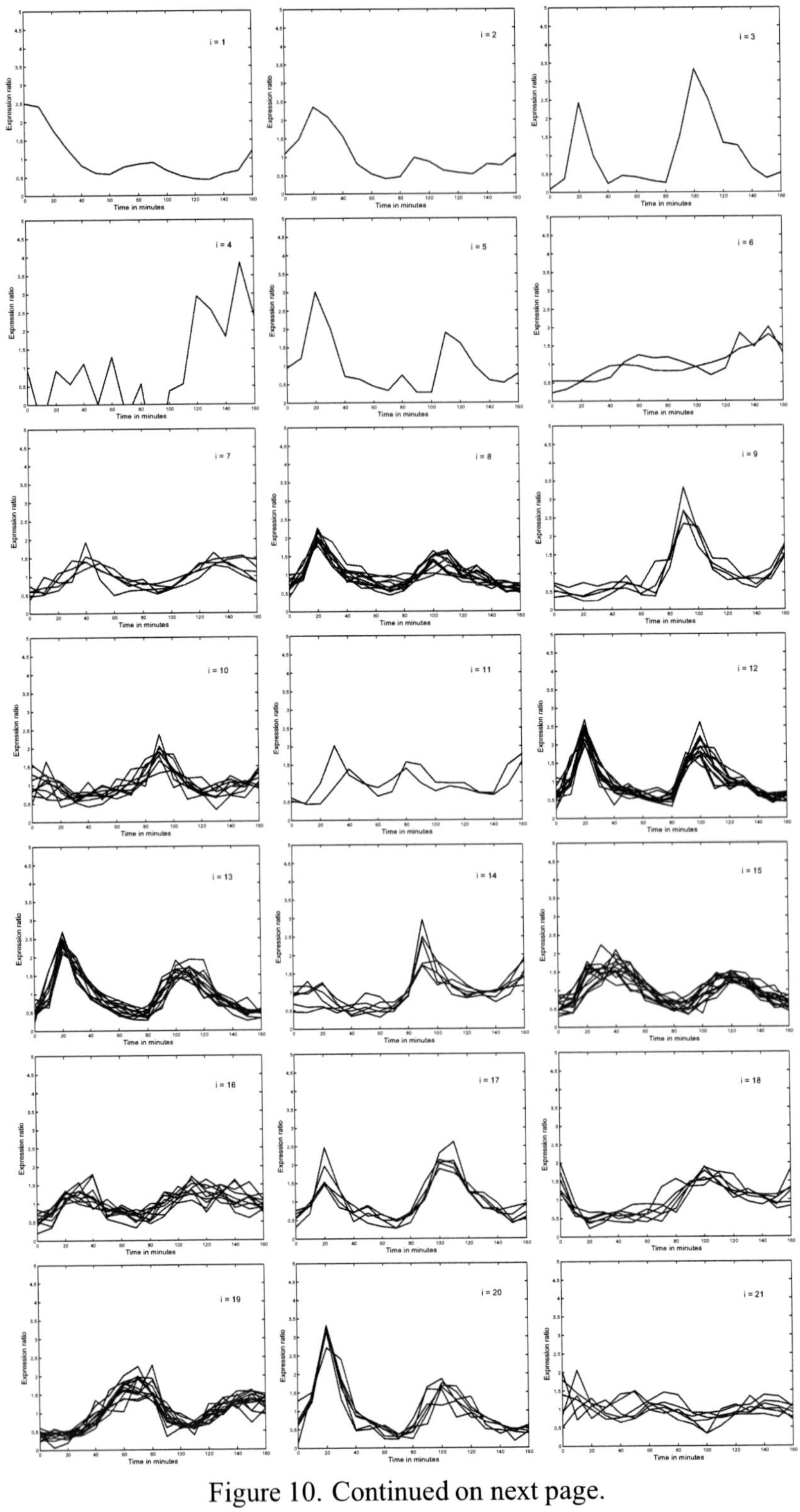

Figure 10. Continued on next page.

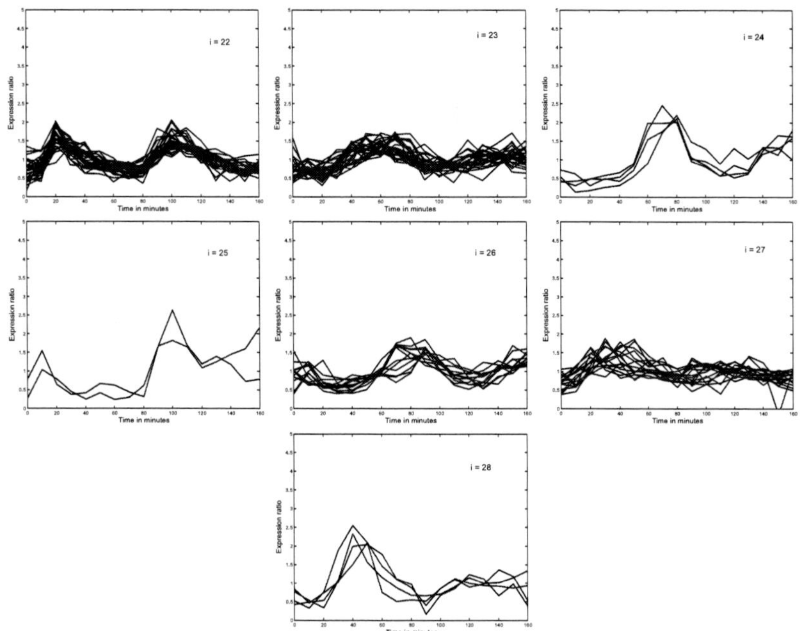

Figure 10. Different clusters obtained using PAAC on the 221 gene temporal expression profiles from Table 1 of [4], where $k = 28$.

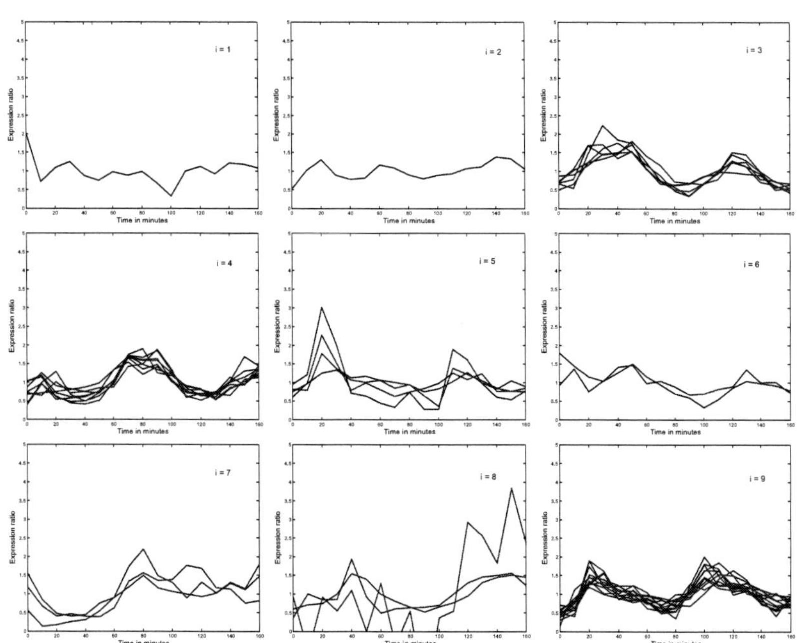

Figure 11. Continued on next page.

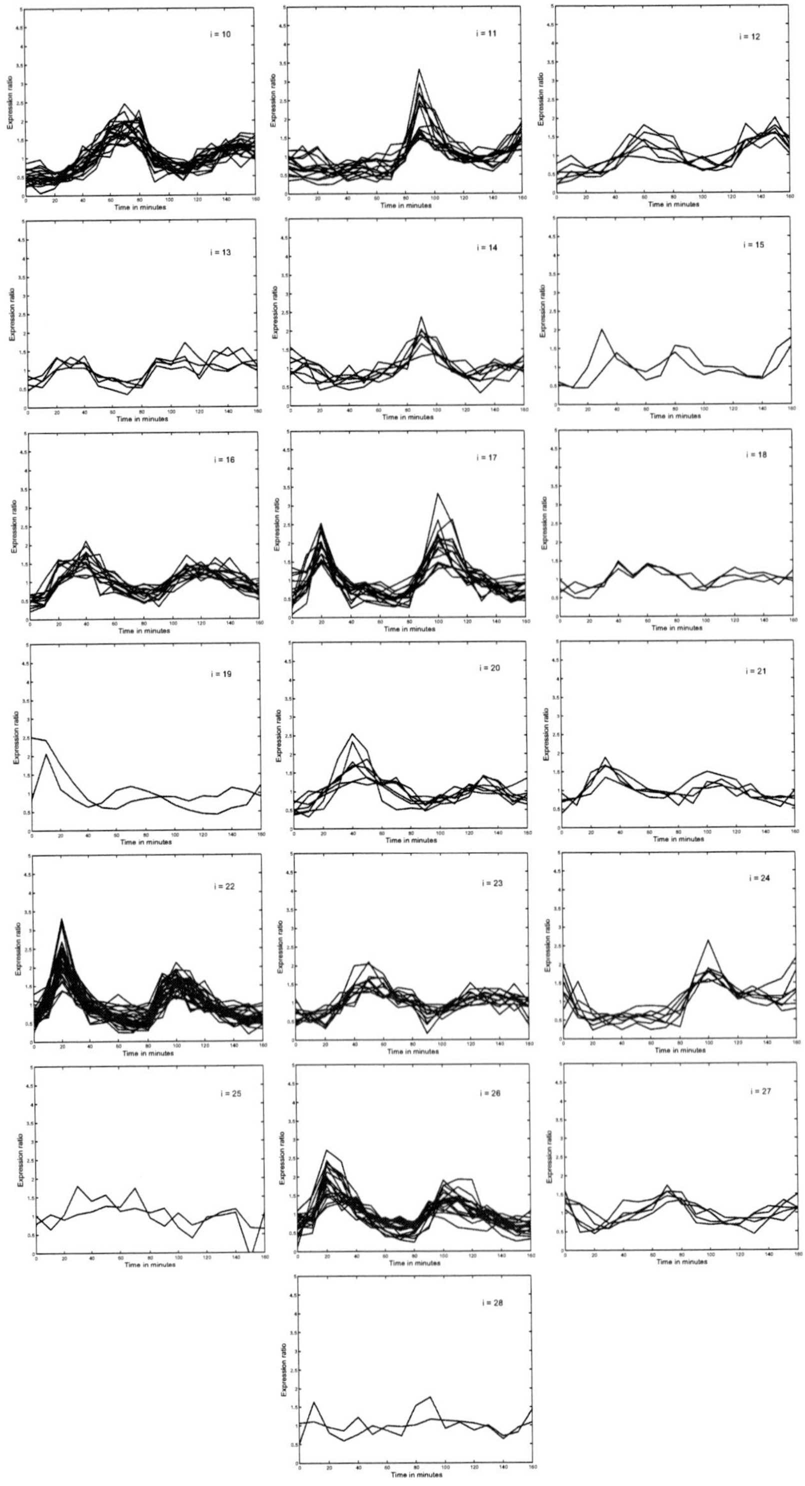

Figure 11. Different clusters obtained using the Pearson correlation coefficient on the 221 gene temporal expression profiles from Table 1 of [4], where $k = 28$.

4. Conclusion

Clustering gene expression time series data offers many computational challenges. Various works have been done is this direction, and some of the issues inherent to the time-series

Table 3. Genes from Table 1 of [4], clustered using PAAC, where $k = 28$

PAAC	Gene	Phase	PAAC	Gene	Phase	PAAC	Gene	Phase	PAAC	Gene	Phase
1	YBR067c/TIP1	Early G_1	13	YNL102W/CDC17	Late G_1	19	YLR353W/BUD8	M	23	YGL255W/ZRT1	G_2
2	YGL055W/OLE1	Early G_1	13	YOR074C/CDC21	Late G_1	19	YMR001C/CDC5	M	23	YJL137c/GLG2	G_2
3	YDL227C/HO	Late G_1	13	YKL113C/RAD27	Late G_1	19	YGL021W/ALK1	M	23	YCR073c/	G_2
4	YAL001C/TFC3	S	13	YLR383W/	Late G_1	19	YLR131c/ACE2	M	23	YDR389w/SAC7	G_2
5	YJL115W/ASF1	S	13	YML060W/OGG1	Late G_1	19	YOR025W/HST3	M	23	YKL068W/NUP100	G_2
6	YDL198C/SHM1	G_2	13	YIL140W/SRO4	S	20	YGR109C/CLB6	Late G_1	23	YGR092W/DBF2	M
6	YDR146c/SWI5	M	13	YAR008W/	S	20	YNL289W/PCL1	Late G_1	23	YOR058C/ASE1	M
7	YKL049C/CSE4	G_2	14	YDL181W/INH1	Early G_1	20	YLR313C/	Late G_1	23	YPL242C/	M
7	YKL048C/ELM1	G_2	14	YML110C/	Early G_1	20	YPR018W/RLF2	Late G_1	23	YCL037c/SRO9	M
7	YER069w/"ARG5,6"	G_2	14	YIL009W/FAA3	Early G_1	20	YPL153C/SPK1	Late G_1	23	YKL130C/	M
7	YJR112W/NNF1	G_2	14	YBR083w/TEC1	Early G_1	20	YBR070c/	Late G_1	23	YNL053W/MSG5	M
7	YPR167C/MET16	M	14	YPL187W/MF(alpha)1	Late G_1	21	YJR159W/SOR1	G_2	23	YIL162W/SUC2	M
8	YFL008W/SMC1	Late G_1	14	YJR148W/TWT2	Late G_1	21	YBR104w/YMC2	G_2	23	YDL048c/STP4	M
8	YMR078C/CHL12	Late G_1	15	YPL127C/HHO1	Late G_1	21	YLR014c/PPR1	G_2	23	YHR152w/SPO12	M
8	YPL209C/IPL1	Late G_1	15	YLL021w/SPA2	Late G_1	21	YOR274W/MOD5	G_2	23	YKL129C/MYO3	M
8	YGR152C/RSR1	Late G_1	15	YBL063w/KIP1	S	21	YDR464w/SPP41	G_2	24	YPL058C/PDR12	Early G_1
8	YIL159W/	Late G_1	15	YDR113c/PDS1	S	21	YLL046c/RNP1	G_2	24	YBR038w/CHS2	G_2
8	YNL233W/	Late G_1	15	YDR356w/NUF1	S	22	YER111c/SWI4	Early G_1	24	YGL116W/CDC20	M
8	YKL045W/PRI2	Late G_1	15	YEL061c/CIN8	S	22	YOR373W/NUD1	Early G_1	24	YGR143W/SKN1	M
8	YNL312W/RFA2	Late G_1	15	YGR140W/CBF2	S	22	YKL092C/BUD2	Early G_1	25	YLR286C/CTS1	Late G_1
8	YML061C/PIF1	Late G_1	15	YHR172W/	S	22	YMR199W/CLN1	Late G_1	25	YGL089C/MF(alpha)2	Late G_1
8	YLR233C/EST1	Late G_1	15	YLR045c/STU2	S	22	YKL042W/	Late G_1	26	YBR200w/BEM1	Early G_1
8	YOR026W/BUB3	S	15	YNL126W/	S	22	YLR212C/TUB4	Late G_1	26	YBL023c/MCM2	Early G_1
9	YDL179w/	Early G_1	15	YPR141C/KAR3	S	22	YPL241C/CIN2	Late G_1	26	YBR202w/CDC47	Early G_1
9	YLR079w/SIC1	Early G_1	15	YJR006W/HUS2	S	22	YDR507c/GIN4	Late G_1	26	YEL032w/MCM3	Early G_1
9	YJL157C/FAR1	Early G_1	15	YER001w/MNN1	S	22	YGL027C/CWH41	Late G_1	26	YLR395C/COX8	Early G_1
9	YKL185W/ASH1	Early G_1	15	YER003c/PMI40	S	22	YJL173C/RFA3	Late G_1	26	YMR256c/COX7	Early G_1
10	YJL194W/CDC6	Early G_1	16	YNL225C/	Late G_1	22	YNL262W/POL2	Late G_1	26	R281W/YOR1	Early G_1
10	YLR274W/CDC46	Early G_1	16	YDR224c/HTB1	Late G_1	22	YDL101C/DUN1	Late G_1	26	YOR316C/COT1	Late G_1
10	YPR019W/CDC54	Early G_1	16	YDR225w/HTA1	Late G_1	22	YLR234W/TOP3	Late G_1	26	YCR042c/TSM1	M
10	YHR005c/GPA1	Early G_1	16	YGL200C/EMP24	Late G_1	22	YML021C/UNG1	Late G_1	26	YOR229W/	M
10	YGR183C/QCR9	Early G_1	16	YNL272C/SEC2	Late G_1	22	YLR382C/NAM2	Late G_1	26	YDL138W/RGT2	M
10	YLR273C/PIG1	Early G_1	16	YDR488c/PAC11	S	22	YJL196C/	Late G_1	26	YIL167W/	M
10	YLL040c/	Early G_1	16	YBL002w/HTB2	S	22	YBR073w/RDH54	Late G_1	27	YNR016C/ACC1	Early G_1
10	YHR038W/	Late G_1	16	YBL003c/HTA2	S	22	YKL101W/HSL1	Late G_1	27	YBR160w/CDC28	Late G_1
10	YAL040C/CLN3	M	16	YKL067W/YNK1	S	22	YKL165C/	Late G_1	27	YBR252w/DUT1	Late G_1
11	YDR277c/MTH1	S	16	YER118c/SSU81	S	22	YLL002W/	Late G_1	27	YBR278w/DPB3	Late G_1
11	YML091C/RPM2	S	17	YJL074C/SMC3	Late G_1	22	YLR457C/NBP1	Late G_1	27	YDR297w/SUR2	Late G_1
12	YDL127w/PCL2	Late G_1	17	YGR041W/	Late G_1	22	YPL057C/SUR1	Late G_1	27	YDL155W/CLB3	S
12	YPR120C/	Late G_1	17	YNL082W/PMS1	Late G_1	22	YPL124W/NIP29	Late G_1	27	YFR037C/	S
12	YDL003W/RHC21	Late G_1	17	YOL090W/MSH2	Late G_1	22	YBR275c/RIF1	S	27	YPL016W/SWI1	S
12	YAR007C/RFA1	Late G_1	17	YHR153c/SPO16	Late G_1	22	YCR065w/HCM1	S	27	YDL093W/PMT5	S
12	YBL035c/POL12	Late G_1	18	YCR005c/CIT2	Early G_1	22	YDL197C/ASF2	S	27	YKR001C/SPO15	S
12	YBR088c/POL30	Late G_1	18	YCL040w/GLK1	Early G_1	22	YKL127W/PGM1	S	27	YER016w/BIM1	S
12	YDL164C/CDC9	Late G_1	18	YLR258W/GSY2	Early G_1	22	YDL095W/PMT1	S	27	YER017c/AFG3	S
12	YML102W/	Late G_1	18	YNL173C/	Late G_1	23	YLR210W/CLB4	S	27	YPR111W/DBF20	G_2
12	YPR175W/DPB2	Late G_1	18	YOR317W/FAA1	Late G_1	23	YDR150w/NUM1	M	27	YOR188W/MSB1	G_2
12	YDR097C/	Late G_1	18	YNL073W/MSK1	S	23	YMR198W/CIK1	S	27	YJL092W/HPR5	G_2
12	YLR032w/RAD5	Late G_1	19	YIL106W/MOB1	G_2	23	YBL052c/	S	27	YKL032C/IXR1	G_2
12	YML027W/YOX1	Late G_1	19	YCL014w/BUD3	G_2	23	YIL126W/STH1	S	28	YMR190C/SGS1	S
12	YMR179W/SPT21	Late G_1	19	YGR108W/CLB1	M	23	YCR035c/	S	28	YIR017C/MET28	S
13	YJL187C/SWE1	Late G_1	19	YPR119W/	M	23	YIL050W/	G_2	28	YHR086w/NAM8	S
13	YPL256C/CLN2	Late G_1	19	YBR138c/	M	23	YBL097w/	G_2	28	YJR137C/	S
13	YMR076C/PDS5	Late G_1	19	YHR023w/MYO1	M	23	YJL099W/CHS6	G_2			
13	YER070w/RNR1	Late G_1	19	YOL069W/NUF2	M	23	YJR076C/CDC11	G_2			
13	YLR103c/CDC45	Late G_1	19	YJR092W/BUD4	M	23	YCR084c/TUP1	G_2			

data have been addressed, while some others yet provide the grounds for further studies and warrant the development of new computational tools. The current research on clustering time-series data and some of the methods from a wide range of clustering approaches proposed in various studies are discussed and presented. The methods differ substantially, each offering its own possibilities along with its own pros and cons. Most of the proposed methods for clustering time-series data are suitable for any data, but only a few have incorporated the idea of performing profile alignment before applying a distance measure. An experimental comparison has been made with three of the methods, and the results show:

(i) Using agglomerative clustering while applying profile alignment before computing the similarity between expression profiles can produce superior results, when compared to that of the Pearson and Spearman correlation similarity measures.

(ii) A variant of the $\mathcal{I}$-index can make a trade-off between minimizing the number of useful clusters and keeping the distinctness of individual clusters.

(iii) The biological significance of the results with profile alignment was confirmed by computing the classification accuracy in pre-clustered yeast data, and was found to be over 83%, which is very high considering PAAC is an unsupervised leaning approach.

There are quite a few issues which have been unresolved to date and which deserve particular attention. The computation of the mean of a cluster using the profile-alignment approach still remains an open issue, since only a greedy approach has been proposed so far. Further studies are needed to develop an appropriate method that can efficiently perform alignment of multiple profiles so that a cluster mean can be computed after aligning all profiles in a cluster at the same time, and hence removing the effect of ordering of profiles during the alignment. Also, the effects of alignment in the partitional iterative methods (e.g., EM, regression, *k*-means, SOM, etc.) are yet to be investigated. Moreover, once a suitable multiple alignment scheme is developed, it would be interesting to study the impact of that approach when applied to the entire dataset.

REFERENCES

[1] A. Bari and L. Rueda. A New Profile Alignment Method for Clustering Gene Expression Data. In proceedings of *19th Conference of the Canadian Society for Computational Studies of Intelligence (AI 2006)*, Quebec City, LNCS, Vol 4013:86–97, Springer, 2006.

[2] A. Brazma and J. Vilo. Gene expression data analysis. *FEBS Lett.*, **480**:17–24, 2000.

[3] L. Bréhélin. Clustering Gene Expression Series with Prior Knowledge. In *Lecture Notes in Computer Science*, volume 3692, pages 27–38, October 2005.

[4] R. J. Cho, M.J.Campbell, E. A. Winzeler, L. Steinmetz, A. Conway, L. Wodicka, T. G. Wolfsberg, A. E. Gabrielian, D. Landsman, D. J. Lockhart, R. W. Davis. A genome-wide transcriptional analysis of the mitotic cell cycle. *Molecular Cell*, **2**(1):65–73, 1998.

[5] A. Conesa, M. J. Nueda, A. Ferrer, and M. Talon. maSigPro: a method to identify significantly differential expression profiles in time-course microarray experiments. *Bioinformatics*, 2006, 22(9): 1096 - 1102.

[6] S. Chu, J. DeRisi, M. Eisen, J. Mulholland, D. Botstein, P. Brown, and I. Herskowitz. The transcriptional program of sporulation in budding yeast. *Science*, **282**:699–705, 1998.

[7] S. Déjean, P. G. P. Martin, A. Baccini, and P. Besse. Clustering Time-Series Gene Expression Data Using Smoothing Spline Derivatives. *EURASIP J Bioinform Syst Biol.*, 2007, 2007:70561.

[8] S. Drăghici. *Data Analysis Tools for DNA Microarrays*. Chapman & Hall, 2003.

[9] R. Duda, P. Hart, and D. Stork. *Pattern Classification*. John Wiley and Sons, Inc., New York, NY, 2nd edition, 2000.

[10] M. Eisen, P. Spellman, P. Brown, and D. Botstein. Cluster analysis and display of genome-wide expression patterns. In *Proc. Natl Acad. Sci.*, volume 95, pages 14863–14868, USA, 1998.

[11] J. Ernst, G. J. Nau and Z. Bar-Joseph. Clustering Short Time Series Gene Expression Data. *Bioinformatics*, 2005 21(Suppl 1):i159–i168; doi:10.1093/bioinformatics/bti1022.

[12] A. P. Gasch and M. B. Eisen Exploring the conditional coregulation of yeast gene expression through fuzzy k-means clustering. *Genome Biology*, **3** (11) (2002) 0059.1–0059.22.

[13] K. Guillemin, N. Salama, L. Tompkins, and S. Falkow. Cag pathogenicity island-specific responses of gastric epithelial cells to Helicobacter pylori infection. Proc. *Natl Acad. Sci* USA, **99**:15136.15141, 2002.

[14] J. A. Hartigan, Clustering Algorithms, John Wiley and Sons, 1975.

[15] W. H. Heijne, R. H. Stierum, M. Slijper, P. J. van Bladeren and B. van Ommen Toxicogenomics of bromobenzene hepatotoxicity: a combined transcriptomics and proteomics approach. *Biochem. Pharmacol.*, **65**, 857–875, 2003.

[16] L. Heyer, S. Kruglyak, and S. Yooseph. Exploring expression data: identification and analysis of coexpressed genes. *Genome Res.*, **9**:1106–1115, 1999.

[17] R. Hogg and A. Craig. *Introduction to Mathematical Statistics.* MacMillan, fifth edition, 1995.

[18] J. Hwang and S. Peddada Confidence interval estimation subject to order restrictions. *Ann. Statist.*, **22**, 67–93, 1994.

[19] V. Iyer, M. Eisen, D. Ross, G. Schuler, T. Moore, J. Lee, J. Trent, L. Staudt, Jr. J. Hudson, and M. Boguski. The transcriptional program in the response of human fibroblasts to serum. *Science*, **283**:83–87, 1999.

[20] Z. Bar-Joseph, G. Gerber, T. Jaakkola, D. Gifford and I. Simon. Continuous representations of time series gene expression data. *Journal of Computational Biology*, 2003, 10(3-4): 341–356.

[21] E. Lobenhofer, L. Bennett, P. Cable, L. Li, P. Bushel and C. Afshari Regulation of DNA replication fork genes by 17betaestradiol. *Molec. Endocrin.*, **16**, 1215–1229, 2002.

[22] U. Maulik and S. Bandyopadhyay. Performance Evaluation of Some Clustering Algorithms and Validity Indices. *IEEE Transactions on Pattern Analysis and Machine Intelligence*, **24**(12):1650–1654, 2002.

[23] C.Moller-Levet, F.Klawonn, K.-H.Cho, and O. Wolkenhauer. Clustering of unevenly sampled gene expression time-series data. *Fuzzy sets and Systems*, Volume 152, Issue 1, 16, 2005, Pages 49–66.

[24] S. Peddada, K. Prescott and M. Conaway Tests for order restrictions in binary data. *Biometrics*, **57**, 1219–1227, 2001.

[25] S. Peddada, E. Lobenhofer, L. Li, C. Afshari, C. Weinberg, and D. Umbach. Gene selection and clustering for time-course and dose-response microarray experiments using order-restricted inference. *Bioinformatics*, **19**(7):834–841, 2003.

[26] T. Petrie Probabilistic functions of finite state Markov chains. *Ann. Math. Statist*, **40**, 97–115, 1969.

[27] M. Ramoni, P. Sebastiani and I. Kohane. Cluster analysis of gene expression dynamics. *Proc. Natl Acad. Sci. USA*, **99**(14), 9121–9126, 2002.

[28] J. Ramsay and B. Silverman. *Functional Data Analysis*, Springer, New York, NY, USA, 2nd edition, 2005.

[29] L. Rueda and A. Bari. Clustering Temporal Gene Expression Data with Unequal Time Intervals. *2nd International Conference on Bio-Inspired Models of Network, Information, and Computing Systems*, To appear. 2007.

[30] A. Schliep, A. Schonhuth and C. Steinhoff. Using hidden Markov models to analyze gene expression time course data. *Bioinformatics*, 2003, 19, I264–I272.

[31] G. Sherlock. Analysis of large-scale gene expression data. *Curr. Opin. Immunol.*, **12**:201–205, 2000.

[32] P. T. Spellman, G. Sherlock, M. Q. Zhang, V. R. Iyer, K. Anders, M. B. Eisen, P. O. Brown, D. Botstein and B. Futcher Comprehensive identification of cell cycleregulated genes of the yeast saccharomyces cerevisiae by microarray hybridization. *Mol Biol Cell* **9** (1998) 3273–3297.

[33] P. Tamayo, D. Slonim, J. Mesirov, Q. Zhu, S. Kitareewan, E. Dmitrovsky, E. Lander, and T. Golub. Interpreting patterns of gene expression with self-organizing maps: Methods and application to hematopoietic differentiation. *Proc. Natl. Acad. Sci.* **96**(6): 2907–2912. 1999.

[34] S. Tavazoie, J. Hughes, M. Campbell, R. Cho, and G. Church. Systematic determination of genetic network architecture. *Nat. Genet.* **22**: 281–285. 1999.

[35] G. Zhu1, P. T. Spellman, T. Volpe, P. O. Brown, D. Botstein, T. N. Davis and B. Futcher. Two yeast forkhead genes regulate cell cycle and pseudohyphal growth. *Nature*, **406**:90–94, 2000.

In: Oligonucleotide Array Sequence Analysis
Editors: M.K. Moretti and L.J. Rizzo, pp. 123-147
ISBN: 978-1-60456-542-3

Chapter 4

RECENT ADVANCES IN MICROBIAL DISCOVERY USING METAGENOMICS, DNA MICROARRAY, BIOSENSORS, MOLECULAR SUBTYPING, AND PHAGE RECOMBINANT PROBES

Sufian F. Al-Khaldi[1,*], Magdi M. Mossoba[2], Betsy Jean Yakes[2], Eric Brown[1], Devang Sharma[1,3] and Santina Carnazza[4]

[1]Division of Microbiology, Office of Regulatory Science (ORS) and
[2]Division of Analytical Chemistry, ORS, Center for Food Safety and Applied Nutrition, US Food and Drug Administration
[3] Fischell Department of Bioengineering, A. James Clark School of Engineering, University of Maryland, College Park
[4]Dipartimento di Scienze Microbiologiche, Genetiche e Molecolari, Università di Messina, Italy

ABSTRACT

Much interest has been generated over the past few years aimed at microbial discovery using state-of–the-art technologies. Microbial communities of uncultured microorganisms are no longer an obstacle for the advance or understanding of 99% of microorganisms. Metagenomics, or culture-independent genomic analysis, has joined DNA microarray technology in recent years to discover many bacterial gene functions and identification in the environment. Biosensors and molecular subtyping as well as phage recombinant probe technologies continue to advance our knowledge in detecting and classifying bacterial isolates. These molecular tools are mainly dependent on DNA or protein sequence analysis. Our abilities to use these tools are helping us to detect and trace bacterial populations found in nature.

---

* E-mail address: Sufian.Alkhaldi@fda.hhs.gov. Correspondence concerning this article should be addressed to Dr. Sufian F. Al-Khaldi, 3E-014 Office, HFS-712, DNA Microarray Lab, Division of Microbiology, 5100 Paint Branch Parkway, College Park, MD 20740-3835

Metagenomic Strategies for Uncultured Microbes

Many bacterial genomes sequenced lately were recovered from cultured bacteria which represent 1% of total microbial communities [4]. This limitation of exploring the DNA sequences and missing 99% of non-cultured bacteria represents a big gap in our knowledge on bacterial strains of microbial communities. Because vast amounts of bacterial strains do not live or exist in isolation in their natural environments, metagenomic approaches - culture-independent genomic analysis- of DNA sequences can help to address the real representation of full genetic and metabolic potential. Therefore, allowing the bio-discovery of gene functions associated with natural population can lead to a) establishment of gene inventories b) discovery of new natural products c) understanding microbial group dynamics in extreme environments d) discovery of new viruses (bacteriophages).

Although different microbial communities were studied using PCR analysis, PCR has shown some biases in primer binding and amplification of specific DNA regions hindering the real representation of microbial diversity [70]. Metagenomic strategies of microbial habitat, however, are more powerful tools for profiling DNA sequences from complex mixtures of uncultured bacterial cells.

Metagenomic DNA Strategies

There are many approaches that can be used in metagenomic investigations. Almost all investigated strategies require some form of cloning and sequencing of the DNA fragments. Cloning different DNA fragments from microbial communities is followed by several possible strategies including:

1. Screening clones for the phenotype. This is usually followed by determination of phylogenetic origins of cloned DNA [10].
2. Screening clones for specific phylogenetic anchors such as 16S RNA or predetermined genes. This is followed by sequencing the entire clones and trying to find other genes residing near the anchor genes [7].
3. Random sequencing of the cloned fragments, identifying the interesting genes, and finally reconstruction of the genomic map [7, 8, 11, 12].
4. Using small DNA fragments to build DNA microarray chips followed by hybridizations with known reference bacterial strains [77], Figure 1.

Metagenomic in Bacterial Communities

In the last five years, DNA sequence technologies have seen rapid improvements. Because of these improvements, it is feasible now to sequence entire metagenomes of complex bacterial communities. The complexity of bacterial communities due to microheterogeneity makes the metagenomic DNA sequence reconstructions simple representatives of real life. Since first attempts at conducting metagenomic sequencing started in viral communities in the ocean [12, 87] and human feces [11, 87], it is now feasible to sequence vast amounts of culture independent bacterial communities from different microbial environments.

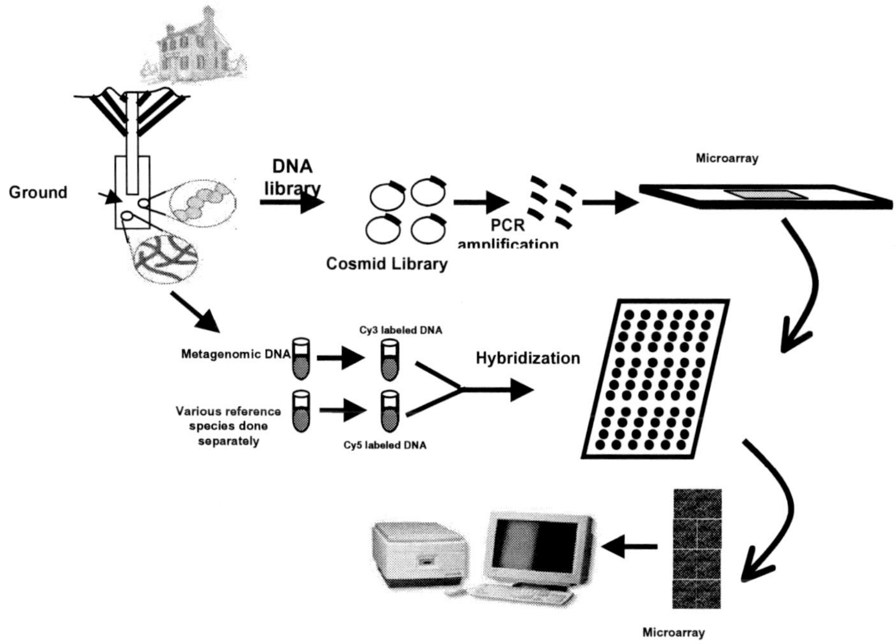

Figure 1. DNA was isolated from groundwater microbial communities to build cosmid DNA library. The DNA fragments were fabricated on the chemically treated slide to build the DNA chip. Reference known bacterial strains mixed with unknown samples were labeled and hybridized to the DNA chip revealing the DNA profiling of the bacterial water communities.

Venter et al. [87] and Tyson et al [86] were first to conduct large-scale metagenomic DNA sequencing. The DNA, in one study for example [87], was isolated from environmental genomes and inserted into cosmid libraries using shotgun cloning. The cloned DNA was sequenced without the need to culture or isolate the bacterial cells. As a result, seawater samples from whole-genome shotgun sequences were collected and further analyzed revealing at least 1800 genomic species including 148 previously unknown bacterial phylotypes. More than 1.2 million genes, including 782 new rhodopsin-like photoreceptors, were identified. In addition, six plasmids larger than 100-kb were discovered as well as putative genes of arsenate, mercury, copper, and cadmium resistance sequences. Putative conjugal transfer systems and plasmid transfer genes were also found suggesting the presence of DNA transfer mechanisms between bacterial strains in the ocean. The discovery of these gene sequences exposed the presence of substantial oceanic microbial diversities. Similarly, Tyson et al. [86] used random DNA shotgun sequencing from natural acidophilic biofilm. A total of 76.2 million base pairs were sequenced and around 103, 462 genes (high-quality reads) were identified.

Recently, microarray platforms were used to screen 1-kb metagenomic libraries [77]. The DNA 1-kb fragments from 672 cosmid metagenomic library clones were amplified using PCR, Figure 1. The PCR fragments were printed on chemically treated slides and hybridized with labeled DNA. The DNA hybridization mixture containing a labeled Cy5-dCTP environmental DNA and labeled Cy3-d-CTP bacterial reference strains DNA created comparative genomic hybridizations (CGH). Positive results were based on the ratios which are greater than one of Cy5/Cy3. This strategy was able to profile each clone in the metagenomic library by using known reference strains.

Metagenomic in Virus Communities

Metagenomic analysis of genome is important in the discovery of viruses and bacteriophages. Bacteriophages take part in diversity and population structures of microbial communities in the human gut. There are approximately 500 human intestinal microbial species with around 40 predominant species in the total population [51, 52].

The presence of bacteriophages was documented in many environments where bacteria live in soil, the intestines of animals and humans, and sea and drinking waters. Since bacteriophages have complex diversity and many of their hosts have not been cultured and therefore not yet discovered, metagenomic approaches were great discovery genomic tools. Bacteriophages isolated from human feces using metagenomic analyses of an uncultured viral community were found, naturally, to be unrelated to any known viruses [11]. A total of 1200 viral genotypes, not including bacterial strains, were recognized revealing the novelty of these microscopic lives. Most of these bacteriophages were found to be siphophages.

Metagenomic research was instrumental in finding the virus causing honey bee colony collapse disorder (CCD) [21]. In fact, CCD has caused a loss of 50-90% of colonies in beekeeping operations in the US. The devastating effect of this disorder has triggered mass fear of economic losses in agricultural farm productions. Metagenomic analysis together with different molecular techniques was used to study the CCD hives for three years revealing that CCD in bees, was caused, most likely, by Israeli acute paralysis virus (IAPV). This breakthrough conclusion highlighted the importance of metagenomic research as instruments capable of finding the cause of unknown complex diseases especially when microorganisms cannot be cultured.

Finally, it is paramount to recognize that metagenomic DNA analyses have opened the door to the discovery of many bacterial species as well as viruses. Discovering the black box with complex microbial environments, metagenoic research represents instrumental tool in the microbial investigation. Once again [1], high throughput screening technologies will be valuable in unlocking many of the unknowns' microbial reservoirs. The future will be filled with new discoveries of genes which will be used to improve our daily lives.

Application of DNA Microarray Technology

The vast multitude of sequenced genomes and the rapid increase in their discovery has increased the output of genomic DNA data. Recently, DNA microarray technology has emerged as a leader in this analysis, testing gene expression for thousands of genes in a single experiment. In a typical DNA microarray assay, a series of oligonucleotide sequences known as "oligoprobes" are selected to identify genes spotted onto a chemically modified glass slide, creating a "chip". DNA collected from a test sample would be amplified and then fluorescently labeled for hybridization on the chip.

Along with its application to gene expression studies, DNA microarray technology has emerged as an indispensable technique for microbial identification and classification in different environments. Recent concern for accidental and intentional contamination of food has called to develop of microarray technology to quickly and selectively characterize microbes, even in cases where only a small concentration of pathogens is present. Another important development in DNA microarray has been its clinical application, identifying and characterizing pathogenic microbes and viruses for diagnoses of infectious diseases in humans.

Gene Expression and Microbial Identification in Food Bacterial Pathogens

Gene expression of virulence and resistance factors in bacterial isolates is an area of high interest for public health and food safety. The study of certain genes provides insight into the interrelationship between virulence and fitness genes. By selecting for unique sets of genes, it is possible to characterize and serotype bacterial strains, as well. Previously, the go-to method for detecting multiple genes has been multiplex PCR combined with gel electrophoresis. Multiplex PCR utilizes more than one set of primers to simultaneously amplify multiple DNA sequences.

A multiplex PCR method was developed to test for virulence genes of *Escherichia coli* O157:H7, a major food pathogen. This method detected *slt*-I and –II toxin structural genes, the *eaeA* gene whose product mediated the adherence of *E. coli* O157:H7 to host cells, and genes involved in the biosynthesis of O157 and H7 antigens. The combination of genes selected successfully identified *E. coli* O157:H7 in bovine feces [1].

This general method has its limitations. The number of different amplifications is restricted due to primer-primer interactions. Also, nonspecific amplification causes multiple bands and problems in identifying PCR product due to DNA fragment size similarity. The application of DNA microarray technology provides a solution to these problems.

A highly sensitive and flexible microarray scheme was developed to identify and genotype *E. coli* O157:H7. Various oligonucleotide sequences were selected to represent four genes in wild-type and mutant *E. coli* O157:H7. They were: *stl*-I and –II, *eaeA*, and *hylA*. The method was successful in genotyping four pure cultures of *E. coli* O157:H7 and one isolate of O91:H2. The hybridization results were 32-fold more sensitive than the comparable multiplex PCR-gel electrophoresis assay [2, 16].

DNA microarray is not only used to test gene expression of bacterial isolates but also to serotype bacterial strains. A recent study of *Staphylococcus aureus* isolated from cattle utilized DNA microarray technology to study 144 virulence and resistance genes in a single experiment. Its efficiency also allowed 128 isolates from Germany and Switzerland to be tested much faster than any other method would allow. The majority of isolates were closely related clonal complexes 8, 25, and 97 (34.4% together) or were related to the bovine strain RF122 (48.4%) [50]. The simultaneous characterization and serotyping of *Salmonella enterica* isolates was also recently achieved using DNA microarray. Whole genomic DNA was fluorescently labeled with Cy5 and hybridized to the chip. A total of 109 oligoprobes detected genes encoding virulence factors, antibiotic resistance factors, and serotype-specific flagella and somatic antigens [47].

The ability to quickly and precisely identify bacteria is especially important in cases of food outbreaks. *Yersinia enterocolitica* is a pathogen able to grow in refrigerated and high salt conditions. It is typically transmitted through consumption of contaminated pork, water, and milk. Myers et al. [55] recently developed a DNA microarray chip using four virulence genes (*virF, ail, yst, blaA*) and a positive control encoding 16S ribosomal components. The chip successfully identified five genes from adulterated pasteurized whole milk, and the detection limit was 1,000 CFU per hybridization. The ultimate goal of DNA microarray chip is to selectively detect multiple food contaminants in a single assay with high sensitivity and low false-positive occurrence.

Possible Clinical Applications of DNA Microarray

A rather general response of any physician examining a patient with high fever in the emergency room would be to take a blood sample and send it to the bacteriology laboratory. Analysis would take two to three days to confirm an infection, to identify a pathogen, and to determine its antibiotic susceptibility profile [76]. The latter is especially important with the increasing number of multiple-drug resistant microbes.

The application of DNA microarray technology provides for highly sensitive and rapid characterization and subtyping of pathogens resistant to multiple antibiotics. Majtan et al [46] recently developed a novel *Salmonella*-specific DNA microarray chip for determining virulence factors, prophage sequences, and antimicrobial resistance genes. Preliminary screening of 24 *Salmonella* clinical isolates was done by susceptibility testing, plasmid profiling, and class 1 integron PCR. The microarray analysis characterized and localized the genetic markers, while the antimicrobial resistance genes were also detected and confirmed with sequencing. All strains revealed resistance to two or more antibiotics, some even showing resistance to up to nine drugs.

Pseudomonas aeruginosa infections, in particular, need fast and reliable methods for antibiotic susceptibility testing in order to achieve therapy improvement. Weile et al [90] have developed a DNA microarray chip and amplification protocol for genotyping multiple-drug resistant *P. aeruginosa*. DNA isolation, target gene amplification, fluorescence labeling, fragmentation, and hybridization could all be completed in less than five hours. There was an 87.8% coverage of relevant resistance genes used in therapy of critically ill patients calculated from genotype-phenotype comparisons.

Adenoviruses infect the respiratory tract of humans and cause acute respiratory disease. There is an absence of effective therapeutic or alternative prophylactic treatment for the infection. Lopez-Campos et al [42] recently developed a simple, cost-effective microarray assay for detection and serotyping of adenoviruses.

In order for these applications to be clinically significant, there must be an initial understanding of the patient's condition and at least a suspicion of the bacterial species causing the infection. If no such information can be determined, broad-range microarray techniques can be applied.

Schrenzel [76] designed a method to select for pathogens causing blood infection. These pathogens were identified from the past ten years' records at the Geneva University Hospitals, by analyzing more than 12,000 isolates. About 1000 species were considered and short regions of their ribosomal RNA genes were compared. A combination of highly conserved and highly variable oligonucleotide probes were chosen to design the microarray. Francois et al. [30] have identified 21 genera and species responsible for approximately 80% of bacteremic episodes. Bacterial cultures isolation validated the microarray further by fluorescently labeling and hybridizing immediately after nucleic acid extraction.

The advantages of microarray technology are clear, yet certain limitations prevent its acceptance as a routine diagnostic technique. Broad-range microarray, in particular, is too complex and costly to integrate as a diagnostic technique. Even with a broad range, the microarray is not cost effective like culturing [76]. The application of this technology can be integrated where it is most helpful. Zhu et al [94] have developed a base stacking hybridization method that uses microarray to rapidly identify clinical isolates without nucleic acid extraction. The technique was completed in less than two hours and matched with biochemical identification for 150 out of 152 clinical strains. Currently, it would be most effective to reserve DNA microarray technology to specific validated applications.

Looking Forward – The Future of DNA Microarray

This section has covered many of the recent applications of DNA microarray technology to bacterial detection in contaminated foods, an area especially important to counter-bioterrorism and public health. The development of this area has been the emerging trend in the literature over the past three to five years. In the future, DNA microarray technology may be used more for understanding the interrelationship between bacteria and the host environment they occupy as pathogens or normal flora.

Palmer et al. [60] used DNA microarray technology to study the bacterial inhabitants of fourteen human infants' colons. They reported comprehensive analysis of bacteria residing in the human intestine. A previous study of adult intestinal microbiota showed the abundance of Gram-positive bacteria, Firmicutes and Actinobacteria, and Bacteroidetes. The recent infant study showed similar prevalence of these two categories of bacteria, but also a major appearance of Proteobacteria, a group of facultative anaerobes. It is speculated that they may be early inhabitants of the colon to clear oxygen content and allowing the anaerobes to colonize. A comparison of the infants' microbiota showed a high degree of species-level variations, but no discernable pattern of abundant species or temporal modes of acquisition. Two fraternal twins were also compared to show that environment may play a greater role than genetics in determining microbial profiles [60].

Characterization and Detection of DNA Hybridization by Using Metal Nanoparticles (NP) and Surface-Enhanced Vibrational Spectroscopy

The successful development of new microanalytical biosensing assays must meet several challenging requirements that include specificity of biological recognition probes and sensitivity of detection. The present discussion will be limited to the sequence-specific recognition of capture DNA, a process that is based on the hybridization of a strand of nucleic acid (probe) to a complementary target sequence. Fluorescent dyes traditionally used for the detection of DNA microarrays (gene chips) suffer from drawbacks such as broad emission peaks which limit multiplexing, quenching of fluorescence and photobleaching [27, 39]. Advances in the development of DNA-nanoparticle (NP) bioconjugates have led to their use as novel analytical research tools for the characterization and detection of DNA [17, 39, 44] in an attempt to overcome these limitations.

There has been an increased interest in applying metal, particularly gold and silver, NPs to the detection of DNA hybridization and to other assays. This is because NPs significantly enhance the sensitivity of physicochemical detection methods including those based on surface-enhanced vibrational spectroscopy, such as surface-enhanced infrared absorption (SEIRA) and surface-enhanced Raman scattering (SERS). These applications are possible due to the fact that the surface of metal NPs as well as the surface of metal island films (with nanometric asperities) can be successfully functionalized via an organic linker with stable receptors, such as oligonucleotides suitable for hybridization in aqueous buffers. In one conjugation technique, thiol or disulfide molecules with appropriate endgroups are used to form gold-bound thiolate, self-assembled monolayers (SAMs) on metal surfaces [93], thus making it possible to design substrates for tethering molecular recognition receptors such as ssDNA probes.

NPs were first reported to tether to thiol-modified DNA by self-assembly in 1996 [3, 48]. These advances in NP-nucleic acid bioconjugates have since been applied to biosensing and have led to the development of many novel assays with a number of different detection methods. With respect to vibrational spectroscopy, there is a paucity of SEIRA publications on monolayer formation and/or detection of DNA hybridization. By contrast, there is a large body of SERS literature on the detection of binding reactions, including those between nucleic acid targets and probes. A few examples of recent studies are presented next as an overview of the application of NPs and surface-enhanced vibrational spectroscopy to DNA hybridization.

SERS is an emerging technology that can overcome many of the challenges associated with fluorescence detection. SERS signals occur from the combination of electromagnetic [31, 37, 75] and chemical [54, 59, 88] enhancement effects that allow for signals up to 10^{14} times larger than normal Raman scattering [38]. While many theories have been proposed to elucidate the mechanisms underlying these effects, electromagnetic enhancement is recognized to be from enhanced local electric fields experienced by the Raman active molecule and the formation of surface plasmons at the metal surface [32], while chemical enhancement is attributed to an increased polarizability of the molecule due to adsorption at the metal surface and charge-transfer [61].

Hairpin DNA structures have traditionally been used in molecular beacon systems with fluorescence detection. However, by using a plasmonic-based nanoprobe called a molecular sentinel [89], a DNA hairpin loop with a Raman-active molecule at one end and a thiol moiety at the other linked to a metal NP can be used as a SERS-beacon in solution. The middle circular section of the hairpin consisted of the DNA sequence probe; the two straight arm sections had complementary sequences in order to form a hairpin configuration. Under normal conditions, the very close proximity of the two arms, and hence that of the NP (SERS surface) to the label (Raman-active molecule), induces a strong SERS signal upon excitation with a laser light. However, this SERS signal disappears upon hybridization with the complementary target DNA sequence. This decrease in SERS signal as the loop of the hairpin opens up upon formation of a DNA duplex occurs from the two arms physically moving away from each other thus separating the NP from the Raman label and decreasing the SERS enhancement effects [89].

Another study deals with multiplexed SERS assays carried out with Au NP probes functionalized with oligonucleotides containing Raman-active labels that exhibit distinctive spectroscopic fingerprints [17]. A microarray of 48 amine-modified capture ssDNA probes were immobilized on a commercially available glass slide coated with a hydrophilic polymer functionalized with succinimidyl ester groups. The tethered capture probes, which consisted of 15-nucleotide strands each, were incubated with a 30-nucleotide target. The overhanging region of the target sequence was subsequently hybridized to a probe consisting of 13-nm Au NPs with an adlayer of 15-nucleotide complementary strands containing Raman-active labels. To carry out measurements, the Au NPs were augmented by a Ag metal shell in order to allow for electromagnetic enhancement and achieve SERS detection of microarrayed spots. Multiplexed detection of oligonucleotide targets was possible with this SERS assay due to the unique spectral signature of the six different Raman-active labels. The potential of this assay to differentiate single nucleotide polymorphisms was demonstrated, and, most importantly, a detection limit of 20 femtomolar was reported [17].

Another recent advance in SERS-based DNA detection is the bio-bar-code amplification method of Mirkin and coworkers [36]. In this system, a bar-code DNA sequence that serves as a unique identifier for a target sequence is detected, rather than the target itself, thus allowing a route for PCR-less target amplification. Through a combination of NPs for SERS signal amplification and magnetic microparticles for separation of unreacted elements, this assay was reportedly able to detect an oligonucleotide associated with anthrax lethal factor in the 500 zeptomolar concentration range [36]. Other methods in DNA detection by SERS that show promise include gene diagnostics via hybridization of probes to complementary DNA-coated, silver chips [22] and the detection of tagged PCR products [49].

Another alternative to fluorescent labeling has been the use of surface-enhanced infrared reflection absorption (SEIRA) spectroscopy. In this SEIRA effect, IR radiation generates oscillating dipole moments in metal nanostructures, which in turn produce an external electromagnetic field that can interact with molecules on or near the metal surface; this interaction leads to the enhancement of IR vibrations perpendicular to the metal surface [58]. In one example, grazing-angle external-reflection Fourier-transform infrared (IR) microspectroscopy and Au NPs were used to detect hybridization [44]. NP-labeled complementary DNA targets were hybridized to thiol-modified capture single stranded DNA probes chemisorbed onto slides coated with Au island films. The surface of the slide was passivated with mercaptohexanol to reduce the number of non-covalent interactions between bases in ssDNA probes and the Au surface. Non-selective binding of cDNA targets was reported to be dependent on the size of NPs and the target concentration, and nearly complete slide surface coverage was observed with a 20% probe density, 10 nM target concentration and 5-nm diameter NPs. IR spectra for ssDNA, dsDNA and NP-dsDNA conjugates were measured; however, only those observed for the latter exhibited characteristic spectral features indicating the necessity of NP conjugation. For instance, bands at 1126 and 1222 cm^{-1} attributed to the relatively intense symmetric and antisymmetric stretching vibrations of the phosphodiester DNA backbone were reported to suggest that the DNA duplex axis makes a significant projection along the surface normal.

Spectral enhancements for nucleic acids or their constituents were also reported in several SEIRA studies [5, 28, 35, 53, 67, 71]. In an external-reflection IR study to characterize SAMs [66], a thiol-modified polythymine deoxyribonucleotide $(dT)_{25}$ probe was immobilized on a Au slide. The data indicated that at a lower coverage, much of the $(dT)_{25}$ probe laid nearly flat on the surface, with a substantial fraction of the thymine bases chemisorbed; while at a higher surface density, the probe was attached via the thiol group and at possibly one or two bases. An observed IR band at 1714 cm^{-1} was attributed to carbonyl groups in thymine bases, while IR features between 1600 and 1550 cm^{-1} were consistent with carbonyl vibration modes in thymine bases chemisorbed on the Au slide.

In conclusion, NPs as signal amplification tags have been used in DNA biosensing applications and offer advantages over fluorescence-based detection methods due to improved multiplexing capabilities and signal stability. As such, the complementary techniques of SEIRA and SERS are coming to the forefront of DNA detection. Most promising are heterogeneous SERS assays, which include immobilization of capture probes on a solid substrate (chip-based) and provide high selectivity with sensitivities that are similar to those of conventional molecular biology techniques. SERS assays have also been shown to support multiplexed DNA detection, where unique Raman molecules can be used to simultaneously interrogate multiple targets of interest. Reflection and absorption mode SEIRA are increasingly being used as surface analytical techniques. For food pathogen, medical, defense, environmental and other applications, the full potential of applying NPs and surface-enhanced vibrational spectroscopy (including multispectral imaging) to gene detection in high throughput screening on a microarray platform has yet to be reached, and advances in this field are expected.

Recent Advances in the Molecular Subtyping of Bacterial Strains

Molecular genetic techniques (i.e., pulsed field gel electrophoresis (PFGE) and single nucleotide polymorphism (SNP) analysis) that pinpoint subtle genotypic differences between closely related strains of a species have allowed for significant advances in strain identification and differentiation. Whether the molecular subtyping involves the tracking of nosocomial outbreaks, identifying the reservoirs of foodborne infections, or the discrimination of plant pathogenic strains through an agricultural field commodity, the association of a specific genotype with a particular outbreak of disease cluster greatly enhances our understanding of the epidemiological principles that govern the spread of a bacterium.

Molecular subtyping methods that yield unique genotypes among strains have revolutionized the way we characterize strain diversity and scrutinize pathogen outbreaks. These techniques offer numerous advantages over classical typing methods including averting the need for stable monocultures of bacterial cells, increased assay sensitivity, and significantly faster reaction times [68]. Molecular subtypes, stemming from DNA fingerprints or polymorphic DNA sequences, can provide substantial genetic information that distinguishes closely related strains. As an example, a single hypervariable gene sequence can offer dozens of potentially polymorphic nucleotide substitution capable of discriminating among highly homogeneous pathogenic bacterial strains. As a rule, strain identification methods can be broken down into two basic conventions: (i) those that fingerprint the chromosome into polymorphic size fragments using gel electrophoretic techniques and oligonucleotide primers or restriction endonucleases (e.g., PFGE or ribotyping); and (ii) those that target specific DNA sequences and discriminate based upon the individual nucleotide substitutions retained within a particular strain (e.g., multilocus sequence typing (MLST) or SNP analysis). Emanating historically from a need for the high resolution study of metazoan populations, DNA-based strain subtyping approaches, whether they be fragmentation based or sequence based, have transformed radically the way we differentiate strains and study their phylogenetic relationships [15, 41].

Molecular subtyping methodologies have been widely effective for the differentiation of many outbreak clusters and for high resolution delineation of discrete evolutionary clades (i.e., the hierarchical classification of species based on evolutionary ancestry of strains). A wide spectrum of mounting epidemiological challenges, however, has begun to expose the various "Achilles' heels" of these once highly touted subtyping methodologies. Newly emerging bacterial pathogens (e.g., the non-O157 strains of hemorrhagic E. coli) as well as several re-emerging classical pathogens (e.g., the group I salmonellae) associated with the food supply are now present with unusual population structures, many of which confound conventional subtyping strategies. *Salmonella* Enteritidis, for example, has re-emerged in poultry as a highly virulent clone with little known genetic variation partitioning any two strains [72]. Such genetic homogeneity stymies current subtyping protocols designed for differentiating *Salmonella* strains.

Here, several novel and recent advances in PFGE and DNA sequence-based subtyping schemes are presented that hold promise in countering the aberrant genetic challenges forged within certain foodborne pathogen populations. Specifically, conventional *Salmonella* PFGE schemes, largely ineffective in discriminating *S.* Enteritidis and other clonal strain lineages, have been enhanced by the addition of several more informative restriction enzymes, markedly improving method resolution. Additionally, the slow substitution rates of housekeeping gene sequences have proven incapable of providing the discriminatory power among genetically uniform clades of salmonellae. Thus, other more rapidly evolving or hypervariable gene sequences are now being examined for their utility in discriminating closely related strains. The availability of full-length genome sequences of two or more strains from the same serovar have provided extensive utility in allowing the pinpointing of loci with greater numbers of sequence polymorphisms.

Enhanced PFGE-Based Subtyping Schemes for Clonal Serovars of the Salmonellae

PFGE is applied for the separation of large linear pieces of DNA up to 10 Mb in length. PFGE relies on rare-cutting restriction enzymes which macro-restrict genomic digestions for input to be loaded onto agarose gels. As such, PFGE can successfully resolve significantly larger fragments than conventional electrophoresis thereby yielding banding patterns with fewer band classes easily discernable among strains. This unique property of the technique has made it useful as a molecular epidemiological approach for detecting outbreak clusters of bacterial pathogens based on common PFGE banding patterns or fingerprints [84].

PFGE has been reported widely as a useful approach for differentiating strains of *Salmonella* serovars often associated with foodborne illness. The technique has greatly improved the accuracy of pinpointing bacterial sources of foodborne outbreaks while significantly reducing the amount of time required to complete an investigation [83]. While PFGE remains the subtyping "gold standard" for many strains and species, however, a few serovars continue to confound PFGE-based subtyping strategies.

Conventional PFGE protocols call for the restriction enzymes *Xba*I and *Bln*I (and sometimes *Spe*I) to resolve *Salmonella* strains. Unfortunately, these particular enzymes lack the discriminatory power to amass the genotypic signatures required to partition many *S.* Enteritidis strains and certain *E. coli* O157:H7 strains–to name two examples. Regarding the former, *S.* Enteritidis comprises a limited number of highly clonal genotypes with a single electrophoretic type (ET) accounting for nearly 80% of the ETs for this serovar [72]. These data reinforce the notion that the vast majority of *S.* Enteritidis strains are derived from a single endemic clone irrelevant of host or geography. In order to overcome this limitation, several public health and food safety laboratories have collaborated to provide enhanced dicriminatory PFGE protocols for a more effective subtyping of these particular pathogens [92].

Specifically, improvement of the discriminatory power of PFGE for highly clonal salmonellae has relied on two strategies. First, several additional macrorestriction enzymes were selected that provided a relatively even distribution and a moderately informative number of cut sites within the genomes of *S.* Typhimurium strain LT2 , *S.* Typhi CT18, *Escherichia coli* K12 and *E. coli* O157:H7 (www.rebase.com). These additional enzymes provided discriminatory PFGE fingerprints among strains that were previously unresolvable using conventional *Xba*I/*Bln*I protocols. The second strategy has been to combine individual enzyme datasets into a single supermatrix thus allowing for a combined analysis of up to six different enzymes simultaneously. This approach has had a substantial effect in differentiating highly homogeneous strains of *S.* Enteritidis. As an example, concatenated binary band class data for six enzymes derived for a clonal strain set of *S.* Enteritidis derived from poultry sources was subjected to a combined single cluster analysis. The dendrogram that resulted from this simultaneous analysis was highly resolved, yielding numerous previously unidentified branching patterns at the tips of the tree which translated into unique PFGE pattern types for nearly every strain included in the analysis (Figure 2). Additionally, a series of streamlined three-enzyme combined protocols have been developed for several notably clonal *Salmonella* serovars such as *S.* Enteritidis, Hadar, Kentucky, and Heidelberg. These strategies capture most of the six-enzyme discriminatory capability while allowing for a more technically streamlined approach. A *Sfi*I/*Pac*I/*Not*I combination, for example, was found to actually exceed corresponding values in the six-enzyme analysis for *S.* Enteritidis indicating that the diversity present in six-enzyme analyses is captured by certain three-enzyme combinations [92].

In addition to markedly improving the discriminatory potential of PFGE when subtyping closely related enteric strains, it has been observed that six enzyme combined PFGE data matrices can be used to infer fairly accurate genetic relationships among strains, an attribute of six enzyme analysis that is particularly helpful in the absence of any relevant epidemiologic findings [24]. Deriving accurate genetic relatedness of individuals from homogeneous strain populations allows for assignment of molecular epidemiological relatedness among strains. Studies focusing on these PFGE- based estimates of genetic similarity have shown great

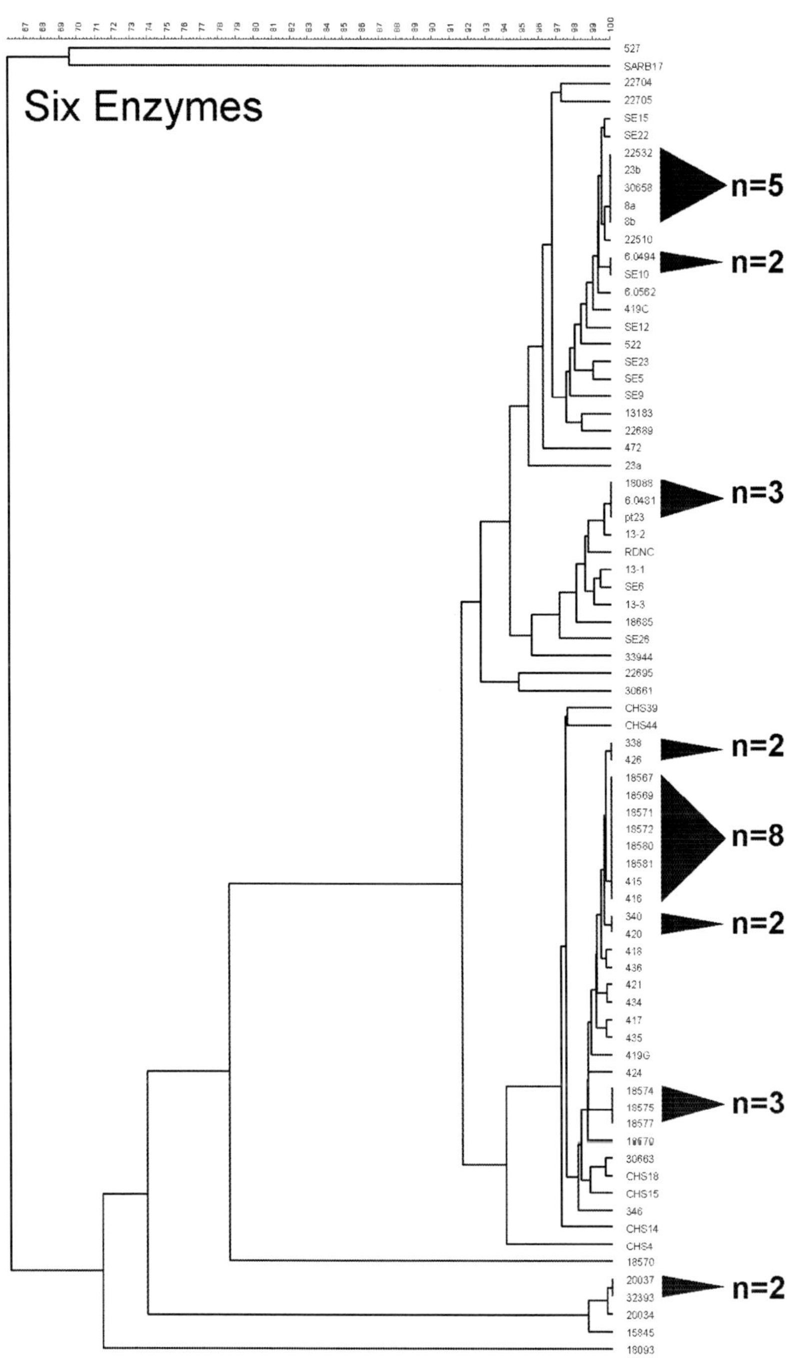

Figure 2. Combined six-enzyme dendrogram of *S.* Enteritidis strains derived from poultry. The tree shown depicts the enhanced discriminatory power resulting from the simultaneous clustering of PFGE data from six enzymes (*Xba*I, *Bln*I, *Spe*I, *Sfi*I, *Pac*I, and *Not*I). Shaded cones to the right of the terminal branches denote polytomies within each dendrogram, while adjacent *n* numbers reveal the strain totals composing their respective polytomies. A scale depicting percent divergence is presented above the dendrogram.

promise in linking potential sources of infection (e.g., feedstuffs) with outbreak strains of *E. coli* O157:H7 in cattle as well as demonstrating a strong genetic correlation with geographical distribution of O157:H7 strains from common cattle herds [25, 26].

Combined three and six enzyme schemes presented here should serve as useful alternatives for surmounting the problem of extreme strain homogeneity among pathogens. Specifically, in cases where strains are tightly linked both ecologically and temporally, potentially resulting from clonal expansion, the simultaneous multi-enzyme approach can allow for an epidemiologically useful partitioning of strains. Indeed, combined data strategies such as those noted above highlight PFGE analysis as a continued essential and informative subtyping method for the molecular epidemiological investigation of foodborne and clinical enteric pathogens.

Enhanced DNA-Sequence Based Subtyping for Clonal Serovars of the Salmonellae

Arguably, the most precise and potent method for bacterial strain discrimination remains the nucleotide sequence itself. Nucleotide sequences have been instrumental in discerning evolutionary relationships and assigning genotypes between disparate lineages of bacteria. DNA sequencing has been applied commonly for discriminating individual genotypes at the strain level [41]. Nucleotide sequence results offer several advantages over fingerprinting methods. Given the analysis of an effectively variable locus, larger numbers of polymorphic sites are usually found to be unique to a given strain. Additionally, results are portable and easily transferred between laboratories without the use of elaborate or expensive image-based databases. Finally, genotypic conclusions drawn from DNA sequence data are unambiguous and do not rely on a secondary measurements such as base pair size to make determinations.

MLST is perhaps the most well known and widely used strain subtyping method for direct sequence comparisons. MLST distinguishes strains or clusters of strains based on nucleotide sequence differences or distinct allelic types of conserved housekeeping (HK) genes [45]. Numerous loci are now available that are known to yield reliable phylogenetic clusters of strains for several enteric species (www.mlst.net). Moreover, the simultaneous analysis of multiple HK genes scattered around the chromosome often yields a snapshot of the general taxonomic or evolutionary relationships of the strains or species in question. These types of binned sequence or allele data are also amenable to cladistic analysis, a classification method in which the members of a cluster have been grouped together on the basis of a more recent common ancestor with each other than with the other members of any other group. Taken together, the advantages of MLST have made it a highly effective method, in general, for subtyping, and the method has been used to subtype populations of a variety of bacterial species including *Salmonella enterica, E. coli, Vibrio cholerae, Listeria monocytogenes*, and *Mycobacterium spp*. [82].

Much like DNA fragment-based subtyping methodologies, one caveat associated with MLST analysis is the inability to differentiate highly clonal lineages of enteric bacteria such as the aforementioned serovars of *Salmonella*. This shortcoming lies in part with the choice of loci being analyzed in a conventional MLST approach. Housekeeping genes represent more evolutionarily stable and slowly evolving portions of a bacterial genome and, as such, are prone to substantially slower rates of substitution. The subsequent lack of substantive nucleotide variation at the DNA level prevents an effective discriminatory scheme for pathogens like *Salmonella Enteritidis* derived from only a few genetic clones.

While an ever expanding arsenal of automated subtyping technologies, such as sequence tiling on DNA microarray, ESI-TOF mass spectrometry, and whole-genome pyrosequencing offer higher throughput approaches for assessing nucleotide diversity between strains, much of the success of these techniques is belied by the choice of genomic targets. With multiple whole-genome sequences now readily available in the public database for many notable enteric pathogens including *S. Enteritidis*, it has become possible to comparatively seek out those sequences that are under more relaxed selective constraints allowing them to evolve at more rapid rates between strains. This provides a more promising strategy for harvesting a greater number of nucleotide polymorphisms between any two strains. Two places in the bacterial genome where attention is being focused include (i) the relatively anonymous hypervariable ORFs that dapple the genome amongst more conserved HK gene loci and (ii) the intergenic segments of unexpressed DNA that separate many of the ORFs in the enteric genome. For some *Salmonella* serovars, strategies that target hypervariable loci are beginning to yield success. A recent publication demonstrated an effective typing scheme for *Salmonella Typhymurium* clone DT104 by simultaneously analyzing a collection of intergenic non-transcribed sequences [34]. In addition, numerous hypervariable loci have now been identified bioinformatically in *S. Enteritidis* through whole-genome comparison of syntenic ORFs between the PT4 and PT8 genomes in GenBank. Several examples of these targets and their inter-genome sequence divergence values are listed in Table 1. Taken together, these novel genome targeting strategies should facilitate more effective typing schemes capable of culling the subtle genomic differences that distinguish strains residing amongst troublesome clonal pathogen complexes.

Table 1. Examples of hypervariable loci shared between *S. Enteritidis* strains PT4 and PT8

Locus	SE PT4 ORF[a]	SE PT8 Contig[b]	Length (bp)	Putative Function	%PT4/PT8 similarity
map	00632	1419	828	methionine aminopeptidase	97.5
nfnB	01472	2007	651	dihydropteridine reductase	97.1
ybdN	01552	1972	1233	3'- phosphoadenosine	97.7
yliH	02208	2037	381	cytoplasmic protein	97.9
yliJ	02230	786	486	glutathione S-transferase	97.4
lonH	02574	1463	1758	protease	97.3
ycgB	03280	318	1035	cytoplasmic protein	97.7
ynfI	04161	412	612	anaerobic dehydrogenase	97.2
ydgP	04274	832	618	Na+ transport NADH:ubiquinone	97.9
ynhG	04426	342	999	lysM domain	97.9
celD	04581	1017	840	transcriptional repressor	97.9
csgG	05067	273	831	transcriptional regulator (curly)	97.6

[a]S. *Enteritidis* strain PT4 genome was available from www.sanger.ac.uk/Projects/Salmonella/
[b]S. *Enteritidis* strain PT8 genome was available in contig format from http://www.salmonella.org/genomics/sen.dbs.

Recombinant Phage Probes in Microbial Diagnostics

Development of systems for monitoring of the environment and food for biological threat agents is a challenge because it requires environmentally stable, long lasting, sensitive and specific diagnostic probes capable of tight selective binding of pathogens in unfavourable conditions. To respond to the challenge, large financial investments and extensive collaborative efforts of specialists in different areas of science and technology are required. In the last years, probe technology is being revolutionized by utilizing methods of combinatorial chemistry and directed molecular evolution. In particular, phage display is recently identified [64, 65] as a new technique for development of diagnostic probes which may meet the strong criteria —fastness, sensitiveness, accuracy, and inexpensiveness— for biological monitoring. Phage display is an exponentially growing research area, and numerous reviews covering different aspects of it have been published [20, 29, 79].

Phage Display Technology

Phage display technology is a powerful tool for selection of short random peptides with high affinity to target structures of interest. It is most fully developed in the Ff class of filamentous bacteriophage, which includes three wild-type strains: f1, M13, and fd. The outer coat of the phage is composed of thousands of subunits of the major coat protein pVIII, which form a tube encasing the viral DNA. At the tips of the phage are several copies of each of the minor coat proteins, pIII, pVI, pVII, and pIX. To create a phage display library, degenerate synthetic oligonucleotides are spliced in-frame into one of the coat protein genes, usually VIII or III, so that the "foreign" peptides encoded by those sequences are fused to the major coat protein pVIII or to the terminus pIII protein, respectively, and thereby displayed in several copies on the exposed surface of the virions. These foreign peptides are arranged in a repeating pattern on the phage capsid, which may thus serve as an interacting scaffold to constrain them into a particular conformation, creating a defined organic surface structure (defined "landscape" by Smith and Petrenko, [79]) that varies from one phage clone to the next. A landscape library is a huge population of such phage, encompassing billions of clones with different surface structures and biophysical properties.

Recombinant peptides specifically binding a target of interest can be selected from random peptidic libraries (usually from 8- to 15-mer), by a process of affinity selection known as "biopanning". Briefly, a simple method for biopanning involves incubating the library with the target (either a single receptor tethered to a solid support or intact cells for selection of tissue and cell targeting proteins) to allow phage displaying a complementary protein to the target to bind. Non-binding phage are then washed away and those that are bound – usually a tiny minority – are eluted in a solution that loosens target-peptide bonds without destroying phage infectivity. The eluted phages are propagated simply by infecting fresh bacterial host cells, yielding an amplified eluate that can serve as input to another round of affinity selection. Successive rounds of biopanning enrich the pool of phage with clones that specifically bind the target. Phage clones from the final eluate (typically after 2–4 rounds of selection) are propagated and characterized individually. The amino acid sequences of the peptides responsible for binding the target receptor are determined simply by ascertaining the corresponding DNA coding sequence in the phage genome.

Recombinant Phage Probes

The proteins displayed on phage surface range from short amino acid sequences to antibody fragments, enzymes, cDNA and hormones. Recently, the possibility of using landscape phage as detection probes in place of antibodies has been presented [13, 62, 64, 65]. Recombinant phage probes with high specificity and selectivity for a wide range of targets have been selected from phage display libraries [40, 62, 63, 69, 74].

Advanced bioselective sensors may meet prerequisite for isolation, concentration of the agents and their immediate real-time detection. The majority of rapid detection biosensors described in the literature has utilized antibodies as bioreceptors [43]. However, while sensitive and selective, antibodies have numerous disadvantages for use as diagnostic biodetectors in biological monitoring, including high cost of production, low availability, great susceptibility to environmental conditions [78] and the need for laborious immobilization methods to sensor substrates [65].

An effective alternative to antibodies may be the short peptides affinity-selected from random phage-displayed peptide libraries for specific, selective binding to biological targets [80]. Recombinant phage-displayed peptides can recognize and bind specific targets, such as cell surface receptors. Thus, they can act as antibody surrogates, possessing distinct advantages including durability, stability, standardization and low-cost production, while achieving equivalent specificity and sensitivity [63, 65]. A selected phage itself can be used as a probe in a detection device, without chemical synthesis of the displayed peptide or fusion to a carrier protein. For example, to be used in an automated fluorescence based sensing assay [33] or FACS [85], phage, exposing thousands of reactive amino-groups, can be conjugated with fluorescent labels and, in this format, successfully compete with antibody-derived probes. The use of antibodies as diagnostic probes outside of a laboratory may be problematic because they are often unstable in severe environmental conditions. Environmental monitoring requires stable probes, such as landscape phage, that carry thousands of foreign peptides on their surfaces, are as specific and selective as antibodies, and can operate in non-controlled conditions. Filamentous phages are probably the most stable natural nucleoproteins capable of withstanding high temperatures (up to 80°C), denaturing agents (6–8 mol/L urea), organic solvents [e.g., 50% alcohol or acetonitrile], mild acids (pH 2), and alkaline solutions. The thermostability of a landscape phage probe was recently examined in comparison with a monoclonal antibody specific for the same target [14]. They were both stable for greater than six months at room temperature, but at higher temperatures the antibody degraded more rapidly than the phage probe. Phage retained detectable binding ability for more than 6 weeks at 63°C, and 3 days at 76°C. These results confirm that phage probes are highly thermostable and can function even after exposure to high temperatures during shipping, storage and operation.

Phage-derived probes inherit the extreme robustness of the wild-type phage and, in addition, allow fabrication of bioselective materials by self-assemblage of phages or their composites on metal, mineral, or plastic surfaces [65]. The recombinant phage probes appear to be highly amenable to simple immobilization through physical adsorption directly to the sensor surface, thus providing another engineering advantage while maintaining biological functionality [18, 19, 57, 80]. This property allows phage to be used as a recognition element in biosensors by using a method to immobilize phage on the sensor surface simpler than Langmuir–Blodgett and molecular assembling technologies [56].

Applications

Numerous phage applications have been proposed, including the detection of small molecules [73] receptors [6], bacterial spores [13], and whole-cell epitopes [18, 19, 23, 57, 64, 80, 81, 91]. In particular, this technology represents a powerful tool for the selection of peptides binding to specific motifs on whole cells since it is a non-targeted strategy, which also enables the identification of surface structures that may not have been considered as targets or have not yet been identified [9].

In our studies, we used a whole-cell phage display approach to isolate phage probes specifically binding to surface of bacterial cells [18, 19]. We demonstrated, along with the other published data, that phage clones show promise as probes for biosensor applications. In fact, these phage probes could be used to build micro-biosensor systems in which biological sensing element are the selected recombinant phage. Our experience shows that recombinant phage selected by phage display technology may be the most suitable for obtaining long lasting specific probes against *L. monocytogenes*, *P. aeruginosa* and other bacterial pathogens.

Recombinant phage selected by phage-display may find application as biosorbent and diagnostic probe for monitoring bacterial cells by various devices in which antibodies have been used to date. The potential advantages of phage probes include: the simplicity of manipulation of the phage libraries, their great variability, high binding affinity, great stability and the possibility to select probes to targets of different nature, also to small molecules or toxic compounds or immunosupressants against which it is difficult to raise natural antibodies. For these properties they may be exploited for development of bioaffinity sensors, whose essential elements are probes that specifically recognize and selectively bind target structures and, as parts of the analytical platform, generate a measurable signal. For example, they may be used for separation and purification of bacteria prior to their identification with polymerase chain reaction, immunoassays, flow cytometry, or other methods. Furthermore, they may find application as bio-recognition elements of real-time biosensor devices. Recombinant phage selected by phage-display selectively recognize and specifically bind complex structures, such as bacterial cells, thus they can be used to develop rapid diagnostic arrays. In fact, traditional diagnostic systems usually involve a multi-step detection method with the use of labelled secondary antibody. Phage-displayed detection microsystems could be considered one-step, simultaneously bind and identify the target microorganism, with no need of further characterization steps. Furthermore, the nature of the bioreceptor phage holds potential utilization for development against any bacterium, virus or toxin to which a corresponding recombinant phage could be affinity-selected for. Therefore, different phage clones could be isolated specifically binding to isolated proteins, enzyme or inorganic material, as well as to different microbial species, thus, with the same microsystem different enzymes, toxins and pathogens might be detected, by performing several different parallel assays in real-time, within the same miniaturized substrate, in a single run. This could ultimately translate to a much lower cost per test. Much of the promise of these microarrays relies in their small dimensions, which reduce sample and reagent requirements and reaction times, while increasing the amount of data available from a single assay. Through the use of different labels in parallel, such as different specific phage probes, multiple tests could be simultaneously performed on the same microarray in a single step, so that standardizing data from multiple separate experiments is unnecessary and truly meaningful comparisons can be made.

The development of highly sensitive and accurate field-usable devices for detection of multiple biological agents could have a number of applications in biomedical field as well as in monitoring of agro-food pathogens and detection of biological warfare agents.

ACKNOWLEDGEMENT

Betsy Jean Yakes was supported in part by an appointment with the Research Participation Program at the Center for Food Safety and Applied Nutrition administered by the Oak Ridge Institute for Science and Education through an interagency agreement between the U.S. Department of Energy and the U.S. Food and Drug Administration.

REFERENCES

[1] Al-Khaldi, S. F., S. A. Martin, A. Rasooly, and J. D. Evans. 2002. DNA microarray technology used for studying foodborne pathogens and microbial habitats: minireview. *J AOAC Int* **85**:906-10.

[2] Al-Khaldi, S. F., M. E. Mossoba, K. M. Myers, K. E. Herold, S. J. Hart, and A. Terray. 2006. Using micro-and nono-technology in bacterial isoaltion and identfication from complex food matrices, p. 135-167. *In* P. D. Flaps (ed.), New Development in Nurition Research. Nova Science Publishers, Inc, New York.

[3] Alivisatos, A. P., K. P. Johnsson, X. Peng, T. E. Wilson, C. J. Loweth, M. P. Bruchez, Jr., and P. G. Schultz. 1996. Organization of 'nanocrystal molecules' using DNA. *Nature* **382**:609-11.

[4] Amann, R. I., W. Ludwig, and K. H. Schleifer. 1995. Phylogenetic identification and in situ detection of individual microbial cells without cultivation. *Microbiol Rev* **59**: 143-69.

[5] Aroca, R. F., D. J. Ross, and C. Domingo. 2004. Surface-enhanced infrared spectroscopy. *Appl Spectrosc* **58**:324A-338A.

[6] Balass, M., Y. Heldman, S. Cabilly, D. Givol, E. Katchalski-Katzir, and S. Fuchs. 1993. Identification of a hexapeptide that mimics a conformation-dependent binding site of acetylcholine receptor by use of a phage-epitope library. *Proc Natl Acad Sci U S A* **90**:10638-42.

[7] Beja, O., L. Aravind, E. V. Koonin, M. T. Suzuki, A. Hadd, L. P. Nguyen, S. B. Jovanovich, C. M. Gates, R. A. Feldman, J. L. Spudich, E. N. Spudich, and E. F. DeLong. 2000. Bacterial rhodopsin: evidence for a new type of phototrophy in the sea. *Science* **289**:1902-6.

[8] Beja, O., M. T. Suzuki, E. V. Koonin, L. Aravind, A. Hadd, L. P. Nguyen, R. Villacorta, M. Amjadi, C. Garrigues, S. B. Jovanovich, R. A. Feldman, and E. F. DeLong. 2000. Construction and analysis of bacterial artificial chromosome libraries from a marine microbial assemblage. *Environ Microbiol* **2**:516-29.

[9] Bishop-Hurley, S. L., F. J. Schmidt, A. L. Erwin, and A. L. Smith. 2005. Peptides selected for binding to a virulent strain of Haemophilus influenzae by phage display are bactericidal. *Antimicrob Agents Chemother* **49**:2972-8.

[10] Brady, S. F., C. J. Chao, and J. Clardy. 2002. New natural product families from an environmental DNA (eDNA) gene cluster. *J Am Chem Soc* **124**:9968-9.

[11] Breitbart, M., I. Hewson, B. Felts, J. M. Mahaffy, J. Nulton, P. Salamon, and F. Rohwer. 2003. Metagenomic analyses of an uncultured viral community from human feces. *J Bacteriol* **185**:6220-3.

[12] Breitbart, M., P. Salamon, B. Andresen, J. M. Mahaffy, A. M. Segall, D. Mead, F. Azam, and F. Rohwer. 2002. Genomic analysis of uncultured marine viral communities. *Proc Natl Acad Sci U S A* **99**:14250-5.

[13] Brigati, J., D. D. Williams, I. B. Sorokulova, V. Nanduri, I. H. Chen, C. L. Turnbough, Jr., and V. A. Petrenko. 2004. Diagnostic probes for Bacillus anthracis spores selected from a landscape phage library. *Clin Chem* **50**:1899-906.

[14] Brigati, J. R., and V. A. Petrenko. 2005. Thermostability of landscape phage probes. *Anal Bioanal Chem* **382**:1346-50.

[15] Brown, E. W. 2001. Molecular epidemiology: a practical approach. Oxford University Press, Oxford.

[16] Call, D. R., F. J. Brockman, and D. P. Chandler. 2001. Detecting and genotyping Escherichia coli O157:H7 using multiplexed PCR and nucleic acid microarrays. *Int J Food Microbiol* **67**:71-80.

[17] Cao, Y. C., R. Jin, and C. A. Mirkin. 2002. Nanoparticles with Raman spectroscopic fingerprints for DNA and RNA detection. *Science* **297**:1536-40.

[18] Carnazza, S., C. Foti, G. Gioffrè, F. Felici, and S. Guglielmino. 2007a. Specific and selective probes for Pseudomonas aeruginosa from phage-displayed random peptide libraries. *Bios Bioelectron* in press.

[19] Carnazza, S., G. Gioffrè, F. Felici, and S. Guglielmino. 2007b. Recombinant phage probes for Listeria monocytogenes. *J Phys Condensed Matter* **19**:395011.

[20] Cortese, R., P. Monaci, A. Nicosia, A. Luzzago, F. Felici, G. Galfre, A. Pessi, A. Tramontano, and M. Sollazzo. 1995. Identification of biologically active peptides using random libraries displayed on phage. *Curr Opin Biotechnol* **6**:73-80.

[21] Cox-Foster, D. L., S. Conlan, E. C. Holmes, G. Palacios, J. D. Evans, N. A. Moran, P. L. Quan, T. Briese, M. Hornig, D. M. Geiser, V. Martinson, D. vanEngelsdorp, A. L. Kalkstein, A. Drysdale, J. Hui, J. Zhai, L. Cui, S. K. Hutchison, J. F. Simons, M. Egholm, J. S. Pettis, and W. I. Lipkin. 2007. A metagenomic survey of microbes in honey bee colony collapse disorder. *Science* **318**:283-7.

[22] Culha, M., S. D., L. R. Allain, and T. Vo-Dinh. 2003. Surface-enhanced Raman scattering substrate based on a self-assembled monolayer for use in gene diagnostics. *Anal. Chem.* **75**:6196-6201.

[23] Cwirla, S. E., E. A. Peters, R. W. Barrett, and W. J. Dower. 1990. Peptides on phage: a vast library of peptides for identifying ligands. *Proc Natl Acad Sci U S A* **87**:6378-82.

[24] Davis, M. A., D. D. Hancock, T. E. Besser, and D. R. Call. 2003. Evaluation of pulsed-field gel electrophoresis as a tool for determining the degree of genetic relatedness between strains of Escherichia coli O157:H7. *J Clin Microbiol* **41**:1843-9.

[25] Davis, M. A., D. D. Hancock, T. E. Besser, D. H. Rice, C. J. Hovde, R. Digiacomo, M. Samadpour, and D. R. Call. 2003. Correlation between geographic distance and genetic similarity in an international collection of bovine faecal Escherichia coli O157:H7 isolates. *Epidemiol Infect* **131**:923-30.

[26] Davis, M. A., D. D. Hancock, D. H. Rice, D. R. Call, R. DiGiacomo, M. Samadpour, and T. E. Besser. 2003. Feedstuffs as a vehicle of cattle exposure to Escherichia coli O157:H7 and Salmonella enterica. *Vet Microbiol* **95**:199-210.

[27] Diamandis, E. P., and T. K. Christopoulos. 1996. *Immunoassay*. Academic Press, San Diego.

[28] Dovbeshko, G. I., V. I. Chegel, N. Y. Gridina, O. P. Repnytska, Y. M. Shirshov, V. P. Tryndiak, I. M. Todor, and G. I. Solyanik. 2002. Surface enhanced IR absorption of nucleic acids from tumor cells: FTIR reflectance study. *Biopolymers* **67**:470-86.

[29] Felici, F., A. Luzzago, P. Monaci, A. Nicosia, M. Sollazzo, and C. Traboni. 1995. Peptide and protein display on the surface of filamentous bacteriophage, p. 149–83. *In* M. R. El-Gewely (ed.), *Biotechnology Annual Review*, vol. 1. Elsevier Science BV, Amsterdam.

[30] Francois, P., Y. Charbonnier, J. Jacquet, D. Utinger, M. Bento, D. Lew, G. M. Kresbach, M. Ehrat, W. Schlegel, and J. Schrenzel. 2006. Rapid bacterial identification using evanescent-waveguide oligonucleotide microarray classification. *J Microbiol Methods* **65**:390-403.

[31] Garrell, R. L. 1989. Surface-enhanced Raman spectroscopy. *Anal. Chem.* **61**:401A-411A.

[32] Gersten, J., Nitzan, A. 1980. Electromagnetic theory of enhanced Raman scattering by molecules adsorbed on rough surfaces. *L. Phys. Chem.* **73**:3023-3037.

[33] Goldman, E. R., M. P. Pazirandeh, J. M. Mauro, K. D. King, J. C. Frey, and G. P. Anderson. 2000. Phage-displayed peptides as biosensor reagents. *J Mol Recognit* **13**:382-7.

[34] Hu, H., R. Lan, and P. R. Reeves. 2006. Adaptation of multilocus sequencing for studying variation within a major clone: evolutionary relationships of Salmonella enterica serovar Typhimurium. *Genetics* **172**:743-50.

[35] K. Ataka, M. O. 1999. In situ infrared study of cytosine adsorption on gold electrodes. *J. Electroanal. Chem.* **460**:188-196.

[36] Kasili, P. M., J. M. Song, and T. Vo-Dinh. 2004. Optical sensor for the detection of caspase-9 activity in a single cell. *J Am Chem Soc* **126**:2799-806.

[37] Kneipp, K., H. Kneipp, I. Itzkan, R. R. Dasari, and M. S. Feld. 2002. Surface Enhanced Raman Scattering and Biophysics. *J. Phys.: Condens. Matter* **14**:R597.

[38] Kneipp, K., H. Kneipp, I. Itzkan, R. R. Dasari, and M. S. Feld. 1999. Ultrasensitive chemical analysis by Raman spectroscopy. *Chem. Rev.* **99**:2957-2975.

[39] Kong, H., D. V. Volokhov, J. George, P. Ikonomi, D. Chandler, C. Anderson, and V. Chizhikov. 2007. Application of cell culture enrichment for improving the sensitivity of mycoplasma detection methods based on nucleic acid amplification technology (NAT). *Appl Microbiol Biotechnol* **77**:223-32.

[40] Kouzmitcheva, G. A., V. A. Petrenko, and G. P. Smith. 2001. Identifying diagnostic peptides for lyme disease through epitope discovery. Clin Diagn Lab Immunol 8:150-60.

[41] Lawrence, J. G., H. Ochman, and D. L. Hartl. 1991. Molecular and evolutionary relationships among enteric bacteria. *J Gen Microbiol* **137**:1911-21.

[42] Lopez-Campos, G., M. Coiras, J. P. Sanchez-Merino, M. R. Lopez-Huertas, I. Spiteri, F. Martin-Sanchez, and P. Perez-Brena. 2007. Oligonucleotide microarray design for detection and serotyping of human respiratory adenoviruses by using a virtual amplicon retrieval software. *J Virol Methods* **145**:127-36.

[43] Luppa, P. B., L. J. Sokoll, and D. W. Chan. 2001. Immunosensors--principles and applications to clinical chemistry. *Clin Chim Acta* **314**:1-26.

[44] M.L. Sauthier, R. L. C., C.B. Gorman, S. Franzen. 2002. Nanoparticle layers assembled through DNA hybridization: Characterization and optimization. *Langmuir* **18**:1825-1830.

[45] Maiden, M. C., J. A. Bygraves, E. Feil, G. Morelli, J. E. Russell, R. Urwin, Q. Zhang, J. Zhou, K. Zurth, D. A. Caugant, I. M. Feavers, M. Achtman, and B. G. Spratt. 1998. Multilocus sequence typing: a portable approach to the identification of clones within populations of pathogenic microorganisms. *Proc Natl Acad Sci U S A* **95**:3140-5.

[46] Majtan, T., L. Majtanova, J. Timko, and V. Majtan. 2007. Oligonucleotide microarray for molecular characterization and genotyping of Salmonella spp. strains. *J Antimicrob Chemother* **60**:937-46.

[47] Malorny, B., C. Bunge, B. Guerra, S. Prietz, and R. Helmuth. 2007. Molecular characterisation of Salmonella strains by an oligonucleotide multiprobe microarray. *Mol Cell Probes* **21**:56-65.

[48] Mirkin, C. A., R. L. Letsinger, R. C. Mucic, and J. J. Storhoff. 1996. A DNA-based method for rationally assembling nanoparticles into macroscopic materials. *Nature* **382**:607-9.

[49] Monaghan, P. B., K. M. McCarney, A. Ricketts, R. E. Littleford, F. Docherty, W. E. Smith, D. Graham, and J. M. Cooper. 2007. Bead-based DNA diagnostic assay for chlamydia using nanoparticle-mediated surface-enhanced resonance Raman scattering detection within a lab-on-a-chip format. *Anal. Chem.* **79**:2844-2849.

[50] Monecke, S., P. Kuhnert, H. Hotzel, P. Slickers, and R. Ehricht. 2007. Microarray based study on virulence-associated genes and resistance determinants of Staphylococcus aureus isolates from cattle. *Vet Microbiol* **125**:128-40.

[51] Moore, W. E., E. P. Cato, and L. V. Holdeman. 1978. Some current concepts in intestinal bacteriology. *Am J Clin Nutr* **31**:S33-42.

[52] Moore, W. E., and L. V. Holdeman. 1974. Human fecal flora: the normal flora of 20 Japanese-Hawaiians. *Appl Microbiol* **27**:961-79.

[53] Moses, S., S. H. Brewer, L. B. Lowe, S. E. Lappi, L. B. Gilvey, M. Sauthier, R. C. Tenent, D. L. Feldheim, and S. Franzen. 2004. Characterization of single- and double-stranded DNA on gold surfaces. *Langmuir* **20**:11134-40.

[54] Moskovits, M. 1985. Surface-enhanced spectroscopy *Rev. Mod. Phys.* **57**:783-826.

[55] Myers, K. M., J. Gaba, and S. F. Al-Khaldi. 2006. Molecular identification of Yersinia enterocolitica isolated from pasteurized whole milk using DNA microarray chip hybridization. *Mol Cell Probes* **20**:71-80.

[56] Nanduri, V., I. B. Sorokulova, A. M. Samoylov, A. L. Simonian, V. A. Petrenko, and V. Vodyanoy. 2007. Phage as a molecular recognition element in biosensors immobilized by physical adsorption. *Biosens Bioelectron* **22**:986-92.

[57] Olsen, E. V., I. B. Sorokulova, V. A. Petrenko, I. H. Chen, J. M. Barbaree, and V. J. Vodyanoy. 2006. Affinity-selected filamentous bacteriophage as a probe for acoustic wave biodetectors of Salmonella typhimurium. *Biosens Bioelectron* **21**:1434-42.

[58] Osawa, M. 1997. Dynamic processes in electrochemical reactions studied by surface enhanced infrared absorption spectroscopy (SEIRAS) *Bull. Chem. Soc. Jpn.* **70**:2861-2880.

[59] Otto, A., J. Timper, J. Billmann, G. Kovacs, and I. Pockrand. 1980. Surface roughness induced electronic Raman scattering. *Surf. Sci.* **92**:L55-L57.

[60] Palmer, C., E. M. Bik, D. B. Digiulio, D. A. Relman, and P. O. Brown. 2007. Development of the Human Infant Intestinal Microbiota. *PLoS Biol* **5**:e177.

[61] Persson, B. N. J. 1981. On the theory of surface-enhanced Raman scattering. *Chem. Phys. Lett.* **82**:561-565.

[62] Petrenko, V. A., and G. P. Smith. 2000. Phages from landscape libraries as substitute antibodies. *Protein Eng* **13**:589-92.

[63] Petrenko, V. A., G. P. Smith, X. Gong, and T. Quinn. 1996. A library of organic landscapes on filamentous phage. *Protein Eng* **9**:797-801.

[64] Petrenko, V. A., and I. B. Sorokulova. 2004. Detection of biological threats. A challenge for directed molecular evolution. *J Microbiol Methods* **58**:147-68.

[65] Petrenko, V. A., and V. J. Vodyanoy. 2003. Phage display for detection of biological threat agents. *J Microbiol Methods* **53**:253-62.

[66] Petrovykh, D. Y., H. Kimura-Suda, L. J. Whitman, and M. J. Tarlov. 2003. Quantitative analysis and characterization of DNA immobilized on gold. *J Am Chem Soc* **125**: 5219-26.

[67] R. Aroca, R. B. 1999. Surface enhanced vibrational spectra of thymine. *Vibrational Spectroscopy* **19**:11-21.

[68] Relman, D. A., and D. H. Persing. 1996. *PCR protocol for emerging infectious diseases.* ASM Press, Washington, DC.

[69] Romanov, V. I., D. B. Durand, and V. A. Petrenko. 2001. Phage display selection of peptides that affect prostate carcinoma cells attachment and invasion. *Prostate* **47**: 239-51.

[70] Romermann, D., J. Warrelmann, R. A. Bender, and B. Friedrich. 1989. An rpoN-like gene of Alcaligenes eutrophus and Pseudomonas facilis controls expression of diverse metabolic pathways, including hydrogen oxidation. *J Bacteriol* **171**:1093-9.

[71] S. Pronkin, T. W. 2003. Time-resolved in situ ATR-SEIRAS study of adsorption and 2D phase formation of uracil on gold electrodes. *J. Electroanal. Chem.* **550-551**:131-147.

[72] Saeed, A. M., S. T. Walk, M. Arshad, and T. S. Whittam. 2006. Clonal structure and variation in virulence of Salmonella enteritidis isolated from mice, chickens, and humans. *J AOAC Int* **89**:504-11.

[73] Saggio, I., and R. Laufer. 1993. Biotin binders selected from a random peptide library expressed on phage. *Biochem J* **293**:613-6.

[74] Samoylova, T. I., V. A. Petrenko, N. E. Morrison, L. P. Globa, H. J. Baker, and N. R. Cox. 2003. Phage probes for malignant glial cells. *Mol Cancer Ther* **2**:1129-37.

[75] Schatz, G. C. 1984. Theoretical Studies of Surface Enhanced Raman Scattering. *Acc. Chem. Res*. **17**:370-376.

[76] Schrenzel, J. 2007. Clinical relevance of new diagnostic methods for bloodstream infections. *Int J Antimicrob Agents* **30** Suppl 1:S2-6.

[77] Sebat, J. L., F. S. Colwell, and R. L. Crawford. 2003. Metagenomic profiling: microarray analysis of an environmental genomic library. *Appl Environ Microbiol* **69**:4927-34.

[78] Shone, C., P. Wilton-Smith, N. Appleton, P. Hambleton, N. Modi, S. Gatley, and J. Melling. 1985. Monoclonal antibody-based immunoassay for type A Clostridium botulinum toxin is comparable to the mouse bioassay. *Appl Environ Microbiol* **50**:63-7.

[79] Smith, G. P., and V. A. Petrenko. 1997. Phage Display. *Chem Rev* **97**:391-410.

[80] Sorokulova, I. B., E. V. Olsen, I. H. Chen, B. Fiebor, J. M. Barbaree, V. J. Vodyanoy, B. A. Chin, and V. A. Petrenko. 2005. Landscape phage probes for Salmonella typhimurium. *J Microbiol Methods* **63**:55-72.

[81] Stratmann, J., B. Strommenger, K. Stevenson, and G. F. Gerlach. 2002. Development of a peptide-mediated capture PCR for detection of Mycobacterium avium subsp. paratuberculosis in milk. *J Clin Microbiol* **40**:4244-50.

[82] Sukhnanand, S., S. Alcaine, L. D. Warnick, W. L. Su, J. Hof, M. P. Craver, P. McDonough, K. J. Boor, and M. Wiedmann. 2005. DNA sequence-based subtyping and evolutionary analysis of selected Salmonella enterica serotypes. *J Clin Microbiol* **43**:3688-98.

[83] Swaminathan, B., T. J. Barrett, and P. Fields. 2006. Surveillance for human Salmonella infections in the United States. *J AOAC Int* **89**:553-9.

[84] Swaminathan, B., T. J. Barrett, S. B. Hunter, and R. V. Tauxe. 2001. PulseNet: the molecular subtyping network for foodborne bacterial disease surveillance, United States. *Emerg Infect Dis* **7**:382-9.

[85] Turnbough, C. L., Jr. 2003. Discovery of phage display peptide ligands for species-specific detection of Bacillus spores. *J Microbiol Methods* **53**:263-71.

[86] Tyson, G. W., J. Chapman, P. Hugenholtz, E. E. Allen, R. J. Ram, P. M. Richardson, V. V. Solovyev, E. M. Rubin, D. S. Rokhsar, and J. F. Banfield. 2004. Community structure and metabolism through reconstruction of microbial genomes from the environment. *Nature* **428**:37-43.

[87] Venter, J. C., K. Remington, J. F. Heidelberg, A. L. Halpern, D. Rusch, J. A. Eisen, D. Wu, I. Paulsen, K. E. Nelson, W. Nelson, D. E. Fouts, S. Levy, A. H. Knap, M. W. Lomas, K. Nealson, O. White, J. Peterson, J. Hoffman, R. Parsons, H. Baden-Tillson, C. Pfannkoch, Y. H. Rogers, and H. O. Smith. 2004. Environmental genome shotgun sequencing of the Sargasso Sea. *Science* **304**:66-74.

[88] Vo-Dinh, T. 1998. Surface-Enhanced Raman Spectroscopy Using Metallic Nanostructures. *Trends Anal. Chem*. **17**:557-582.

[89] Wabuyele, M. B., M. Culha, G. D. Griffin, P. M. Viallet, and T. Vo-Dinh. 2005. Near-field scanning optical microscopy for bioanalysis at nanometer resolution. *Methods Mol Biol* **300**:437-52.

[90] Weile, J., R. D. Schmid, T. T. Bachmann, M. Susa, and C. Knabbe. 2007. DNA microarray for genotyping multidrug-resistant Pseudomonas aeruginosa clinical isolates. *Diagn Microbiol Infect Dis* **59**:325-38.

[91] Yu, J., and G. P. Smith. 1996. Affinity maturation of phage-displayed peptide ligands. *Methods Enzymol* **267**:3-27.

[92] Zheng, J., C. E. Keys, S. Zhao, J. Meng, and E. W. Brown. 2007. Enhanced subyping scheme for *Salmonella* Enteritidis. . *Emerg Infect Dis* **13**:1932-1935.

[93] Zhong, C. J., R. C. Brush, J. Anderegg, and M. D. Porter. 1999. Organosulfur monolayers at gold surfaces: Reexamination of the case for sulfide adsorption and implications to the formation of monolayers from thiols and disulfides. *Langmuir* **15**:518-525.

[94] Zhu, L. X., D. Wang, G. B. Zhang, D. Jiang, Z. W. Zhang, Q. Zhang, K. Mitchelson, and J. Cheng. 2007. Development of a base stacking hybridization-based microarray method for rapid identification of clinical isolates. *Diagn Microbiol Infect Dis* **59**: 149-56.

In: Oligonucleotide Array Sequence Analysis
Editors: M.K. Moretti and L.J. Rizzo, pp. 149-174

ISBN: 978-1-60456-542-3

Chapter 5

STATISTICAL ANALYSIS OF TILING-PATH MICROARRAYS

Xiangfeng Wang*

Department of Molecular, Cellular and Developmental Biology,
Yale University, New Haven, CT, 06520-8104

ABSTRACT

Genomic tiling-path microarray has become an economical and prompt way to efficiently investigate the functional DNA elements in many forms, such as coding/non-coding transcripts, transcription factor binding sequences, DNA content variation, DNA methylation, histone methylation, acetylation and other numerous essential epigenetic events that contribute to gene transcription regulation and chromatin organization. With the increasing of tiling array application in many fields of biology, various methods related to tiling array processing and analysis have been developed and evolved from simple assessments to more scientific model-based mathematical algorithms; however, a unified and regularized procedure must be established. This chapter provides a general review of the application and statistical methodology involved in tiling array analysis from the following aspects:

Firstly, from an initial report of human ENCODE project, the importance of the application of tiling array in post-genomics era will be stressed. Secondly, some basic classifications of representative tiling array platforms with different definitions and several algorithms for selecting an optimal tile path of probes to cover a reference genome during tiling array design will be introduced. The third part relates to the methods and technical issues in preprocessing of tiling array raw data including standard within- and between-slide normalizations, spatial smoothing normalizations of unevenly distributed probe signals across array surface, background correction methods and some ChIP-chip, array-CGH specific normalizations methods. The fourth part will be focused on how to statistically infer the interrogated genomic regions with significantly enriched tiling array signals. A number of well-recognized statistical models and algorithms will be introduced, such as hidden Markov model (HMM), Wilcoxon signed-rank test, Model-based Analysis of Tiling-arrays (MAT) and dynamic programming algorithm, etc. in the last part, a brief introduction of the tiling array analysis software and online resources will be given.

---

* Phone: 203-508-4833 E-mail: Xiangfeng.Wang@yale.edu

INTRODUCTION: VERSATILE APPLICATIONS OF TILING ARRAYS

The development of genomic tiling-path design in microarray technology has provided an unprecedented opportunity to investigate the functional DNA elements in an extremely efficient fashion. These successes in utilizing high-through tiling array have greatly broadened our understanding of biological sciences and have proven the core role of tiling arrays in many fields of genomics research. Construction of tiling array demands a large set of short oligonucleotide probes, which are either overlapped or spaced at certain bases, to create a pathway spanning the reference genome sequences. To monitor the activities of transcribed elements encoded in DNA or the enrichment of designated genomic regions with specific epigenetic modifications, tiling arrays will be hybridized with fluorescent-labeled RNA or DNA pooled from coding/non-coding transcripts samples or chromatin immunoprecipitation (ChIP) assays. From the array signals we could recognize the interrogated products by mapping hybridized probes back to the target genome according to their genomic coordinates. During the past decade, many computational biologists have been dedicated to the development of mathematical models and methods for best analyzing and interpreting tiling array results in many aspects, including optimal design of tiling probes across an ideal genomic path, tiling array data preprocessing, statistical inference of interested genomic regions based on hybridization signals and how tiling array data could be visualized, stored and integrated to current genome browsers for public online queries. In the first part of this chapter, from a recent report on analyzing functional elements in 1% of the human genome by the ENCODE project [1], we will see how tiling array technology fulfills its critical functions in the post-genomics era.

1. Application of Tiling Arrays in the ENCODE Project

The Encyclopedia of DNA Elements (ENCODE) project, whether in animal or plant, involves hundreds of groups all over the world using hundreds of ways of experimental and computational techniques to explore all possible functional elements embedded in DNA and to decipher the inheritable genetic information contained in all possible forms. The first assignment of ENCODE is to evaluate predicted exon/intron structures that are still lacking of full length cDNA confirmation. Applications of tiling arrays hybridized with 3'-polyadenylated RNA transcripts have been proven to be an effective way to accurately determine the boarders of genic and intergenic territories, junctions of neighboring exons and introns, as well as the uncharacterized un-translated regions (UTR) (Figure 1A & 1B). These utilizations in several model organism genomes such as human, mouse, *Arabidopsis* and rice, have facilitated in the construction of a scoring system for existing gene annotation and integrating the tiling array information into the popular genome browsers as universal experimental evidence for annotated gene models.

The second assignment is to continue to uncover the missed protein-coding genes by current computational prediction methods. Initial tiling array analysis on hybridization data with total or Poly-A RNA products showed an unexpected proportion of transcribe sequences distributed in the 1% of the human genome according to ENCODE's report [1]: nearly 15% of the bases were identifiable to bear transcriptional activity, but 63% of them were located outside existing annotated exons; for those non-exon transcriptional fragments, one third in the intergenic regions indicates the percentage of novel independent transcripts while the other two thirds in the intronic regions present a rough estimate of possible alternatively spliced exons or incorrect annotation as well. Another amazing finding was that nearly one third of the annotate gene loci have antisense transcriptional signals which is still unexplained and universally found in other model organisms conducted by genome wide tiling array experiments.

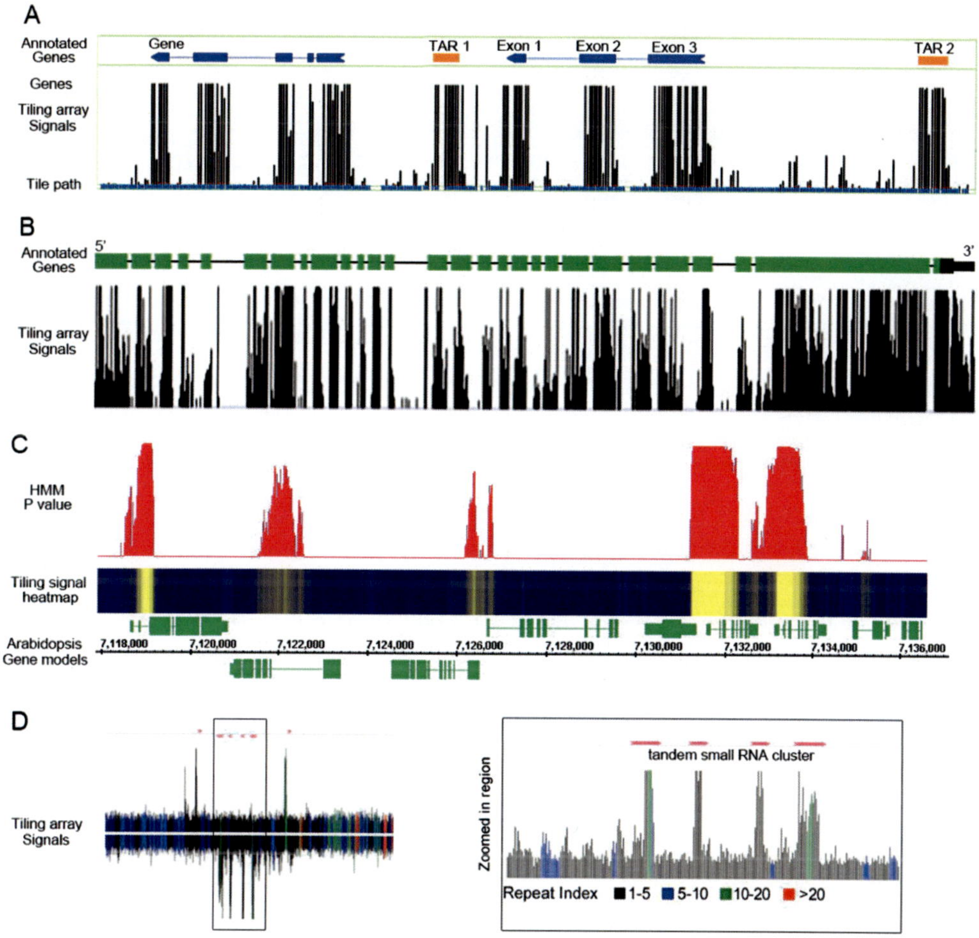

Figure 1. Illustrations of tiling arrays for mRNA analysis, noncoding RNA analysis and ChIP-on-chip experiments. A) An average resolution 46 bp tiling array used for an experiment confirms the predicted gene structures, as the blue gene models shown. The probes located in genes' transcribed regions present abundant array signals while the probes located in intron and intergenic regions have no hybridization signals. From the signal clusters we could correspond and confirm the predicted genes' structures. Additionally, the intergentic signals may indicate the potential novel transcriptional units, as yellow bars shown. B) Resolution 5 bp tiling array, the higher resolution of tiling array, the smaller exons could be identified. C) High resolution (5 bp) tiling arrays used for non-coding RNA transcripts analysis. Four small non-coding RNA transcriptional activities were reflected from the tiling array signals. D) Tiling arrays in ChIP-on-chip experiment for detecting histone H3 lysine acetylations. Pink peaks are distributions of P values calculated by Hidden Markov Model. Second track is a tiling array signal heat map, in which yellow intensive regions represents higher ChIP-enriched regions. All the figures are re-edited from recently published rice and Arabidopsis tiling array project at Yale Univeristy.

Thirdly, another important subject of ENCODE project is to characterize the regulatory *cis*-elements across the genome by ChIP-on-chip assays to profile the binding sites of the specific transcription factors. In terminological "ChIP-on-chip", the first "ChIP" is the abbreviation of "chromatin immunoprecipitation", which means using specific antibodies to pull down the target proteins and then extract the DNA fragments wrapped on them for subsequent microarray hybridization. The second "chip" is specifically referred to tiling-path microarrays since the exact positions of most of regulatory factors' binding sites usually are unknown and an unbiased fashion is needed for the examination. Because the chromatin immunoprecipitated DNA fragments are around 500 bp in length, the resolution of the ChIP tiling array is not stringent and usually every region of 100bp to 200 bp is tiled by 3 to 5 successive probes to interrogate a chromatin immunoprecipitated DNA fragment.

The forth objective in ENCODE is to profile the epigenetic landscape in the genome including the methylated DNA regions; an array of histone modifications such as H3 lysine 4 mono- di- and tri-methylations (H3K4me1, H3K4me2, H3K4me3), H3 lysine 9 acetylation (H3K9ac), H3 lysine 27 methylations (H3K27me) *etc.*; various histone unit variants and other forms of chemical modifications (Figure 1C). The comprehensive epigenomics data could be associated with various known featured genomic elements and yield numerous sub-branches of epigenetics related subjects. For example, mapping epigenetic modification to genome and protein-coding genes will be helpful to understand how they contribute to chromosomal organization and how they distribute within a transcriptional unit; correlation analysis of epigenetically modified genes and their expression level will provide evidence how epigenetic modifications influence gene transcription at initiation or elongation phases from a genome-basis survey; integrated information of enriched epigenetic modification sites and transcription factor binding sites will facilitate the prediction of transcription start sites (TSSs) and distal specific regulatory sequences; finally, how multiple modifications interact and result in combinatory effect on gene transcription is also under investigation.

Those remarkable results partially illustrated from the ENCODE project has evidently shown powerful potency of the application of tiling array in genomics, transcriptomics and epigenomics related fields and provided novel knowledge into biological sciences.

2. Low-Resolution Tiling Array for Comparative Genomics Hybridization

Comparative genomics hybridization analysis, usually abbreviated as CGH, has been developed as a substituent technique for fluorescence *in situ* hybridization (FISH) and its greatest improvement is that the amounts of deleted or duplicated DNA fragments could be quantitatively measured and accurately positioned in the genome at kilo-base level accuracy. Since the tiling array is also the central technology for conducting a CGH experiment, this technology is also called array-CGH. However, since the duplicated or deleted genomic regions that cause a meaningful genetic variation usually ranges from several kilo base pairs to several mega base pairs and the resolution of the tiling array for CGH experiment becomes unimportant. Thus, in a typical array-CGH experiment, the probes are usually prepared from PCR amplification and then printed onto the slides.

To examine the changes of copy number in DNA content, the test DNA samples from tumor cells labeled with one dye are often co-hybridized with a reference DNA samples from normal cell labeled with another dye, and the ratio of the two dyes' fluorescence intensities will be used to estimate the amounts of gained or lost DNA copies in the test sample.

3. High-Resolution Tiling Array in Non-Coding RNA Analysis

Using tiling array to globally identify the non-coding RNAs in genome is still a new challenge, since the mature microRNAs are usually 20~22 nt in length and other longer small RNA population are usually below 100 bp, the ncRNA tiling array must reach a extremely high resolution for better hybridization. Combining the tiling array data with known non-coding RNA elements and structural RNA has provided universal transcription evidence for this important population. Mapping the non-coding RNA signals to the genome is helpful for analyzing the genomic organization of non-coding RNAs and how they impact the target genes in the neighborhood. A recent pilot tiling array with an extreme high resolution of 5 bp in a representative region on rice chromosome 10 has been conducted with hybridization of pooled processed non-messenger RNA [2]. By this experiment, the high-resolution tiling array is proven to be an effective way for large-scale screening of small non-coding elements and the same paper also proposed an interesting assumption that some of the mature small RNA might be originated from the antisense 5' end, probably the opposite strand of UTR, and co-transcribed with their target genes and then processed to be functional forms to fulfill their regulatory roles [2].

Optimal Design of Tiling-Path Microarray

Tiling-path microarray has been involved in many aspects of genomics research fields, from experimental genome re-annotation, comparative genomics hybridization, gene alternative splicing, transcription binding sites and genome epigenetic modification profiling, etc. For different goals of conducting a tile path analysis, the strategies of designing tiling array may vary a lot to fit different requirements as the studied objectives raise, and the most important parameter to discriminate different types of tiling array is called "resolution". Resolution is defined as the average or median range length for each probe to cover a genomic interval. As Figure 2A shows, for a standard tile path design, the probes usually arranged in an overlapped manner or spaced with roughly equal distances. For the overlapped design, the resolution is calculated as the probe length minus the overlapped portion, while for the spaced design, it's the sum of probe length and average distance length. Higher resolution provides higher detection accuracy, denser signals and larger coverage of all possible genomic loci. As a matter of fact, due to the current probe capacity of a microarray and considering the economical reason, it's unrealistic to make every base pair in the genome to be covered. Additionally, the DNA sequences that lack good sequence nature for hybridization or will potentially produce non-specific hybridization must be skipped to avoid false positive signal. Therefore, how to generate a maximum optimal tile path becomes a mathematical problem that needs proper computational algorithms to dynamically select the probe set along the continuous genome sequence. This section will be focused on several classic methods for designing a tiling-path microarray. First of all, we will introduce some basic classifications of tiling arrays categorized based on their various resolutions, different tile path arrangements as well as the research purposes.

1. Basic Classifications of Tiling-Path Microarrays

The first introduced member in tiling array family is based on PCR-generated genomic DNA fragments, which are spotted on glass slides as probes. Because the probes are generated from amplified PCR products or cloned DNA to construct the tile path, their average lengths range usually hundreds to several kilo- base pairs and have the lowest resolution comparing to other types of tiling array. At this resolution, the probe is not capable to discriminate the sources of similar genomic loci, however, its non-specific nature well agrees with the needs for investigating the genome duplication and copy numbers. Therefore, the PCR-fragments-based tiling array becomes a favorable tool in the Comparative Genomics Hybridization experiments (CGH) and usually contains 10,000 to 50,000 probes per chip to represent a genome.

The other types of tiling arrays have higher resolution and different sources of producing the short oligonucleotide probes, which are *in situ* synthesized on array surface or spotted onto glass slide by an array printing machine. A typical high density tiling array usually contains about 100,000 to several millions of probes on each chip. Todd Mockler defined four types of high-density tiling array that based on different design strategy early published in 2005 [3].

True-whole-genome tiling array design (Figure 2A and 2B), in which the short oligo probes are partially overlapped or spaced with quite small interval, provides the highest resolution and the most comprehensive coverage of the genome. The true genome-wide tiling arrays hybridized with mRNA samples have been widely used in many organisms to facilitate the computational genome annotation and to identify the potential novel transcriptional elements that have been missed by current annotation. A decent resolution allows an elaborate comparison of tiling array signals with the computer-predicted genes, and a good match could reflect the structure of exons and introns. The true whole-genome tiling arrays have also been employed in profiling the epigenetic landscape and chromosomal organization of a genome by hybridized with the fragmented DNA products from a series of chromatin immunoprecipitation assay, such as DNA methylations, histone modifications and nucleosome occupancy *etc*.

Quasi-whole-genome tiling array design (Figure 2C), in which longer probes are spaced by longer intervals, provides a little bit lower resolution and are usually coupled with ChIP assay to map the transcription factor binding sites and to roughly identify the potential genomic positions of upstream regulatory cis-elements.

Exon-scanning tiling array design (Figure 2D), is an annotation-dependent array design, in which the probes are selected right reside the junctions of predicted exons and introns. Exon-scanning tiling arrays have been used to discover the alternatively spliced transcripts and characterize their potential iso-forms in different tissues and samples.

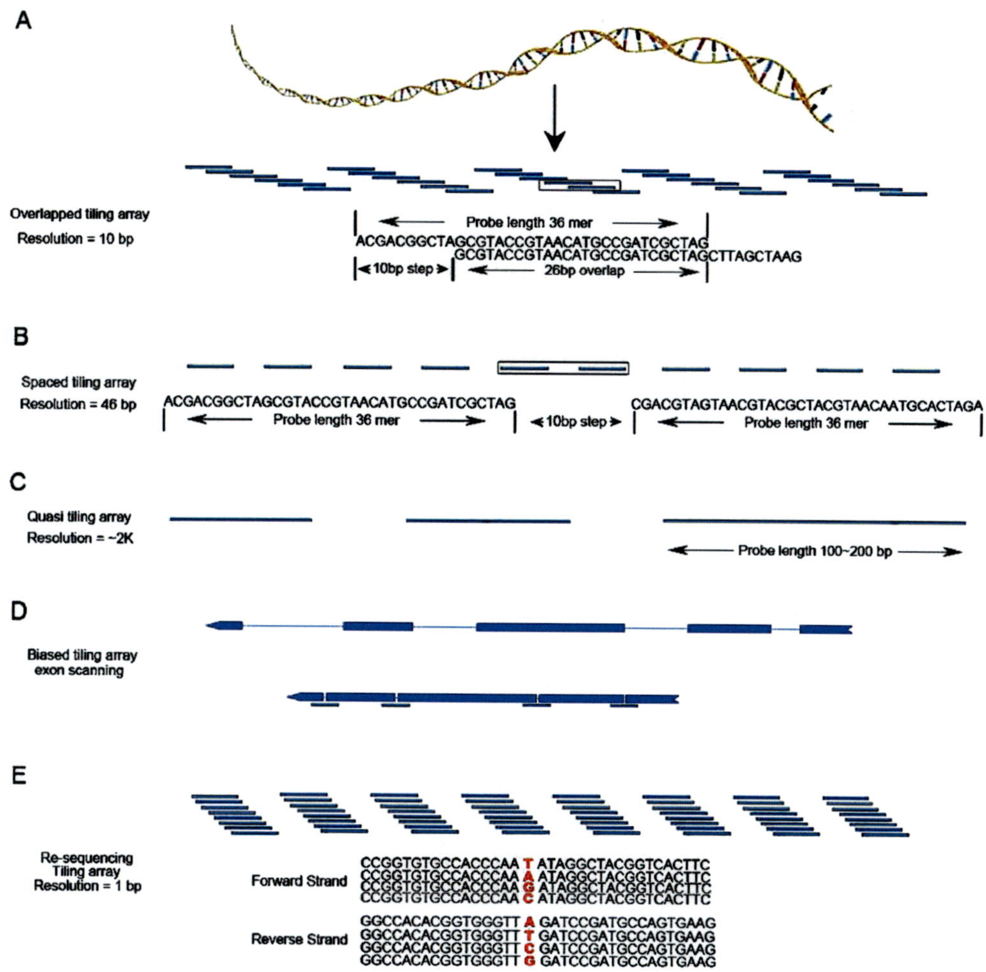

Figure 2. Four types of tiling-path microarrays with different resolution for different purposes. A) True overlapped tiling arrays whose resolution is the size of backward step. Usually hybridized with mRNA transcripts used for experimental validation of genome annotation and identification of novel transcriptional units. B) True spaced tiling arrays whose resolution is the sum of probe length and average interval length. Usually used in ChIP-on-chip experiment. Since the chromatin immunoprecipitated DNA fragments often ranges from 500 bp to 2000 bp, the resolution is not highly required. C) Quasi tiling arrays in which the probes are generated from PCR products, their resolutions usually range hundreds base pairs. This type of tiling array is often used for mapping transcription factors. D) Exon scanning tiling arrays in which the probes designed at the junction of adjacent introns and exons, for detecting alternative splicing isoforms. E) Re-sequencing tiling arrays whose resolution is 1 bp, are usually used for validating SNP sites. Eight probes with eight different nucleotides are designed and the perfect matched one reflects the real SNP nucleotide.

Re-sequencing tiling array design (Figure 2E), has the highest resolution at the top of all other types of tiling array, in which the tiling-step is one nucleotide for the two neighboring probes and the tile path spans from two directions to represent the double DNA strands. Re-sequencing tiling array has been utilized to experimentally confirm the Single Nucleotide Polymorphisms (SNP) in the genome. Four probes as a set with the four forms of nucleotides in the middle of each probe are used to differentiate the signals from a genomic locus bearing a SNP candidate site, and the one with the highest value among the four probes will represent the existing nucleotide.

For the exon-scanning tiling array and re-sequencing tiling array that are based on annotated genome information, there is no much room for a design to select the probes because all the probes should be located right at the interrogated position. However, for the true whole genome tiling array, the design and selection of oligo probes will be focused on how to find a maximally optimized tile path that offers the highest resolution and coverage with lowest artificial hybridizations.

2. Difference in Designing Tiling Arrays and Traditional Gene Arrays

The strategy of designing a traditional gene-level microarray and genomic tiling-path microarray varies in several aspects regarding different purposes and factors that need to be taken into account during the selection of optimal probes. In gene-level oligonucleotide or cDNA microarray design, the probes are selected within the coding regions of annotated gene models and the central principle of probe selection algorithm is to produce probes with highest specificity and maximum sensitivity to the target genes. Usually one or more probes are chosen to present each gene in a gene-level microarray, and the main purpose is to profile the expression pattern of the target genes and to screen the differentially expressed genes across multiple conditions.

As the purpose of tiling array is to interrogate the genome transcription activity in an unbiased fashion, the core of tiling array design is to find out an optimal tile path that provides the maximum coverage of genomic sequence, allows most flexible range of DNA hybridization affinity but furthest avoids cross-hybridization from non-specific oligonucleotides. An efficient tiling array probe selection algorithm must take more factors into consideration than in a gene microarray design, such as how to skip a wide range of repetitive sequence region, how to index the repetitive frequency of oligonucleotides probes, how to dynamically adjust tiling-step, how to select the optimal probe within a step range based on probe specificity, GC composition, Tm value and sequence complexity, and how to process the computation with the minimum system requirements *etc*.

3. Probe Selection Algorithm in Tiling Array Design

As a general initial step in tiling array design, sequence-comparison-based screening of repetitive sequences is carried out using certain of repeat identification software, among which RepeatMasker (http://www.repeatmasker.org/) is the most widely used program [4]. RepeatMasker searches for the non-unique portions of genome sequence against a species-specifiable repeat database that stores all known repetitive elements such as long LTR (terminal repeat) transposons, LINE/SINE (long/short interspersed nuclear element) repeats, other types of transposons, tandem repeat elements and low-complexity sequences. Averagely, nearly 45% of the human genomic DNA will be recognized as repetitive sequences and fatherly masked from the genome sequence [5].

The first step of repeat masking procedure creates numerous partitions of non-repetitive genomic sequence, but the potential of cross-hybridization still remains at short oligonucleotide level because of the existence of gene family or multiple copy genes. To maximally eliminate the cross-hybridization, several solutions and algorithms for indexing the similarity of probe sequence have been proposed, basically classified as alignment-based approach and frequency counting approach. Bertone *et al.* presented an alignment-based approach by adopting a BLAST-like procedure to look up substrings of a certain length of oligonucleotide in the reference genome sequence [6]. To find out the optimal tile path in the non-repetitive sequence partitions, they proposed a scoring function to evaluate all possible tile paths and used a dynamic programming algorithm to implement this procedure. Their methods resulted in a web-interface tiling array design tool "OligoTile" at http://tiling.gersteinlab.org/ [6].

Compared with the alignment-based approach, the fundamental of frequency counting way is to calculate similarity frequency of oligonucleotide in an entire genome without masking the repetitive sequences. A recent publication introduced a *uniqueness score "U"* as a parameter for measuring the repeat frequency of a given oligonucleotide probe, and a greedy algorithm to design the genome-wide tiling array based on computing *U* score [7]. In this strategy, the reference genome sequence is first subdivided into equal-sized windows, and then minimum unique substrings within each window are determined for calculating the *U* score using an FMindex data structure. The biggest advantage of this *uniqueness score U* computing- based probe selection approach is that a small portion of non-repetitive sequences can be retrieved from the previously filtered sequences by RepeatMasker program and more probes containing important information within the repetitive genomic regions can be retained in this tiling array design. The source code of tiling array design developed by Stefan et al., which use *uniqueness score U* as the core of the algorithm is accessible via http://www.ebi.ac.uk/~graef/arraydesign/ [7].

NORMALIZATION METHODS IN TILING ARRAY ANALYSIS

1. General Tiling Microarray Data Preprocessing Pipeline

Within the recent decade of the application of microarray technology in biology research, a great number of statistical models and methods have been developed for processing and analyzing microarray data. Those methods usually are classified and organized in an appropriate workflow in accordance with the types of systematic errors that need to be removed from the raw microarray data, and this procedure is called "Normalization". To normalize the data from dual-channel cDNA or oligonucleotide microarrays, the first step of normalization is to eliminate the variance of hybridization intensity in red and green labels, which is caused by the different strength of dye fluorescence of Cy3 and Cy5. The most common method in dealing with this problem is to use a *LOWESS (or LOESS) normalization* process to balance the ratio values from Cy3 channel and Cy5 channel by fitting a curve to the local neighborhood of data points and then adjust the each probe's intensity [8]. From a MA plot, the trend of the distribution of ratio values of Cy3 versus Cy5 after LOESS process should be centered at zero indicating the bias of different fluorescence dye has been removed. Another normalization procedure on a single slide is to remove the variance introduced by the inconsistent printing efficiency of print-tip needle. This kind of experimental error appears to be a periodical pattern that the observed overall hybridization intensities from different sub-arrays show various strength. Print-tip normalization can effectively remove the variance.

Besides the two normalization methods that work within a slide, *Scale normalization* and *Quantile normalization* are usually performed to eliminate the experimental variance between multiple slides. *Scale normalization* is the simplest way to adjust the microarray data from multiple experiments to have a consistent median value or average value which is called a "baseline" [9]. *Quantile normalization* is commonly used in single-channel high-density oligonucleotide microarray e.g. Affymetrix GeneChip, which is developed based on a basic assumption that the gene expression abundances in all the samples are nearly identical [10]. Quantile algorithm is performed on logarithmized probe intensities and then transforms data from all samples to have an absolutely consistent distribution.

In processing of tiling microarray data, some normalization methods share the algorithms in traditional gene-level DNA microarray analysis, but in several aspects, some of them are improved or modified to meet the requirements of high-throughput calculation and take the unique feature of tiling microarray into consideration. The following flowchart (Figure 3) illustrate a proposed procedure while processing a batch of data from tiling array experiment and each issue will be separately discussed in details.

2. Tiling Array Raw Data Extraction and Organization

Unlike the traditional oligo or cDNA microarray which is usually constructed by Microarray printing machine, tiling microarrays are usually produced on *in situ* oligonucleotide synthesizer platform, such as Affymetrix, NimbleGen. The maximum advantage of *in situ* synthesized microarray is its huge capacity that a single slide could contain 200,000 to 2,000,000 short oligonucleotide probes and thus provides the minimum consumption of a set of chips but reach a decent coverage of the genome sequence. A set of tiling array experiments typically cost over a hundred of slides and generates raw microarray data in the amount of dozens to hundreds of gigabytes. A dataset of this size can hardly be processed slide by slide using graphics-interfaced software on a personal computer, and therefore several tiling array processing software provide a function of raw data extraction and organization as the initial step.

TileMap, which is a tool developed for Affymetrix tiling array analysis, includes an independent sub-program to import the raw data from Affymetrix arrays and convert multiple standard Affymetrix format (*.CEL) files into a single standard TileMap recognizable text file [11]. Another software called NMPP, which is developed for processing NimbleGen tiling array data, also provides a functional module for parsing and integrating raw data that automatically collects probe intensities from all chips in accordance with a microarray design file and an experiment description file. As a result, all the chips will be integrated into a TAB-delimited table as the only input file to NMPP pipeline [12]. This procedure largely reduces the redundancy of microarray outcome and increases the efficiency of batch processing in multiple operation systems.

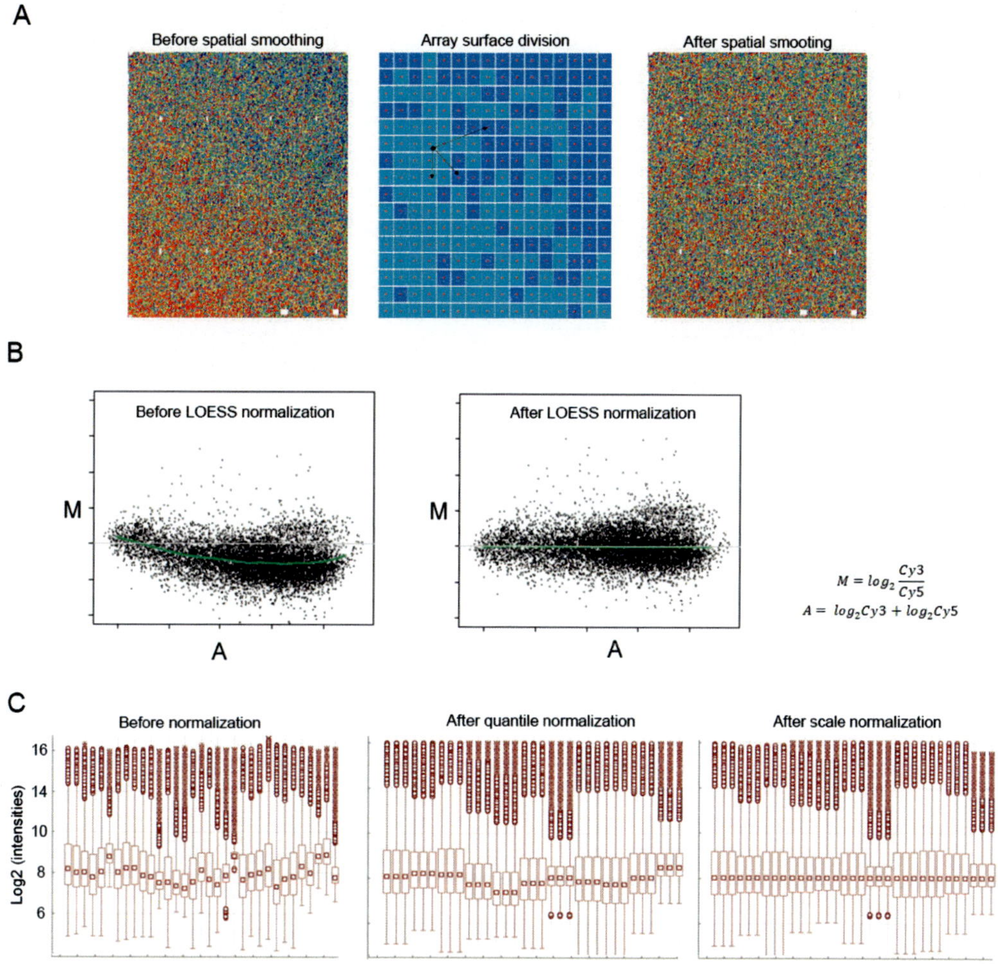

Figure 3. A standard tiling array data normalization procedure including four major steps: A). *Spatial smoothing normalization* to remove spatial bias. B) *LOESS normalization* to remove dye bias in dual-channel ChIP-on-chip experiments. C) Between-slide normalization including *Quantile normalization* and *Scale normalization* within replicates and between samples, respectively.

3. Spatial Normalization of Tiling Microarray

As the major within-slide experimental artifact on *in situ* synthesized microarrays, spatial effect means the unbalanced distribution of hybridization signals across the surface of microarray. From the scanned microarray image, a gradient change in signal from one area of the slide to the other is observed and this artifact is usually originated from uneven washing during microarray hybridization. Several methods have been proposed to correct spatial bias of tiling array before any other steps of normalization are carried out.

In a recent publication, Neuvial et al. defined two distinct types of spatial artifacts that were ubiquitously found in high density tiling array-based CGH data [13]. The first type of spatial bias is called "local spatial bias" which means some discrete area on the array surface present significant higher or lower signal than the global signals. The other type is "continuous spatial gradient" which means the entire array surface presents a continuous uneven trend of signal distribution from a part of slide to the other. They therefore proposed a spatial normalization method to characterize the trend of the spatial bias across an array's surface and subsequently correct the affected portions.

The heart of their method dealing with local spatial bias is a segmentation algorithm (*seg*) that consists of three major steps [13]. Firstly, the overall spatial gradient trend across the array surface is estimated by two-dimensional *LOESS* regression (*2dLoess*); secondly a so-called Neighborhood Expectation Maximization (NEM) algorithm is applied to segment the array surface into spatial areas among which each area has similar signal trend; thirdly, the regions affected by spatial bias is identified.

In most of the cases, the local spatial bias and continuous spatial gradients simultaneously exist on the same slide. Therefore, a combination of two algorithms (*seg+2dLoess*) is implemented in the software package MANOR in R environment: first the spatial segmentation algorithm to identify the potential areas of local spatial bias, and in the second step, the two dimensional LOESS regression is applied to correct the continuous spatial gradients [13]. This method is universal to all *in situ* synthesized high-density oligonucleotide microarrays and should be integrated in the processing of all forms of tiling array data.

To address the same issue on NimbleGen tiling arrays, software NMPP used a distance-weighted smoothing algorithm to eliminate the spatial artifact, which is similar with background correction procedure in Affymetrix MAS 5.0 [14, 15]. First, the entire array surface is segmented to *n* x *n* sub-squares (*n*=16 is the default), and then, a set of *n* x *n* weight factors for each probe are computed based on the distances from its position to the centers of all sub-squares (See formula A).

$$Weight_{i,j}(x,y) = \frac{1}{(x_i - x_j)^2 + (y_i - y_j)^2 + SF}$$

where $Weight_{i,j}$ is the weight factor of squared distance between oligo *i* to the center of zone *j*. and *SF* is a small *smooth factor* to avoid the denominator being zero, default $SF = 100$.

In the following step, the smoothing procedure is performed using a linear regression model as shown below:

$$\log_2(FG_i^{'}(x,y)) = \log_2(FG_i(x,y)) + \frac{1}{\sum_{J=1}^{J} W_{i,j}(x,y)} \sum_{J=1}^{J} W_{i,j}(x,y)\left\{\left[\log_2(FG_{baseline}) - \log_2(M(FG_{z(j)}))\right]\right\}$$

where $FG_i^{'}$ and FG_i is the smoothed and original foreground intensity of oligo *i*, respectively. $M(FG_{z(j)})$ is the median foreground intensity of zone *j*. $FG_{baseline}$ is the baseline intensity and $W_{i,j}$ is the distance weighted factor of oligo *i* to zone *j*. *x* and *y* is the coordinate of oligo *i* on the processed array. Figure 3 demonstrates the effect before and after spatial smoothing.

4. Background Correction of Tiling Microarray

The difference between spatial normalization and background correction is that the former is working on the unevenly distributed foreground intensities but latter is working on the overall background intensities. The purpose of background correction is to estimate the proportion of background noise and accordingly subtracted from the foreground signals. Background subtraction is essential for an appropriate comparison of gene expression or probe-level intensity comparison. For example, the observed expression intensity of a gene under condition *a* and *b* is I_a and I_b, and background noise for the two experiments are B_a and B_b. If the background noise is not considered, the measured expression ratio is I_a/I_b; however, if we subtract the background noise from the observed intensity, the expression ratio should be another value $(I_a - B_a)/ (I_b - B_b)$. For Affymetrix microarrays, there are no spiked-in probes as negative control for estimating the distribution of background noise and the proportion of the background noise must be computed from an overall distribution of probe signals.

Affymetrix Microarray Suite 5.0 provides a "Background Subtraction" function to estimate a local background value for each probe [15]. In MAS 5.0's algorithm, the first step is to divide the array into a number of equally spaced sub-zones and the lowest 2% of hybridization intensities within each zone is regarded as the background value. In the second step, the MAS calculate a weighting factor for each probe based on its distances to the centers of all divided sub-zones, and subsequently a weighted local background value is computed and assigned to each probe. In the third step, the local background noise for each probe is subtracted from the original intensity and a background-noise adjusted intensity is calculated for each probe. For those probes whose intensities are lower than the computed local background noise, a minimum noise value that is calculated from standard deviation of the probe intensities within sub-zone, will substitute probe intensity.

Academic software RMA (Robust Multi-array Average), which is developed for processing and analyzing Affymetrix microarray data, has proposed a totally different way to estimate the background noise from the information of all the PM (perfect match) probes on the slide, based on the assumption that the observed intensity *O* is composed of a normally distributed background noise *N* and an exponentially distributed signal *S*, and the parameters *u o* and *a* can be inferred from all PM probes on a slide [16].

$$O = N + S, N \sim N(\mu, \sigma^2), S \sim \text{Exp}(\alpha)$$

Given this model, the observed intensities can be adjusted by replacing them with the expected true signal $E(s|O=o)$

$$E(s|O=o) = a + b\frac{\phi\left(\frac{a}{\sigma}\right) - \phi\left(\frac{o-a}{\sigma}\right)}{\Phi\left(\frac{a}{\sigma}\right) - \Phi\left(\frac{o-a}{\sigma}\right) - 1}$$

where $\alpha = o - \mu - \sigma^2\alpha$, ϕ and Φ are the standard normal density and distribution functions, respectively [10]. This procedure ensures that no negative intensity values are produced since S cannot be greater than O which is reflected in the expected value $E(s|O=o)$ [16].

The custom designed microarray allows the user to include a considerable amount of negative control oligos that selected from unexpressed genomic sequence to construct a global background distribution. NMPP uses the same assumption that the background noise is normally distributed while the foreground signal is exponentially distributed as described in RMA [16]. However the background noise is not a typical normal distribution, and thus NMPP first makes a transformation of the original distribution of negative control oligos to exclude the false negative control who exhibit high hybridization intensities for unpredictable reasons [14]. By this means, a normal distribution of background noise is constructed and then a threshold at 95% confidence level will be calculated for each given tissue to determine a gene's transcription activity is "on" or "off".

5. Normalization across Replicate Microarrays

Parallel replicate hybridization is usually conducted during microarray experiment, but produces systematic artifacts mixed with real biological difference. To eliminate this kind of technical bias in a tiling microarray experiment, a so-called *Quantile Normalization* is commonly recognized to deal with high density oligonucleotide microarrays. Unlike the *LOESS Normalization* that works on the log-transformed ratio of Cy3 and Cy5, or the *Scale Normalization* that needs to choose an arbitrary baseline, *Quantile Normalization* algorithm assumes that the probe intensities have same distribution across all the replicate arrays, and will directly work on absolute intensity or the value after log-transformation in three major steps. Given an n columns multiplying p rows dataset X_{raw} from an experiment of n replicate arrays containing p probes for each, the first step is to sort each column according to the probe's intensities and get n sorted columns that form a new matrix X_{sort}; secondly, calculate an average or median value for each row in X_{sort} and assign the value to each probe in each column; thirdly, rearrange the probes in each column to their original order as in X_{raw}, and finally a normalized dataset X_{norm} is achieved.

Quantile Normalization has been proven to be applicable to all kinds of microarray design and to be robust in dealing with the non-biological variance resulted from replicate experiments, despite a little concern that it will also probably destroy or shrink the real biological relevance to a certain content. Some popular software integrate *Quantile Normalization* as the major option for processing the multi-array data, such as RMA, MBEI, TileMap, dChip, NMPP, and Affymetrix's Tiling Array Suit *etc.*

6. ChIP-on-chip Specific Normalization Using Input Control

One unique feature of NimbleGen tiling array products that are different from Affymetrix tiling array is that NimbleGen tiling array allows two samples simultaneously hybridized to one slide using Cy3 and Cy5 fluorescence labels. This feature makes it feasible to statistically identify the investigated genomic loci by comparing the chromatin-immunoprecipitated DNA versus the reference input as control, typically sonicated genomic DNA. A "mock" control is also commonly used to hybridize against the input DNA, in which the sample only contains IP dilution without the specific antibody. By this means, the background noise resulted from environmental factors could be largely reduced and the accuracy of identifying a true positive investigated genomic sites is increased.

To process the two-color ChIP-on-chip data, several standard methods for normalizing the gene expression microarray have been introduced in ChIP-on-chip experiments such as *LOESS Normalization, Quantile Normalization*. However some concerns have been raised due to the different property of ChIP-on-chip experiment and traditional gene expression array. For example, all the normalization methods developed for gene expression microarray are all based on the assumption that most of the genes' expression level remains unchanged and the difference of overall distributions of gene expression intensities in different samples is considered to be caused by non-biological factors. In a recent publication, Peng *et al.* demonstrated an asymmetric distribution of the log-ratios of ChIP value versus input DNA value to prove the impropriety to impose the standard gene microarray normalization methods on ChIP-on-chip data. Either *LOESS Normalization* or *Quantile Normalization* will force the signals from ChIP sample and control sample to follow a consistent distribution, but will eliminate the real variation in the experiments to some degree. Peng *et al.* presented a novel normalization which is specifically designed for ChIP-on-chip data and their method is proven to be both of great effectiveness on ChIP-on-chip experiments with or without "mock control" [17].

The optimal normalization procedure for two-dye ChIP-on-chip data requires three steps: first, using mock control data to separate the real hybridization signal from background. Comparing the log-ratio of ChIP-on-chip data versus the log-ratio of mock control data indicates that the shape of the mock control data could well describe the distribution of noise background and the mission of mock control normalization is to properly separate the bubbled portion, which is supposed to be the real binding signals. However, the traditional subtraction by establishing an arbitrary threshold line that parallels the diagonal will cause the loss of partial true positive signals as well as the import of false negative signals. Therefore, the optimal solution would be a weighted subtraction based on the control [17].

As we have introduced, a conventional way to eliminate the intensity-dependent dye-bias is to perform *LOESS Normalization* by fitting a regression throughout the body of data points. However, Peng *et al* has proven it's less effective to be working on ChIP-on-chip data and proposed a new solution by performing the line fitting on the "first order differences" of the intensities based on the chromosomal position instead of the original data, because of the less differences in values between the neighboring probes [17]. The outcome of the chromosomal position-based data transformation is a rotation of the data points, which resulted in a roughly symmetric distribution centering at M=0 line. The angle of the rotation is determined by the slope of sigma (M) versus sigma (A). After the rotation done, the following step is a standard *LOESS Normalization* based on data value to remove the remaining systematical bias. The authors suggested the combinatory procedure of rotation and *LOESS Normalization* is also effective in dealing with between slide variations.

7. Array-CGH Data Normalization

For a typical array-CGH experiment, it's usually conducted on dual-channel microarray platform that the test samples and reference samples are labeled by two fluorescence for each spot and the ratio values is calculate to infer the "single copy gains and losses". Calum

MacAculay's group defined four sources of systematic variations, namely Spatial bias, Plate bias, intensity bias and background bias (subtracted before taking log ratio) , included in array CGH data, and a procedure of three-step normalizations is proposed to remove the biases [18].

As we I have discussed previously, the first kind of variation, "spatial bias", which is supposed to be caused by different efficiency of each print-tip or uneven washing during the hybridization experiments will be removed by the *Spatial Normalization* as the first step. The spatial trend firstly estimated from the neighboring probes within *n* rows by *m* columns square and then the biased part is subtracted from the original value of each spot. The second type of variation, "plate bias" is originated from difference of micro-plates because different clones produced in different micro-plates may have experienced slightly different physical conditions during PCR or in subsequent purification. Plate bias is estimated from the log ratios of the spots within each group and will be subtracted from "Spatial" normalized values. "Intensity bias" is similar with the intensity-dependent dye-bias as we mentioned in two-color gene expression arrays and could be clearly identified by MA plots. The traditional way is to perform *LOESS Normalization* as the third step to balance the intensities from red and green channel.

Statistical Models in Tiling Array Analysis

Since the invention of tiling arrays, many computational biologists have been dedicated to build up suitable statistical models for better interpreting the tiling array data and mining meaningful biological relevance. Some aspects are mainly focused on how to properly identify the interrogated genomic regions that are usually represented by the probe clusters of enriched hybridization signals, or called "peaks". To discriminate the significant peaks of signal probes from the background noise of non-hybridized probes, several state-of-the-art mathematical models have been introduced into tiling array analysis. Among them, Wilcoxon Signed-Rank Test and Hidden Markov Model (HMM) are two most representative methods that are used by Affymetrix tiling array analysis suite (TAS) [19] and some of other classical academic software. In this section, our discussion of statistical inference of interested genomic regions by tiling array is mainly around three frequently used methods.

1. Pair-Wise Test in a Scanning Mode

A standard ChIP-tiling array experiment usually contains two samples, a ChIP-enriched DNA sample and input genomic DNA sample as control, which are labeled with different dye and hybridized to the same array (NimbleGen array) or hybridized to two arrays (Affymetrix array). The most direct way for judging the difference of the hybridization intensities of a probe between two samples is to calculate a ratio of ChIP-enriched sample versus control sample and use the fold change to evaluate whether a probe is hybridized with ChIP-DNA or not. However, this method is lacking of statistical sense and always has problem in defining a threshold and combining neighboring probes to form an interested region. Therefore, a simple and more statistical solution for judging a designated genomic region being hybridized or not is to perform a pair-wise test within a certain length of moving bin along the chromosome.

This strategy was adopted by Keles et al. [20]. The first step of their method is to perform a one-side Welch's *t-test* for each probe *i* on N_1 replicates of ChIP-enriched sample 1 and N_2 replicates of control sample 2, as the following formula denoted:

$$T_i = \frac{\overline{X}_{i,1} - \overline{X}_{i,2}}{\sqrt{\frac{\sigma_{i,1}^2}{N_{i,1}} - \frac{\sigma_{i,2}^2}{N_{i,2}}}}$$

Where $\overline{X}_{i,1}$ and $\overline{X}_{i,2}$ are probe *i*'s average $\log_2$ ratio of ChIP sample 1 and control sample 2, respectively; $\sigma_{i,1}^2$ and $\sigma_{i,2}^2$ are the variances for probe *i* across the i$N_{i,1}$ and $N_{i,2}$ replicates in two samples; and T_i is the Welch t-statistics for probe *i*.

In the second step, an averaged t-statistics within a sliding window is calculated and called scan-statistics, as the following formula shown:

$$T_i^* = \frac{1}{w} \sum_{i=i}^{i+w-1} T_i$$

Where T_i^* is denote as *scan-statistics* which is the mean of the T statistics of *w* probes within a window starting from the ith probe.

However, Keles's method allows only two sample's comparison and each sample must have multiple replicates. To resolve this problem, the authors of TileMap, Ji and Wong, developed a similar method called Moving Window Average (MA) to combine the signals from neighboring probes [11], and the average t-statistics m_i is calculated as:

$$m_i = \frac{1}{(2w+1)} \sum_{k=i-w}^{i+w} T_k$$

Where m_i is the t-statistics for i^{th} probe averaged from $2w+1$ probes within a moving window.

The difference between TileMap and Keles's method is that TileMap uses a modified formula to compute the t-statistics, and in this way, the multiple sample comparison is allowed.

2. Wilcoxon Signed-Rank Test

Wilcoxon signed-rank test is a non-parametric hypothetical testing for measuring the significance of the difference of two related and paired samples [21]. Unlike the pair-wise *t-test* that requires the data to fit a normal distribution, Wilcoxon signed-ranked test is an alternative choice for the data whose distribution is unknown and quantity is low. As its name indicated, Wilcoxon test uses the ranks of the value instead of the original value as the *t-test* does, by this way it could largely reduced the influence of outlying data points, which causes biased and larger variances during the *t-test*.

Affymetrix integrated Wilcoxon test to its Microarray Analysis Suite (MAS) for examining the changes in gene expression level between two samples from a set of PM and MM 25-mer probes within one gene, usually 20 pairs, instead of only one probe's information across limited replicates in traditional gene array. Obviously, the Wilcoxon test from summarized information of multiple probes and their background noise could be more robust and greatly increase the accuracy of the comparison. Affymetrix further introduced this method into tiling array analysis and its first application was presented in a mapping of three transcription factors (Sp1, cMyc, and p53) along human chromosomes 21 and 22 by ChIP-tiling array assays [22]. In this work, the Wilcoxon test was performed on the probe set within ±500 bp range across an probe, who hybridization intensity has been normalized and transformed by $\log_2(\max(PM\text{-}MM,1))$. As a result, a P value will be calculated and assigned to the probe in the center. Wilcoxon test is carried out in a one-sided manner, because the examination only interests in the cases that the values in treated sample are positively higher than those in control sample and furthermore, the lower P value than a designated threshold will reject the null hypothesis that the two populations are equal. Hereby, we will discuss the details of Wilcoxon test algorithm based on the introduction in *Affymetrix Statistical Algorithm Document* for their software [15].

Assuming we have n probes within a 1000 bp sliding window under two set of experiments, one is from control sample $b_i = (b_1, b_2, \ldots, b_n)$ and the other one is from treated sample $e_i = (e_1, e_2, \ldots, e_n)$. Both b_i and e_i are the difference of a pair PM_i and MM_i probes calculated from $\log_2(\max(PM_i\text{-}MM_i, 1))$. The first thing is to calculate an absolute difference $|d_i|=|e_i\text{-}b_i|$ for each probe and then sort $|d_i|$in increasing order. If all the $|d_i|$ within the 1000 bp window are zero, a P value of 0.5 will be output for the central probe. Secondly, assign ranks to $|d_i|$ and if two $|d_i|$ are equal, assign a tied rank for each of them. Thirdly, set the ranks back in the original order as the probes in the genome and assign to each such rank a "+" sign when $e_i\text{-}b_i > 0$ and a "-" sign when $e_i\text{-}b_i < 0$. So we have signed ranks S_i for n probes in a sliding window and the sum of S_i is calculated by $S = \sum_{i=1}^{n} S_i$. The next step is to compute the p value for the probes in a designated region in two ways and the first way is for the region that contains less than 12 probes.

$$p(S) = 2^{-n} \sum_{i-1}^{2^n} u(S_i > S) + 0.5u(S_i = S)$$

Where $n < 12$ and u is a logical expression: when $S_i > S$, $u = 1$ otherwise $u = 0$; and if $S_i = S$, $u = 1$ otherwise $u = 0$. And 2^n is all possible results of assigning ranks for n probes.

In another words, this equation means for all signed ranks higher than S, we add a weight of 1 and those equal to S, we add 0.5, and the final ratio of the added weights versus the number of all possible signed ways is the final P value, which is between 0 and 1.

Finally we could use the P value to compare with a designated threshold (i.e. 0.05, 0.01, 0.001) to determine whether we reject the null hypothesis or not. If there is no significant difference between e_i and b_i, the P value should be around 0.5; if the e_i is significantly higher than b_i, the *P* value should be close to 0 and the lower of *P* value, the higher significance of the difference for $e_i > b_i$ in one-sided manner.

For the number of probe pairs is greater or equal to 12, Affymetrix gives a different way to calculate the P value based on the assumption that S follows a standard normal distribution with mean being 0 and variance being 1, as shown below.

$$p(S) = 1 - f(S')$$

Where S' is calculated as:

$$S' = \frac{S - \frac{n(n+1)}{4}}{\sqrt{n(n+1) * \frac{2n+1}{24} - vt}}$$

Where vt is a term modifying the variance for ties and calculated as:

$$vt = \sum_{k=1}^{t} \frac{b_k(b_k^2 - 1)}{48}, k = 1{:}t$$

Where *t* is the number of tie groups; b_k is the number of ties in the tie group *k*.

Affymetrix has integrated the Wilcoxon one side signed-rank test in their tiling analysis software TAS [19] as a default method to identify the genomic regions of interet or the ChIP-enriched regions.

3. Hidden Markov Model (HMM) Inference

HMM, as a powerful statistical model, has been successfully applied to model consecutive events in many area of computational biology, such as prediction of gene structure, detection of core motif in regulatory sequences, identification of homologous proteins, discovery of non-coding RNA based on structural alignment and many other aspects of biological pattern recognition [23].

A standard hidden Markov process describes a consecutive event with two fundamental elements, of which one is *observed sequence* and the other is an invisible *state path* that may continually transit from one state to another. The challenge of building HMM model is to determine the underlying hidden states from the observed sequence [23]. Regarding tiling array experiment, the *observed sequence* represents the tiling probes with observed values of hybridization intensities, while the hidden states may represent the interrogated genomic regions of interest but the specific meanings may vary depending on the design and hybridized samples in the microarray experiments. In transcriptional activity monitoring tiling assay, the *state path* is composed of transcribed regions (exons, UTRs) and non-transcribed regions (Intergenic DNA, introns). In ChIP-Chip assay, the *state path* is composed of "non-enriched regions" and "ChIP-enriched regions" that may represent transcription factor binding sites, methylated genomic DNA or histone modification regions.

Besides the two above elements, an HMM model also contains *initial probabilities* indicating the chance of a certain state occurs first at the start of a hidden Markov chain; *transition probabilities* indicating the chance of moving from current state to a new state; and *emission probabilities* indicating the probability distribution for each state own. Therefore, a full HMM model λ including five parameters could be described as [24]:

$$\lambda = (\Omega_X, \Omega_O, A, B, \pi)$$

Where $\Omega_X = \{q_1,...,q_N\}$ represents the set of N possible states
$\Omega_O = \{v_1,...,v_M\}$ represents the set of M observed sequence
$A = \{a_{ij}\}, a_{ij} = p(X_{t+1} = q_j \mid X_t = q_i)$ denotes the transition probabilities
$B = \{b_{ik}\}, b_{ik} = p(O_t = v_k \mid X_t = q_i)$ denotes the emission probabilities
$\pi = \{\pi_i\}, \pi_i = p(X_1 = q_i)$ denotes the initial probabilities

Basically, the fundamental outcome of an HMM model is to generate the best hidden state path Ω_X with the highest probability from the observed sequence Ω_O and all probability parameters A, B and π.

In a recent publication, Li et al. proposed an HMM approach for analyzing the ChIP-on-chip data to identify the *p53* binding sites in human genome [25]. They assumed the path has two hidden states, *ChIP-enriched state* and *non-enriched state* and among the K probes in total, there are J probes associated with *p53* binding regions. Consequently, the initial probabilities could be defined as J/K for ChIP-enriched state and 1-J/K for non-enriched state; and the transition probabilities could be defined as J/K for moving from one state to new state and 1-J/K for staying in current state. The emission probabilities was defined based on an empirical distribution: $N\ (\mu_i + 2\sigma_i, (1.5\sigma_i)^2)$ for ChIP-enriched state while $N\ (\mu_i + \sigma_i^2)$for non-enriched state.

However, another publication immediately following Li's work introduced a more advanced tool – *TileMap* which was also developed based on HMM inference but more features were integrated to compensate the limitations in Li's method and other previous work [11]. Compared with Li's HMM model, the greatest advantage of *TileMap* is that the author utilized Hierarchical empirical Bayes model to estimate the necessary parameters for HMM model and at the same time computed probe-level test-statistics for each oligo in its first step. In the second step, *TileMap* conducts HMM inference on each probe's test-statistics instead of the normalized hybridization intensity. Another important innovation in TileMap is a process called Unbalanced Mixture Subtraction (UMS) that provides the function of modeling HMM parameters and controls the false discovery rate (FDR) during the multiple-sample comparison [11].

Besides the use of HMM model in detection of binding sites or epigenetic modification region from ChIP-tiling array experiments, some other computational biologists have shown their effort in using HMM model to score and characterize the transcriptional structures, especially for those unknown transcription units who are lacking of typical structure and cannot be predicted merely based on nucleotide sequences. Toyoda et al presented a sophisticated HMM model to predict the structure of unknown genes from Arabidopsis transcription tiling array, and their work was implemented as program called ARTADE (http://omicspace.riken.jp/ARTADE/) [26]. In their HMM model, they defined three hidden states of the Markov chain, namely exon-state, introns-state and Intergenic-state, and the procedure was carried out in two main steps: firstly all possible transcribed regions with significant tiling array signals were identified by a binomial distribution test; secondly, a HMM model combined array signals with nucleotide information was applied on both strands to estimate the best exon/intron structure.

Another work from Stjernqvist et al. shows the ability of using HMM model to analyze the array-CGH data. Their method is called continuous-index hidden Markov model and Monte Carlo EM algorithm was used to estimate the necessary HMM parameters [27].

TILING ARRAY DATA ANALYSIS SOFTWARE AND RESOURCES

1. Affymetrix Tiling Analysis Software

Affymetrix Company has been dedicated to develop compatible software platform to facilitate the best use of its tiling microarray products. Tiling Analysis Software (TAS) version 1.1 could be freely accessed and provides the maximum capability to process and analyze tiling array data. The TAS workflow encloses several functional modules that meet all the requirements for a typical tiling array analysis procedure. The original input files for TAS are *.CEL files in which each CEL file contains probe intensities from a scanned image of a hybridized array, and the other one are BPMAP (Binary Probe Map) files in which each probe's X-Y coordinate on the array has been corresponded to its original genomic position. To start tiling array analysis project, a TAG file (Tile Analysis Group) will be created to contain the all required information, such as specified CEL data files and BPMAP file, normalization options and other necessary parameters. The following part introduces some selected important function modules.

Normalization After an analysis project has been initiated, all the arrays selected in the first step will be Quantile normalized across all the probe's intensities and the user is allowed to select normalization performed across all the groups at one time or normalization performed group by group.

Probe Analysis To do the probe-level analysis, a *Bandwidth* parameter should be firstly setup to determine the size of a sliding window. Then, the intensities from all the probes within each 2*BW+1 will be used for computing a signal value and a P-value assigned to the central probe of the window. Selection of bandwidth size depends on the type of experiments and the resolution of tiling array. For a 35-bp tiling array hybridized with mRNA transcripts, the suggested bandwidth size is 50 bp which is roughly equal to the half of the average length of exons. For ChIP-on-chip histone modification analysis, the *bandwidth* could be adjusted to

~500, since the DNA fragments in ChIP products are usually around 1 Kbp in length. The yielded result of probe signals and P values will be stored in *.BAR files and *.CHP files

Interval Analysis This part of analysis in TAS aims at summarizing the interested genomic regions based on the probe signals and P-value computed from *Probe Analysis* and the *.BAR files are used as input. Three important parameters should be determined for a proper characterization of a detected genomic region. The first parameter is a P value threshold that TAS uses 0.001 and 0.00001 as default to discriminate the high and low significant probes. Another two parameters *max_gap* and *min_run* should be setup to determine the smallest distance of two neighboring probes should be merged as one region and how small of a detected interval is allowed to be considered as a possible candidate region, respectively. Like the Bandwidth in probe analysis part, determination of a proper *max_gap* and *min_run* largely depends on the type of tiling array assay. For example, in histone modification data analysis, ~150 bp, the length of DNA fragment wrapped on a histone could be considered as a minimum unit for choosing the *max_gap* and *min_run* parameters. As a final result, the detected regions will be exported to a *.BED file containing the start and stop positions of each computed genomic interval. BED file can be used as input for Affymetrix's Integrated Genome Browser and UCSC's genome browser to visualize the tiling array signal simultaneously with annotated genome information.

*Statistical Testing*TAS performs statistical testing on judging the significant level of a the probe set within each sliding window in two ways, one-sample analysis and two-sample analysis. One-sample analysis procedure is suitable for one channel experiments while both PM and MM information would be necessarily considered. TAS will conduct a Wilcoxon signed-rank test on PM and MM oligo pairs within the sliding bin and a pseudo-median value will be calculated as the estimated signal intensity. Two-sample analysis is suitable for PM-only analysis and the Wilcoxon signed-rank test will be carried out on a pair of treatment sample versus control sample.

TAR can finally derive a report containing all detected genomic regions and generate a summary of statistics for a tiling analysis project. TAS's outcome is readily imported into Affymetrix's Integrated Genome Browser (IGB) tool for visualization against genomic annotations.

2. Integrated Genome Browser (IGB)

Integrated Genome Browser is an online application provided by Affymetrix to visualize and explore the tiling signals mapping with the updated genome annotation [27]. IGB is developed by Java in order to meet the requirements of multiple operation system and Java 2 Runtime Environment (JRE) and Java 2 Standard Edition SDK (J2SE SDK) must be installed before running the application under windows.

Launching IGB program will initially open a blank window and users need to load the specified genome annotation data from DAS server (Distributed Annotation Sever) which is maintained by NetAffy Analysis Center from Affymetrix. Besides the NetAffy provided genome information, IGB is also a powerful tool to display other format of annotation files including GFF format, PSL format and BED format generated by different annotation resources. To graphically display the tiling array signals, *.BAR file or *.CHP from the TAS software should be specified and IGB provides several panels with an array of parameters to manipulate the display of graphs for a better view effect.

Another convenient feature of IGB is its more flexible interaction with other popular genome browser, which largely expands its capability to explore genome resource, such as UCSC Genome Browser (http://genome.ucsc.edu) and GBrowser from the Generic Model Organism Database, which could be installed on local computer station.

3. TileMap

Another well acknowledged academic tiling array analysis tool package is called *TileMap* which is developed by Ji and Wong from the Department of Statistics in Harvard and Stanford University, respectively [11]. As we have introduced in previous part, TileMap basically utilizes hidden Markov Model (MA) to infer the interested genomic regions from the tiling array signals. Other basic functions of TileMap includes *Quantile normalization*, repetitive sequences filtering, probe-level summary and statistical inference of enriched genomic loci. TileMap is developed by C and can be run under UNIX system by a command-line based interface. TileMap package is composed of four individual programs:

*tilemap_importaffy*is used to extract the probe intensities from Affymetrix data files. Version 3 of *.CEL files and *.BPMAP files are needed as standard input. During the import, the user is also allowed to choose whether filter the repeat sequences or not.

*tilemap_norm*is used to perform *Quantile normalization* or the normalization procedure could be also included in *importaffy* program.

tilemap is the main program that fulfills the identification of signal peaks by two steps, firstly compute the probe-level statistics based on a hierarchical empirical Bayes model, and secondly infer the interested genomic regions by applying HMM or Moving Window Average (MA). There are three major output files after *tilemap* processing done: *.sum files store the statistics for each probe; *.bed files store the detected regions and *.reg files store the sorted regions according to their significant level. *.bed files are also compatible to UCSC genome browser for visualization.

tilemap_extract is used to extract probe-level statistics of a specified region and to store the data in tab-delimited text file.

Since the *tilemap* programs are designed to run in command line mode, the user defined parameters must be setup before the execution. *.cmpinfo file is edited to indicate primary experiment design and how to do the comparison. Importaffy_parameter_file and norm_parameter_file is edited to specify the normalization procedure. And the tilemap_parameter_file is edited to configure the procedure how to carry out a statistical identification of interested genomic regions based on normalized tiling array signals. The last extract_parameter_file is used to store the parameters for extracting the data for visualization.

4. Model Based Analysis of Tiling Array (MAT)

Liu's group from Harvard School of Public Health developed a "Model based Analysis of Tiling array" (MAT) tool for analyzing the Affymetrix tiling array data for ChIP-on-chip experiments [28]. One unique feature of MAT is that, besides the previous described factors involved in processing tiling array, it also takes probe sequence composition and array copy number into account. MAT also provides a function to calculate the false discovery rate (FDR) and generate adjusted P-value.

MAT is developed by Python (http://chip.dfci.harvard.edu/~wli/MAT) and runs under multiple operation systems. Besides the standard Affymetrix *.CEL data files and *.BPMAP genome annotation files, MAT also requires a RepeatMasker outcome file containing the chromosomal coordinates of the repetitive sequences. MAT exports a similar result format as TAS does, a *.BAR file that can be loaded into Affymetrix's Integrated Genome Browser for visualization and a *.BED file that stores the identified genomic regions of interest and is also compatible to UCSC Genome browser.

5. NMPP NimbleGen Microarray Processing Pipeline

NMPP package is a bundle of user-customized tools that helps the researchers to process self-designed NimbleGen tiling microarray data, which was developed by Perl [12]. NMPP has been compiled to be executable programs running on Windows MS-DOS mode and any version of Linux system. As we said earlier, NMPP works in an integrative processing manner: namely, to collect a large amount of raw expression data from multiple microarray experiments, to visualize array surface probe signal distribution, to perform single-array background normalizations and multi-array normalizations. NMPP package is composed of seven sub-programs, of which the following introduced four programs responsible for basic tiling array data preprocessing while other three are related to expression analysis for traditional gene-level arrays:

ParseRawdata program reads multiple microarray experiments information from an "experiment description file" and an "oligo design file". All the probe intensity data from all the arrays will be extracted and integrated into a single TAB-separated output file, in which each array occupies a column.

SpatialSmooth program is used to remove array surface spatial variation introduced from uneven washing or scanning during the hybridization experiments. User needs to define a number of the division of the entire array surface, which is 256 as the default number, but the more the zone numbers the longer time the smoothing procedure will cost.

SpatialDraw program is developed to visualize the array spatial effect by Scalable Vector Graphics (SVG image) Perl library before and after the smoothing procedure. Similar with the *SpatialSmooth* program, the *SpatialDraw* program also needs a user-defined parameter to initially divide the array surface into *n* zones (default is 256).

Quantile normalization is used to remove between-slide variation. NMPP provides two ways to carry out the procedure, one-step *Quantile normalization* and the two-step *Quantile normalization*. The one-step way is to normalize all the chips at one time, enforcing the distributions of all the slides to be absolutely the same, which is a more robust method that removes all the variations during replicate experiments and the differences from the RNA sample preparation, but at the expense that some real biological differences might be smoothed to a certain content. The latter method divides the procedure into two steps: first, perform *Quantile normalization* on replicate slides within each tissue, and then perform global normalization to adjust all tissues to an average baseline.

REFERENCES

[1] The ENCODE Project Consortium Identification and analysis of functional elements in 1% of the human genome by the ENCODE pilot project. *Nature*. 2007 Jun 14;447(7146):799-816.

[2] Li L, Wang X, Sasidharan R, Stolc V, Deng W, He H, Korbel J, Chen X, Tongprasit W, Ronald P, Chen R, Gerstein M, Wang Deng X. Global identification and characterization of transcriptionally active regions in the rice genome. *PLoS ONE*. 2007 Mar 14;2(3):e294

[3] Mockler TC, Chan S, Sundaresan A, Chen H, Jacobsen SE, Ecker JR. Applications of DNA tiling arrays for whole-genome analysis. *Genomics*. 2005 Jan;85(1):1-15.

[4] http://www.repeatmasker.org/

[5] Houldsworth J, Chaganti RS. Comparative genomic hybridization: an overview. *Am J Pathol*. 1994 Dec;145(6):1253-60.

[6] Bertone P, Trifonov V, Rozowsky JS, Schubert F, Emanuelsson O, Karro J, Kao MY, Snyder M, Gerstein M. Design optimization methods for genomic DNA tiling arrays. *Genome Res*. 2006 Feb;16(2):271-81.

[7] Gräf S, Nielsen FG, Kurtz S, Huynen MA, Birney E, Stunnenberg H, Flicek P. Optimized design and assessment of whole genome tiling arrays. *Bioinformatics*. 2007 Jul 1;23(13):i195-204.

[8] Smyth GK, Speed T. Normalization of cDNA microarray data. *Methods*. 2003 Dec;31(4):265-73.

[9] Yang YH, Dudoit S, Luu P, Lin DM, Peng V, Ngai J, Speed TP. Normalization for cDNA microarray data: a robust composite method addressing single and multiple slide systematic variation. *Nucleic Acids Res*. 2002 Feb 15;30(4):e15

[10] Bolstad BM, Irizarry RA, Astrand M, Speed TP. A comparison of normalization methods for high density oligonucleotide array data based on variance and bias. *Bioinformatics*. 2003 Jan 22;19(2):185-93

[11] Ji H, Wong WH. TileMap: create chromosomal map of tiling array hybridizations. *Bioinformatics*. 2005 Sep 15;21(18):3629-36.

[12] Wang X, He H, Li L, Chen R, Deng XW, Li S. NMPP: a user-customized NimbleGen microarray data processing pipeline. *Bioinformatics*. 2006 Dec 1;22(23):2955-7.

[13] Neuvial P, Hupé P, Brito I, Liva S, Manié E, Brennetot C, Radvanyi F, Aurias A, Barillot E. Spatial normalization of array-CGH data. *BMC Bioinformatics*. 2006 May 22;7:264.

[14] *Affymetrix® Microarray Suite software* (MAS 5.0)

[15] *Statistical Algorithms Description Document*, 2002 Affymetrix, Inc.

[16] Irizarry RA, Hobbs B, Collin F, Beazer-Barclay YD, Antonellis KJ, Scherf U, Speed TP. Exploration, normalization, and summaries of high density oligonucleotide array probe level data. *Biostatistics*. 2003 Apr;4(2):249-64.

[17] Peng S, Alekseyenko AA, Larschan E, Kuroda MI, Park PJ. Normalization and experimental design for ChIP-chip data. BMC Bioinformatics. 2007 Jun 25;8:219

[18] Khojasteh M, Lam WL, Ward RK, MacAulay C. A stepwise framework for the normalization of array CGH data. *BMC Bioinformatics*. 2005 Nov 18;6:274.

[19] *Affymetrix Tiling Analysis Software* - Version 1, Build 15, 12/23/2005

[20] Keleş S, van der Laan MJ, Dudoit S, Cawley SE. Multiple testing methods for ChIP-Chip high density oligonucleotide array data. *J Comput Biol*. 2006 Apr;13(3):579-613

[21] Wilcoxon, F. (1945). Individual comparisons by ranking methods. *Biometrics*, **1**, 80-83

[22] Cawley S, Bekiranov S, Ng HH, Kapranov P, Sekinger EA, Kampa D, Piccolboni A, Sementchenko V, Cheng J, Williams AJ, Wheeler R, Wong B, Drenkow J, Yamanaka M, Patel S, Brubaker S, Tammana H, Helt G, Struhl K, Gingeras TR. Unbiased mapping of transcription factor binding sites along human chromosomes 21 and 22 points to widespread regulation of noncoding RNAs. *Cell*. 2004 Feb 20;116(4):499-509

[23] Eddy SR. What is a hidden Markov model? *Nat Biotechnol*. 2004 Oct;22(10):1315-6. Review

[24] Eddy SR. Profile hidden Markov models. *Bioinformatics*. 1998;14(9):755-63. Review

[25] Li W, Meyer CA, Liu XS. A hidden Markov model for analyzing ChIP-chip experiments on genome tiling arrays and its application to p53 binding sequences. *Bioinformatics*. 2005 Jun;21 Suppl 1:i274-82

[26] Toyoda T, Shinozaki K. Tiling array-driven elucidation of transcriptional structures based on maximum-likelihood and Markov models. *Plant J*. 2005 Aug;43(4):611-21

[27] Stjernqvist S, Rydén T, Sköld M, Staaf J. Continuous-index hidden Markov modelling of array CGH copy number data. *Bioinformatics*. 2007 Apr 15;23(8):1006-14. Epub 2007 Feb 19

[28] *Affymetrix Integrated Genome Browser* (IGB)

[29] Johnson WE, Li W, Meyer CA, Gottardo R, Carroll JS, Brown M, Liu XS. Model-based analysis of tiling-arrays for ChIP-chip. *Proc Natl Acad Sci U S A*. 2006 Aug 15;103(33):12457-62

In: Oligonucleotide Array Sequence Analysis
Editors: M.K. Moretti and L.J. Rizzo, pp. 175-207

ISBN: 978-1-60456-542-3

Chapter 6

APPLICATION OF MICROARRAY IN RAT MODEL OF ALCOHOLIC AND NONALCOHOLIC LIVER DISEASE

Meena R. Sharma[1*], Mahesh C. Sharma[2] and Amin A. Nanji[3]
[1] School of Medicine, University of Pennsylvania, Philadelphia PA 19104 and
[2]Department of Surgery, Drexel College of Medicine, Drexel University
Philadelphia PA 19104
[3] Dalhousie University School of Medicine, Halifax, Canada B3H 1V8

ABSTRACT

Alcoholic and nonalcoholic liver diseases are increasingly recognized conditions that may progress to end-stage liver disease. Nonalcoholic liver disease (NAFLD) and alcoholic liver disease (ALD) share similar pathological characteristics despite the etiological difference between two diseases. In both diseases, the first manifestation of injury is the accumulation of fat within hepatocytes (steatosis), followed by the development of necrosis and inflammation that leads to cirrhosis. Although NAFLD is a metabolic disease associated with obesity and insulin resistance, it is impossible to distinguish alcoholic from nonalcoholic steatohepatitis on histological grounds. This suggests that common pathogenic mechanisms are involved in both conditions. In addition to alcohol consumption, genetic factors, dietary fatty acids, gender and viral infection potentiate the severity of alcoholic liver injury. A high fat diet in rat and mice models reproduces many of the features of NAFLD found in humans. Female rats fed with fish oil and dextrose developed fatty liver, necrosis and inflammation. In an intragastric infusion rat model of ALD, saturated fatty acids (medium chain triglycerides) are protective, whereas unsaturated fatty acids (palm oil and fish oil) promote ALD. The rats fed with palm oil and ethanol developed only fatty liver, whereas rats fed with fish oil and ethanol developed fatty liver, severe necrosis and inflammation. In ALD and NAFLD, female rats developed more severe liver injury than males. In humans, significant gender differences in susceptibility to ALD are observed. The proposed molecular mechanism through which polyunsaturated fatty acid and ethanol develops pathogenesis includes enhanced lipid peroxidation-induced oxidative stress, mitochondrial dysfunction, endotoxin, and proinflammatory cytokines. To understand the effect of dietary fatty acids and gender on the development of more severe

* E-mail address: prasunsh@mail.med.upenn.edu. Phone: 215.823.4439. FAX: 215.823.5994. Correspondence to: Meena R. Sharma, Ph.D., School of Medicine, University of Pennsylvania, A-414 VA Medical Center, Woodland Avenue, Philadelphia, PA 19104

ALD and NAFLD in female rats, we performed affymetrix microarray based expression profiling of genes. A large number of genes showed significant changes in female livers compared to males. We employed EASE and Ingenuity computational pathway analysis tools to analyze the microarray data. Ingenuity computational pathway analysis tool identifies specific regulatory networks of genes operative in promoting liver injury. These networks allowed us to identify genes involved in different signaling pathways for a better understanding of the role of cytokines in ALD in response to different types of fatty acids, we used inflammatory cytokines pathway specific superarray. Many cytokine genes underwent significant changes in the liver of rats fed with palm oil plus ethanol, and fish oil and ethanol, as compared to animals fed with medium chain triglycerides and ethanol. Genes listed in this article that significantly increased or decreased (expression two-fold or more) were assigned to pathological functional groups and reviewed in relevance to established hypotheses of NAFLD and ALD mechanisms of action in female liver injury.

Key Words: alcoholic liver disease, nonalcoholic liver disease, Ingenuity computational pathway, biological theme.

Overview of Chapter

The discovery of the human genome has demonstrated its implications in biological research. High-performance techniques were needed to define the biological functions of these newly identified genes. Microarray technology allows the investigators to screen thousands of genes at one time, providing a comparative genes profile of the diseased and normal tissue (Brown and Botstein, 1999; Pollack et al., 1999). Microarray experiments are replacement of conducting thousands of northern blots, therefore avoiding variability among experiments. Different types of microarrays have been used in a number of studies to identify the global gene expression pattern and to elucidate the mechanisms of different disease (Schena et al., 1995).

DNA Microarray

Varieties of DNA microarray experiment is a multistep process requiring specific equipment, and expertise in many areas including molecular biology, image analysis, computing, and statistics. Microarray analysis has a broad range of applications in biological field that involve different types of probes and/or targets.

A DNA microarray consists of an orderly arrangement of thousands of DNA fragments, representing the genes of an organism, immobilized on a solid support. The solid support can be a glass slide or membrane. The fragments of DNA are called as probes. Robotic spotters are used to place thousands of spots on one slide, allowing biologists to analyze virtually every gene present in a genome.

To assess the gene expression, RNA is synthesized to labeled cDNA (complementary DNA) or cRNA. The labeled target cDNA or cRNA is hybridized to the immobilized probes on a chip. The labeled target binds to specific gene spot and is detected by chemiluminescent, fluorescent, or radioactive methods. The signal produced at each spot represents the expression level of message in the RNA sample. Two major types of commercial microarrays, oligonucleotide arrays (Affymetrix) and cDNA arrays (BD Biosciences and others), are available for many species.

Oligonucleotide Arrays

Oligonucleotide arrays also known as GeneChip (Affymetrix) are comprised of 25 base pair long sequences. These DNA sequences are selectively designed to have little or no cross-reactivity with other genes so that non-specific hybridization will be minimized. To avoid non-specific hybridization, the perfect Match/Mismatch (PM/MM) probe strategy is used. In this strategy, a second probe, that is identical to the first except for a mismatched base at its centre, is placed next to the first probe. Any background hybridization with the MM probe is subtracted from the PM probe signal.

cDNA Arrays

In the cDNA array, probes are large pieces of DNA that are complementary to the genes of interest. cDNA probes can be generated either from a commercially available cDNA library or alternatively, specific genes from genomic DNA can be amplified using PCR, representing the entire genome of an organism on the array. cDNA probes can be mechanically spotted onto a glass slide, normally in duplicate to serve as controls.

Comparison of Arrays

cDNA arrays do not need probe design which is required for oligonucleotide arrays. In the cDNA array, the entire genome of an organism can be represented on the array. This makes cDNA arrays more useful for the analysis of gene expression on a global level. In contrast, oligonucleotide arrays due to their PM/MM probe design and specific fluorescence detection are more useful for monitoring expression levels of a spliced transcript. Probes for each exon can be printed on the array and heterogeneous transcripts that would all bind to the same cDNA probe will bind differentially to each oligonucleotide probe.

Pathway Directed Microarray

Pathway directed DNA microarrays are the ideal tools for studying gene expression associated with a biological pathway or disease state. GEArrays are spotted on nylon membrane. These arrays, with a low density of genes, easy target labeling method and a chemiluminescent detection system, provide a very valuable tool in microarray applications. No special equipment is required for the GEArray processing. The array images can be captured using a gel documentation system having a cooled CCD camera or X-ray film. Array data analysis can be performed through the use of SuperArray's GEArray® Expression Analysis Suite web-based software. This software provides analysis tools including scatter plots and hierarchical clustering.

The use of the techniques described above focus on the applications of affymetrix and pathways specific SuperArray's GEArray microarrays in various stages of alcoholic and nonalcoholic liver disease. Profiling of gene expression in alcoholic and non alcoholic liver disease has been demonstrated in both human and animal models (Arlinde et al., 2004; Deaciuc et al., 2004a; Deaciuc et al., 2004b; Deaciuc et al., 2005; Hoffman and Tabakoff, 2005; Lewohl et al., 2001; Sikela et al., 2006; Tadic et al., 2002; Yin et al., 2007). However published studies do not define the role of dietary fat and liver pathology and gender in the severity of ALD and NAFLD. The pathology in ALD and NAFLD ranges from fatty liver alone to progressive fibrosis, cirrhosis, and eventually hepatocellular cancer (Forgione et al., 2007). All these conditions can be induced by feeding mice and rats diets that are high in fat or sucrose alone or with ethanol (Deaciuc et al., 2004a; Gregoire et al., 2002; Koteish and Diehl, 2001). Although the amount of dietary fat and its accumulation in the liver play an important role in the pathogenesis of ALD and NAFLD, the type of dietary fat is also an important factor. Studies in experimental ALD have highlighted the role of polyunsaturated fatty acids in promoting ethanol-induced liver injury. In contrast, no evidence of liver injury is seen in rats fed a diet containing mainly saturated fatty acids (Nanji et al., 2001b; Nanji et al., 1997). We describe the use of affymetrix microarrays and superarray to understand the role of dietary fatty acids and gender in modifying the severity of alcoholic and nonalcoholic disease by utilizing the generation of gene expression profiles, and the construction of gene networks.

Introduction of ALD and NAFLD

ALD is a major chronic illness and a cause of mortality around the world. Alcohol has been identified as a direct hepatoxin and various pathways have been implicated but the contribution to liver injury is not clear. Prolonged exposure of alcohol can lead to progressive fibrosis, cirrhosis, and eventually hepatocellular cancer (Yip and Burt, 2006). In addition to ALD, non alcoholic disease (NAFLD) is emerging as one of the most common chronic liver diseases in the western world. The histologic features of NAFLD range from fatty liver alone to nonalcoholic steatohepatitis (NASH). NASH is characterized by hepatocellular necrosis, inflammation, and fibrosis (Wilfred de Alwis and Day, 2007). Initially in both diseases, fat accumulation in hepatocytes leads to the development of fatty liver (steatosis). An emerging body of evidence suggests that the pathological mechanisms of advanced ALD and NAFLD are very similar (Figure1) and involve cytokine and oxidative stress–mediated injury.

Pathogenesis

1) Mechanism of Fat Accumulation in the Liver

Hepatic steatosis develops as a consequence of dysfunction of several metabolic pathways. The exact mechanisms leading to steatosis are not fully understood. The role of defective lipid metabolism and fatty acid oxidation in both ALD and NAFLD has been elucidated. In both diseases, excessive triglyceride accumulation in hepatocytes triggers liver damage. Previous studies suggest that increased influx of free fatty acids in hepatocytes is due to an increase in fatty acid synthesis, a decrease in fatty acid oxidation, increased transport of fatty acids from the adipocytes to the liver, and decreased transport of fatty acids from the liver (Day, 2006).

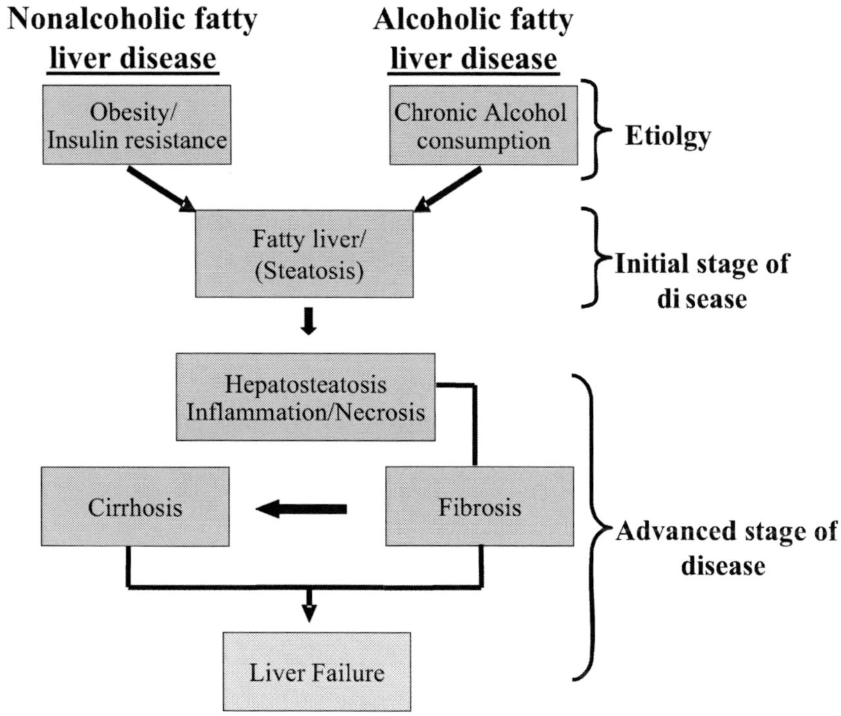

Figure 1. Schematic diagram shows the overlapping mechanism in pathogenesis of ALD and NAFLD.

In NAFLD, insulin resistance is likely the most important factor related to the increase in free fatty acid (FFA) influx into the liver; thereby increasing hepatic triglyceride synthesis. Hyperinsulinaemia and hyperglycaemia also promote lipogenesis by upregulating transcription factors such as sterol regulatory binding protein-1c (SREBP-1c) and carbohydrate response element binding protein. Activation of SREBP-1c increases malonyl-CoA synthesis which inhibits FFA oxidation, thereby facilitating hepatic triglyceride accumulation (Canbay et al., 2007; Yin et al., 2007).

2) Steatohepatitis

The mechanisms through which the fatty acids promote alcoholic liver injury include enhanced oxidative stress, production of endotoxin, and increased expression of proinflammatory cytokines. Tumor necrosis factor-alpha has been implicated in the pathogenesis of alcohol-induced liver injury in human beings, as well as in animals (Nanji et al., 2001a). Increased hepatic cellular fibronectin is an early response to injury, and an increased alpha-smooth muscle actin concentration suggests hepatic stellate cell activation, which may result in excessive collagen deposition and subsequent fibrosis (Miyahara et al., 2000).

The cause of non-alcoholic steatohepatitis remains elusive, but most investigators suggested that steatosis requires a second hit to induce development of inflammation, fibrosis, or necrosis (Day, 2002a; Day, 2002b). Once hepatic steatosis is established, it is thought that the second hit is caused by oxidative stress, mitochondrial dysfunction, gut-derived lipopolysaccharide and adipocytokines, which promote hepatocellular damage, inflammation and progressive liver disease (Pessayre et al., 2002).

FACTORS CONTRIBUTING TO NAFLD AND ALD

Many factors are responsible for potentiating ALD and NAFLD. These include among many causes, dietary fatty acids, and gender. These are discussed below.

Dietary Factors

Dietary fat has been shown to play an important role in the pathogenesis of alcoholic liver disease. Polyunsaturated fatty acids (e.g., omega-6 and omega-3 fatty acids) potentiate the severity of alcoholic liver injury whereas saturated fatty acids (medium chain triglycerides) are protective (Nanji et al., 2001a; Nanji et al., 2001b; Tsukamoto et al., 1986). The deleterious effects of polyunsaturated fatty acids are thought to be mediated through increased oxidative stress (lipid peroxidation), whereas the mechanisms of the protective effects of saturated fatty acids is not clear. One of the mechanisms by which unsaturated fatty acids cause oxidative stress is through induction of the free radical-generating enzyme CYP2E1. The level of induction of CYP2E1 correlates with the degree of lipid peroxidation, which, in turn, is linked to the severity of liver injury (Morimoto et al., 1993; Nanji et al., 2001a; Nanji et al., 1994).

Gender

Physiological differences between males and females exist in a variety of tissues, including the liver. The rodent liver has provided an excellent model for investigating gender differences in the complex interaction of hormonal, developmental, and tissue-specific control of gene expression. Sex-specific patterns of gene expression occur in liver for several cytochrome P450 enzymes involved in steroid and xenobiotic metabolism (Carulli et al., 2006; Shapiro et al., 1995; Sharma et al., 1998; Shimizu and Ito, 2007). In humans, hepatic gender differences exist in pathogenesis of many diseases, such as differences in alcohol metabolism (Enomoto et al., 2004; Muller, 2006; Sato et al., 2001) and incidence of nonalcoholic fatty liver disease (NAFLD) (Kotoh et al., 2007). Women develop ALD at a lower intake of alcohol than men. Recent studies have demonstrated that estrogen increases gut permeability to endotoxin and up-regulates endotoxin receptors on Kupffer cells, leading to an increased production of TNF-α in the rat model (Thurman, 1998). Estrogen has many roles that seem to be protective against fatty liver (Sato et al., 2001). In the intragastric infusion rat model, rats fed with unsaturated fat (fish oil) and ethanol or dextrose the degree of liver injury was greater in female than in male rats (Nanji et al., 2001a). One of the molecular mechanisms responsible for the gender differences in the susceptibility of the liver to alcohol and fish oil may be related to the differential expression of hepatic genes, especially those regulated by hormones.

Experimental Design

Animal Model: Male and age-matched female Wistar rats were fed by continuous infusion of a liquid diet through permanently implanted gastric tubes, as described previously (Tsukamoto et al., 1990). The rats were given their total nutrient intake by intragastric infusion. The diet provided 250 kcal/kg/day and was modified to include a salt and vitamin mix (Dyets, Bethlehem, PA) (French, 1993). The source of protein was lactalbumin. The percentage of total calories derived from fat was 35%. The diet was supplemented with choline and methionine. The amount of ethanol was modified to maintain high levels of blood ethanol (150-300 mg/dl) throughout the day. The amount of ethanol was initially 10 g/kg/day and was increased up to 16 g/kg/day as tolerance developed. Rats were fed different dietary fats to produce increasing degrees of liver injury: Medium chain triglycerides plus dextrose or ethanol, palm oil plus dextrose or ethanol, and fish oil plus dextrose or ethanol. All control animals were pair-fed the same diet as ethanol-fed rats except that ethanol was isocalorically replaced by dextrose. The care of animals and euthanasia procedures were in accordance with the National Institutes of Health criteria for use of laboratory animals. The measurement of pathological and biochemical parameters has been described previously (Nanji et al., 1995; Nanji et al., 1993).

In our experiments, rats fed with medium chain triglycerides plus dextrose or ethanol develop no pathological changes. Palm oil plus ethanol-fed rats develop fatty liver only. Female rats fed with fish oil plus ethanol develop more severe injury (fatty liver, inflammation and necrosis) than male rats. The female rats fed fish oil and dextrose (FFD) also develop NAFLD showing fatty liver, necrosis and inflammation (Nanji et al., 2001a) whereas male rats fed the same diet show no evidence of pathological changes. Therefore, FFD rats provide an excellent model to study the complex interaction of gene expression in NAFLD.

Microarray Analysis

Affymetrix Microarray Analysis

Microarray analysis was performed in liver of rats fed with unsaturated fat (fish oil) and ethanol (developed ALD) or fish oil plus dextrose (developed NAFLD) using the Affymetrix Rat Genome U34A array (Affymetrix Inc.), which contained a total number of 8799 probe sets as described previously (Sharma et al., 2006; Sharma et al., 2007). Microarray analysis was performed according to manufacturer's instructions. Briefly, total RNA prepared from the individual rat liver sample was used for cDNA synthesis using a T7-(dT)24 primer (Amersham Biosciences). The cDNA was cleaned and used for biotin-labeled cRNA synthesis through in vitro transcription using the MEGAscript T7 in Vitro Transcription Kit (Ambion Inc., TX). Each biotin-labeled and fragmented cRNA sample (~ 10 μg) was individually hybridized to a Rat Genome U34A Array, which contained a total number of

8799 probe sets, at 45 °C for 16h. The array was washed twice and stained with a streptavidin-phycoerythrin conjugate in the Affymetrix Fluidics Station.

Microarray Data Analysis: The fluorescence images of hybridized GeneChip array were scanned with a GeneArray Scanner (Hewlett-Packard, Santa Clara, CA) and the hybridization signal intensity was analyzed with MAS 5.0 software (Affymetrix, Inc). In this study, the fold-change analysis method was used to select differentially expressed genes in response to diet. The increase or decrease in mRNA level was considered reliable if the difference in hybridization signal intensity between a male and female sample yielded at least a two fold change (>2.0 or <-2.0). Data sets were analyzed using GeneSpring software. The differentially expressed up-regulated or down regulated genes in female rats in reference to male were further analyzed for relevant biological information. Only genes showing >2 or <-2 fold change in female rats were subjected to an extensive literature search and correlated with pathological conditions in ALD and NAFLD of female rat liver based on their reported functions in the literature.

Inflammatory Cytokines Pathway Specific Super Array

To understand the role of inflammatory cytokines in the pathogenesis of ALD that is seen in rats fed palm oil and ethanol, and fish oil and ethanol respectively, we used rat inflammatory cytokines and receptors microarray (ORN-011) from SuperArray (SuperArray,Bioscience Corporation, Frederick, MD). These arrays contained 96 proinflammatory cytokines and receptors genes along with non-specific genes; plasmid c-DNA and housekeeping genes that were used for normalization purposes. The complete list of genes is available on www.superarray.com. The study described here was designed to profile the expression of proinflammatory genes in rats fed with saturated fatty acids (medium chain triglycerides) compared to rats fed with palm oil (developed only fatty liver). The rats fed with palm oil were compared to rats fed with fish oil (developed fatty liver, necrosis and inflammation) to investigate the role of inflammatory genes relevant in transcriptional shift from fatty liver to necrosis and inflammation in pathogenesis of ALD. Synthesis of cDNA and cRNA was carried out with the True labeling-Amp linear RNA amplification Kit (SuperArray, Bioscience Corporation) according to the manufacturer's instructions. The biotinylated cRNAs from different rats were hybridized overnight to separate membranes at 60°C, washed with SSC/SDS solutions, incubated with the avidin--alkaline phosphatase conjugate and exposed to a chemiluminescent substrate. Arrays were exposed on X-ray film. Images were scanned and genes spots were normalized with house keeping gene GAPDH. Comparison of the intensities of the genes in control vs experimental membranes and as well as among the rats fed MCT plus ethanol, palm oil plus ethanol and fish oil and ethanol was achieved using the web-based GEArray Expression Analysis Suite software from SuperArray Bioscience Corporation (http://geasuite.superarray.com).

Biological theme analysis: Expression Analysis Systematic Explorer (EASE) software application was used to determine biologically relevant themes in the lists of genes differentially expressed in the female versus male liver. These gene lists were analyzed for themes using Gene Ontology (GO) cellular compartment, molecular function, biological process, and biochemical function (provided by NCBI). The data sets were analyzed with the categorical over-representation function of EASE. All significant categories resulting from each list produced the over-represented biological themes.

Pathway analysis: Biologically relevant pathways were constructed using Ingenuity Pathways analysis (IPA) application (www. Ingenuity.com). The two lists of differentially regulated genes in female rats, were combined together and the final list containing Affymetrix identifiers and associated fold-change values were used for IPA analysis. Genes which had a greater than 2.0 fold change were included in this analysis. These genes, known as focus genes, were used as the starting point to generate the biological network. The program computes a score for each network according to the fit of the network to the set of focus genes.

Analysis of Microarray Data in ALD

To understand the molecular mechanisms involved in the increased severity of ALD seen in female rats compared to males fed with fish oil and ethanol, we designed a study to profile the expression of hepatic genes in female rats compared to male rats in response to an ethanol-fish oil diet.

We have previously shown that the liver of female rats fed with fish oil and ethanol (FFE) for 4 weeks showed more severe fatty liver, necrosis and inflammation, than the male liver. The molecular mechanism for increased severity of ALD proposes the role of oxidative stress, endotoxin, chemokines and cytokines in alcohol-mediated injury. We, therefore, have chosen to restrict this discussion to the genes promoting fat infiltration and inflammation in the liver.

Genes Up-regulated in Female Rats

We identified 840 up regulated genes in female rats fed with fish oil plus ethanol compared to males fed the same diet. The Expression Analysis Systematic Explorer (EASE) was used to identify important biological themes overrepresented among genes that were differentially expressed. The EASE score is calculated by penalizing (removing) one gene within the given category from the list and calculating the resulting Fisher exact probability for that category. It therefore represented the upper boundary of the distribution of jackknife Fisher exact probabilities and has advantages in terms of penalizing the significance of categories supported by few genes (Hosack et al., 2003). Table 1 summarizes some of the most significant themes, representing expression of genes encoding proteins involved mostly in metabolic functions, with scores that were identified using this analysis. Genes involved in catalytic activity, lipid and alcohol metabolism, and mitochondrial functions were maximally expressed.

Out of 840 genes that were upregulated in females, we only considered the genes that showed a greater than 2- fold increase (Table 2). The genes in Table 2 encode for proteins that belong to various functional groups such as lipid metabolism, inflammatory response, cell cycle regulation, oxidative stress, signal transduction, and estrogen dependent metabolism.

Table 1. EASE Biological Theme Analysis : Upregulated Genes in Female Rat Liver Fed with Fish Oil and Ethanol

System	Gene Category	EASE score*
GO Molecular Function	catalytic activity	2.04E-16
GO Molecular Function	proteasome endopeptidase activity	2.00E-13
GO Cellular Component	proteasome complex (sensu Eukarya)	1.40E-12
GO Cellular Component	cytoplasm	1.67E-12
GO Biological Process	lipid metabolism	1.78E-11
GO Biological Process	metabolism	3.34E-11
GO Cellular Component	proteasome core complex (sensu Eukarya)	8.95E-11
GO Biological Process	carboxylic acid metabolism	4.82E-09
GO Biological Process	organic acid metabolism	4.82E-09
GO Cellular Component	intracellular	7.81E-09
GO Cellular Component	endoplasmic reticulum	2.39E-08
GO Biological Process	ubiquitin-dependent protein catabolism	3.52E-08
GO Biological Process	alcohol metabolism	7.22E-08
GO Biological Process	modification-dependent protein catabolism	9.39E-08
GO Molecular Function	oxidoreductase activity	9.43E-08
GO Cellular Component	mitochondrion	9.61E-08
GO Biological Process	sterol metabolism	6.31E-07
GO Biological Process	steroid metabolism	6.91E-07
GO Biological Process	catabolism	1.07E-06
GO Biological Process	sterol biosynthesis	2.85E-06
GO Biological Process	cholesterol metabolism	6.81E-06
GO Biological Process	fatty acid metabolism	8.83E-06
GO Biological Process	lipid biosynthesis	1.58E-05
GO Biological Process	cholesterol biosynthesis	3.58E-05
GO Biological Process	macromolecule catabolism	9.54E-05
GO Molecular Function	hydrolase activity	0.000107927
GO Biological Process	coenzyme and prosthetic group metabolism	0.000117097
GO Biological Process	protein catabolism	0.00012616
GO Cellular Component	cytosol	0.000162173
GO Biological Process	proteolysis and peptidolysis	0.000190915
GO Biological Process	physiological process	0.000238165
GO Biological Process	steroid biosynthesis	0.000263902
GO Molecular Function	oxidoreductase activity\, acting on CH-OH group of donors	0.000722191
GO Molecular Function	ligase activity	0.000934865
GO Biological Process	coenzyme metabolism	0.000973853
GO Cellular Component	microbody	0.001076288
GO Cellular Component	peroxisome	0.001076288
GO Biological Process	tricarboxylic acid cycle	0.00111923
GO Molecular Function	oxidoreductase activity\, acting on the CH-OH group of donors\, NAD or NADP as acceptor	0.00151487
GO Molecular Function	endopeptidase activity	0.001676576
GO Molecular Function	lyase activity	0.001832968

Table 1. Continued

System	Gene Category	EASE score*
GO Molecular Function	peptidase activity	0.00243268
GO Molecular Function	intramolecular isomerase activity\, transposing C=C bonds	0.002623977
GO Biological Process	energy derivation by oxidation of organic compounds	0.002785749
GO Molecular Function	transferase activity\, transferring acyl groups	0.003350237
GO Molecular Function	oxidoreductase activity\, acting on the CH-CH group of donors	0.005015584
GO Biological Process	main pathways of carbohydrate metabolism	0.005697589
GO Biological Process	isoprenoid metabolism	0.00581835
GO Biological Process	energy pathways	0.005946143

* EASE score is the upper bound of the distribution of Jacknife Fisher exact probabilities.
This table is adapted from Sharma et al 2007 with permission of Experimental and Molecular Pathology

Table 2. Differentially Expressed Genes in Livers of Female Rats in ALD and NAFLD

Category/Function		ALD		NAFLD	
		Up-regulated	Down-regulated	Up-regulated	Down-regulated
	Gene Identifiers	Male: Female Ratio		Male: Female Ratio	Male: Female Ratio
Fatty liver					
SREBP regulated genes					
Cd36 antigen	29184	2.67			
N-acylsphingosine amidohydrolase (acid ceramidase)	84431	2.77			
malic enzyme 1	24552	2.33			
histidine ammonia lyase	29301	2.48			
3-hydroxy-3-methylglutaryl-Coenzyme A synthase 1	29637	3.70			
stearoyl-Coenzyme A desaturase 1	246074	11.67			
enoyl-Coenzyme A, hydratase	171142	2.51			
sterol-C4-methyl oxidase-like	140910	2.41			
cytosolic acyl-CoA thioesterase 1	50559	7.89		23.18	
stearoyl-Coenzyme A desaturase 2	83792	2.14			
mitochondrial acyl-CoA thioesterase 1	192272	3.84		3.48	
enoyl coenzyme A hydratase 1	64526	2.24			
pyruvate dehydrogenase phosphatase isoenzyme 2	246311	2.36			
fatty acid synthase	50671	2.19			
sterol carrier protein 2	25541	3.74			
PPAR regulated genes					
fatty acid binding protein 1	25598	21.00			
farensyl diphosphate synthase	83791	3.39		2.22	
cytochrome P450, 7a1	25428	3.43			
Cbp/p300-interacting transactivator	114490		-2.2		-4.5

Table 2. Continued

Category/Function		ALD		NAFLD	
		Up-regulated	Down-regulated	Up-regulated	Down-regulated
	Gene Identifiers	Male: Female Ratio		Male: Female Ratio	Male: Female Ratio
Inflammation					
pregnancy upregulated non-ubiquitously expressed CaM kinase	29660	3.10			
Kupffer cell receptor	114598	2.54			
ferredoxin 1	29189	2.38			
S100 calcium-binding protein A8 (calgranulin A)	116547	2.93			
prolactin receptor	24684	4.42			
aminopeptidase B	81761	3.87			
guanylate binding protein 2, interferon-inducible	171164	4.08			
amino-terminal enhancer of split	29466	2.52			
CD48 antigen	245962	2.01			
Fc receptor, IgG, low affinity III	116591	2.42			
guanylate binding protein 2, interferon-inducible	171164	4.08			
MHC class Ib RT1.S3	294228	2.40			
similar to interferon regulatory factor 7	293624	4.79			
Cd36 antigen	29184	4.62			
lipopolysaccharide binding protein	29469	3.49			
Gro	81503	17.48			
K-kininogen	25087	2.95			
benzodiazepin receptor	24230	2.11			
insulin-like growth factor 1	24482	2.21			
histidine decarboxylase	24443			38.67	
alpha-1-inhibitor III; plasma proteinase inhibitor	65279				-20.51
Oxidative Stress					
NF-E2-related factor 2	83619	2.86			
Heme oxygenase 1	24451			3.08	
Heme oxygenase 2	79239	2.35			
aminolevulinic acid synthase 1	65155	2.20			
glycerol-3-phosphate dehydrogenase 2	25062	8.60			
glutathione peroxidase 2	29326	2.77			
peroxisomal membrane protein Pmp26p (Peroxin-11)	85249	3.14			
NADH dehydrogenase, mitochondrial genome	854269				-2.23

Table 2. Continued

Category/Function		ALD		NAFLD	
		Up-regulated	Down-regulated	Up-regulated	Down-regulated
	Gene identifiers	Male: Female Ratio		Male: Female Ratio	Male: Female Ratio
Cell cycle					
achaete-scute complex homolog-like 2 (Drosophila)	24209	3.75			
cell division cycle 2 homolog (S.pombe)-like 1	252879	5.02			
X-rayrepair complementing defective repair in Chinese hamster cells 5	363247	7.63			
DNA primase, p49 subunit	246327	8.40			
dentatorubral pallidoluysian atrophy	29515	7.83			
similar to RA175	363058	2.59			
similar to deoxycytidyl transferase	316344	3.79			
signal transduction					
HRAS like suppressor	24913	2.03			
regulated endocrine-specific protein 18	50561	4.03			
signal transducer and activator of transcription 1	25124	12.61			
somatostatin receptor 1	25033	2.56			
growth factor receptor bound protein 14	58844	2.05			
purinergic receptor P2Y, G-protein coupled 2	29597	2.18			
protein tyrosine phosphatase, non-receptor type 6	116689	2.03			
TBP-interacting protein 120A	117152	2.17			
proline rich 2	286988	1.87			
similar to Serine/threonine kinase 38	361813	4.72			
similar to nuclear ATP/GTP-binding protein	290986	5.31			
similar to archvillin	361256	2.07			
interleukin 6 signal transducer	25205				-5.77
Estrogen regulated					
Cytochrom P450 15-beta gene	25011	75.5		40.00	
similar to Alcohol sulfotransferase	361511	2.10			
signal transducer and activator of transcription 1	25124	12.61			
Rat cytochrome P-450(M-1) gene, exon 9	203693		-12.1		-8.51
R.norvegicus CYP3A1 gene for cytochrome P450 PCN1	56037				-2.49
Rat 3-beta-hydroxysteroid dehydrogenase	24470				-12.84
sulfotransferase, estrogen preferring	25355				-17.01

Genes Down-regulated in Female Rats

Total 295 genes were down regulated and EASE analysis of these genes is presented in Table 3. Biological theme analysis categorized these genes as being included mainly as those involved in oxidoreductase activity, nucleic acid metabolism, chaperone and transcriptional processes. Of the affected genes, only few genes showed a greater than two-fold down regulation.

Table 3. EASE Biological Theme Analysis: down-regulated genes in female liver

System	Gene Category	EASE score*
GO Cellular Component	endoplasmic reticulum	7.08E-08
GO Molecular Function	oxidoreductase activity	1.24E-07
GO Cellular Component	vesicular fraction	8.83E-07
GO Cellular Component	microsome	8.83E-07
GO Cellular Component	intracellular	9.59E-06
GO Cellular Component	cytoplasm	1.87E-05
GO Biological Process	metabolism	5.66E-05
GO Molecular Function	chaperone activity	6.26E-05
GO Biological Process	amino acid and derivative metabolism	0.000100165
GO Biological Process	amine metabolism	0.000204852
GO Cellular Component	membrane fraction	0.000342928
GO Cellular Component	endoplasmic reticulum lumen	0.001769152
GO Molecular Function	electron transporter activity	0.001883121
GO Cellular Component	cell fraction	0.002114396
GO Biological Process	physiological process	0.002375494
GO Biological Process	amino acid metabolism	0.002819564
GO Biological Process	organic acid metabolism	0.00525025
GO Biological Process	carboxylic acid metabolism	0.00525025
GO Biological Process	amino acid biosynthesis	0.006425313
GO Biological Process	electron transport	0.007149855
GO Biological Process	amine biosynthesis	0.008048286
GO Molecular Function	catalytic activity	0.008828216
GO Molecular Function	oxidoreductase activity\, acting on paired donors\, with incorporation or reduction of molecular oxygen	0.011849921
GO Biological Process	protein biosynthesis	0.015361282
GO Cellular Component	mitochondrion	0.015911612
GO Molecular Function	oxidoreductase activity\, acting on the aldehyde or oxo group of donors	0.0170461
GO Molecular Function	oxidoreductase activity\, acting on CH-OH group of donors	0.020559708
GO Molecular Function	unspecific monooxygenase activity	0.021685797
GO Molecular Function	oxidoreductase activity\, acting on paired donors\, with incorporation or reduction of molecular oxygen\, reduced flavin or flavoprotein as one donor\, and incorporation of one atom of oxygen	0.025083545
GO Biological Process	heterocycle metabolism	0.025684829
GO Biological Process	glutamine family amino acid metabolism	0.0345118
GO Biological Process	peptidyl-amino acid modification	0.03803068

Table 3. Continued

System	Gene Category	EASE score*
GO Biological Process	serine family amino acid metabolism	0.049200457
GO Biological Process	biosynthesis	0.056075163
GO Molecular Function	oxidoreductase activity\, acting on the CH-OH group of donors\, NAD or NADP as acceptor	0.05611378

* EASE score is the upper bound of the distribution of Jacknife Fisher exact probabilities.
This table adapted from Sharma et al. 2007 with permission of Experimental and Molecular Pathology

Identification of Biologically Relevant Networks in ALD

IPA global functional analysis showed that maximum numbers of genes were differentially modulated in lipid metabolism, cellular growth and proliferation, cancer, signaling, and cellular function maintenance. As pointed out previously, biologically relevant networks were constructed from the list of genes that were differentially expressed. The pathway shown in 2 and 3 were identified the most significant pathways and contained 35 focus genes, with a highly significant score of 49.

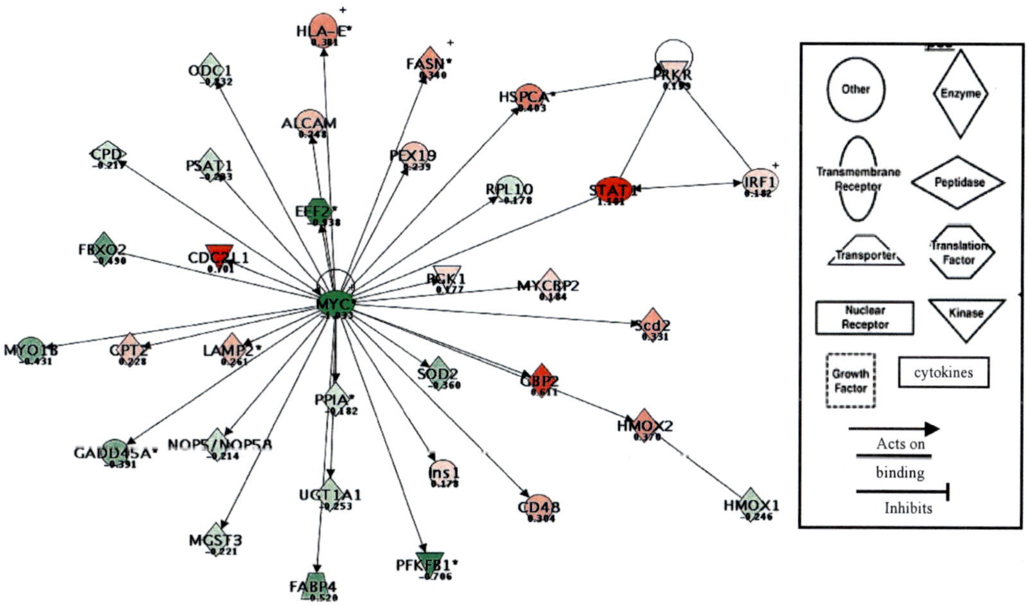

Figure 2. Ingenuity pathway analysis constructed a network of genes regulated in response to fish oil and ethanol diet in female rat liver. The network is displayed graphically as nodes (genes/gene products) and edges (the biological relationships between the nodes). The node color intensity indicates the expression levels of genes; red, up regulated; green, down regulated. Different shapes of Nodes represent the functional class of the gene product. Edges describe the nature of the relationship between the nodes (A; activation, B; binding; and I; inhibition). A total of 35 differentially expressed "focus" genes were displayed into the network with main function in transcription regulation, cell growth, proliferation, cell death, and apoptosis via myc signaling. (Figures 2-7 adopted from Sharma et al 2006 and 2007 with the permission of the Experimental and molecular pathology).

As shown in Figure 2, these pathways analysis identified genes that are mostly participants in Myc and Stat pathways. Myc has well-established roles in cell growth and proliferation and may have an important role in the predisposition to hepatocarcinogenesis (Chan et al., 2004; Feitelson, 2004). Estradiol upregulates c-Myc gene in aggressive form of breast cancer (McNeil et al., 2006). Figure 3 shows modulation of genes primarily involved in IGF-1 signaling, amino acid metabolism and protein synthesis. Many studies have shown that IGF-1 acts synergistically with estrogen to stimulate breast cancer cells (Chong et al., 2007). In our study expression of the c-Myc and IGF-1 related genes in Figures 2 and 3 clearly demonstrates that estrogen-regulated genes play an important role in promoting liver injury in female rats in ALD. Figure 4 shows the genes that are predominantly regulated by SREBP, PPARα and HNF4α implicating their role in the lipid metabolism.

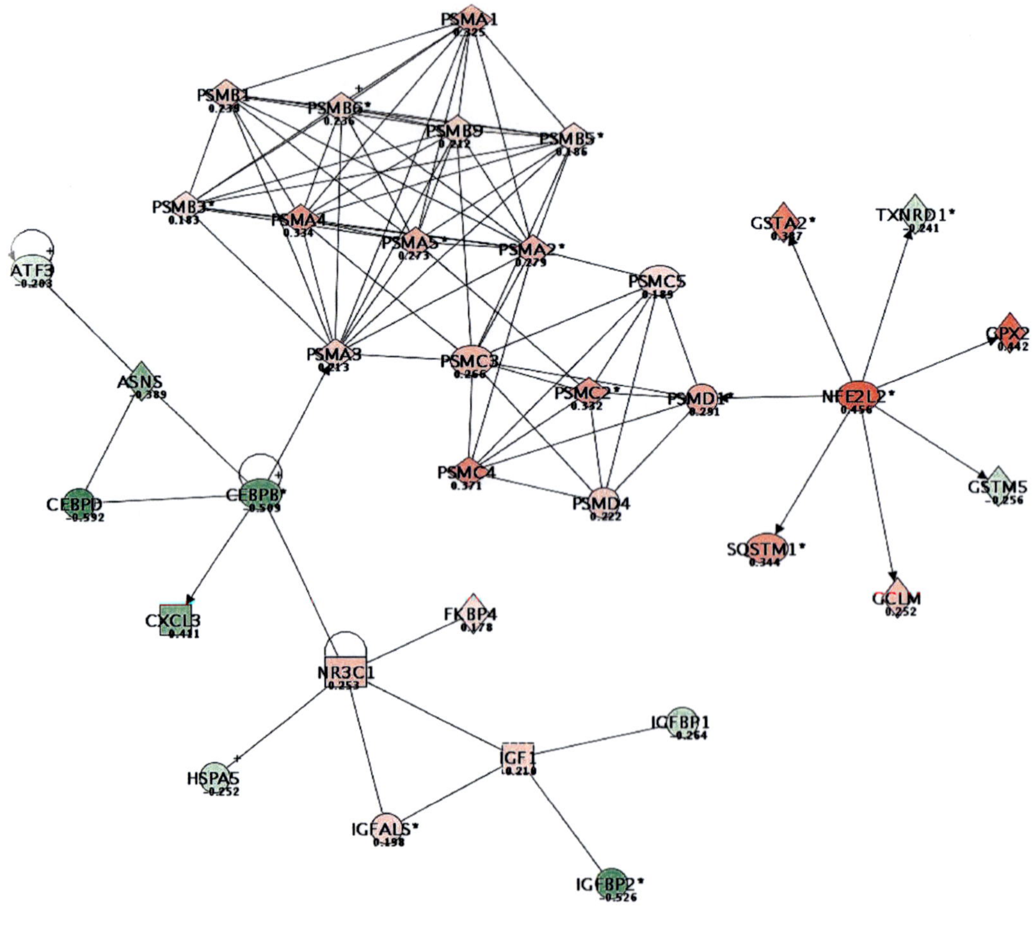

Figure 3. Ingenuity pathway constructed a network of 35 differentially expressed "focus" genes with main function in amino acid metabolism, protein synthesis and tissue morphology.

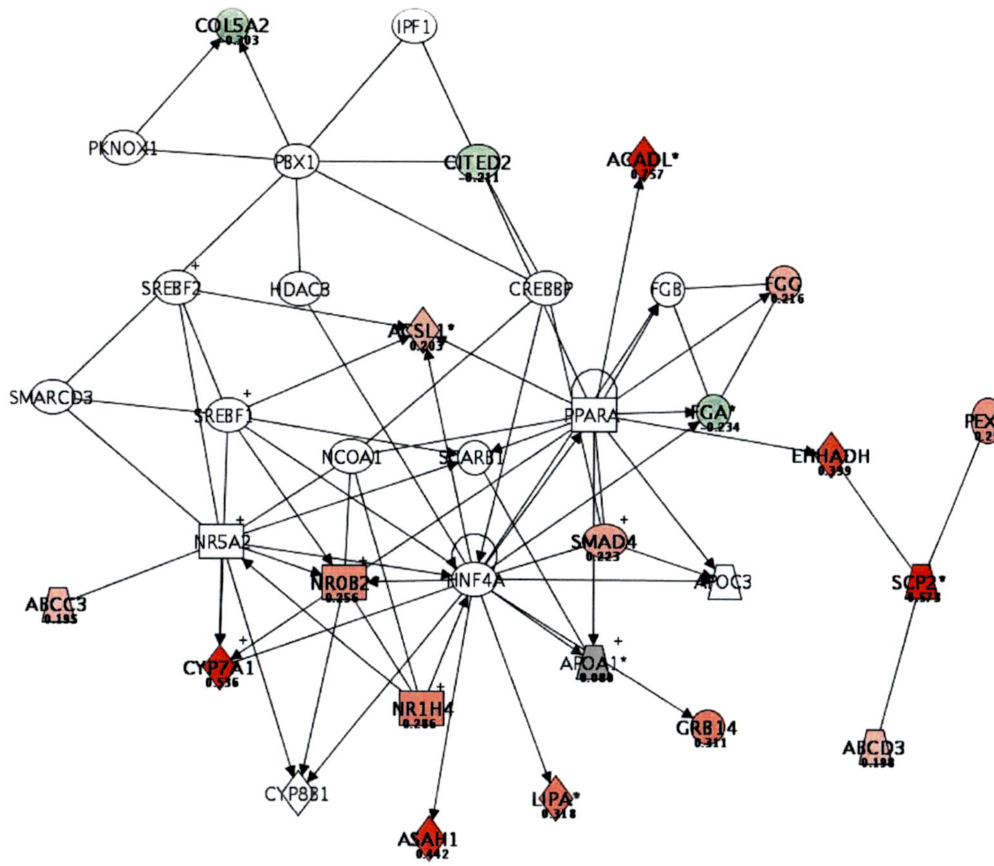

Figure 4. Ingenuity pathway constructed a network of 18 differentially expressed "focus" genes with main function in lipid metabolism, genetic disorder and hematological disease.

Analysis of Microarray Data in NAFLD

To understand how dietary fatty acids influence the development of NAFLD and NASH, we took advantage of our findings which showed that female rats fed with fish oil and dextrose develop pathological changes typical of NASH (Nanji, 2004). We therefore performed a microarray experiment to profile the expression of hepatic genes in female rats compared to male rats in response to dextrose and fish oil diet.

Up-regulated Genes in Female Rats

We identified 149 genes that were up-regulated in female versus male rats. The genes of most significant themes that are over represented in female rats were involved in lipid and fatty acid metabolism, chaperone activity, and mitochondrial function. Of the 149 genes, we considered only 13 genes that showed a greater than 2.0-fold up regulation (Table 2). The genes listed in Table 2 encode for proteins that belong to various functional groups involved in anti-oxidant function, hepatic carcinogenesis, transport and protein folding, the inflammatory response, liver regeneration, xenobiotic metabolism and obesity.

Down-regulated Genes in Female Rats

Thirty genes were down regulated in female versus male rats and biological theme analysis categorized these genes as being involved mostly in nucleic acid metabolism and transcriptional processes. Table 2 summarizes the highly differentially expressed 15 genes (> 2 fold) which encode for proteins that are involved in carbohydrate metabolism, inflammatory responses, lipid metabolism, transcription regulation and xenobiotic metabolism.

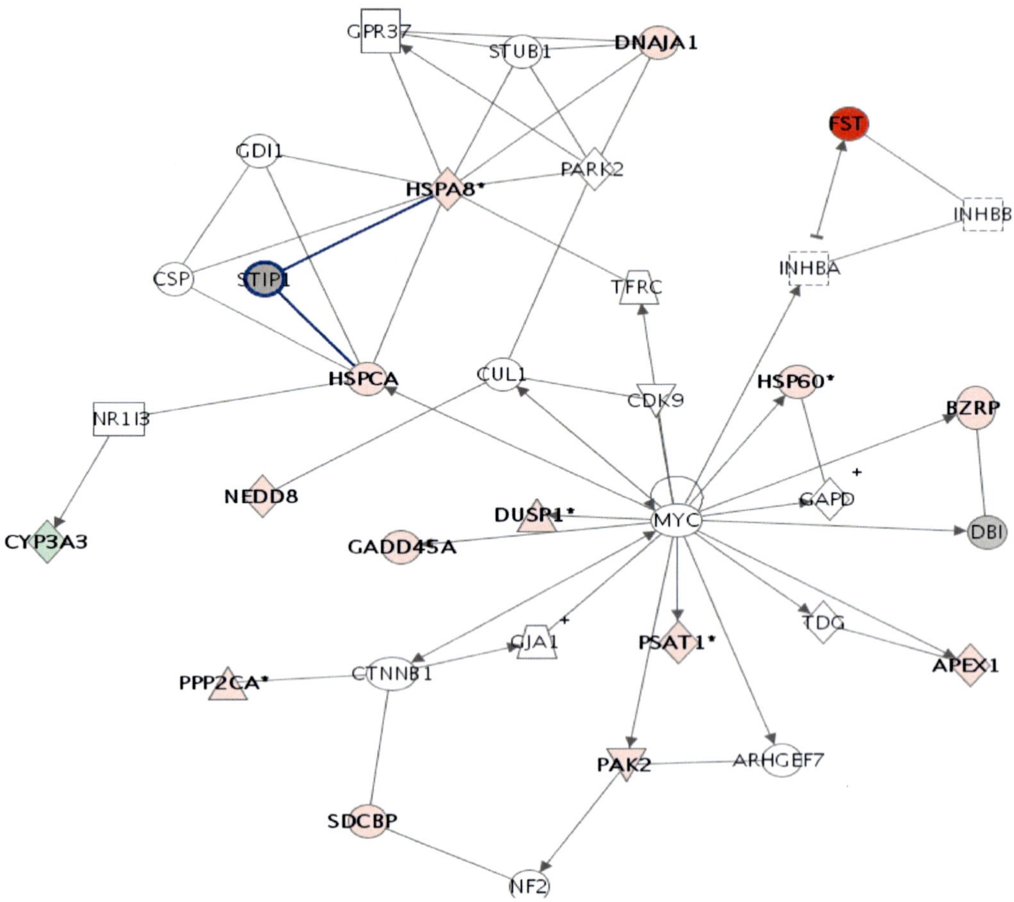

Figure 5. Ingenuity pathway analysis identified a network of genes regulated in response to fish oil and dextrose diet in female rat liver. The network contained 15 focus genes with a highly significant score of 22. These "focus" genes are: apurinic/apyrimidinic endonuclease 1 (APEX1), benzodiazepine receptor (BZRP) DnaJ-like protein (DNAJA1), heat shock protein 1, alpha (HSPCA), neural precursor cell expressed, developmentally down-regulated gene 8 (NEDD8), protein phosphatase 2a (PPP2CA), heat shock protein 60 (HSP60), dual specificity phosphatase 1 (DUSP1), growth arrest and DNA-damage-inducible 45 alpha (GADD45A), heat shock protein 8 (HSP8), p21 (CDKN1A)-activated kinase 2 (PAK2), phosphoserine aminotransferase 1 (PSAT1), syntenin (SDCBP), and follistatin (FST).

Networks in NAFLD

The pathway shown in Figure 5 contained 15 focus genes, with a highly significant score of 22. The pathways in Figure 5 identified genes that are participants in MAPK, PI3K/AKT, Myc and TGF-β signaling pathways which promote cell death, DNA damage, cell proliferation and post transcriptional modification of proteins. Figure 6 showed primarily modulation of genes of the PPAR signaling and lipid metabolism pathways. Figure 7 shows the genes of JAK/STAT and MAPK pathways that are predominantly regulated by cytokines implicating their role in the inflammatory response and apoptosis in NAFLD. We only included networks with significant scores and therefore exclude the possibility that focus genes in a given network are found together due to random chance. A score of 2 or higher has a greater than 99% likelihood of not being generated by random chance alone.

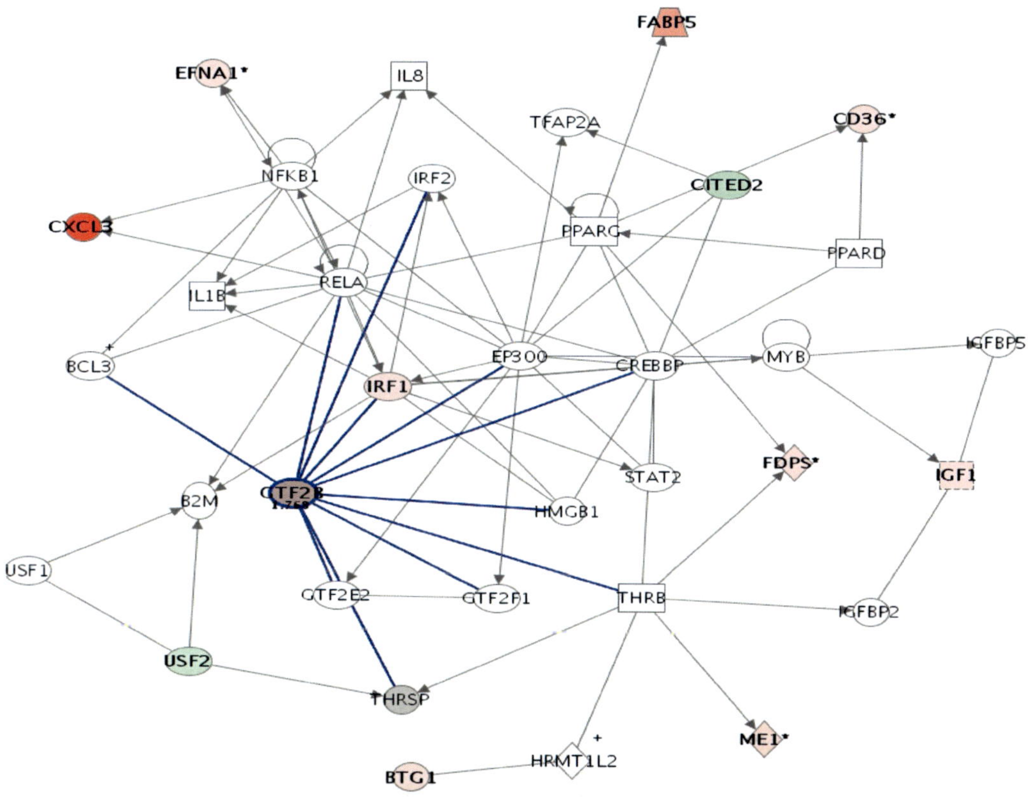

Figure 6. Ingenuity pathway analysis shows a network of 12 differentially expressed "focus" genes which include fatty acid binding protein 5 (FABP5), farensyl diphosphate synthase (FDPS), cd36 antigen (CD36), insulin-like growth factor 1 (IGF1), malic enzyme 1 (ME1) whose main functions are in fatty acid metabolism,and transcription regulation.

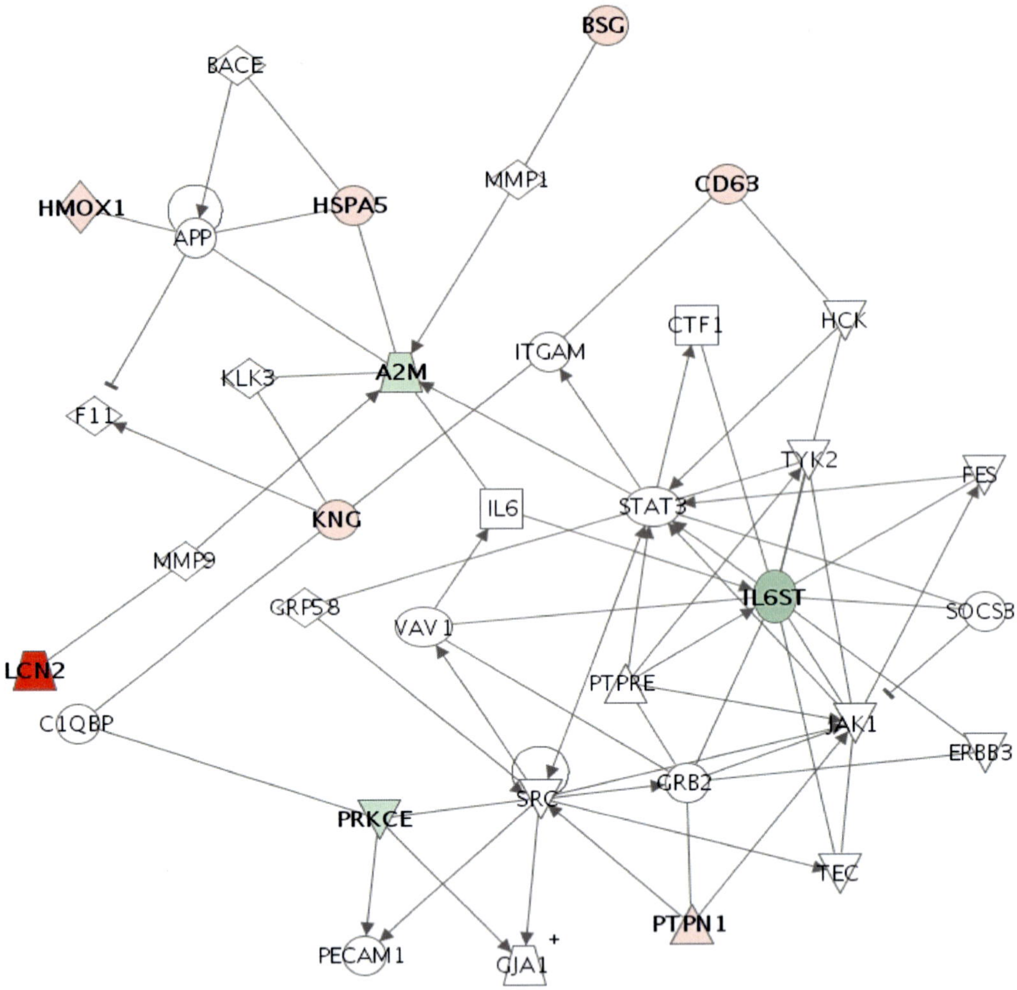

Figure 7. Ingenuity pathway analysis shows a network of 10 differentially expressed "focus" genes with main functions in immune response, cell death and development.

Functional Classification of Genes in Pathogenesis of the ALD and NAFLD

1) Genes Relevant to Promoting Fatty Liver

The progression of ALD is characterized by the initial appearance of fatty liver, followed by inflammation, necrosis and apoptosis, fibrosis and, then, in some individuals cirrhosis (Tilg and Day, 2007). An increase in the expression of genes regulating fatty acid synthesis, coupled with suppression of genes for fatty acid oxidation and transport from the liver, have been reported in a number of animal models of chronic ethanol exposure (Nagy, 2004). Chronic ethanol exposure also affects the activity of several transcription factors that coordinate the expression of genes involved in fatty acid metabolism. One of these transcription factors, sterol regulatory element binding protein (SREBP), primarily coordinates the regulation of genes involved in fatty acid synthesis (Eberle et al., 2004). SREBP also regulates PPAR, which in turn regulates the expression of genes involved in fatty acid oxidation.

(i) Role of SREBP

In our study many of the genes up-regulated in fatty acid metabolism are modulated by sterol regulatory element binding proteins (SREBPs). These genes include acetyl-CoA carboxylase, fatty acid synthase, stearolyl –CoA desaturase 1 and 2, sterol carrier protein 2, malic enzyme, and ATP citrate lyase, all of which are important in hepatic triglyceride synthesis (Table 2) (Guay et al., 2007). High-fat diets induce SREBP1 expression in ICAM-1 deficient mice; a transgenic mouse model that is very sensitive to a high-fat-diet-induced fatty liver (Gregoire et al., 2002). SREBP1 levels are increased in hepatoma cell lines, exposed to ethanol, and also in mice receiving an ethanol-containing diet for four weeks (You et al., 2002). The increase in SREBP is associated with increased expression of genes involved in the regulation of hepatic lipogenesis, as well as the accumulation of triglyceride in the liver (Ji et al., 2006; Repa et al., 2000). Moreover, ethanol, endoplasmic reticulum stress, shear stress, and cytokines have been reported to upregulate the lipogenic gene expression in hepatocytes and/or vascular endothelial cells by activation of SREBPs (Diomede et al., 2001; Werstuck et al., 2001).

Intragastric ethanol feeding in mice induces expression of unfolded protein response/endoplasmic reticulum (UPR/ER) stress response genes. Alcohol-induced ER stress can activate sterol regulatory element-binding protein (SREBP)-1c and SREBP-2, which contribute to the accumulation of triglyceride and cholesterol (Esfandiari et al., 2005; Kaplowitz and Ji, 2006).

(ii) Role of PPARα

The genes encoding peroxisomal, microsomal, and certain mitochondrial fatty acid metabolizing enzymes in liver are transcriptionally regulated by PPAR α (Rao and Reddy, 2001). PPARs have also been implicated as regulators of inflammatory responses (Cabrero et al., 2002; Fujii, 2005). Many genes in ALD and NAFLD are involved in PPAR signaling pathways. Fatty acid binding protein 2 (Fabp2), farnesyl diphosphate synthase (Fdps), and Cyp7a1 are upregulated in ALD (Table 2). Fabp2 plays a key intermediary role in the transport of fatty acids to hepatocytes. Chronic ethanol feeding to mice decreases hepatic PPAR-α DNA binding activity. Decreased PPAR binding is associated with decreased expression of PPAR-dependent genes, with the exception of fatty acid binding protein, which, in contrast is induced with ethanol feeding (Fischer et al., 2003). Fdps is also regulated by PPARα, and in the rat, is involved in the control of the sterol biosynthesis pathway (Le Jossic-Corcos et al., 2004). Importantly, treatment of these chronic ethanol-fed mice with WY14643, a PPAR agonist, restores PPAR DNA binding activity and prevents the development of chronic ethanol-induced steatosis (Fischer et al., 2003). The activation of PPARα by Clofibrate, a PPAR-α-activating ligand, can prevent not only the pathological manifestations of alcohol-induced perturbations in fatty acids metabolism but also alcohol-induced biochemical and histological inflammatory changes in liver (Nanji et al., 2004).

Although many genes in NAFLD are involved in PPAR signaling pathways, the most dramatic changes were observed with lipopolysaccharide-binding protein (LBP), farnesyl diphosphate synthase (Fdps), and Cbp/p300-interacting transactivator (CITED2) (Table 2).

Induction of lipopolysaccharide-binding protein (LBP), an acute phase protein produced mainly by hepatocytes, plays a key intermediary role in the activation of cells by endotoxin. Endotoxin levels were increased in female rats fed with fish oil and dextrose (Nanji et al., 2001a). Fdps is also upregulated in ALD, suggesting a overlapping role of sterol biosynthesis pathway in promoting fatty liver. Downregulation of CITED2 in NAFLD, a coactivator of PPAR α causes impaired fatty acid oxidation and increased fat storage (Tien et al., 2004).

2) Cytokines and Inflammatory Response

Role of Kupffer Cells

Chronic ethanol consumption is characterized by an increase in the expression of a number of inflammatory mediators, including cytokines, reactive oxygen, nitrogen species, and chemokines (McClain et al., 2002; Nanji et al., 2001a). One of the primary mechanisms for increased production of inflammatory mediators in response to ethanol exposure is due to activation of Kupffer cells in the liver (Jarvelainen et al., 2000; Nanji, 2002). Ablation of Kupffer cells prevents the development of fatty liver and inflammation in rats chronically exposed to ethanol (Jarvelainen et al., 2000). The upregulation of the Kupffer cell receptor in our study indicates that activation of Kupffer cells occurred in response to alcohol. Additional mechanisms of Kupffer cell activation during chronic ethanol exposure may involve activation of immune mechanisms in response to oxidative damage to hepatic proteins and phospholipids (Mottaran et al., 2002). As seen in Table 2, many genes involved in the immune response were upregulated.

Other markers whose increase is consistent with the presence of an inflammatory response include upregulation of Interferon induced Guanylate binding protein 2 (gbp2) (Neun et al., 1996), Kupffer cell receptor (Kclr), and prolactin receptor all of which have been shown to be involved in inflammatory processes (Crunkhorn et al., 2004) and were up-regulated in livers of FFE rats (Table 2). Aminopeptidase B, is increased after chronic intake of ethanol (Mayas et al., 2001) and is a putative marker of T lymphocytes activation (Belhacene et al., 1993). S100A8 proteins are also upregulated in the acute phase of inflammation (Ikemoto et al., 2007).

In NAFLD, growth regulated oncogene (gro), histidine decarboxylase (HDC), and benzodiazepine receptor genes, which are reported to be involved in inflammatory processes, were up-regulated in livers of FFD rats (Table 2). The chemokine CXC receptor family of proteins including GRO, are increased after stimulation by IL-1 or TNF-α (De Ceuninck et al., 2004; Kono et al., 2001; Lorenzetti et al., 2002) both cytokines were up-regulated in NASH and are capable of inducing HDC in the liver (Endo et al., 1995). Other markers which are expressed in an inflammatory response include upregulation of benzodiazepine receptor expression and down regulation of alpha 1-Inhibitor III (Abraham et al., 1990; Olofsson et al., 2002).

Superarray Cytokines Gene Expression Profile

We compared gene profiles of the rats fed MCT plus ethanol, palm oil plus ethanol and fish oil and ethanol with the control rats using GEArray Expression It is clear from Figure 8 that cytokine receptor like protein CYRL, FGF activating protein, IL-10 receptor alpha, IL-1 receptor1, IL-1 receptor accessory protein, IL-4 receptor, IL-6 receptor, TGF beta 1 induced transcript 4 were expressed in rats fed with MCT, palm oil or fish oil and ethanol. We did not observe any significant difference in expression level of these genes in MCT plus dextrose vs MCT plus ethanol fed rats supporting the idea that saturated fatty acids are hepatoprotective. In comparison to MCT and ethanol, rats fed with palm oil and ethanol showed a significant change in anti-inflammatory cytokines IL-10 receptor alpha and IL-4R. These cytokine receptors were up-regulated in the liver of palm oil and ethanol fed rats. When compared palm oil to fish oil, the most significant change was observed in cytokine IL-10 receptor alpha, IL-4R, TGF-βi4, FGF receptor activating protein1, and IL-6st. It is well established that Interleukin-10 (IL-10) is a potent anti-inflammatory cytokine. The interaction of IL-10 with its receptor activates the tyrosine kinases Jak1 and Tyk2, which are associated with the IL-10R1 and IL-10R2, respectively (Ding et al., 2001). It is evident from the figure 8 that anti-inflammatory cytokine IL-10 receptor alpha is down regulated but other cytokines are upregulated in female rats fed with fish oil and ethanol. Increased expression of IL-10 controls inflammatory processes by suppressing the expression of proinflammatory cytokines such as IL-6, chemokines, adhesion molecules, as well as antigen-presenting and costimulatory molecules in inflammatory, and T cells (Asadullah et al., 2003). Therefore, it is reasonable to speculate that decreased IL-10 expression may contribute to increased expression of IL-6st and IL-4R resulting in more severe injury in ALD in female than male rats. This observation is in the line of previous reports which show that chronic alcohol exposure inhibits the expression of IL-10 (Hill et al., 2002). In patients with acute alcoholic hepatitis, expression of IL-10 was below the detection limit in the serum (Sanchez Perez et al., 2006) suggesting an important role of IL-10 in hepatoprotective mechanism. IL-4 induces apoptosis of human hepatocytes through IL-4 receptor binding, STAT6 activation, decreased mitochondrial membrane potential, and increased caspase activation (Aoudjehane et al., 2007). In rat models, IL-4 can induce lethal hepatitis mediated by hepatocyte apoptosis (Guillot et al., 2001). In recent study of drug-induced hepatitis, IL-10 and IL-4 help to prevent over expression of IL-6 and other potential hepatotoxic factors (Bourdi et al., 2007). In animal models of fatty liver, developed in leptin deficient ob/ob mice, ethanol-fed mice, and mice fed a high-fat diet, IL-6 treatment decreased steatosis and normalized serum aminotransferase. IL-6 treatment in vivo resulted in part from an increase in mitochondrial beta oxidation of fatty acid and an increase in hepatic export of triglyceride and cholesterol (Nonogaki et al., 1995). IL-6 treatment increased hepatic peroxisome proliferator-activated receptor (PPAR) alpha and decreased liver and serum tumor necrosis factor (TNF) alpha (Van Bladel et al., 1991), therefore, ameliorates liver injury, suggesting the therapeutic potential of IL-6 in treating human fatty liver disease (Hong et al., 2004; Memon et al., 2001; Nonogaki et al., 1995; Sanchez Perez et al., 2006; Zhang et al., 2006; Zhang et al., 2005). Other cytokine receptors were increased in an inflammatory response include upregulation of IL-1 receptor1, IL-1 receptor accessory protein in ALD (Chen et al., 2005; Hoek and Pastorino, 2002; Koo et al., 2002; Valles et al., 2003). Permanent alcohol abuse may lead to chronic liver injury with deleterious sequelae such as liver cirrhosis and hepatocellular carcinoma. Mechanisms of fibrogenesis include recruitment of inflammatory cells at the site of injury and cytokine mediated activation of hepatic stellate cells (HSC) with accumulation of interstitial collagens. IL-6 produced during the acute phase response, alone or in conjunction with other cytokines, could play an important role in liver fibrogenesis by inducing the expression of collagen, fibronectin, and transforming growth factor-beta (Greenwel et al., 1995). Transforming growth factor-beta represents the main profibrogenic cytokine in liver fibrosis and other fibroproliferative disorders (Bedossa and Paradis, 1995; Chevallier et al., 1995). In parallel, TGF-beta triggers hepatocytes to undergo apoptosis, thereby inducing HSC proliferation and generation of a collagenous matrix (Breitkopf et al., 2005; George et al., 2003; Gnainsky et al., 2007; Prosser et al., 2006; Senties-Gomez et al., 2005; Spee et al., 2006; Wang et al., 2006; Wiercinska et al., 2006; Xu et al., 2004; Zhuge and Cederbaum, 2006). An increase in IL-6st and TGF-βi4 in our study suggests a shift in fibrogenic phenotypes in female liver in response to fish oil and alcohol.

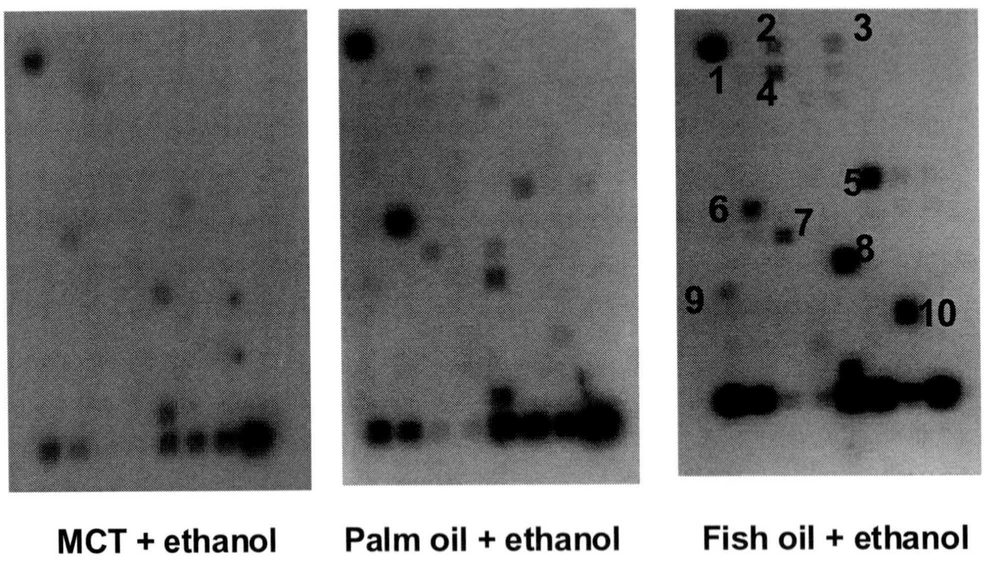

Figure 8. Inflammatory cytokines and receptors microarray (ORN-011) probed with cRNA of rats fed with medium chain triglycerides plus ethanol, palm oil plus ethanol and fish oil and ethanol and exposed on X-ray film. The scanned images of microarray showing differential expression of genes (1- peptidylprolyl isomerase A, 2-bone morphogenetic protein 2, 3-caspase 1, 4- CCAAT/enhancer binding protein, 5- FGF activating protein, 6- IL-10 receptor alpha, 7- IL-1 receptor 1, 8- IL-4 receptor, 9- IL-6 receptor, and 10- TGF beta 1 induced transcript 4).

Genes Relevant to Enhanced Oxidant Stress

Mitochondrial dysfunction is one factor that contributes to oxidative stress in NAFLD and ALD (Angulo and Lindor, 2001; Pessayre et al., 2004). Mitochondrial enzymes acyl-CoA thioesterase1 (MTE1) and cytosolic acyl CoA thioesterase1 (Cte1) (Table 2) were markedly up-regulated in livers of female rats of NAFLD and ALD. These enzymes are expressed in tissues having high rates of fatty acid oxidation and regulated by PPAR-α, fatty acid and insulin (Bostrom et al., 2004; Dongol et al., 2007; Durgan et al., 2006). Impaired mitochondrial function which is suggested by increased levels of the mitochondrial enzyme, glycerol 3 phosphate dehydogenase 2 (Table 2), contributes to both oxidative stress and inflammatory injury. ROS can also induce a variety of genes encoding proteins that possess anti-oxidant properties. Of particular note in this study was the upregulation of hemeoxygenase-1, 2, glutathione peroxidase 2, and down regulation of MT1A which can operate as a defense mechanism against tissue injury (Cantoni et al., 1991; Rizzardini et al., 1998). The MT1A gene encodes a protein responsible for cell detoxification (Yu et al., 2003). NF-E2-related factor 2 (Nrf2) regulates the induction of genes encoding antioxidant proteins and phase 2 detoxifying enzymes. Nrf2 is a key factor for cytoprotection against carcinogens and pro-inflammatory agents (Kobayashi and Yamamoto, 2005; Rim and Kozak, 2002).

In NAFLD decreased levels of the mitochondrial enzyme, NADH dehydrogenase (Table 2) is a marker for mitochondrial dysfunction (Lu et al., 2000). The upregulation of hemeoxygenase-1 (HO-1) can protect liver against injury (Dorman et al., 2004).

Cell Cycle Regulatory Genes

Fatty acids are not only a source of energy but also regulate molecular events that govern gene expression (Sabin et al., 2007). It is evident from table 2 that many genes upregulated are transducers of signals to cells for preserving vital functions. In (Figure 2) genes involved in cell death, apoptosis, and proliferation, these were regulated by myc and stat signaling pathways, further confirming that cell cycle regulatory genes play important role in pathogenesis of liver injury. The role of MAPK signaling is well documented in hepatocyte apoptosis (Czaja, 2002). Although, many genes that are modulated in female rat liver promote liver injury in response to fish oil and alcohol, the most highly upregulated genes were DNA primase, dentatorubral pallidoluysian atrophy, X-ray repair complementing defective repair, achaete-scute complex homolog like 2, cell division cycle 2 homolog all of which contribute to DNA repair, replication, cell death and apoptosis. It should be emphasized that mitochondria and oxidative stress may also play key roles in apoptosis in ALD.

Conclusion

Our studies have defined the gene expression profiles in alcoholic and non-alcoholic liver diseases. The studies confirm some of the pathways implicated in ALD and NASH but also identify novel genes and pathways not identified previously.

Acknowledgment

This work was supported by National Institutes on Alcohol abuse and alcoholism grant AA12892.

REFERENCES

Abraham, L.J., A.D. Bradshaw, B.R. Shiels, W. Northemann, G. Hudson, and G.H. Fey. 1990. Hepatic transcription of the acute-phase alpha 1-inhibitor III gene is controlled by a novel combination of cis-acting regulatory elements. *Mol Cell Biol*. **10**:3483-91.

Angulo, P., and K.D. Lindor. 2001. Insulin resistance and mitochondrial abnormalities in NASH: a cool look into a burning issue. *Gastroenterology*. **120**:1281-5.

Aoudjehane, L., P. Podevin, O. Scatton, P. Jaffray, I. Dusanter-Fourt, G. Feldmann, P.P. Massault, L. Grira, A. Bringuier, B. Dousset, S. Chouzenoux, O. Soubrane, Y. Calmus, and F. Conti. 2007. Interleukin-4 induces human hepatocyte apoptosis through a Fas-independent pathway. *Faseb J*. **21**:1433-44.

Arlinde, C., W. Sommer, K. Bjork, M. Reimers, P. Hyytia, K. Kiianmaa, and M. Heilig. 2004. A cluster of differentially expressed signal transduction genes identified by microarray analysis in a rat genetic model of alcoholism. *Pharmacogenomics J*. **4**:208-18.

Asadullah, K., W. Sterry, and H.D. Volk. 2003. Interleukin-10 therapy--review of a new approach. *Pharmacol Rev*. **55**:241-69.

Bedossa, P., and V. Paradis. 1995. Transforming growth factor-beta (TGF-beta): a key-role in liver fibrogenesis. *J Hepatol*. **22**:37-42.

Belhacene, N., B. Mari, B. Rossi, and P. Auberger. 1993. Characterization and purification of T lymphocyte aminopeptidase B: a putative marker of T cell activation. *Eur J Immunol*. **23**:1948-55.

Bostrom, M., S.E. Alexson, B. Lundgren, B.D. Nelson, and J.W. DePierre. 2004. The expression of cytosolic and mitochondrial type II acyl-CoA thioesterases is upregulated in the porcine corpus luteum during pregnancy. *Prostaglandins Leukot Essent Fatty Acids*. **71**:319-27.

Bourdi, M., D.P. Eiras, M.P. Holt, M.R. Webster, T.P. Reilly, K.D. Welch, and L.R. Pohl. 2007. Role of IL-6 in an IL-10 and IL-4 double knockout mouse model uniquely susceptible to acetaminophen-induced liver injury. *Chem Res Toxicol*. **20**:208-16.

Breitkopf, K., S. Haas, E. Wiercinska, M.V. Singer, and S. Dooley. 2005. Anti-TGF-beta strategies for the treatment of chronic liver disease. *Alcohol Clin Exp Res*. 29:121S-131S.

Brown, P.O., and D. Botstein. 1999. Exploring the new world of the genome with DNA microarrays. *Nat Genet*. **21**:33-7.

Cabrero, A., J.C. Laguna, and M. Vazquez. 2002. Peroxisome proliferator-activated receptors and the control of inflammation. *Curr Drug Targets Inflamm Allergy*. **1**:243-8.

Canbay, A., L. Bechmann, and G. Gerken. 2007. Lipid metabolism in the liver. *Z Gastroenterol*. **45**:35-41.

Cantoni, L., C. Rossi, M. Rizzardini, M. Gadina, and P. Ghezzi. 1991. Interleukin-1 and tumour necrosis factor induce hepatic haem oxygenase. Feedback regulation by glucocorticoids. *Biochem J*. **279** (Pt 3):891-4.

Carulli, L., A. Lonardo, S. Lombardini, G. Marchesini, and P. Loria. 2006. Gender, fatty liver and GGT. *Hepatology*. **44**:278-9.

Chan, K.L., X.Y. Guan, and I.O. Ng. 2004. High-throughput tissue microarray analysis of c-myc activation in chronic liver diseases and hepatocellular carcinoma. *Hum Pathol*. **35**:1324-31.

Chen, W.X., G.Y. Xu, C.H. Yu, P. Wang, M. Yu, P. Xu, Y.S. Liu, and Y.M. Li. 2005. Correlation of polymorphism in the interleukin-1 receptor antagonist gene intron 2 with alcoholic liver disease. *Hepatobiliary Pancreat Dis Int.* **4**:41-5.

Chevallier, M., V. Paradis, and P. Bedossa. 1995. [Liver fibrosis]. *Ann Pathol.* **15**:372-9.

Chong, Y.M., A. Subramanian, A.K. Sharma, and K. Mokbel. 2007. The potential clinical applications of insulin-like growth factor-1 ligand in human breast cancer. *Anticancer Res.* **27**:1617-24.

Crunkhorn, S.E., K.E. Plant, G.G. Gibson, K. Kramer, J. Lyon, P.G. Lord, and N.J. Plant. 2004. Gene expression changes in rat liver following exposure to liver growth agents: role of Kupffer cells in xenobiotic-mediated liver growth. *Biochem Pharmacol.* **67**: 107-18.

Czaja, M.J. 2002. Induction and regulation of hepatocyte apoptosis by oxidative stress. *Antioxid Redox Signal.* **4**:759-67.

Day, C.P. 2002a. NASH-related liver failure: one hit too many? *Am J Gastroenterol.* **97**:1872-4.

Day, C.P. 2002b. Non-alcoholic steatohepatitis (NASH): where are we now and where are we going? *Gut.* **50**:585-8.

Day, C.P. 2006. Non-alcoholic fatty liver disease: current concepts and management strategies. *Clin Med.* **6**:19-25.

De Ceuninck, F., L. Dassencourt, and P. Anract. 2004. The inflammatory side of human chondrocytes unveiled by antibody microarrays. *Biochem Biophys Res Commun.* **323**:960-9.

Deaciuc, I.V., D.E. Doherty, R. Burikhanov, E.Y. Lee, A.J. Stromberg, X. Peng, and W.J. de Villiers. 2004a. Large-scale gene profiling of the liver in a mouse model of chronic, intragastric ethanol infusion. *J Hepatol.* **40**:219-27.

Deaciuc, I.V., X. Peng, N.B. D'Souza, S.I. Shedlofsky, R. Burikhanov, I.V. Voskresensky, and W.J. de Villiers. 2004b. Microarray gene analysis of the liver in a rat model of chronic, voluntary alcohol intake. *Alcohol.* **32**:113-27.

Deaciuc, I.V., Z. Song, and C.J. McClain. 2005. Lessons from large-scale gene profiling of the liver in alcoholic liver disease. *Hepatol Res.* **31**:187-92.

Ding, Y., L. Qin, D. Zamarin, S.V. Kotenko, S. Pestka, K.W. Moore, and J.S. Bromberg. 2001. Differential IL-10R1 expression plays a critical role in IL-10-mediated immune regulation. *J Immunol.* **167**:6884-92.

Diomede, L., D. Albani, M. Bianchi, and M. Salmona. 2001. Endotoxin regulates the maturation of sterol regulatory element binding protein-1 through the induction of cytokines. *Eur Cytokine Netw.* **12**:625-30.

Dongol, B., Y. Shah, I. Kim, F.J. Gonzalez, and M.C. Hunt. 2007. The acyl-CoA thioesterase I is regulated by PPARalpha and HNF4alpha via a distal response element in the promoter. *J Lipid Res.* **48**:1781-91.

Dorman, R.B., M.L. Bajt, A. Farhood, J. Mayes, and H. Jaeschke. 2004. Heme oxygenase-1 induction in hepatocytes and non-parenchymal cells protects against liver injury during endotoxemia. *Comp Hepatol.* **3** Suppl 1:S42.

Durgan, D.J., J.K. Smith, M.A. Hotze, O. Egbejimi, K.D. Cuthbert, V.G. Zaha, J.R. Dyck, E.D. Abel, and M.E. Young. 2006. Distinct transcriptional regulation of long-chain acyl-CoA synthetase isoforms and cytosolic thioesterase 1 in the rodent heart by fatty acids and insulin. *Am J Physiol Heart Circ Physiol.* **290**:H2480-97.

Eberle, D., B. Hegarty, P. Bossard, P. Ferre, and F. Foufelle. 2004. SREBP transcription factors: master regulators of lipid homeostasis. *Biochimie*. **86**:839-48.

Endo, Y., M. Nakamura, Y. Nitta, and K. Kumagai. 1995. Effects of macrophage depletion on the induction of histidine decarboxylase by lipopolysaccharide, interleukin 1 and tumour necrosis factor. *Br J Pharmacol*. **114**:187-93.

Enomoto, N., Y. Takei, S. Yamashina, K. Ikejima, S. Suzuki, T. Kitamura, and N. Sato. 2004. [Gender difference in alcoholic liver injury]. *Nihon Arukoru Yakubutsu Igakkai Zasshi*. **39**:163-7.

Esfandiari, F., J.A. Villanueva, D.H. Wong, S.W. French, and C.H. Halsted. 2005. Chronic ethanol feeding and folate deficiency activate hepatic endoplasmic reticulum stress pathway in micropigs. *Am J Physiol Gastrointest Liver Physiol*. **289**:G54-63.

Feitelson, M.A. 2004. c-myc overexpression in hepatocarcinogenesis. *Hum Pathol*. **35**: 1299-302.

Fischer, M., M. You, M. Matsumoto, and D.W. Crabb. 2003. Peroxisome proliferator-activated receptor alpha (PPARalpha) agonist treatment reverses PPARalpha dysfunction and abnormalities in hepatic lipid metabolism in ethanol-fed mice. *J Biol Chem*. **278**:27997-8004.

Forgione, A., L. Miele, C. Cefalo, G. Gasbarrini, and A. Grieco. 2007. Alcoholic and nonalcoholic forms of fatty liver disease. *Minerva Gastroenterol Dietol*. **53**:83-100.

French, S.W. 1993. Nutrition in the pathogenesis of alcoholic liver disease. *Alcohol Alcohol*. **28**:97-109.

Fujii, H. 2005. [Anti-inflammatory action of PPARs]. *Nippon Rinsho*. **63**:609-15.

George, J., N. Pera, N. Phung, I. Leclercq, J. Yun Hou, and G. Farrell. 2003. Lipid peroxidation, stellate cell activation and hepatic fibrogenesis in a rat model of chronic steatohepatitis. *J Hepatol*. **39**:756-64.

Gnainsky, Y., Z. Kushnirsky, G. Bilu, Y. Hagai, O. Genina, H. Volpin, R. Bruck, G. Spira, A. Nagler, N. Kawada, K. Yoshizato, D.P. Reinhardt, T.A. Libermann, and M. Pines. 2007. Gene expression during chemically induced liver fibrosis: effect of halofuginone on TGF-beta signaling. *Cell Tissue Res*. **328**:153-66.

Greenwel, P., M.J. Iraburu, M. Reyes-Romero, N. Meraz-Cruz, E. Casado, J.A. Solis-Herruzo, and M. Rojkind. 1995. Induction of an acute phase response in rats stimulates the expression of alpha 1(I) procollagen messenger ribonucleic acid in their livers. Possible role of interleukin-6. *Lab Invest*. **72**:83-91.

Gregoire, F.M., Q. Zhang, S.J. Smith, C. Tong, D. Ross, H. Lopez, and D.B. West. 2002. Diet-induced obesity and hepatic gene expression alterations in C57BL/6J and ICAM-1-deficient mice. *Am J Physiol Endocrinol Metab*. **282**:E703-13.

Guay, C., S.R. Madiraju, A. Aumais, E. Joly, and M. Prentki. 2007. A role for ATP-citrate lyase, malic enzyme and pyruvate/citrate cycling in glucose-induced insulin secretion. *J Biol Chem*.

Guillot, C., H. Coathalem, J. Chetritt, A. David, P. Lowenstein, E. Gilbert, L. Tesson, N. van Rooijen, M.C. Cuturi, J.P. Soulillou, and I. Anegon. 2001. Lethal hepatitis after gene transfer of IL-4 in the liver is independent of immune responses and dependent on apoptosis of hepatocytes: a rodent model of IL-4-induced hepatitis. *J Immunol*. **166**: 5225-35.

Hill, D.B., N.B. D'Souza, E.Y. Lee, R. Burikhanov, I.V. Deaciuc, and W.J. de Villiers. 2002. A role for interleukin-10 in alcohol-induced liver sensitization to bacterial lipopolysaccharide. *Alcohol Clin Exp Res*. **26**:74-82.

Hoek, J.B., and J.G. Pastorino. 2002. Ethanol, oxidative stress, and cytokine-induced liver cell injury. *Alcohol*. **27**:63-8.

Hoffman, P., and B. Tabakoff. 2005. Gene expression in animals with different acute responses to ethanol. *Addict Biol*. **10**:63-9.

Hong, F., S. Radaeva, H.N. Pan, Z. Tian, R. Veech, and B. Gao. 2004. Interleukin 6 alleviates hepatic steatosis and ischemia/reperfusion injury in mice with fatty liver disease. *Hepatology*. **40**:933-41.

Hosack, D.A., G. Dennis, Jr., B.T. Sherman, H.C. Lane, and R.A. Lempicki. 2003. Identifying biological themes within lists of genes with EASE. *Genome Biol*. **4**:R70.

Ikemoto, M., H. Murayama, H. Itoh, M. Totani, and M. Fujita. 2007. Intrinsic function of S100A8/A9 complex as an anti-inflammatory protein in liver injury induced by lipopolysaccharide in rats. *Clin Chim Acta*. **376**:197-204.

Jarvelainen, H.A., C. Fang, M. Ingelman-Sundberg, T.A. Lukkari, H. Sippel, and K.O. Lindros. 2000. Kupffer cell inactivation alleviates ethanol-induced steatosis and CYP2E1 induction but not inflammatory responses in rat liver. *J Hepatol*. **32**:900-10.

Ji, C., C. Chan, and N. Kaplowitz. 2006. Predominant role of sterol response element binding proteins (SREBP) lipogenic pathways in hepatic steatosis in the murine intragastric ethanol feeding model. *J Hepatol*. **45**:717-24.

Kaplowitz, N., and C. Ji. 2006. Unfolding new mechanisms of alcoholic liver disease in the endoplasmic reticulum. *J Gastroenterol Hepatol*. **21** Suppl 3:S7-9.

Kobayashi, M., and M. Yamamoto. 2005. Molecular mechanisms activating the Nrf2-Keap1 pathway of antioxidant gene regulation. *Antioxid Redox Signal*. **7**:385-94.

Kono, H., T. Uesugi, M. Froh, I. Rusyn, B.U. Bradford, and R.G. Thurman. 2001. ICAM-1 is involved in the mechanism of alcohol-induced liver injury: studies with knockout mice. *Am J Physiol Gastrointest Liver Physiol*. **280**:G1289-95.

Koo, H.N., S.H. Hong, H.J. Jeong, E.H. Lee, N.G. Kim, S.D. Choi, K.W. Ra, K.S. Kim, B.K. Kang, J.J. Kim, J.G. Oh, and H.M. Kim. 2002. Inhibitory effect of Artemisia capillaris on ethanol-induced cytokines (TNF-alpha, IL-1alpha) secretion in Hep G2 cells. *Immunopharmacol Immunotoxicol*. **24**:441-53.

Koteish, A., and A.M. Diehl. 2001. Animal models of steatosis. *Semin Liver Dis*. 21:89-104.

Kotoh, K., M. Nakamuta, M. Fukushima, S. Morizono, M. Enjoji, and H. Nawata. 2007. Fertile females with nonalcoholic fatty liver disease (NAFLD) have higher levels of ALT than postmenopausal females: implications for the influence of fertility on NAFLD. *Hepatogastroenterology*. **54**:224-8.

Le Jossic-Corcos, C., G.M. Pastori, S. Duclos, Y. Kawabe, T. Pineau, and P. Bournot. 2004. Peroxisome proliferator-activated receptor alpha (PPARalpha) activators induce hepatic farnesyl diphosphate synthase gene expression in rodents. *J Steroid Biochem Mol Biol*. **88**:203-11.

Lewohl, J.M., P.R. Dodd, R.D. Mayfield, and R.A. Harris. 2001. Application of DNA microarrays to study human alcoholism. *J Biomed Sci*. **8**:28-36.

Lorenzetti, B.B., F.H. Veiga, C.A. Canetti, S. Poole, F.Q. Cunha, and S.H. Ferreira. 2002. Cytokine-induced neutrophil chemoattractant 1 (CINC-1) mediates the sympathetic component of inflammatory mechanical hypersensitivitiy in rats. *Eur Cytokine Netw.* **13**:456-61.

Lu, F., M. Selak, J. O'Connor, S. Croul, C. Lorenzana, C. Butunoi, and B. Kalman. 2000. Oxidative damage to mitochondrial DNA and activity of mitochondrial enzymes in chronic active lesions of multiple sclerosis. *J Neurol Sci.* **177**:95-103.

Mayas, M.D., M.J. Ramirez-Exposito, M.J. Garcia, M. Ramirez, and J.M. Martinez-Martos. 2001. [Influence of alcohol on brain aminopeptidases. An in vitro study]. *Rev Neurol.* **32**:1031-40.

McClain, C.J., D.B. Hill, Z. Song, I. Deaciuc, and S. Barve. 2002. Monocyte activation in alcoholic liver disease. *Alcohol.* **27**:53-61.

McNeil, C.M., C.M. Sergio, L.R. Anderson, C.K. Inman, S.A. Eggleton, N.C. Murphy, E.K. Millar, P. Crea, J.G. Kench, M.C. Alles, M. Gardiner-Garden, C.J. Ormandy, A.J. Butt, S.M. Henshall, E.A. Musgrove, and R.L. Sutherland. 2006. c-Myc overexpression and endocrine resistance in breast cancer. *J Steroid Biochem Mol Biol.* **102**:147-55.

Memon, R.A., A.H. Moser, J.K. Shigenaga, C. Grunfeld, and K.R. Feingold. 2001. In vivo and in vitro regulation of sterol 27-hydroxylase in the liver during the acute phase response. potential role of hepatocyte nuclear factor-1. *J Biol Chem.* **276**:30118-26.

Miyahara, T., L. Schrum, R. Rippe, S. Xiong, H.F. Yee, Jr., K. Motomura, F.A. Anania, T.M. Willson, and H. Tsukamoto. 2000. Peroxisome proliferator-activated receptors and hepatic stellate cell activation. *J Biol Chem.* **275**:35715-22.

Morimoto, M., A.L. Hagbjork, A.A. Nanji, M. Ingelman-Sundberg, K.O. Lindros, P.C. Fu, E. Albano, and S.W. French. 1993. Role of cytochrome P4502E1 in alcoholic liver disease pathogenesis. *Alcohol.* **10**:459-64.

Mottaran, E., S.F. Stewart, R. Rolla, D. Vay, V. Cipriani, M. Moretti, M. Vidali, M. Sartori, C. Rigamonti, C.P. Day, and E. Albano. 2002. Lipid peroxidation contributes to immune reactions associated with alcoholic liver disease. *Free Radic Biol Med.* **32**:38-45.

Muller, C. 2006. Liver, alcohol and gender. *Wien Med Wochenschr.* **156**:523-6.

Nagy, L.E. 2004. Molecular aspects of alcohol metabolism: transcription factors involved in early ethanol-induced liver injury. *Annu Rev Nutr.* **24**:55-78.

Nanji, A.A. 2002. Role of Kupffer cells in alcoholic hepatitis. *Alcohol.* **27**:13-5.

Nanji, A.A. 2004. Another animal model for nonalcoholic steatohepatitis: how close to the human condition? *Am J Clin Nutr.* **79**:350-1.

Nanji, A.A., A.J. Dannenberg, K. Jokelainen, and N.M. Bass. 2004. Alcoholic liver injury in the rat is associated with reduced expression of peroxisome proliferator-alpha (PPARalpha)-regulated genes and is ameliorated by PPARalpha activation. *J Pharmacol Exp Ther.* **310**:417-24.

Nanji, A.A., F. Fogt, and B. Griniuviene. 1995. Alterations in glucose transporter proteins in alcoholic liver disease in the rat. *Am J Pathol.* **146**:329-34.

Nanji, A.A., K. Jokelainen, M. Fotouhinia, A. Rahemtulla, P. Thomas, G.L. Tipoe, G.L. Su, and A.J. Dannenberg. 2001a. Increased severity of alcoholic liver injury in female rats: role of oxidative stress, endotoxin, and chemokines. *Am J Physiol Gastrointest Liver Physiol.* **281**:G1348-56.

Nanji, A.A., K. Jokelainen, G.L. Tipoe, A. Rahemtulla, and A.J. Dannenberg. 2001b. Dietary saturated fatty acids reverse inflammatory and fibrotic changes in rat liver despite continued ethanol administration. *J Pharmacol Exp Ther.* **299**:638-44.

Nanji, A.A., U. Khettry, S.M. Sadrzadeh, and T. Yamanaka. 1993. Severity of liver injury in experimental alcoholic liver disease. Correlation with plasma endotoxin, prostaglandin E2, leukotriene B4, and thromboxane B2. *Am J Pathol.* **142**:367-73.

Nanji, A.A., D. Zakim, A. Rahemtulla, T. Daly, L. Miao, S. Zhao, S. Khwaja, S.R. Tahan, and A.J. Dannenberg. 1997. Dietary saturated fatty acids down-regulate cyclooxygenase-2 and tumor necrosis factor alfa and reverse fibrosis in alcohol-induced liver disease in the rat. *Hepatology.* **26**:1538-45.

Nanji, A.A., S. Zhao, S.M. Sadrzadeh, A.J. Dannenberg, S.R. Tahan, and D.J. Waxman. 1994. Markedly enhanced cytochrome P450 2E1 induction and lipid peroxidation is associated with severe liver injury in fish oil-ethanol-fed rats. *Alcohol Clin Exp Res.* **18**:1280-5.

Neun, R., M.F. Richter, P. Staeheli, and M. Schwemmle. 1996. GTPase properties of the interferon-induced human guanylate-binding protein 2. *FEBS Lett.* **390**:69-72.

Nonogaki, K., G.M. Fuller, N.L. Fuentes, A.H. Moser, I. Staprans, C. Grunfeld, and K.R. Feingold. 1995. Interleukin-6 stimulates hepatic triglyceride secretion in rats. *Endocrinology.* **136**:2143-9.

Olofsson, P., N. Nordquist, C. Vingsbo-Lundberg, A. Larsson, C. Falkenberg, U. Pettersson, B. Akerstrom, and R. Holmdahl. 2002. Genetic links between the acute-phase response and arthritis development in rats. *Arthritis Rheum.* **46**:259-68.

Pessayre, D., B. Fromenty, and A. Mansouri. 2004. Mitochondrial injury in steatohepatitis. *Eur J Gastroenterol Hepatol.* **16**:1095-105.

Pessayre, D., A. Mansouri, and B. Fromenty. 2002. Nonalcoholic steatosis and steatohepatitis. V. Mitochondrial dysfunction in steatohepatitis. *Am J Physiol Gastrointest Liver Physiol.* **282**:G193-9.

Pollack, J.R., C.M. Perou, A.A. Alizadeh, M.B. Eisen, A. Pergamenschikov, C.F. Williams, S.S. Jeffrey, D. Botstein, and P.O. Brown. 1999. Genome-wide analysis of DNA copy-number changes using cDNA microarrays. *Nat Genet.* **23**:41-6.

Prosser, C.C., R.D. Yen, and J. Wu. 2006. Molecular therapy for hepatic injury and fibrosis: where are we? *World J Gastroenterol.* **12**:509-15.

Rao, M.S., and J.K. Reddy. 2001. Peroxisomal beta-oxidation and steatohepatitis. *Semin Liver Dis.* **21**:43-55.

Repa, J.J., E.G. Lund, J.D. Horton, E. Leitersdorf, D.W. Russell, J.M. Dietschy, and S.D. Turley. 2000. Disruption of the sterol 27-hydroxylase gene in mice results in hepatomegaly and hypertriglyceridemia. Reversal by cholic acid feeding. *J Biol Chem.* **275**:39685-92.

Rim, J.S., and L.P. Kozak. 2002. Regulatory motifs for CREB-binding protein and Nfe2l2 transcription factors in the upstream enhancer of the mitochondrial uncoupling protein 1 gene. *J Biol Chem.* **277**:34589-600.

Rizzardini, M., M. Zappone, P. Villa, P. Gnocchi, M. Sironi, L. Diomede, C. Meazza, M. Monshouwer, and L. Cantoni. 1998. Kupffer cell depletion partially prevents hepatic heme oxygenase 1 messenger RNA accumulation in systemic inflammation in mice: role of interleukin 1beta. *Hepatology.* **27**:703-10.

Sabin, M.A., C.E. Stewart, E.C. Crowne, S.J. Turner, L.P. Hunt, G.I. Welsh, M.J. Grohmann, J.M. Holly, and J.P. Shield. 2007. Fatty acid-induced defects in insulin signalling, in myotubes derived from children, are related to ceramide production from palmitate rather than the accumulation of intramyocellular lipid. *J Cell Physiol.* **211**:244-52.

Sanchez Perez, M.J., E. Gonzalez-Reimers, F. Santolaria-Fernandez, M.J. de la Vega-Prieto, A. Martinez-Riera, P.A. Gonzalez, E. Rodriguez Rodriguez, and M.C. Duran-Castellon. 2006. Lipid peroxidation and serum cytokines in acute alcoholic hepatitis. *Alcohol Alcohol.* **41**:593-7.

Sato, N., K.O. Lindros, E. Baraona, K. Ikejima, E. Mezey, H.A. Jarvelainen, and V.A. Ramchandani. 2001. Sex difference in alcohol-related organ injury. *Alcohol Clin Exp Res.* **25**:40S-45S.

Schena, M., D. Shalon, R.W. Davis, and P.O. Brown. 1995. Quantitative monitoring of gene expression patterns with a complementary DNA microarray. *Science.* **270**:467-70.

Senties-Gomez, M.D., F.J. Galvez-Gastelum, E. Meza-Garcia, and J. Armendariz-Borunda. 2005. [Hepatic fibrosis: role of matrix metalloproteases and TGFbeta]. *Gac Med Mex.* **141**:315-22.

Shapiro, B.H., A.K. Agrawal, and N.A. Pampori. 1995. Gender differences in drug metabolism regulated by growth hormone. *Int J Biochem Cell Biol.* **27**:9-20.

Sharma, M.C., A.K. Agrawal, M.R. Sharma, and B.H. Shapiro. 1998. Interactions of gender, growth hormone, and phenobarbital induction on murine Cyp2b expression. *Biochem Pharmacol.* **56**:1251-8.

Sharma, M.R., R. Polavarapu, D. Roseman, V. Patel, E. Eaton, P.B. Kishor, and A.A. Nanji. 2006. Transcriptional networks in a rat model for nonalcoholic fatty liver disease: a microarray analysis. *Exp Mol Pathol.* **81**:202-10.

Sharma, M.R., R. Polavarapu, D. Roseman, V. Patel, E. Eaton, P.B. Kishor, and A.A. Nanji. 2007. Increased severity of alcoholic liver injury in female verses male rats: A microarray analysis. *Exp Mol Pathol.*

Shimizu, I., and S. Ito. 2007. Protection of estrogens against the progression of chronic liver disease. *Hepatol Res.* **37**:239-47.

Sikela, J.M., E.J. Maclaren, Y. Kim, A. Karimpour-Fard, W.W. Cai, J. Pollack, R. Hitzemann, J. Belknap, S. McWeeney, R.T. Kerns, C. Downing, T.E. Johnson, K.J. Grant, B. Tabakoff, P. Hoffman, C.C. Wu, and M.F. Miles. 2006. DNA microarray and proteomic strategies for understanding alcohol action. *Alcohol Clin Exp Res.* **30**:700-8.

Spee, B., B. Arends, T.S. van den Ingh, B. Brinkhof, H. Nederbragt, J. Ijzer, T. Roskams, L.C. Penning, and J. Rothuizen. 2006. Transforming growth factor beta-1 signalling in canine hepatic diseases: new models for human fibrotic liver pathologies. *Liver Int.* **26**:716-25.

Tadic, S.D., M.S. Elm, H.S. Li, G.J. Van Londen, V.M. Subbotin, D.C. Whitcomb, and P.K. Eagon. 2002. Sex differences in hepatic gene expression in a rat model of ethanol-induced liver injury. *J Appl Physiol.* **93**:1057-68.

Thurman, R.G. 1998. II. Alcoholic liver injury involves activation of Kupffer cells by endotoxin. *Am J Physiol.* **275**:G605-11.

Tien, E.S., J.W. Davis, and J.P. Vanden Heuvel. 2004. Identification of the CREB-binding protein/p300-interacting protein CITED2 as a peroxisome proliferator-activated receptor alpha coregulator. *J Biol Chem.* **279**:24053-63.

Tilg, H., and C.P. Day. 2007. Management strategies in alcoholic liver disease. *Nat Clin Pract Gastroenterol Hepatol.* **4**:24-34.

Tsukamoto, H., M. Matsuoka, and S.W. French. 1990. Experimental models of hepatic fibrosis: a review. *Semin Liver Dis.* **10**:56-65.

Tsukamoto, H., S.J. Towner, L.M. Ciofalo, and S.W. French. 1986. Ethanol-induced liver fibrosis in rats fed high fat diet. *Hepatology*. **6:**814-22.

Valles, S.L., A.M. Blanco, I. Azorin, R. Guasch, M. Pascual, M.J. Gomez-Lechon, J. Renau-Piqueras, and C. Guerri. 2003. Chronic ethanol consumption enhances interleukin-1-mediated signal transduction in rat liver and in cultured hepatocytes. *Alcohol Clin Exp Res*. **27**:1979-86.

Van Bladel, S., C. Libert, and W. Fiers. 1991. Interleukin-6 enhances the expression of tumor necrosis factor receptors on hepatoma cells and hepatocytes. *Cytokine*. **3**:149-54.

Wang, J.H., R.G. Batey, and J. George. 2006. Role of ethanol in the regulation of hepatic stellate cell function. *World J Gastroenterol*. **12**:6926-32.

Werstuck, G.H., S.R. Lentz, S. Dayal, G.S. Hossain, S.K. Sood, Y.Y. Shi, J. Zhou, N. Maeda, S.K. Krisans, M.R. Malinow, and R.C. Austin. 2001. Homocysteine-induced endoplasmic reticulum stress causes dysregulation of the cholesterol and triglyceride biosynthetic pathways. *J Clin Invest*. **107**:1263-73.

Wiercinska, E., L. Wickert, B. Denecke, H.M. Said, J. Hamzavi, A.M. Gressner, M. Thorikay, P. ten Dijke, P.R. Mertens, K. Breitkopf, and S. Dooley. 2006. Id1 is a critical mediator in TGF-beta-induced transdifferentiation of rat hepatic stellate cells. *Hepatology*. **43**:1032-41.

Wilfred de Alwis, N.M., and C.P. Day. 2007. Genetics of alcoholic liver disease and nonalcoholic fatty liver disease. *Semin Liver Dis*. **27**:44-54.

Xu, X.B., X.S. Leng, X. Yang, and Z.P. He. 2004. [Obstruction of TGF-beta1 signal transduction can decrease the process of hepatocellular carcinoma in mice induced by CCl4/ethanol]. *Zhonghua Yi Xue Za Zhi*. **84**:1122-5.

Yin, H.Q., M. Kim, J.H. Kim, G. Kong, K.S. Kang, H.L. Kim, B.I. Yoon, M.O. Lee, and B.H. Lee. 2007. Differential gene expression and lipid metabolism in fatty liver induced by acute ethanol treatment in mice. *Toxicol Appl Pharmacol*. **223**:225-33.

Yip, W.W., and A.D. Burt. 2006. Alcoholic liver disease. *Semin Diagn Pathol*. **23**:149-60.

You, M., M. Fischer, M.A. Deeg, and D.W. Crabb. 2002. Ethanol induces fatty acid synthesis pathways by activation of sterol regulatory element-binding protein (SREBP). *J Biol Chem*. **277**:29342-7.

Yu, J., H.Y. Zhang, Z.Z. Ma, W. Lu, Y.F. Wang, and J.D. Zhu. 2003. Methylation profiling of twenty four genes and the concordant methylation behaviours of nineteen genes that may contribute to hepatocellular carcinogenesis. *Cell Res*. **13**:319-33.

Zhang, X.G., P. Xu, Q. Liu, C.H. Yu, Y. Zhang, S.H. Chen, and Y.M. Li. 2006. Effect of tea polyphenol on cytokine gene expression in rats with alcoholic liver disease. *Hepatobiliary Pancreat Dis Int*. **5:**268-72.

Zhang, X.G., C.H. Yu, Q. Jiang, Y. Zhang, S.H. Chen, and Y.M. Lu. 2005. [Research on cytokines gene express of rats alcoholic liver disease affected by tea polyphenol]. *Zhongguo Zhong Yao Za Zhi*. **30**:847-50.

Zhuge, J., and A.I. Cederbaum. 2006. Increased toxicity by transforming growth factor-beta 1 in liver cells overexpressing CYP2E1. *Free Radic Biol Med*. **41**:1100-12.

In: Oligonucleotide Array Sequence Analysis
Editors: M.K. Moretti and L.J. Rizzo, pp. 209-243
ISBN: 978-1-60456-542-3

Chapter 7

Biological Interpretation in Oligonucleotide Array Sequence Analysis Based on Molecular Networks

Hiromi Sato [*]
Life Science Research Department, Institute of Medicinal Molecular Design, Inc. (IMMD) 5-24-5 Hongo, Bunkyo-ku, Tokyo 113-0033, Japan

Abstract

The availability of information on genome sequences has prompted researchers to clarify the molecular pathogenesis and molecular mechanisms underlying diseases, predict the toxicity of drugs, and realize personalized therapy immediately. The available information can help in achieving these ends; however, the process that would be involved for doing this is not very simple. Under these circumstances, DNA microarray, one of the technologies that employ this information, has become a well-established and useful tool for life scientists, with improvement in its sensitivity and probe design, etc. since the post-genome era. Nevertheless, researchers would be interested in seeking options for obtaining comprehensive gene expression data for their studies through a simple and inexpensive process. Further, even though such comprehensive data are acquired, the difficulties involved in the processing and interpretation of the large quantities of the data remain unresolved. We have therefore developed a life science information-orchestrated platform termed KeyMolnet that focuses on "molecular relations." This system consists of content databases collecting life science information, and a search software that can generate molecular networks based on the contents. When importing experimental and/or clinical data such as DNA microarray data into the system, molecular networks can be generated based on these data. Since the various kinds of life science information collected in the KeyMolnet, such as that on biomolecules, established pathways, diseases and drugs, are mutually related through key molecules, researchers could obtain working hypotheses, suggestions, and new insights from the generated molecular networks based on their data. Here, I demonstrate new approaches toward analyzing and profiling DNA microarray data for medical, pharmaceutical, and biological applications in practice by using molecular networks reflecting molecular mechanisms.

[*] E-mail address: hiromis@immd.co.jp.

ABBREVIATIONS

TZD	=	Thiazolidinediones
Pio	=	pioglitazone
Ros	=	rosiglitazone
Tro	=	troglitazone
GEO	=	Gene Expression Omnibus
GO	=	Gene Ontology
GOA	=	Gene Ontology Annotation
HPRD	=	human protein reference databases
AC	=	adenocarcinoma
BE	=	Barrett's esophagus
KO	=	knockout
WT	=	wildtype

INTRODUCTION

Previously, in investigating molecular mechanisms of the causes of onset and/or development of diseases, drug actions, and changes in biological processes by stimuli, a gene and/or a protein involved were initially required to be identified and isolated. Thus, for instance, processes such as cloning of DNA sequences for subjects and probe design for the northern blot analysis to measure and detect gene expression levels in genetic analysis were inevitable. First of all, the differences between normal and abnormal samples or between samples treated with chemicals and untreated ones had to be compared by thoroughly observing the pathology, responsiveness, gene and protein expression levels, etc., after establishing a suitable experimental model. On the contrary, in the post-genome era, we could first verify all the gene expression in our subjects because a set of probes for all the genome sequences has already been prepared and is easily available. We have experienced the so-called paradigm shift in the approach to life science research more than ever before.

The accomplishment of mapping the entire human genome had brought forth various magical visions such as an immediate resolution of the molecular pathogenesis and molecular mechanisms underlying diseases, a prompt prediction of drug toxicity, and also, an instant realization of personalized therapy. However, such advantageous achievements with multistep tasks are not easily attainable. Subsequently after decoding the entire human genome, the immediate realization of these advances appear to be far-fetched, and the need for patient efforts, appropriate strategy, and sufficient time have been accepted again.

Currently, DNA microarray analysis is one of the most effective applications of DNA sequence information [1]. Probe design and density have been improved year after year, and prices have reduced due to high-volume manufacturing, which led the way to the widespread use of the DNA microarray technology among life scientists [2, 3]. Previously, cDNA sequences such as hundred-mers, which are relatively long nucleotide sequences, were widely used as probes in DNA microarrays. However, oligonucleotide sequences have recently gained popularity among life scientists because high quality homogeneous chips can now be produced in large quantities by some companies, as compared to production using cDNA probes [4]. Thus, due to the improvement in chip quality and high reproducibility, DNA microarray technology is now well established as a general tool. Currently, Affymetrix (Gene Chip®) [5, 6], Agilent Technologies [7, 8], Illumina [9, 10], and GE (CodeLink®; discontinued) [11, 12] are the major manufactures of gene chips, and their products are actively used by scientists all over the world.

Obtaining more than tens of thousands of gene expression data at a time would be an interesting prospect for every researcher; however, once we have obtained this data, we recognized that we were at a loss as to how we can handle such large quantities of data. At this point, bioinformatics and systems biology became conspicuous. Comprehensive gene expression data and their changes were attempted to be understood as the tendency, classification, and/or signatures for a whole organism. In fact, the clustering analysis was actively performed to classify groups that exhibit similar gene expression behavior in several different conditions such as time points [13, 14]. This approach made it possible to classify similar samples that had not been previously elucidated into clinical and/or therapeutic differences at the molecular level. In the application of DNA microarray data classified by hierarchical clustering in diffuse large B cell lymphoma, tumors were more successfully classified into clinically significant cancer subtypes based on gene expression patterns than ever before [15]. In classification of ovarian cancer, which is one of the worst prognoses among gynecological malignancies, hierarchical clustering of gene expression data revealed distinctive groups of tissue samples that correlated to histological and pathological observations and cell-type specific gene expression [16]. Thus, classification of gene expression patterns using DNA microarray technology could take clinical studies one step forward to the molecular level [17-19]. However, the pathological classification based on gene expression patterns using DNA microarray data could not provide an insight into the molecular mechanisms for the cause of onset and/or development of diseases, actions of drugs, and biological changes. This is a challenge that is yet to be resolved.

Formation of molecular networks is essential to living organisms to exert the actions of chemicals, manifest drug toxicity, and lead to biological processes because proteins do not function alone; rather, they form complexes and networks. Molecular networks could be generated as the result of the actions of drugs, causes of onset and/or development of disease, or reflection of biological processes, which could be efficient ways to clarify their molecular mechanisms. Furthermore, their molecular mechanisms could be elucidated, which led to effective ways of evaluating candidates of molecular target of drugs. Although in recent years, information in life sciences, i.e., the accumulation of knowledge, has rapidly and dramatically increased, it has not been organized to be efficiently and effectively utilized in fields other than life sciences despite the fact that this information has immense potential. Therefore, a life science information platform that can generate molecular networks by using organized information was required. In this review, using the life science information platform termed KeyMolnet (Figure 1) [20, 21], I demonstrate and discuss methods of analyzing DNA microarray data, and their applications to medical, pharmaceutical, and life science research based on molecular network-centric understanding.

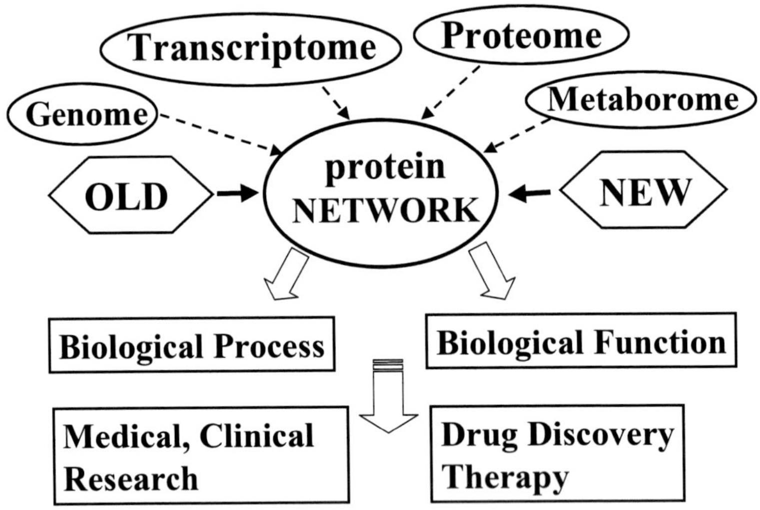

Figure 1. Molecular network-centric strategy.

Oligonucleotide Array Analysis Based on Molecular Networks

Organized Information in KeyMolnet

KeyMolnet is a life science information platform comprising content databases and a search software that can generate molecular networks based on the content information, developed by the Institute of Medicinal Molecular Design, Inc. (IMMD) (http://www.immd.co.jp/; Figure 2) [20, 21]. The content databases are human- and protein-centric databases categorized into the molecular network, disease and drug information that are collected, curated, compiled, and mined manually from literature such as review articles, original papers, and text books. The molecular network information is further devised into core, secondary, and user contents. The core content collects pathway information (well-known, termed, and canonical pathways), from high-level review articles. Each pathway is originally defined based on several review articles, and collected as fixed pathway diagrams whose component is a molecular relation among 2 or 3 proteins and/or small molecules. The core content collects the information that can be regarded as relatively and generally stable, basic, and well established among life scientists because it is derived from high-level review articles. On the other hand, the secondary content comprises accumulated molecular relations collected from original papers. Information collected in the secondary content may be unstable as compared with that in the core content, and could be revised, modified, and/or altered by new findings. The molecular relations in the secondary content can be characterized as detailed and newly obtained information in KeyMolnet. Furthermore, for the most recent information such as current, tentative, virtual, or experimental information, these molecular relations can be added to the core and/or the secondary contents as the user-defined content. Thus, the molecular network information is hierarchical and is designed to have several layers. Molecules are managed in the database by unified IDs such that a molecule with many synonyms can be recognized as a unique molecule by manual verification. The information in the human protein reference databases (HPRD®) is currently integrated with the KeyMolnet contents. HPRD is also a manually curated database collecting protein-protein interactions with domain architecture and post-translational modification information in the human proteome, which is directed and managed by Dr. Pandey of the Johns Hopkins University and the Institute of Bioinformatics [22-25].

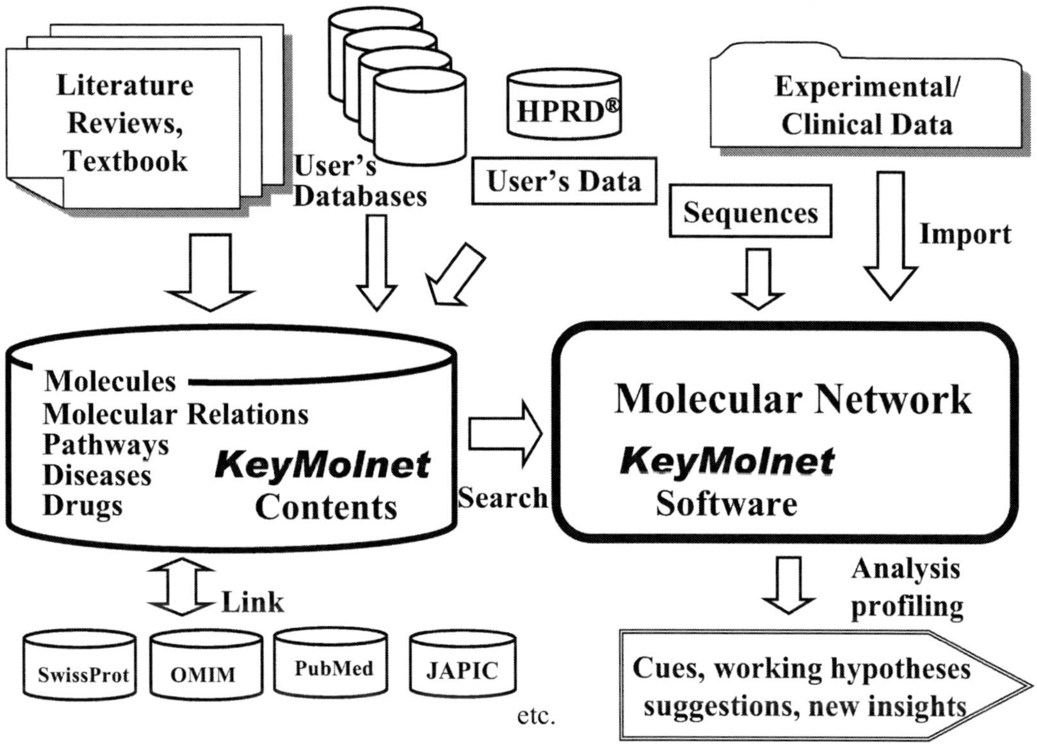

Figure 2. KeyMolnet overview.

Disease information is collected with a focus on molecules termed the Disease Mediating Molecules that are defined as molecules involved in the onset, progression and/or development of a certain disease, and is obtained from Harrison's Internal Medicine textbook (McGraw-Hill Companies, Japanese version by Medical Sciences International, Ltd.) and review articles. Furthermore, pathological information defined as the biological processes occurring in a diseased state is collected with key molecules termed Pathological Molecules. These molecules are managed in the same way as those in the molecular network information content in KeyMolnet.

The third portion of the content is the collection of drug information regarding drugs launched in Japan and the US, and focusing on key molecules in the drugs, termed the Target Molecules. This information is obtained by referring to the drug package insert, PDR (Physician's Desk Reference), and Goodman & Gilman's the pharmacological basis of therapeutics (McGraw-Hill Companies). Target Molecules are also managed in the same way as molecules in the molecular network and disease information content. Therefore, molecules in the molecular network information, Disease Mediating Molecules, and Drug Target

Molecules can be handled in the same way so that information in the 3 categories is interconnected and mediated by key molecules.

The Gene Ontology Annotation (GOA, http://www.ebi.ac.uk/GOA/) [26] and family information (http://au.expasy.org/sprot/) from the Gene Ontology (GO) Consortium [27, 28] by the European Bioinformatics Institute, the Swiss Institute of Bioinformatics, and the Universal protein resource (UniProt, http://www.expasy.uniprot.org/), respectively, are also collected in KeyMolnet. Bio-Event information is also collected using the biological process information in GOA.

Molecules in KeyMolnet are categorized according to proteins, small molecules, complexes, exogenous molecules, and user-defined molecules. Molecular relations are classified into positive and negative direct; positive and negative indirect; and transcriptional activations and repressions in the core, secondary, and HPRD contents. This difference is displayed in network diagrams by using different colors, edges, and nodes and their combinations.

All the information is provided in both Japanese and English.

Network Generation by KeyMolnet

Molecular networks can be generated by the KeyMolnet software in real time upon request by using the KeyMolnet contents in addition to HPRD, GO, and family information. After the search point(s) including molecules and the terms connecting with key molecules such as Disease Mediating and Drug Target Molecules in all databases is set, molecular relations will be searched and obtained from the KeyMolnet databases. If more than one term is used for the search, the searched molecular relations would be mediated by perfectly matched molecules, and molecular networks are generated as a result. Currently, 4 network search algorisms are used alone and/or in combination. The Neighboring Search method is used to search molecular relations one by one based on one or many search term(s) in the up- and/or downstream direction with any number of paths. The Start point to End point Search method is used to search molecular relations in the database with one search term set as the start point(s) and another search point(s) set as the end point(s) via the shortest paths or a specified numbers of paths. The Common upstream/downstream Search method is used to search molecular relations from every pair of search terms until the common upstream and/or downstream molecules are obtained. The Interrelation Search method is used to search minimal molecular networks containing as many search points as possible from multiple search points. The Weighed Search using an annotation assigned to a molecule enables network generation to extract by a certain condition such as subcelluar localization, expression levels in tissues or organs, molecular types, molecular relation types, and content types. For nucleotide or amino acid sequences, networks can be generated based on molecules searched by the blast search. With this method, molecules in species other than humans can be included as search points.

Network Extraction

Using the logical operations on molecular relations, the OR, DIFF, or AND operations can be used to extract all, subtracted, or shared parts of networks, respectively, in case of multiple networks. For example, subtracted parts of networks for multiple molecular networks reflecting molecular mechanisms in different pathological states can be regarded as condition-specific or condition-dependent networks while shared networks can be regarded as condition-independent networks.

Network Summary

A summary on the generated or extracted molecular networks is reported in the following content categories: Pathway, Disease, Pathological Event, Bio-Event, and GO. The number of molecules or molecular relations displayed in the generated or extracted networks and those collected in the contents are calculated, and the Score reveals possible relevance of the generated or extracted networks to the categories by the hypergeometric distribution method [29].

Subcellular Localization Display

A standard network diagram display is used for the even distribution of molecules. However, as a familiar display to life scientists, molecular networks can be displayed in subcellular localization such as a membrane, cytoplasm, nucleus, and mitochondria by using the cellular component information of GOA [28].

Experimental/Clinical Data Import into KeyMolnet

Experimental/clinical data such as DNA microarray data can be imported into KeyMolnet and analyzed [30-32]. Numerical data of gene expression levels and their ratios corresponding with probes or gene IDs are retrieved as is. However, several molecular IDs are required to convert into KeyMolnet IDs to ultimately generate molecular networks. In the process, multiple converting tables using the information in Ensembl (http://www.ebi.ac.uk/ensembl/) [33] and SwissProt are prepared. In the case of data in mice or rats, the data are further converted into KeyMolnet IDs by using the ortholog information provided by Ensembl [33]. The corresponding numerical data are displayed on the molecule as several colors assigned by a range of numerical values.

The network analyses demonstrated in this review were performed using KeyMolnet Contents 8.720071004181711.

APPLICATION OF DNA MICROARRY DATA TO PHARMACOLOGICAL AND TOXICOLOGICAL RESEARCH

Thiazolidinediones (TZDs) are used as insulin sensitizers for treating patients with type 2 diabetes mellitus [34-36]. The common target molecule for TZDs is peroxisome proliferator-activated receptor gamma (PPARγ) [37]. Pioglitazone (Pio) is currently approved in Japan and the United States [38]; however, rosiglitazone (Rog) has not yet been approved in Japan. Troglitazone (Tro) was withdrawn from the market in both Japan and the US in 2000. The toxicological effects of these chemicals are different although the target molecule is the same [38]. One of the reasons could be that due to differences in their chemical structures, the off-targets of the chemicals differ. Specific conformational changes that allow for the recruitment of PPARγ coactivators after ligand binding have been suggested to induce different biological effects [38, 39]. However, details regarding the differential pharmacological and/or toxicological effects, molecular mechanisms, and off-targets of these drugs are not very clear. Therefore, by using published DNA microarray data [40] and network analysis by KeyMolnet, I demonstrate a novel network analysis for clarifying the differential pharmacological and/or toxicological effects of 3 TZDs, namely, Pio, Ros, and Tro, on mouse 3T3-L1 adipocytes.

Data and Data Import

Published DNA microarray data submitted to the Gene Expression Omnibus (GEO, http://www.ncbi.nlm.nih.gov/geo/) [41-44] by Hsiao et al were used (GEO accession #, GDS734) [40]. These researchers used U74A ver. 2 Affymetrix GeneChip® to obtain DNA microarray data from mouse 3T3-L1 adipocytes treated with a vehicle (control, 0.1% DMSO) or one of the following TZDs: 20 μM Tro, 20 μM Pio, or 1 μM Ros [40]. In the network analysis using KeyMolnet, the fold-change values, i.e., ratios of the gene expression data obtained by treatment with each TZD to those obtained by treatment with the controls were imported into KeyMolnet as numerical data. Molecules with p values less than 0.05 were regarded as significant in this analysis.

1. Search for transcription factors other than PPARγ involved in altered gene expression by the TZD treatments

By the Common upstream Search method [20, 32], possible transcription factors regulating genes that were altered in expression by the treatment of Pio, Ros, or Tro were searched. After extracting molecules that were significantly upregulated and downregulated more than 2-fold as compared to those in the controls, networks in the KeyMolnet database were searched using the extracted molecules as the starting points with the molecular relation type condition as "transcriptional regulation" only. The Summary uses scores to demonstrate the relevance of transcription factors to the significantly up- and downregulated genes by each TZD treatment (Table 1). The Summary suggested that "Transcriptional regulation by SMAD" was affected by Pio, Ros, as well as Tro treatments. A recent study suggested that PPARγ activation induces SMAD2 phosphorylation mediated by TGFb1 (Transforming growth factor beta 1) [45]. SMAD is suggested to be involved in the apoptotic effects of pioglitazone. Other transcriptional mechanisms except PPARγ are suggested to be involved in the analysis by the Common upstream Search method; however, the main working mechanisms of the 3 TZDs are also suggested to be similar in the cells (Table 1).

2. Extraction of differential pharmacological and toxicological effects of the 3 TZDs focusing on molecular networks.

Table 1. Possible transcription factors involved in altered gene expression by TZDs

Pathway Summary on the networks searched based on molecules significantly up- and downregulated more than 2-fold by Pio (A), Ros (B), or Tro (C) treatments using the Common upstream Search method. Since the search condition was set as transcriptional regulation only, the generated networks comprised transcriptional regulations. Transcriptional regulations involved in the changes in gene expression by the 3 TZDs with the same target molecule are very similar.

Scores were calculated by the numbers of molecules or molecular relations appearing in the displayed networks and those obtained with each of the terms in the content by the hypergeometric distribution method in Tables 1–6.

	rank	name	score	score(v)	score(c)
A	**rank**	**name**	**score**	**score(v)**	**score(c)**
	1	Transcriptional regulation by SMAD	123.757	0.125	0.352
	2	Transcriptional regulation by RB/E2F	74.501	0.108	0.203
	3	Transcriptional regulation by C/EBP	72.091	0.077	0.321
	4	Transcriptional regulation by RUNX	70.134	0.057	0.513
	5	Transcriptional regulation by STAT	67.397	0.074	0.306
B	**rank**	**name**	**score**	**score(v)**	**score(c)**
	1	Transcriptional regulation by SMAD	121.65	0.127	0.344
	2	Transcriptional regulation by STAT	77.336	0.083	0.329
	3	Transcriptional regulation by C/EBP	65.094	0.074	0.298
	4	Transcriptional regulation by IRF	53.349	0.062	0.284
	5	Transcriptional regulation by RUNX	50.773	0.047	0.41
C	**rank**	**name**	**score**	**score(v)**	**score(c)**
	1	Transcriptional regulation by SMAD	126.859	0.145	0.336
	2	Transcriptional regulation by C/EBP	62.047	0.079	0.274
	3	Transcriptional regulation by RUNX	54.348	0.055	0.41
	4	Transcriptional regulation by STAT	53.43	0.072	0.247
	5	Transcriptional regulation by RB/E2F	47.36	0.093	0.144

Molecular mechanisms of the pharmacological and toxicological actions of chemicals were analyzed, due to which downstream networks were focused, and differential parts of networks reflecting the distinctive molecular networks induced by the actions of Pio, Ros, and Tro were extracted by the DIFF logical operation method. Figure 3 illustrates the analytical flow and steps of the network analysis used for extracting distinctive molecular networks by the Pio, Ros, and Tro treatments.

After the gene expression data were statistically processed, molecules whose gene expression was significantly ($P < 0.05$) upregulated by more than 2-fold in the cells treated with each drug compared to those treated with the controls were extracted (Figure 3, Step 1). Therefore, the 3 datasets of molecules that were significantly upregulated by more than 2-fold by each drug treatment were obtained by the procedure. The numbers of molecules (proteins) extracted by the processes described above in the case of Pio, Ros, and Tro treatments were 135, 143, and 117, respectively. Based on the molecules, networks were generated using the Neighboring Search method with a path limit of 1, and only in the downstream direction (Figure 3, Step 2). Since a network search was performed for each drug treatment, 3 kinds of networks were generated (Step 2 a, b, and c). After generating the 3 kinds of networks, specific parts of the networks for each drug treatment were extracted by the DIFF operation function on the networks. These processes are shown as Step 3 a, b, and c, as shown in the dark regions of the Venn diagram for the 3 generated networks.

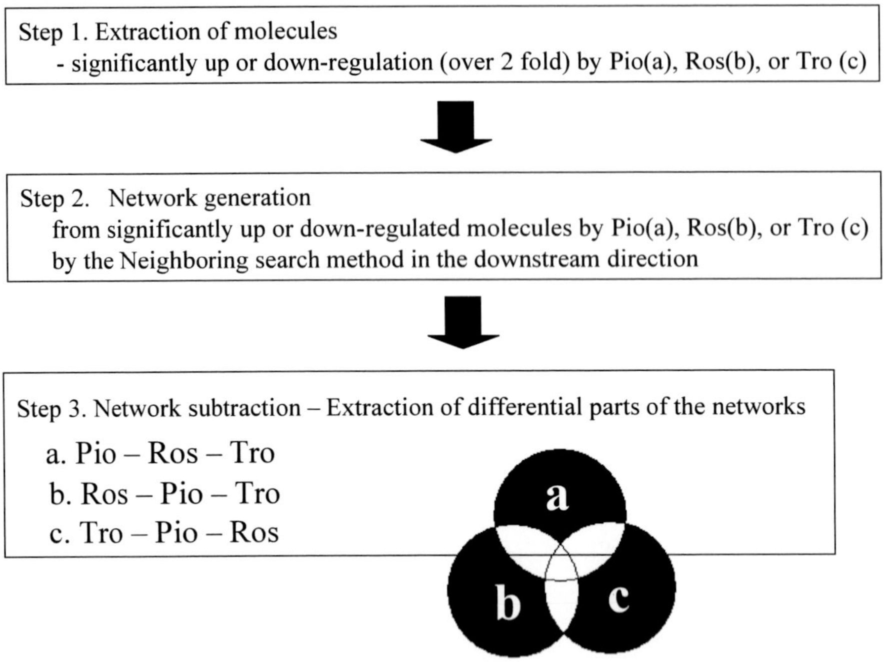

Figure 3. Analytical scheme for the extraction of each TZD-distinctive network. Procedure to extract condition-distinctive molecular networks by the DIFF (subtraction) operation is divided into 3 main steps: extraction of molecules (Step 1), network generation (Step 2), and network subtraction (Step 3). In each step, each operation was repeated 3 times for each of the TZD treatments, namely, Pio (a), Ros (b), and Tro (c). However, since the extraction was divided into up- and down-regulated molecules in each TZD treatment, each operation was repeated 6 times. The division into up- and down-regulated molecules can enable a less complex network analysis.

In this analysis, each generated network is termed the upregulated downstream network because it was generated on the basis of the upregulated molecules induced by the drugs in the downstream direction. The distinctive regions of the networks based on the actions of Pio, Ros, or Tro are partly illustrated in Figure 4A, B, and C. These networks are further displayed by the Subcellular localization Layout using the information of the cellular component of GOA. The white (upper), light blue, white (center), light green, and orange regions represent the extracellular region, membrane, cytoplasm, nucleus, and mitochondria, respectively (Figure 4A, B, and C). In the network diagrams, oval, capsular, and rectangular shapes represent types of molecules, i.e., proteins, complexes, and small molecules, respectively. Edge lines types representing molecular relations are also illustrated in Figure 4.

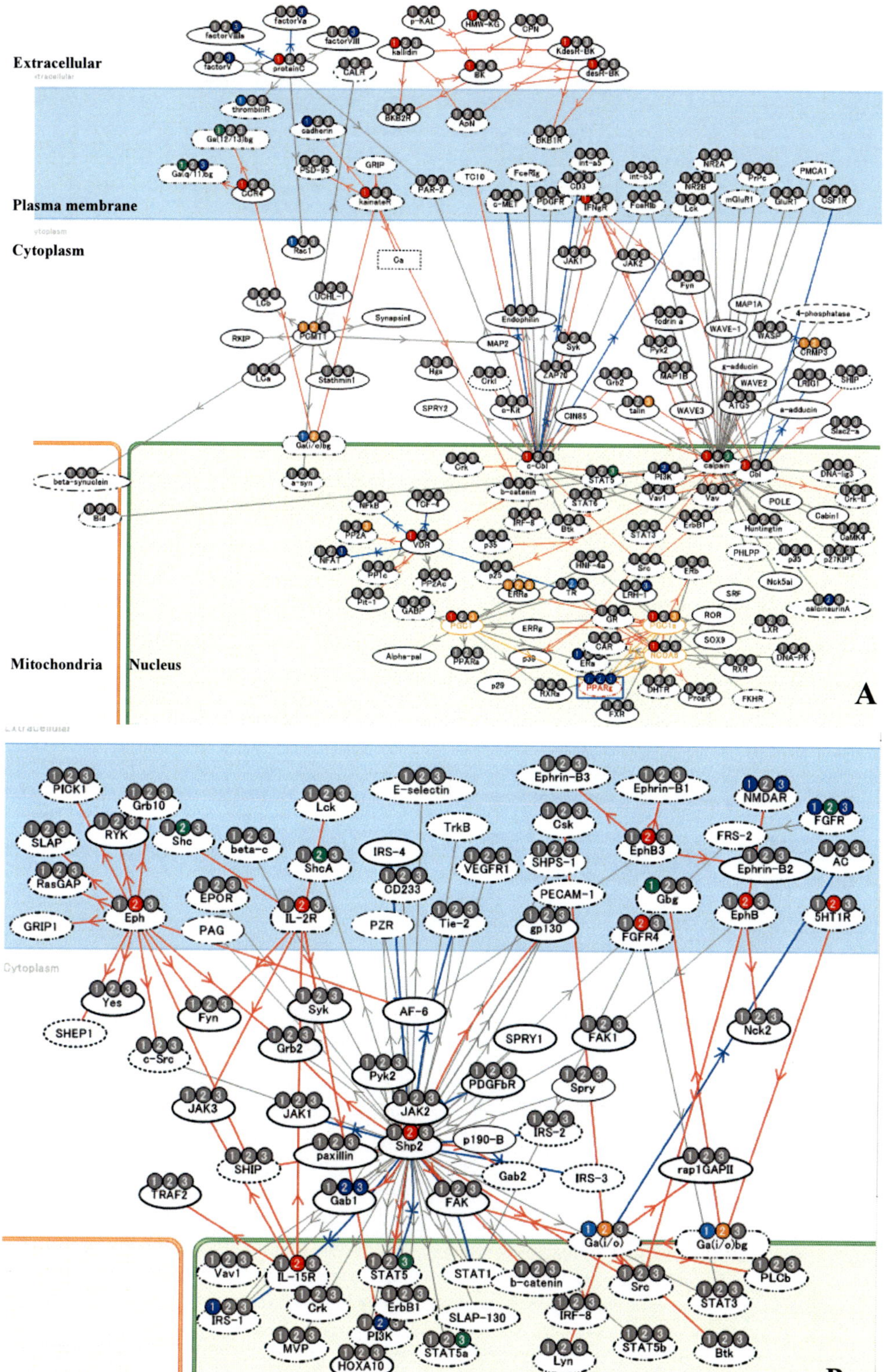

Figure 4. Continued on next page.

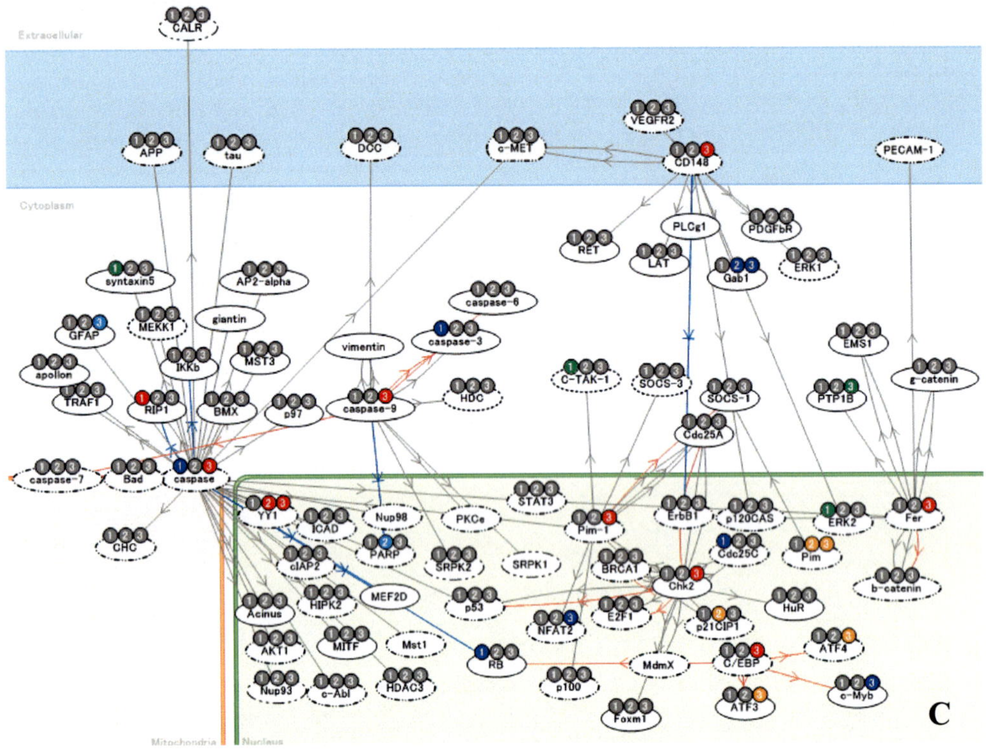

Legends:

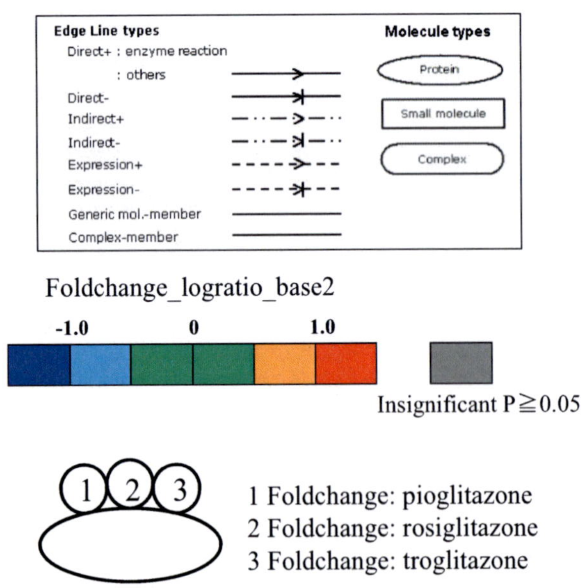

Figure 4. Drug action-distinctive molecular networks. After drug action-distinctive (A, Pio; B, Ros; and C, Tro) molecular networks were extracted by the DIFF operation (Step 3 in Figuer 3), each network was rearranged by the Subcellular localization Layout. The light blue, white (upper, center), light green, and orange areas represent plasma membrane, extracellular region, cytoplasm, nucleus, and mitochondria, respectively. Shapes of nodes representing molecular types and edge types on the networks are represented as Legends, which are used in all molecular networks. The colors of the small circles on the molecules represent the range of fold changes, which is also represented under Legends; however, these colors and ranges can be changed based on different condition settings.

After generating the networks, numerical data representing fold-change values (log ratios, base 2) for each drug treatment were displayed on the networks by color-notation conversion. The colors representing the range of fold change are shown in the figure legend, i.e., red, orange, green, light blue, and dark blue represent upregulation greater than 2-fold and less than 2-fold, no change, downregulation less than 2-fold and greater than 2-fold, respectively. Gray indicates statistically insignificant data ($P \geq 0.05$). Colored circles 1, 2, and 3 on the molecules represent the fold-change values due to Pio, Ros, and Tro treatment, respectively, in Figures 4 and 5. The distinctive parts of the networks generated based on the significantly upregulated molecules induced by the treatment of Pio are illustrated in Figure 4A. Molecules directly interacting with PPARγ (highlighted in red, framed by blue line) such as nuclear receptor coactivator 6 (NCOA6, a synonym for peroxisome proliferator-activated receptor-interacting protein, highlighted in orange) [46] and peroxisome proliferator activated receptor gamma coactivator 1 alpha (PGC1a, highlighted in orange) [47] are upregulated in the network. Therefore, the pharmacological effects of PPARγ activation have a possibility to be enhanced by the co-upregulation of these molecules in the cells.

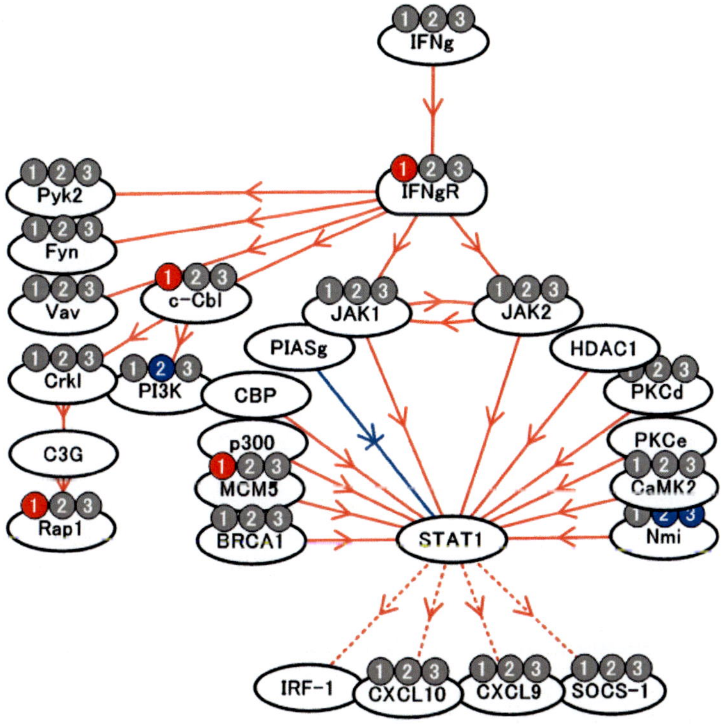

Figure 5. Effects of Pio on "IFNγ signaling pathway". The IFNγ signaling pathway, which is one of the fixed and canonical pathways, was obtained from the KeyMolnet database, and numeric data representing fold-change values were displayed as different colors on the small circles 1, 2, and 3 on the molecules based on different ranges of fold-change values. Legends are the same as those used in Figure 4.

By calculating the numbers of molecules and molecular relations in the generated networks and KeyMolnet databases, Scores representing relevance of the networks to canonical Pathways, Diseases, Pathological Events, Bio-Events and GO terms can be obtained in the Summary. The Summary suggests that the "IFNγ-signaling pathway" may be affected by Pio treatment of the cells (not shown). In fact, when displaying the data revealing fold-change values due to Pio, Ros, and Tro treatments in the pathway, a signal appears to be transduced from the interferon gamma receptor (IFNgR) to Ras-related protein (Rap1) mediated by CBL E3 ubiquitin protein ligase (c-Cbl) in the Pathway by Pio stimulation (Figure 5). IFNgR may be an off-target of Pio; however, additional experiments are required to confirm this supposition. From the Pathological Event Summary exhibiting adverse events, "heart failure" and "edema" are listed in the lower part of the table with respect to Pio treatment (Table 2 A). Heart failure and edema are known to occasionally develop as adverse effects in patients with type 2 diabetes and due to treatment with Pio [38, 48]; results of transcriptome analysis in mouse adipocytes are also consistent with those reported here. On the other hand, cells treated with Ros, JAK1, JAK3, FAK1, STAT1, STAT3, and PI3K may be affected because they are reported to be involved with multiple upregulated molecules such as Shp2, IL-2R, IL-15R, FGFR4, and EphB (Figure 4B) [49-60]. Further experiments other than those in the transcriptome analysis are required to confirm the assumptions obtained from this network analysis. Furthermore, 5-hydroxytryptamine 1 receptor (5HT1R) may be a candidate off-target of Ros because a possible downstream molecule of 5HTR1R, the G protein (Ga (i/o)), was also upregulated in this extracted network (Figure 4B). SH2-containing protein tyrosine phosphatase-2 (Shp2) was reportedly involved in diabetes mellitus [61, 62], and the change in Shp2 gene expression was induced only by Ros in this study (Figure 4B). According to the Summary, the relevance of Pathological Events and Bio-Events to the extracted networks induced by Ros was suggested to be similar to that induced by Pio treatment (Table 2A &B); however, heart failure and edema were not ranked within the top 15 Events (Table 2B). When the DIFF operation was used to extract the distinctive upregulated downstream networks in the Tro-treated cells, the results from the network suggested the induction of apoptosis in these cells (Figure 4C). In fact, as illustrated in Table 2C, both Pathological Events and Bio-Events also suggest that the extracted Tro-induced upregulated downstream network was involved in apoptosis and cell cycle. Based on the relevance of Pathways in the Summary, apoptosis-related pathways such as the "Cytochrome *c* signaling", "Caspase signaling", "ATM signaling", and "BRCA signaling" pathways may be affected in Tro-treated cells (not shown). Additional experiments such as the measurement of caspase activity are required to confirm these; however, of the 3 chemicals, treatment with Tro was strongly suggested to induce apoptosis in the cells. Bae and Song [63] reported that Tro enhanced levels of the proapoptotic proteins Bad and Bax, release of cytochrome *c*, and cleavage of Bid in human hepatoma HepG2 cells, although this transcriptome analysis did not indicate Bad, Bax, or Bid induction in mouse adipocytes. Yin et al. [64, 65] suggested that Tro-induced apoptosis was involved in cell cycle arrest in MCF-7 breast carcinoma cells. Yu et al. [66] also indicated that Tro inhibited cell growth and proliferation and induced apoptosis in hepatocellular carcinoma both in vitro and in vivo. Tro was withdrawn in 2000 in the US and Japan because of its hepatotoxicity in humans [38]. The detailed molecular mechanism of its hepatotoxicity has not yet been elucidated in humans. Bae and Song [63] stated that this may have been due to cytotoxicity resulting from Tro-mediated apoptosis in human hepatocytes.

Table 2. Summary on Pathological Events and Bio-Events for the networks based on the molecules upregulated by TZDs

Relevance of the extracted networks shown in Figures 4A, B, and C to the Pathological Events and Bio-Events is displayed with the scores in A (Pio), B (Ros), and C (Tro), respectively.

A Pathological Event

rank	name	score	score(v)	score(c)
1	Mast-cell Function	33.251	0.054	0.175
2	Cancer	27.561	0.064	0.115
3	Epithelial-Mesenchymal Transition	23.848	0.042	0.157
4	Hypertension	22.107	0.048	0.121
5	Bleeding/Hemostasis	20.733	0.035	0.159
6	Erythropoiesis	20.454	0.029	0.205
7	Steroid Regulation	19.102	0.035	0.143
8	Thrombosis	17.839	0.035	0.131
9	Platelet Function	16.499	0.038	0.109
10	Pulmonary Fibrosis	13.48	0.026	0.131
11	Endoplasmic Reticulum Stress	12.21	0.026	0.116
12	Heart Failure	12.019	0.019	0.158
13	Autophagy	11.949	0.013	0.286
14	Edema	11.401	0.019	0.146
15	Immune System	11.146	0.054	0.06

Bio-Event

rank	name	score	score(v)	score(c)
1	cell communication	92.067	0.479	0.056
2	hemostasis	47.569	0.073	0.193
3	apoptosis	44.652	0.163	0.078
4	blood coagulation	43.756	0.067	0.194
5	cell death	42.37	0.163	0.075
6	regulation of apoptosis	40.7	0.137	0.083
7	wound healing	40.127	0.067	0.172
8	cell differentiation	37.448	0.201	0.059
9	positive regulation of apoptosis	32.713	0.086	0.102
10	response to wounding	31.851	0.102	0.086

Pio

B Pathological Event

rank	name	score	score(v)	score(c)
1	Erythropoiesis	61.896	0.075	0.386
2	Cancer	55.542	0.115	0.149
3	Thrombosis	52.485	0.084	0.226
4	Platelet Function	44.858	0.084	0.173
5	Epithelial-Mesenchymal Transition	36.741	0.066	0.181
6	Bone Regulation	22.992	0.044	0.156
7	Angiogenesis	21.94	0.04	0.17
8	Mast-cell Function	20.263	0.048	0.113
9	Pulmonary Fibrosis	20.143	0.04	0.148
10	Immune System	18.94	0.079	0.063
11	Fibroblast Function	18.791	0.035	0.157
12	Connective Tissue Regulation	16.144	0.031	0.149
13	Thyroid Hormone Regulation	15.523	0.026	0.176
14	Bleeding/Hemostasis	15.457	0.035	0.116
15	Atherogenesis	15.347	0.031	0.137

Ros

Bio-Event

rank	name	score	score(v)	score(c)
1	cell communication	106.052	0.559	0.047
2	cell proliferation	42.329	0.189	0.061
3	organ development	30.653	0.181	0.049
4	regulation of cell proliferation	29.561	0.123	0.065
5	positive regulation of cell proliferation	28.431	0.088	0.087
6	multicellular organismal development	28.294	0.247	0.039
7	regulation of progression through cell cycle	23.052	0.11	0.057
8	positive regulation of enzyme activity	22.539	0.084	0.071
9	negative regulation of progression through cell cycle	20.088	0.053	0.101
10	protein complex assembly	18.887	0.075	0.066

C

Tro

Pathological Event

rank	name	score	score(v)	score(c)
1	Cancer	36.635	0.105	0.103
2	Erythropoiesis	19.503	0.041	0.159
3	Endoplasmic Reticulum Stress	15.078	0.041	0.101
4	Apoptosis	14.972	0.029	0.172
5	Mast-cell Function	14.785	0.047	0.082
6	Pancreatic Exocrine Secretion	14.486	0.029	0.161
7	Gastric Secretion	14.13	0.023	0.235
8	Angiogenesis	14.059	0.035	0.113
9	Epithelial-Mesenchymal Transition	13.352	0.041	0.084
10	Neurologic Disorders	11.368	0.035	0.081
11	Fibroblast Function	10.996	0.029	0.098
12	Esophagus Disease	10.499	0.018	0.214
13	Chemotherapy	10.024	0.029	0.085
14	Sun Exposure	9.389	0.018	0.167
15	Cell Cycle	8.725	0.018	0.143

Bio-Event

rank	name	score	score(v)	score(c)
1	cell death	48.934	0.228	0.057
2	apoptosis	43.304	0.211	0.055
3	cell differentiation	41.075	0.263	0.042
4	cell communication	40.366	0.444	0.028
5	regulation of apoptosis	28.433	0.152	0.05
6	cellular protein metabolic process	25.798	0.257	0.031
7	regulation of progression through cell cycle	22.309	0.123	0.048
8	cell cycle	21.688	0.146	0.041
9	positive regulation of apoptosis	21.645	0.094	0.061
10	induction of apoptosis	17.492	0.076	0.059

The same operation on network analysis focused on significantly downregulated molecules induced by the treatment of Pio, Ros, or Tro. Numbers of significantly downregulated molecules (over 2-fold) by the treatment of Pio, Ros, and Tro were 279, 177, and 268, respectively. The Pathway and Bio-Event Summary (Table 3) revealed that the Pio and Ros treatments may also have downregulated apoptosis (Table 3A and B), but that the downregulated molecules by Tro treatment may not have been involved in apoptosis (Table 3C). Induction of apoptosis is generally suggested by treatment of TZDs including Pio, Ros, and Tro in various types of cells [67-69]. Tro may induce apoptosis in the cells to the greatest degree as compared to the other 2 TZDs in this analysis. This network analysis was performed using the data on the mouse adipocyte 3T3-L1 cells; however, a similar biological response induced by Tro in the mouse adipocytes may be demonstrated in human hepatocytes. Further experiments are required to confirm the suggestions from the network analysis; however, possible detailed actions and molecular mechanisms in the treatment of each TZD could be obtained by network analysis using the oligonucleotide array data.

Table 3. Summary on Pathway and Bio-Event for the networks based on the molecules downregulated by TZDs

Relevance of the extracted networks based on the molecules significantly downregulated by over 2-fold for each TZD treatment (not shown) to the Pathway and Bio-event are displayed with the Scores in A (Pio), B (Ros), and C (Tro), respectively.

A **Pathway**

rank	name	score	score(v)	score(c)
1	Caspase signaling pathway	75.585	0.065	0.326
2	Neurotrophin signaling pathway	13.045	0.02	0.167
3	Tyro3 family signaling pathway	12.972	0.005	1
4	Transcriptional regulation by RB/E2F	11.795	0.038	0.107
5	Catecholamine metabolism	9.709	0.009	0.24
6	Cytochrome c signaling pathway	9.391	0.014	0.164
7	Androgen metabolism	9.342	0.006	0.364
8	Relaxin receptor signaling pathway	8.284	0.008	0.238
9	Insulin receptor signaling pathway	6.878	0.025	0.097
10	Netrin signaling pathway	5.116	0.008	0.143

Bio-Event

rank	name	score	score(v)	score(c)
1	cell cycle	84.079	0.161	0.145
2	cell death	74.544	0.163	0.132
3	cell differentiation	73.594	0.21	0.109
4	regulation of progression through cell cycle	72.011	0.125	0.159
5	apoptosis	68.633	0.154	0.129
6	transcription	49.28	0.185	0.094
7	regulation of apoptosis	47.942	0.116	0.123
8	transcription from RNA polymerase II promoter	41.748	0.118	0.111
9	regulation of transcription	38.451	0.145	0.094
10	cell communication	37.165	0.324	0.066

Pio

B **Pathway**

rank	name	score	score(v)	score(c)
1	PI3K signaling pathway	35.712	0.059	0.217
2	JNK signaling pathway	33.101	0.068	0.152
3	Granzyme signaling pathway	17.965	0.027	0.231
4	Transcriptional regulation by ternary complex factor	15.465	0.023	0.238
5	Prolactin signaling pathway	11.878	0.027	0.113
6	Complement activation pathway	7.635	0.018	0.098
7	MAPK signaling pathway	5.844	0.036	0.041
8	Transcriptional regulation by NFAT	5.42	0.018	0.063

Ros

Bio-Event

rank	name	score	score(v)	score(c)
1	transcription	56.632	0.267	0.058
2	transcription from RNA polymerase II promoter	51.988	0.186	0.075
3	cell differentiation	45.614	0.242	0.054
4	RNA metabolic process	43.888	0.25	0.051
5	cell communication	40.817	0.407	0.036
6	cell death	39.485	0.178	0.061
7	apoptosis	37.201	0.169	0.061
8	positive regulation of transcription	36.746	0.097	0.104
9	regulation of transcription	35.156	0.191	0.053
10	regulation of transcription, DNA-dependent	30.951	0.161	0.055

C

Pathway

rank	name	score	score(v)	score(c)
1	CD44 signaling pathway	72.696	0.083	0.329
2	Tissue kallikrein family	62.858	0.07	0.344
3	G protein (Gq/11) signaling pathway	32.673	0.033	0.385
4	Rho family signaling pathway	23.05	0.047	0.135
5	Complement activation pathway	18.044	0.027	0.195
6	Vitamin D metabolism	12.667	0.017	0.217
7	Transcriptional regulation by NFAT	10.548	0.023	0.109
8	Integrin family	9.011	0.043	0.059
9	Coagulation cascade	7.702	0.013	0.133
10	Transcriptional regulation by ternary complex factor	6.381	0.01	0.143

Tro

Bio-Event

rank	name	score	score(v)	score(c)
1	cell communication	63.505	0.412	0.055
2	cell adhesion	51.351	0.137	0.103
3	response to wounding	46.132	0.115	0.11
4	cell motility	45.867	0.115	0.109
5	organ development	41.065	0.171	0.074
6	response to stress	39.68	0.165	0.074
7	defense response	39.058	0.118	0.093
8	cellular defense response	31.82	0.048	0.187
9	cell migration	28.917	0.064	0.12
10	cell-cell adhesion	28.42	0.056	0.135

Networks Representing the Development of Esophageal Cancer

Esophageal cancer cases are either squamous cell carcinomas or adenocarcinoma (AC) and many etiological factors have been implicated in the development of the disease [70]. Squamous cell carcinoma is believed to be associated with excess alcohol consumption, cigarette smoke, lye ingestion, etc. Further, Barrett's esophagus (BE), which is a chronic gastric reflux and specialized intestinal metaplasia of the normal squamous to columnar epithelium, is believed to be a premalignant transformation and is associated with a 30- to 125-fold increased risk for AC of the esophagus [71]. Barrett's epithelium progresses through a dysplastic stage before developing into AC [71]. Aneuploidy and p53 mutations are detected in the dysplastic epithelium before neoplasm is detectable [70, 72] Cyclooxygenese-2 (COX-2) is reported to be overexpressed early in the neoplastic transformation process of BE, and the overexpression increased significantly in the progression from dysplasia to AC [73]. Since the incidence of AC has increased 3.5-fold over the past 3 decades in the US [74], identification of the biological markers and diagnostic tools for BE progression have been anticipated for the early detection and survival in BE. Therefore, DNA microarray analysis is actively performed for the screening and detection of tumor markers, and the approach to treatment of esophageal cancer [75-77]. In this network analysis, using publicly available DNA microarray data from patients, molecular networks representing abnormal mechanisms due to cancer development were extracted by the AND operation functions of KeyMolnet, if significantly up and/or downregulated networks in both the BE and the AC groups can be regarded as the up and/or downregulated networks from the early to the late stages of the cancer.

Data and Data Import

Further, published DNA microarray data submitted to GEO (GEO accession #, GSE 1420, GDS1312) by Kimchi et al. were used [75]. U133A Affymetrix GeneChip® was used to obtain DNA microarray data in fresh pathological specimens of normal (n = 8), BE (n = 8), and AC (n = 8) in patients with known BE and esophageal AC and who had undergone esophagectomy [75]. In the network analysis using KeyMolnet, the fold-change values, i.e., the ratios of gene expression levels in BE and AC samples to those in normal controls were imported into KeyMolnet as numerical data. Molecules with p values less than 0.01 were regarded as significant data in this analysis.

Networks of Progress of Esophageal Cancer

After significantly up- and downregulated (over 2-fold) molecules were extracted in the BE and AC groups compared to those in the normal group, networks were generated based on these extracted molecules as search points in both the groups. Numbers of the significantly up- and downregulated molecules (over 2-fold) in the BE and AC groups were 294 (upregulated, BE), 612 (upregulated, AC), 201 (downregulated, BE), and 782 (downregulated, AC). Based on these molecules, 4 types of molecular networks were generated by the Interrelation network Search method to search molecular relations including the search points as many as possible in the database. In this step, upregulated and downregulated networks in the BE and AC were generated. With regard to up- and downregulated networks, differential and shared parts of the networks can be extracted by the DIFF and AND operations, respectively. BE-distinctive (up and down) networks can be regarded as early events and/or uninvolved mechanisms in progression of the disease, while AC-distinctive (up and down) networks can be regarded as late events of the disease. Furthermore, shared regions of the 2 networks (up and down) can be regarded as events that occurred throughout the disease. Therefore, possible candidates of prognostic markers could be included in this shared network (up and down). Figure 6 and Table 4 illustrate the shared upregulated network in both BE and AC and the Summary on Pathways, Diseases, and Pathological Events regarding this network, respectively. According to the Pathway Summary (Table 4A), parts of the "Transcriptional regulation by RB/E2F", "CD44 signaling pathway", "Transcriptional regulation by PXR", etc., are included in the extracted network illustrated in Figure 6. When the gene expression data in both BE and AC were imported into whole of the "Transcriptional regulation by RB/E2F" (Figure 7), E2F (E2F3 in this study), which promotes DNA synthesis and cell cycle progression, is significantly upregulated in both the groups; furthermore, it may be transcriptionally activated in both the groups. Yamazaki et al. suggested that increased cell proliferation and decreased apoptosis are associated with adverse prognosis in patients with esophageal squamous cell carcinoma [78]; however, the relationship between E2F3 activation and progression of esophageal adenocarcinoma has not yet been reported. Likewise, when the same data were displayed on the "CD44 signaling pathway" (Figure 8), this pathway is also suggested to be activated in both BE and AC status. Turnar et al. suggested that increased expression of CD44 was significantly associated with shortened patient survival in AC patients with chemoradiotherapy, and that CD44 also may be a valuable prognostic marker in Barrett's-associated adenocarcinoma [79], although CD44 itself was not upregulated in both BE and AC in this study. The possible upregulated CD44 cascade may reflect metastatic potential in this case. Downregulated networks in both BE and AC (not shown) were suggested as the result of possible dysfunction of p53 whose mutations in the gene has been reported in many papers [72, 80].

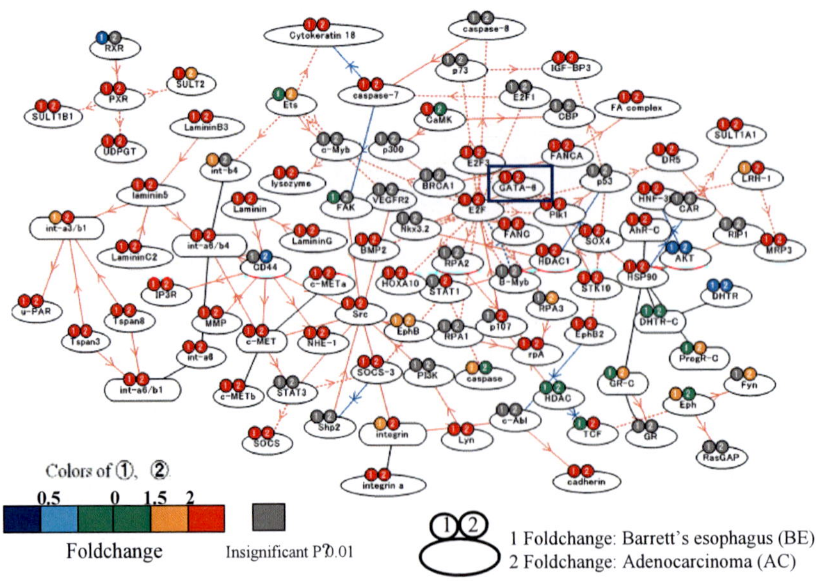

Figure 6. Upregulated network through development from Barrett's esophagus to adenocarcinoma. After extracting molecules whose gene expression levels in the BE and AC groups were significantly (P < 0.01) upregulated more than 2-fold as compared to those in the normal group, networks were generated based on the molecules in each group as search points by the Interrelation network Search method. Next, the shared region of the 2 networks in the 2 groups were extracted by the AND operation. Colors of circles 1 and 2 on the molecules represent the range of fold-change values in BE and AC, respectively, as shown in the figure. The correspondence between the range of fold-changes and colors are also shown in the figure. GATA6, which is suggested by Kimchi et al (75) as a candidate for biomarkers, is indicated with a blue-framed rectangle.

Both analyses in the up- and downregulated networks are taken together, and progression of cell proliferation and repression of apoptosis are suggested to be the consequence of the development of the cancer. The Summary on the Pathological Event presented in Table 4C suggests that angiogenesis and erythropoiesis have already been activated in BE. GATA6 (blue-framed in Fig. 6) suggested by Kimchi et al. (75) as a candidate for biomarkers also appeared in the network (Fig. 6). The Disease Summary reveals that a number of cancer-related molecules (Disease Mediating Molecules in KeyMolnet) are included in the upregulated networks in both BE and AC, as expected. Esophageal cancer is ranked lower in the Disease Summary because fewer molecules related to the disease have been reported as compared to those in other cancers. No remarkable characteristics, except downregulation of apoptosis in Bio-Events (not shown), were suggested in the shared downregulated networks in both BE and AC. However, in this way, the extraction of networks that are up- and/or downregulated through the development of a disease could enable to identify candidates of biomarkers for prognosis, and discuss the mechanisms since molecular networks suggest biological process and/or changes in disease states and biological phenomena.

Table 4. Summary on Pathway, Disease, and Pathological Event for the upregulated networks through from Barrett's esophagus to adenocarcinoma

The Summaries on the extracted network in Figure 6 on Pathway (A), Disease (B) and Pathological Event (C) are presented with the scores.

A	rank	name	score	score(v)	score(c)
	1	Transcriptional regulation by RB/E2F	40.314	0.153	0.081
	2	CD44 signaling pathway	16.991	0.056	0.092
	3	Transcriptional regulation by PXR	14.552	0.032	0.19
	4	Transcriptional regulation by Myb	10.049	0.032	0.087
	4	HGF signaling pathway	10.049	0.032	0.087
B	**Disease**				
	rank	**name**	**score**	**score(v)**	**score(c)**
	1	Gastric Cancer	35.911	0.107	0.121
	2	Epidermolysis Bullosa	18.257	0.033	0.333
	3	Prostate Cancer	16.055	0.049	0.103
	4	Breast Cancer	14.633	0.041	0.119
	5	Ovarian Cancer	14.301	0.041	0.114
	6	Colorectal Cancer	13.924	0.049	0.08
	7	Retinoblastoma	10.972	0.016	0.5
	7	Meningiomas	10.972	0.016	0.5
	9	Hepatocellular Carcinoma	10.818	0.041	0.068
	10	Chronic Myelogenous Leukemia	10.125	0.025	0.143
C	**Pathological Event**				
	rank	**name**	**score**	**score(v)**	**score(c)**
	1	Cancer	41.338	0.139	0.098
	2	Angiogenesis	25.221	0.066	0.151
	3	Erythropoiesis	22.812	0.057	0.159
	4	Endoplasmic Reticulum Stress	18.271	0.057	0.101
	5	Epithelial-Mesenchymal Transition	16.48	0.057	0.084
	6	Cell Cycle	10.125	0.025	0.143
	7	Thrombosis	9.894	0.041	0.06
	8	Fibroblast Function	9.677	0.033	0.078
	9	Chemotherapy	8.894	0.033	0.068
	10	Apoptosis	8.744	0.025	0.103

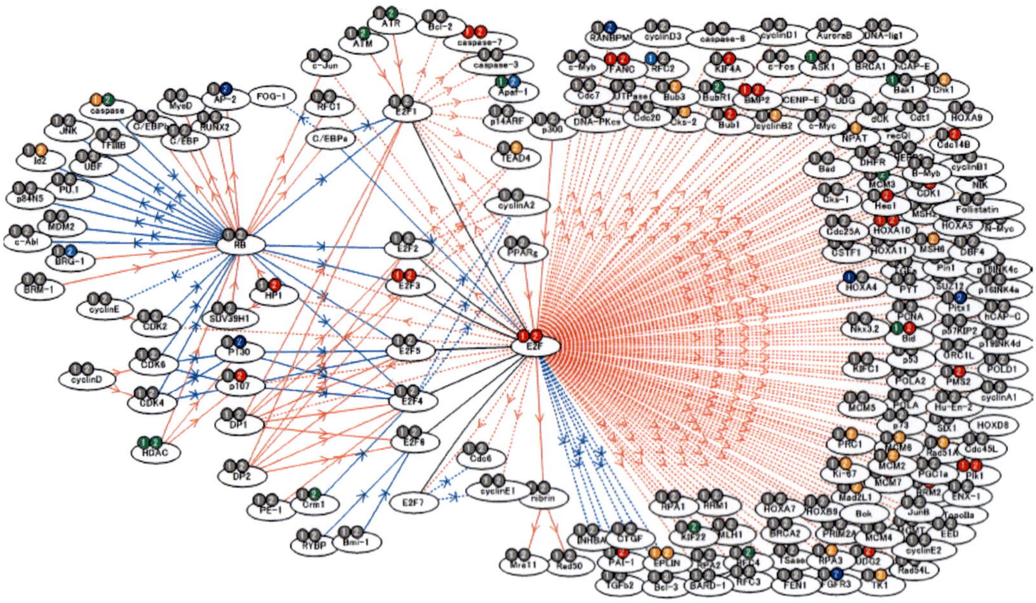

Figure 7. Gene expression in BE and AC on "Transcriptional regulation by RB/E2F". The transcriptional regulation by RB/E2F data that is one of the fixed and canonical pathways was obtained from the KeyMolnet database, and numerical data representing fold-change values in BE and AC were displayed in small circles 1 and 2, respectively, on the molecules after coding them using several colors. The legends are the same as those used in Figure 6.

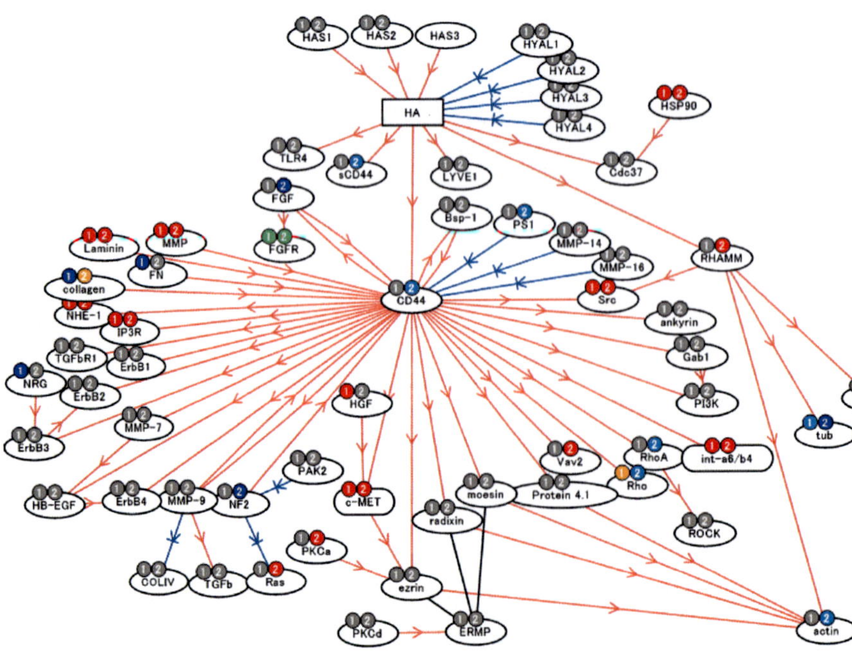

Figure 8. Gene expression in BE and AC on "CD44 signaling pathway". The CD44 signaling pathway data that is one of the fixed and canonical pathways was obtained from the KeyMolnet content, and numerical data representing fold-change values in BE and AC are indicated on small circles 1 and 2, respectively, on the molecules after coding them using several colors. The legends are the same as those used in Figures 6 and 7.

PREDICTION OF MOLECULAR FUNCTION USING OLIGONUCLEOTIDE ARRAY DATA

Changes in phenotypes by targeted disruption of a gene are sometimes difficult to detect. Thus, subtle changes or predictable physiological effects of molecular deficiencies in biological specimens could be clarified using comprehensive gene expression data in gene knockout (KO) mice, or after repression of gene expression by RNAi technology because the sensitivity of signals of gene expression levels in DNA microarray is high, and their data can be covered comprehensively.

Nuclear factor erythroid 2-related factor 2 (Nrf2) is an essential transcription factor for inducible and/or constitutive expression of antioxidants and a group of phase II detoxification enzyme genes, such as glutathione S-transferases (GSTs) and NAD(P)H dehydrogenase, quinone 1 (NQO1). Further, Nrf2 actions are mediated through antioxidant responsive elements (AREs) [81, 82]. Nrf2KO mice exhibit high sensitivity to oxidative stresses such as that observed during hepatotoxicity induced by acetaminophen [83], lung injury by hypoxia [84], and accelerated DNA adduct formation by exposure to diesel exhaust [85]. In this network analysis, by using oligonucleotide array data from livers and spleens of Nrf2 KO and wild-type (WT) mice, functions of Nrf2 are predicted and its influences are discussed.

Data and Data Import

Published DNA microarray data submitted to GEO (GEO accession #; GSE867, GDS514) by Li et al. were used [86]. To obtain DNA microarray data in WT female livers (n = 2), Nrf2 KO female livers (n = 2), WT male livers (n=2), Nrf2 KO male livers (n = 2), WT female spleens (n = 2), and Nrf2 KO female spleens (n = 2), MG-U74A Affymetrix GeneChip® were used [86]. In the network analysis using KeyMolnet, the fold-change values, i.e., the ratios of the gene expression levels in the KO samples to those in the WT controls were imported into KeyMolnet as numerical data. The *t* test/ANOVA was not performed. The fold-change values were used as the average of duplicate data available in the GEO.

Molecular Network-Related Biological Processes Resulted in the Absence of Nrf2

Before network analysis, effects of Nrf2 deficiency on gene expression in molecules related to Nrf2 were observed by searching for molecules directly interacting with Nrf2 and transcriptionally regulated by Nrf2 in the downstream direction, by using the Neighboring Search method. In other words, molecular network search was performed by setting Nrf2 (in red) as the start point with the following conditions: path limit of 1, downstream direction only, and direct relation and/or transcriptional regulation. As illustrated in Figure 9, in Nrf2 KO without any stimulation, only NAD(P)H dehydrogenase1 (quinone1) (NAD(P)HDHq1) was downregulated more than 2-fold in gene expression in the male KO mice livers as compared to those of the male WT mice livers.

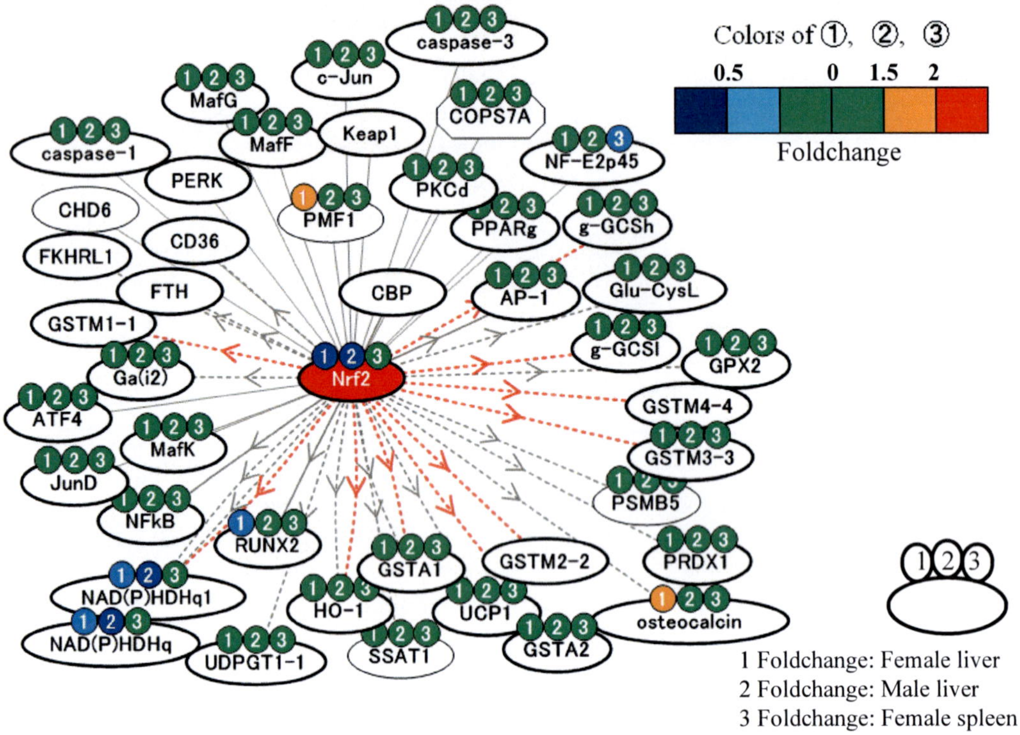

Figure 9. Effects of Nrf2 KO on gene expression of molecules directly related to Nrf2. After setting Nrf2 as a search point (red), molecules directly related to Nrf2 were searched using the Neighboring Search method with the following conditions: downstream direction only, and direct and transcriptional regulations as the types of molecular relations. Numerical data representing fold-changes in the female livers, male livers, and female spleens were indicated using several colors on small circles 1, 2 and 3, respectively, on the molecules as shown in the figure.

By the disruption of Nrf2 without any stimulation, no molecules were upregulated in gene expression more than 2-fold in female livers, spleens, and male livers of the Nrf2 KO mice compared to those of the WT mice. The numbers of molecules downregulated over 2-fold by the Nrf2 KO were 220, 35, and 41 in the female livers, spleens, and the male livers, respectively. Based on these molecules as the starting points, molecular networks were generated by the Interrelation network Search method with a path limit of 2, and molecular relation types as direct relation and transcriptional regulation. In this step, 3 kinds of networks in the female livers, male livers, and female spleens were generated. After generating these 3 molecular networks, for each of the 2 combinations of the 3 networks, the AND and DIFF operations were performed. By these operations, the female- and male-distinctive parts and the shared parts of the networks in both the livers, the liver- and spleen- distinctive parts and shared parts of the networks in the female were extracted. As indicated numbers of molecules downregulated over 2-fold in the female livers, male livers, and female spleens, that in the female livers was predominant among the 3 groups. Therefore, generated molecular networks in the female livers were the largest and most complex, as compared to those in the male livers and female spleens due to the number of molecules set as starting points (not shown). The shared region of networks and molecular relations in the female livers and spleens were not extracted, and those in the female and the male livers comprised small parts of "Hyaluronan cross-linking", "ATM signaling pathway", "Integrin signaling pathway", and "CDK signaling pathway" (not shown). The distinctive networks in the female livers, as compared to the networks in the male livers and female spleens, were complicated but similar to each other, and consisted of cross-talks in many pathways such as the "Insulin receptor signaling pathway", "Transcriptional regulation by SMAD", "Integrin signaling pathway", and "IgE signaling pathway" (not shown). However, risks of hepatocellular carcinoma and hepatitis in the female livers of Nrf2 KO mice were suggested by the Disease Summary (not shown). On the other hand, the distinctive region of the molecular network in male livers is considerably simple, although gene expression levels of molecules in the network were lower in male livers than in female livers, according to the colors of the small circles on the molecules (Figure 10). Therefore, the male liver-distinctive network as compared to the network in the female livers could rather be regarded as more repressive networks.

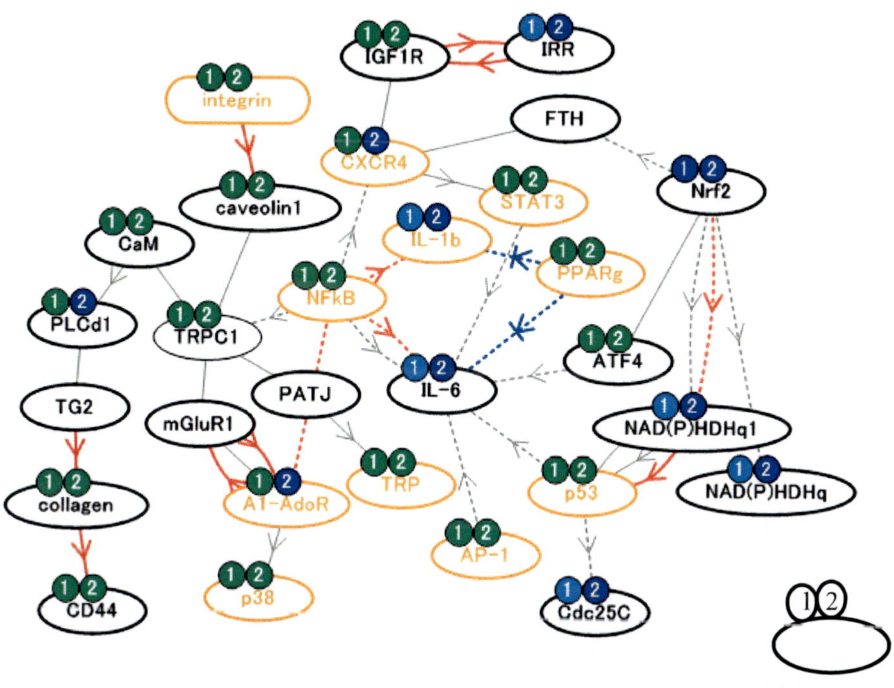

1 Foldchange: Female liver
2 Foldchange: Male liver

Figure 10. Differential part of network in male livers compared to networks in female livers. After extracting molecules whose gene expression levels in the female and male livers of Nrf2 KO mice were downregulated by more than 2-fold as compared to those in the WT mice, networks were generated based on the molecules in each group as search points by the Interrelation network Search method with a path limit of 2. Next, differential regions of the 2 networks in the female and male livers were extracted by the DIFF operation. Molecules with orange highlights in the figure represent molecules annotated with the GO term "response to external stimulus." Colored small circles 1 and 2 on molecules represent the range of fold-change values in the female and male livers, respectively, as shown in the figure. The correspondence between the range of fold-changes and colors are the same as illustrated in Figure 9.

In Figure 10, molecules highlighted in orange indicate molecules annotated with "response to external stimulus (GO:0009605)" in the biological process information of GO. Thus, this molecular network is suggested to be repression or dysfunctions of the host defense system that is ranked the highest in the GO information Summary (Table 5). Furthermore, the distinctive network in the female spleens as compared to networks in the female livers is suggested to be clearly reflected in the "host defense" and "immune system" (Figure 11 and Table 6A). This molecular network comprises the "BCR signaling pathway", "TCR(a/b) signaling pathway", "TNFR-signaling pathway", and "Transcriptional regulation by E2A/EBF/PAX5" (not shown).

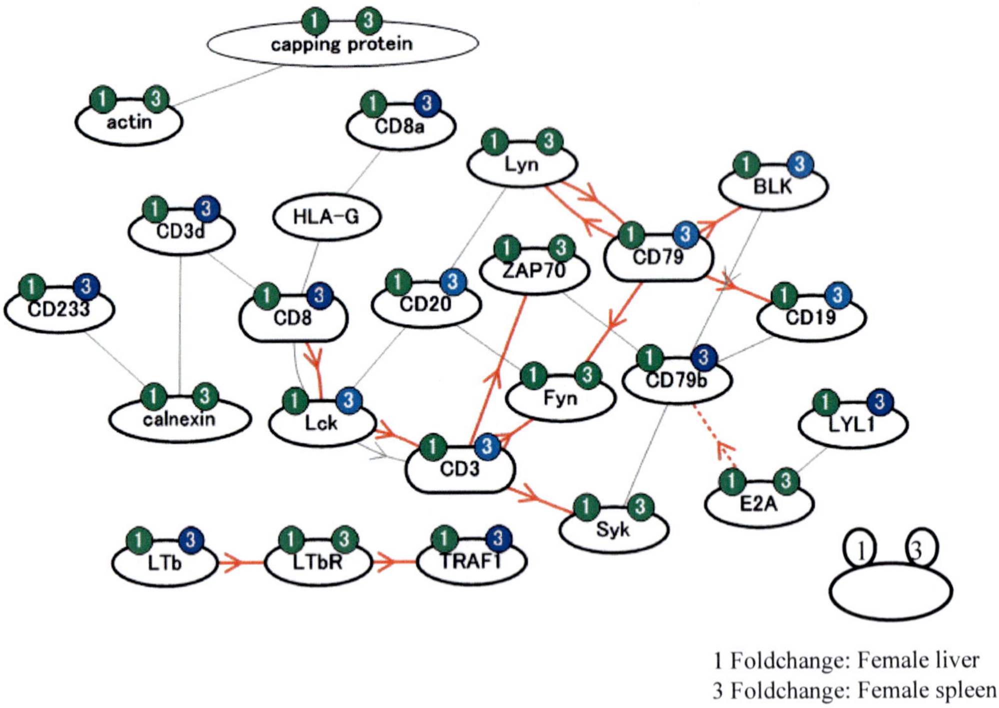

Figure 11. Differential regions of networks in female spleen compared to networks in female livers. The networks were generated and extracted for the female spleens and livers by the same operations and conditions as performed in Figure 10. Colored small circles 1 and 3 on molecules represent the range of fold-change values in female livers and spleens, respectively, as shown in the figure. The correspondence between the range of fold-changes and colors are the same as illustrated in Figure 9.

It is not difficult to predict that mice without Nrf2 would be susceptible to several cancers caused by oxidative stress, as Nrf2 regulates gene expression of phase II-detoxifying enzymes and antioxidant proteins [87, 88]. However, the network Summary on the Pathological Events reveals that the immune system may be strongly influenced by Nrf2 deficiency (Table 6A). Further, varieties of diseases are also suggested to be susceptible by Nrf2 deficiency (Table 6B).

Table 5. Summary of GO terms for differential part of the network in male livers

The Summary and Scores are presented on the differential part of the networks in male livers as compared to those in female livers in Figure 10 regarding the biological process of the GO information.

rank	name	score	score(v)	score(c)
1	response to external stimulus (GO:0009605)	32.715	0.419	0.023
2	response to stimulus (GO:0050896)	29.152	0.581	0.011
3	inflammatory response (GO:0006954)	27.675	0.29	0.035
4	response to stress (GO:0006950)	26.6	0.419	0.016
5	cell communication (GO:0007154)	26.02	0.677	0.008

Table 6. Summary of Pathological Event and Disease for the differential parts of network in female spleens

The Summaries and Scores of the differential parts of network in female spleens as compared to those in female livers are presented in Figure 11 regarding Pathological Event (A) and Disease (B).

A Pathological Event

rank	name	score	score(v)	score(c)
1	Host Defense	33.517	0.207	0.176
2	Immune System	22.992	0.276	0.028
3	Thymocyte Differentiation	15.2	0.103	0.111
4	Thrombosis	14.951	0.138	0.048
5	T cell Differentiation	11.035	0.069	0.133
6	B cell Differentiation	9.911	0.069	0.091
7	Mast-cell Function	9.693	0.103	0.031
8	Biologic Therapy	8.121	0.069	0.049
9	Hypersensitivity	7.001	0.069	0.033

B Disease

rank	name	score	score(v)	score(c)
1	Non-Hodgkin's Lymphoma	18.829	0.138	0.093
2	ZAP70 Tyrosine Kinase Deficiency	16.141	0.069	0.667
2	T Cell Receptor Deficiency	16.141	0.069	0.667
4	Hereditary Elliptocytosis	15.143	0.069	0.5
5	Autosomal Recessive Agammaglobulinemia	13.825	0.069	0.333
6	Inclusion Body Myositis	12.681	0.103	0.063
7	Polymyositis	12.173	0.103	0.056
8	Immunoglobulin-Associated beta Deficiency	8.837	0.034	1
8	CD8 Deficiency	8.837	0.034	1
10	Focal and Segmental Glomerulosclerosis	8.413	0.069	0.054

Next, the molecular mechanisms in the relationships between CD233, CD79b, and CD3d that are downregulated by over 2-fold in the female spleens of the KO mice as compared to WT mice were investigated. Nrf2 was searched using the Start point(s) to End point(s) Search method to determine the molecular relations between the start and the end point(s) in KeyMolnet by setting a path number. This search was performed using the conditions of the shortest path search. The start point Nrf2 is indicated in red while the end points CD233, CD79b, and CD3d are indicated in blue (Figure 12). Relatively simple paths from Nrf2 to the

3 molecules mediated by transcription factors such as NFkB and AP-1 are suggested (Figure 12); however, additional experiments are required to confirm these.

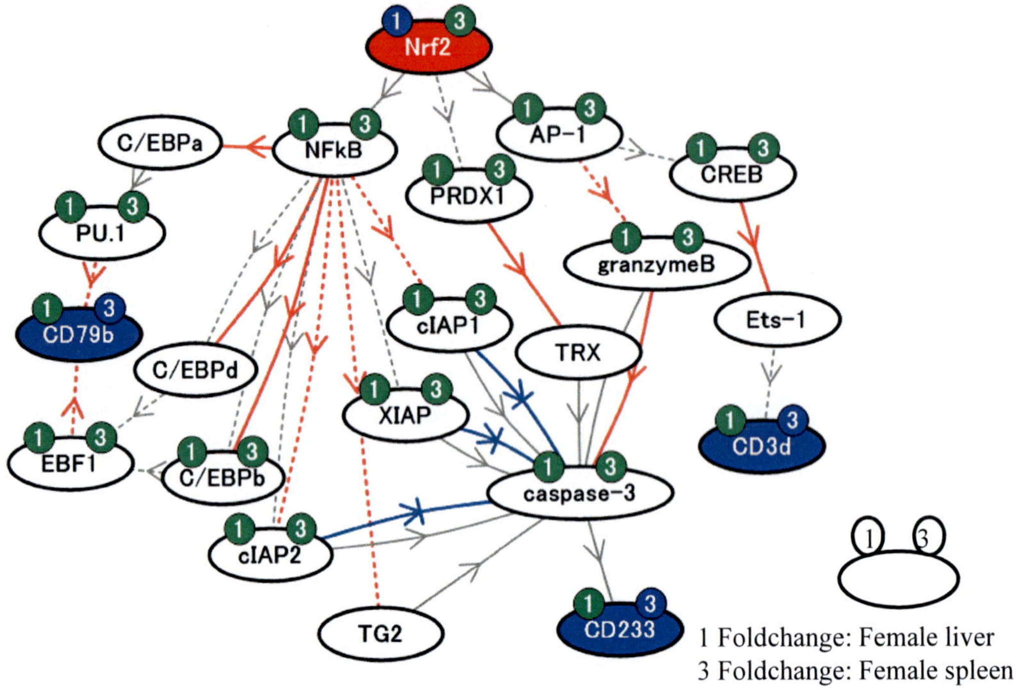

Figure 12. Possible paths from Nrf2 to downregulated molecules in female spleens of Nrf2 KO mice. After setting Nrf2 as the start point (red) and the downregulated molecules induced by the Nrf2 KO as end points (blue), i.e., CD3d, CD233, and CD79b appeared on the network as illustrated in Figure 12, paths from Nrf2 to the end points were searched using the Start point(s) to End point(s) Search method with the shortest path search condition. Colored small circles 1 and 3 on molecules represent the range of fold-changes in the female livers and spleens, respectively, as shown in the figure. The correspondence between the range of fold-changes and colors are the same as shown in Figure 9.

The authors submitted the gene expression data to GEO with a discussion on the relationship between Nrf2 and systemic autoimmune disease [86]. This network analysis by using their data also suggests the possibility of disease involvement in the absence of Nrf2.

Conclusions and Perspectives

Comprehensive gene expression analysis will generally be utilized as an established technique and become a part of research flow in the future. Analysis of huge amounts of gene expression data using hundreds of chips at a time is a difficult transition from large quantities of numerical data to interpretation apart from tendency analysis and/or classification based on gene expression patterns. The DNA microarray analysis is particularly unique for medical, pharmaceutical, and biological uses in that it can obtain clues from the comprehensive information pool in gene expression, such as finding a biomarker or a lead in experimental/clinical research. On the other hand, an oligonucleotide array will further be diverged depending on the aims of research such as SNP detection [89], gene expression profiling, chIp (chromatin immunoprecipitation) studies (ChIp on chip) [90], and copy number analysis [91]. For clinical diagnosis or prediction of prognosis, concentrated and mini-chips focusing on a certain range could be practical and useful. For prediction of toxicity of chemicals, chips to examine several biomarkers in several organs may be effective; however, the detection of important biomarkers is first required. In any case, low-cost, simple, and speedy analyses are desired especially for the realization of personalized therapy.

In this review, interpretation method of large quantities of gene expression data obtained by oligonucleotide array leading to biological processes, phenomena, and functions, based on molecular networks using KeyMolnet was demonstrated. Since molecular networks are essential to perform actions of drugs, manifest adverse events of chemicals, develop diseases, and cause changes in phenomena in living organisms, this analytical method could be a bridge between rows of numerical data using genome sequence information and molecular function using accumulated life science information. Experimental/clinical condition-distinctive networks could enable analysis of the narrowing down of specific molecular mechanisms under certain conditions. Additional experiments are required to confirm the assumptions by generated molecular networks based on gene expression data obtained by DNA microarray; however, working hypotheses and new insights could be obtained from the molecular networks and related life science information.

REFERENCES

[1] Lipshutz RJ, Fodor SP, Gingeras TR, Lockhart DJ. High density synthetic oligonucleotide arrays. *Nat Genet*. 1999;21(1 Suppl):20-4.

[2] Gershon D. Microarray technology an array of opportunities. *Nature*. 2002;416:885-91.

[3] McGall GH, Christians FC. High-density genechip oligonucleotide probe arrays. *Adv Biochem Eng Biotechnol*. 2002;77:21-42.

[4] Mills JC, Roth KA, Cagan RL, Gordon JI. DNA microarrays and beyond: completing the journey from tissue to cell. *Nature Cell Biology*. 2001;3:E175-8.

[5] Dalma-Weiszhausz DD, Warrington J, EY T, CG M. The Affymetrix GeneChip® Platform: An Overview. Kimmel A, Oliver B, editors: Academic Press; 2006.

[6] Heber S, Sick B. Quality assessment of Affymetrix GeneChip data. *OMICS*. 2006;10(3):358-68.

[7] Wolber PK, Collins PJ, Lucas AB, De Witte A, Shannon KW. The Agilent in situ-synthesized microarray platform. Kimmel A, Oliver B, editors: Academic Press; 2006.

[8] Ben-Dor A, Karp R, Schwikowski B, Yakhini Z. Universal DNA tag systems: a combinatorial design scheme. *J Comput Biol*. 2000;7(3-4):503-19.

[9] Shen R, Fan JB, Campbell D, Chang W, Chen J, Doucet D, et al. High-throughput SNP genotyping on universal bead arrays. *Mutat Res*. 2005;573(1-2):70-82.

[10] Fan JB, Gunderson KL, Bibikova M, Yeakley JM, Chen J, Garcia EW, et al. Illumina universal bead arrays. Kimmel A, Oliver B, editors: Academic Press; 2006.

[11] Diez D, Alvarez R, Dopazo A. Codelink: an R package for analysis of GE healthcare gene expression bioarrays. *Bioinformatics*. 2007;23(9):1168-9.

[12] Ramakrishnan R, Dorris D, Lublinsky A, Nguyen A, Domanus M, Prokhorova A, et al. An assessment of Motorola CodeLink microarray performance for gene expression profiling applications. *Nucleic Acids Res*. 2002;30(7):e30.

[13] Sherlock G. Analysis of large-scale gene expression data. *Curr Opin Immunol.* 2000;12(2):201-5.

[14] Shannon W, Culverhouse R, Duncan J. Analyzing microarray data using cluster analysis. *Pharmacogenomics*. 2003;4(1):41-52.

[15] Alizadeh AA, Eisen MB, Davis RE, Ma C, Lossos IS, Rosenwald A, et al. Distinct types of diffuse large B-cell lymphoma identified by gene expression profiling. *Nature*. 2000;403(6769):503-11.

[16] Welsh JB, Zarrinkar PP, Sapinoso LM, Kern SG, Behling CA, Monk BJ, et al. Analysis of gene expression profiles in normal and neoplastic ovarian tissue samples identifies candidate molecular markers of epithelial ovarian cancer. *Proc Natl Acad Sci U S A.* 2001;98(3):1176-81.

[17] Honda M, Kaneko S, Kawai H, Shirota Y, Kobayashi K. Differential gene expression between chronic hepatitis B and C hepatic lesion. *Gastroenterology*. 2001;120(4): 955-66.

[18] Singer S, Socci ND, Ambrosini G, Sambol E, Decarolis P, Wu Y, et al. Gene expression profiling of liposarcoma identifies distinct biological types/subtypes and potential therapeutic targets in well-differentiated and dedifferentiated liposarcoma. *Cancer Res.* 2007;67(14):6626-36.

[19] Moos PJ, Raetz EA, Carlson MA, Szabo A, Smith FE, Willman C, et al. Identification of gene expression profiles that segregate patients with childhood leukemia. *Clin Cancer Res*. 2002;8(10):3118-330.

[20] Sato H, Ishida S, Toda K, Matsuda R, Hayashi Y, Shigetaka M, et al. New approaches to mechanism analysis for drug discovery using DNA microarray data combined with KeyMolnet. *Curr Drug Discov Technol*. 2005;2(2):89-98.

[21] Sato H. Understanding pharmacological mechanisms by using molecular networks - development of KeyMolnet. *Curr Topics in Pharmacology*. 2008;in press.

[22] Peri S, Navarro JD, Amanchy R, Kristiansen TZ, Jonnalagadda CK, Surendranath V, et al. Development of human protein reference database as an initial platform for approaching systems biology in humans. *Genome Res*. 2003;13(10):2363-71.

[23] Peri S, Navarro JD, Kristiansen TZ, Amanchy R, Surendranath V, Muthusamy B, et al. Human protein reference database as a discovery resource for proteomics. *Nucleic Acids Res*. 2004;32(Database issue):D497-501.

[24] Mathivanan S, Periaswamy B, Gandhi TK, Kandasamy K, Suresh S, Mohmood R, et al. An evaluation of human protein-protein interaction data in the public domain. *BMC Bioinformatics*. 2006;7 Suppl 5:S19.

[25] Mishra G, Suresh M, Kumaran K, Kannabiran N, Suresh S, Bala P, et al. Human Protein Reference Database - 2006 Update. *Nucleic Acids Res.* 2006;34 (Database issue): D411-14.

[26] Camon E, Magrane M, Barrell D, Binns D, Fleischmann W, Kersey P, et al. The Gene Ontology Annotation (GOA) project: implementation of GO in SWISS-PROT, TrEMBL, and InterPro. *Genome Res.* 2003;13(4):662-72.

[27] Ashburner M, Ball CA, Blake JA, Botstein D, Butler H, Cherry JM, et al. Gene ontology: tool for the unification of biology. The Gene Ontology Consortium. *Nat Genet*. 2000;25(1):25-9.

[28] Gene Ontology Consortium. Creating the gene ontology resource: design and implementation. *Genome Res*. 2001;11(8):1425-33.

[29] Boyle EI, Weng S, Gollub J, Jin H, Botstein D, Cherry JM, et al. GO:TermFinder-open source software for accessing Gene Ontology information and finding significantly enriched Gene Ontology terms associated with a list of genes. *Bioinformatics.* 2004 Dec 12;20(18):3710-5.

[30] Iwakawa M, Ohno T, Imadome K, Nakawatari M, Ishikawa K, Sakai M, et al. The radiation-induced cell-death signaling pathway is activated by concurrent use of cisplatin in sequential biopsy specimens from patients with cervical cancer. *Cancer Biol Ther.* 2007;6(6):905-11.

[31] Kuzuhara T, Suganuma M, Kurusu M, Fujiki H. Helicobacter pylori-secreting protein Tipalpha is a potent inducer of chemokine gene expressions in stomach cancer cells. J *Cancer Res Clin Oncol.* 2006;133(5):287-96.

[32] Satoh J, Illes Z, Peterfalvi A, Tabunoki H, Rozsa C, Yamamura T. Aberrant transcriptional regulatory network in T cells of multiple sclerosis. *Neurosci Lett.* 2007;422(1):30-3.

[33] Hubbard T, Andrews D, Caccamo M, Cameron G, Chen Y, Clamp M, et al. Ensembl 2005. *Nucleic Acids Res.* 2005;33(Database issue):D447-53.

[34] Sugiyama Y, Murase K, Ikeda H. Mechanisms of thiazolidinedione derivatives for hypoglycemic and insulin sensitizing effects. *Nippon Rinsho.* 2000;58(2):370-5.

[35] Yamanouchi T. Insulin sensitizer drugs--review. *Nippon Rinsho.* 2002;60(Suppl 9): 409-15.

[36] Johnson MD, Campbell LK, Campbell RK. Troglitazone: review and assessment of its role in the treatment of patients with impaired glucose tolerance and diabetes mellitus. *Ann Pharmacother.* 1998;32(3):337-48.

[37] Olefsky JM, Saltiel AR. PPARg and the treatment of insulin resistance. *Trends Metab Endo.* 2000;11(9):362-8.

[38] 38. Yki-Jarvinen H. Thiazolidinediones. *N Engl J Med.* 2004;351(11):1106-18.

[39] Berger JP, Petro AE, Macnaul KL, Kelly LJ, Zhang BB, Richards K, et al. Distinct properties and advantages of a novel peroxisome proliferator-activated protein selective modulator. *Mol Endocrinol.* 2003;17(4):662-76.

[40] Hsiao A, Worrall DS, Olefsky JM, Subramaniam S. Variance-modeled posterior inference of microarray data: detecting gene-expression changes in 3T3-L1 adipocytes. *Bioinformatics.* 2004;20(17):3108-27.

[41] Edgar R, Domrachev M, Lash AE. Gene Expression Omnibus: NCBI gene expression and hybridization array data repository. *Nucleic Acids Research.* 2002;30(1):207-10.

[42] Barrett T, Suzek TO, Troup DB, Wilhite SE, Ngau WC, Ledoux P, et al. NCBI GEO: mining millions of expression profiles-database and tools. *Nucleic Acids Research.* 2005;33(Database Issue):D562-6.

[43] Barrett T, Edgar R. Gene Expression Omnibus (GEO): Microarray data storage. *Methods Enzymol.* 2006;411:352-69.

[44] Barrett T, Troup DB, Wilhite SE, Ledoux P, Rudnev D, Evangelista C, et al. NCBI GEO: mining tens of millions of expression profiles-database and tools update. *Nucleic Acids Research.* 2007;35(Database issue):D760-5.

[45] Redondo S, Ruiz E, Santos-Gallego CG, Padilla E, Tejerina T. Pioglitazone induces vascular smooth muscle cell apoptosis through a peroxisome proliferator-activated receptor-gamma, transforming growth factor-beta1, and a Smad2-dependent mechanism. *Diabetes.* 2005;54(3):811-7.

[46] Zhu Y, Kan L, Qi C, Kanwar YS, Yeldandi AV, Rao MS, et al. Isolation and characterization of peroxisome proliferator-activated receptor (PPAR) interacting protein (PRIP) as a coactivator for PPAR. *J Biol Chem*. 2000;275(18):13510-6.

[47] Zhang Y, Castellani LW, Sinal CJ, Gonzalez FJ, Edwards PA. Peroxisome proliferator-activated receptor-gamma coactivator 1alpha (PGC-1alpha) regulates triglyceride metabolism by activation of the nuclear receptor FXR. *Genes Dev*. 2004;18(2):157-69.

[48] Kahara T, Takamura T, Misaki T, Sakurai M, Takeshita Y, Shimizu A, et al. Relationship between plasma hANP level and pretibial edema by pioglitazone treatment. *Endocr J*. 2005;52(3):373-6. .

[49] Cacalano NA, Johnston JA. Interleukin-2 signaling and inherited immunodeficiency. *Am J Hum Genet*. 1999;65(2):287-93.

[50] Alpdogan O, van den Brink MR. IL-7 and IL-15: therapeutic cytokines for immunodeficiency. *Trends Immunol*. 2005;26(1):56-64.

[51] Rafiq K, Kolpakov MA, Abdelfettah M, Streblow DN, Hassid A, Dell'Italia LJ, et al. Role of protein-tyrosine phosphatase SHP2 in focal adhesion kinase down-regulation during neutrophil cathepsin G-induced cardiomyocytes anoikis. *J Biol Chem*. 2006;281(28):19781-92. .

[52] Tsutsumi R, Takahashi A, Azuma T, Higashi H, Hatakeyama M. Focal adhesion kinase is a substrate and downstream effector of SHP-2 complexed with Helicobacter pylori CagA. *Mol Cell Biol*. 2006;26(1):261-76.

[53] Hart KC, Robertson SC, Kanemitsu MY, Meyer AN, Tynan JA, Donoghue DJ. Transformation and Stat activation by derivatives of FGFR1, FGFR3, and FGFR4. *Oncogene*. 2000;19(29):3309-20.

[54] Moeller ML, Shi Y, Reichardt LF, Ethell IM. EphB receptors regulate dendritic spine morphogenesis through the recruitment/phosphorylation of focal adhesion kinase and RhoA activation. *J Biol Chem*. 2006;281(3):1587-98.

[55] Wu TR, Hong YK, Wang XD, Ling MY, Dragoi AM, Chung AS, et al. SHP-2 is a dual-specificity phosphatase involved in Stat1 dephosphorylation at both tyrosine and serine residues in nuclei. *J Biol Chem*. 2002;277(49):47572-80.

[56] Wang CZ, Su HW, Hsu YC, Shen MR, Tang MJ. A discoidin domain receptor 1/SHP-2 signaling complex inhibits alpha2beta1-integrin-mediated signal transducers and activators of transcription 1/3 activation and cell migration. *Mol Biol Cell*. 2006; 17(6): 2839-52.

[57] Ke Y, Lesperance J, Zhang EE, Bard-Chapeau EA, Oshima RG, Muller WJ, et al. Conditional deletion of Shp2 in the mammary gland leads to impaired lobulo-alveolar outgrowth and attenuated Stat5 activation. *J Biol Chem*. 2006;281(45):34374-80.

[58] Andersson CX, Sopasakis VR, Wallerstedt E, Smith U. Insulin antagonizes interleukin-6 signaling and is anti-inflammatory in 3T3-L1 adipocytes. *J Biol Chem*. 2007;282(13):9430-5.

[59] Sugamura K, Asao H, Kondo M, Tanaka N, Ishii N, Ohbo K, et al. The interleukin-2 receptor gamma chain: its role in the multiple cytokine receptor complexes and T cell development in XSCID. *Annu Rev Immunol*. 1996;14:179-205.

[60] Stirewalt DL, Radich JP. The role of FLT3 in haematopoietic malignancies. *Nat Rev Cancer*. 2003;3(9):650-65.

[61] Seki N, Hashimoto N, Taira M, Yagi S, Yoshida Y, Ishikawa K, et al. Regulation of Src homology 2-containing protein tyrosine phosphatase by advanced glycation end products: the role on atherosclerosis in diabetes. *Metabolism*. 2007;56(11):1591-8.

[62] Lima MH, Ueno M, Thirone AC, Rocha EM, Carvalho CR, Saad MJ. Regulation of IRS-1/SHP2 interaction and AKT phosphorylation in animal models of insulin resistance. *Endocrine*. 2002;18(1):1-12.

[63] Bae MA, Song BJ. Critical role of c-Jun N-terminal protein kinase activation in troglitazone-induced apoptosis of human HepG2 hepatoma cells. *Mol Pharmacol*. 2003;63(2):401-8.

[64] Yin F, Wakino S, Liu Z, Kim S, Hsueh WA, Collins AR, et al. Troglitazone inhibits growth of MCF-7 breast carcinoma cells by targeting G1 cell cycle regulators. *Biochem Biophys Res Commun*. 2001;286(5):916-22.

[65] Yin F, Bruemmer D, Blaschke F, Hsueh WA, Law RE, Herle A. Signaling pathways involved in induction of GADD45 gene expression and apoptosis by troglitazone in human MCF-7 breast carcinoma cells. *Oncogene*. 2004;23(26):4614-23.

[66] Yu J, Qiao L, Zimmermann L, Ebert MP, Zhang H, Lin W, et al. Troglitazone inhibits tumor growth in hepatocellular carcinoma in vitro and in vivo. *Hepatology*. 2006;43(1):134-43.

[67] Thorp E, Kuriakose G, Shah YM, Gonzalez FJ, Tabas I. Pioglitazone increases macrophage apoptosis and plaque necrosis in advanced atherosclerotic lesions of nondiabetic low-density lipoprotein receptor-null mice. *Circulation* 2007;116(19): 2182-90.

[68] Cao LQ, Chen XL, Wang Q, Huang XH, Zhen MC, Zhang LJ, et al. Upregulation of PTEN involved in rosiglitazone-induced apoptosis in human hepatocellular carcinoma cells. *Acta Pharmacol Sin*. 2007;28(6):879-87.

[69] Takenokuchi M, Saigo K, Nakamachi Y, Kawano S, Hashimoto M, Fujioka T, et al. Troglitazone inhibits cell growth and induces apoptosis of B-cell acute lymphoblastic leukemia cells with t(14;18). Acta Haematol. 2006;116(1):30-40.

[70] Mayer RJ. Harrison's Internal Medicine, 16th Edition. New York: McGraw-Hill 2005 [cited. Available from: http://www.accessmedicine.com/content.aspx?aID=62567&searchStr=esophageal+cancer.

[71] Stein HJ, Siewert JR. Barrett's esophagus: pathogenesis, epidemiology, functional abnormalities, malignant degeneration, and surgical management. *Dysphagia*. 1993;8(3):276-88. .

[72] Keswani RN, Noffsinger A, Waxman I, Bissonnette M. Clinical use of p53 in Barrett's esophagus. *Cancer Epidemiol Biomarkers Prev*. 2006;15(7):1243-9.

[73] Shirvani VN, Ouatu-Lascar R, Kaur BS, Omary MB, Triadafilopoulos G. Cyclooxygenase 2 expression in Barrett's esophagus and adenocarcinoma: Ex vivo induction by bile salts and acid exposure. *Gastroenterology*. 2000;118(3):487-96.

[74] Devesa SS, Blot WJ, Fraumeni JF Jr. Changing patterns in the incidence of esophageal and gastric carcinoma in the United States. *Cancer*. 1998;83(10):2049-53.

[75] Kimchi ET, Posner MC, Park JO, Darga TE, Kocherginsky M, Karrison T HJ, Smith KD, Mezhir JJ, Weichselbaum RR, Khodarev NN. Progression of Barrett's metaplasia to adenocarcinoma is associated with the suppression of the transcriptional programs of epidermal differentiation. *Cancer Res*. 2005;65(8):3146-54.

[76] Selaru FM, Zou T, Xu Y, Shustova V, Yin J, Mori Y, et al. Global gene expression profiling in Barrett's esophagus and esophageal cancer: a comparative analysis using cDNA microarrays. *Oncogene*. 2002;21(3):475-8.

[77] Siersema PD. Pathogenesis, diagnosis and therapeutic possibilities of esophageal cancer. *Curr Opin Gastroenterol*. 2007;23(4):456-61.

[78] Yamazaki K, Hasegawa M, Ohoka I, Hanami K, Asoh A, Nagao T, et al. Increased E2F-1 expression via tumour cell proliferation and decreased apoptosis are correlated with adverse prognosis in patients with squamous cell carcinoma of the oesophagus. *J Clin Pathol*. 2005;58(9):904-10.

[79] Turner JR, Torres CM, Wang HH, Shahsafaei A, Richards WG, Sugarbaker D, et al. Preoperative chemoradiotherapy alters the expression and prognostic significance of adhesion molecules in Barrett's-associated adenocarcinoma. *Hum Pathol*. 2000;31(3):347-53.

[80] Fléjou JF, vrcek M. Barrett's oesophagus-a pathologist's view. *Histopathology*. 2007;50(1):3-14.

[81] Itoh K, Chiba T, Takahashi S, Ishii T, Igarashi K, Katoh Y, et al. An Nrf2/small Maf heterodimer mediates the induction of phase II detoxifying enzyme genes through antioxidant response elements. *Biochem Biophys Res Commun*. 1997;236(2):313-22.

[82] Nguyen T, Sherratt PJ, Pickett CB. Regulatory mechanisms controlling gene expression mediated by the antioxidant response element. *Annu Rev Pharmacol Toxicol*. 2003;43:233-60.

[83] Enomoto A, Itoh K, Nagayoshi E, Haruta J, Kimura T, O'Connor T, et al. High sensitivity of Nrf2 knockout mice to acetaminophen hepatotoxicity associated with decreased expression of ARE-regulated drug metabolizing enzymes and antioxidant genes. *Toxicol Sci*. 2001;1:169-77.

[84] Cho HY, Jedlicka AE, Reddy SP, Kensler TW, Yamamoto M, Zhang LY, et al. Role of NRF2 in protection against hyperoxic lung injury in mice. *Am J Respir Cell Mol Biol*. 2002;2:175-82.

[85] Aoki Y, Sato H, Nishimura N, Takahashi S, Itoh K, Yamamoto M. Accelerated DNA adduct formation in the lung of the Nrf2 knockout mouse exposed to diesel exhaust. *Toxicol Appl Pharmacol*. 2001 173(3):154-60.

[86] Li J, Stein TD, Johnson JA. Genetic dissection of systemic autoimmune disease in Nrf2-deficient mice. *Physiol Genomics*. 2004;18(3):261-72.

[87] Iida K, Itoh K, Kumagai Y, Oyasu R, Hattori K, Kawai K, et al. Nrf2 is essential for the chemopreventive efficacy of oltipraz against urinary bladder carcinogenesis. *Cancer Res*. 2004;64(18):6424-31.

[88] Ramos-Gomez M, Kwak MK, Dolan PM, Itoh K, Yamamoto M, Talalay P, et al. Sensitivity to carcinogenesis is increased and chemoprotective efficacy of enzyme inducers is lost in nrf2 transcription factor-deficient mice. *Proc Natl Acad Sci USA*. 2001;98(6):3410-5.

[89] Hacia JG, Fan JB, Ryder O, Jin L, Edgemon K, Ghandour G, et al. Determination of ancestral alleles for human single-nucleotide polymorphisms using high-density oligonucleotide arrays. *Nat Genet*. 1999;22(2):164-7.

[90] Royce TE, Rozowsky JS, Bertone P, Samanta M, Stolc V, Weissman S, et al. Issues in the analysis of oligonucleotide tiling microarrays for transcript mapping. *Trends Genet*. 2005;21(8):466-75.

[91] Friedman JM, Baross A, Delaney AD, Ally A, Arbour L, Armstrong L, et al. Oligonucleotide microarray analysis of genomic imbalance in children with mental retardation. *Am J Hum Genet.* 2006;79(3):500-13.

In: Oligonucleotide Array Sequence Analysis ISBN 978-1-60456-542-3
Editors: M.K. Moretti and L.J. Rizzo, pp. 245-264

Chapter 8

GENERALIZED ANALYSIS OF VARIANCE (ANOVA) MODELS FOR OLIGONUCLEOTIDE MICROARRAY DATA ANALYSIS

***Hua Li*[1] *and Dongxiao Zhu*[2,3]**
[1]Stowers Institute for Medical Research,
1000 E 50th St, Kansas City, MO 64110.
[2]Department of Computer Science, University of New Orleans,
New Orleans, LA 70148, USA.
[3]Research Institute for Children,
200 Henry Clay Ave., New Orleans, LA 70118, USA

Abstract

Generalized ANOVA including fixed and mixed ANOVA are widely used statistical models in analyzing microarray experiments from various platforms. The flexibility of this approach allows investigators to simultaneously assess the contributions of multiple factors and their interactions to gene expression variation. It could account for all sources of variation in the complex experiment and provide easy interpretation of results. In the representative oligonucleotide platform Affymetrix GeneChips, each gene is represented by one or more probe sets (termed sibling probe sets) and each probe set is represented by multiple probes. The complicated many-to-one representation poses critical challenges to data analysts. While summary methods that provide an "average" measure of multiple probes to be the probe set expression score are useful, the mixed model ANOVA analysis using probe level expression data greatly increases sample size and hence increases power in identification of differentially expressed gene. We survey some representative works in probe-level analysis of Affymetrix oligonucleotide data. For genes that are represented by multiple probe sets (sibling probe sets), differentially expressed genes between conditions could be identified using ANOVA by treating the sibling probe sets as a factor along with other interesting factors in the designed experiment. Expression levels of sibling probe sets that are proportional to each other across different conditions are more likely to be replicated measures of the same target gene. The decision on whether to consolidate sibling probe sets was made through statistical testing of the interaction effect in the ANOVA model. In our recent work of applying ANOVA for sibling probe set analysis, results show marked improvement of gene expression estimates and identification of more biologically relevant genes. We demonstrate this advantage by analyzing three publicly available datasets and validate it using Gene Ontology (GO) terms enrichment analysis.

Introduction

DNA microarrays have become widespread for the genome-wide measurement of mRNA level expression. One of the most fundamental and central tasks in any microarray experiment is to identify a list of differentially expressed genes over different conditions. Application of classical statistical methods such as ANOVA has dramatically improved the analysis of differential gene expression. In the early stage of microarray analysis for small scale experiments, researchers often use an arbitrary fold change threshold (such as 2) to screen differential expression. Although the fold change provides easy interpretation of results, its lack of statistical properties quickly drives people to t-test when comparing gene expression for two groups of samples. T-test could generate p-value for statistical significance support, but the relatively small sample size in microarray experiments results in inaccurate estimates of variance of expression values. To improve the test statistics by utilizing parallel information among genes, Tusher et al [1] introduced a method that employs a modified t-test associated with a permutation test to assess the significance between two classes of samples. For complex experiments that involve multiple factors and multiple levels of each factor (e.g. three cell lines, four time points), a model with more flexibility and power is needed. The classical ANOVA model is a suitable framework because it accounts for multiple sources of variation. Kerr et al [2] first proposed using ANOVA in two-color microarray data analysis by fitting a model with all measures across genes, referred to the global ANOVA model. Wolfinger et al [3] showed that the global model could be implemented with a two-step procedure with approximately similar estimates, referred to the gene specific model for analyzing one gene at a time. Chu et al [4] further extended this methodology to high density oligonucleotide microarrays.

High density oligonucleotide microarrays are produced by printing oligonucleotide sequences by synthesizing these squences directly onto the array surface. These oligonucleotide sequences, called probes, are designed to match parts of the sequence of known or predicted open reading frames. Sequences may be longer (60-mer probes such as the Agilent design) or shorter (25-mer probes produced by Affymetrix); longer probes are more specific to individual target genes, shorter probes may be spotted in higher density across the array.

Affymetrix GeneChip arrays, one of the representative and dominate oligonucleotide arrays on the market, use a set of oligonucleotide probes each with a length of 25 bases to represent a gene. A gene, referred to a probe set, is composed of 8 to 20 pairs of these short oligonucleotides to improve specificity of the expression measure. Each probe pair is composed of a perfect match (PM) probe and a mismatch (MM) probe: the former contains exactly 25 nucleotides and the latter differs from the former in that there is one single mismatch at position 13. A PM probe is designed to hybridize only with transcripts from the intended gene (specific hybridization). A MM probe is designed with the intention of measuring only non-specific components of the corresponding PM probe. A single gene is often represented by multiple probe sets on a GeneChip. These multiple probe sets map to the same target gene and share the same annotation. They are designed this way for historical reasons. Some might intend to represent the real biological phenomena such as alternative splicing, some might intend to measure the target gene more specifically and accurately. For example, in the mouse moe4302 chip, there are $45,101$ probe sets corresponding to $25,724$ distinct genes, and 40% of all genes are represented by multiple probe sets, called "sibling probe sets" throughout this chapter (Fig. 1). For these 40% of genes, almost half of them are represented by more than two probe sets on the chip, and some genes even have more than ten probe sets. Similarly in the human hgu133plus2 chip, the total of $28,919$ genes are represented by $54,675$ probe sets on the chip. The built-in redundancy for gene expression measure in Affymetrix GeneChip arrays, i.e., one gene - multiple probe sets and one probe set - multiple probe pairs, has to been considered during data processing and analysis.

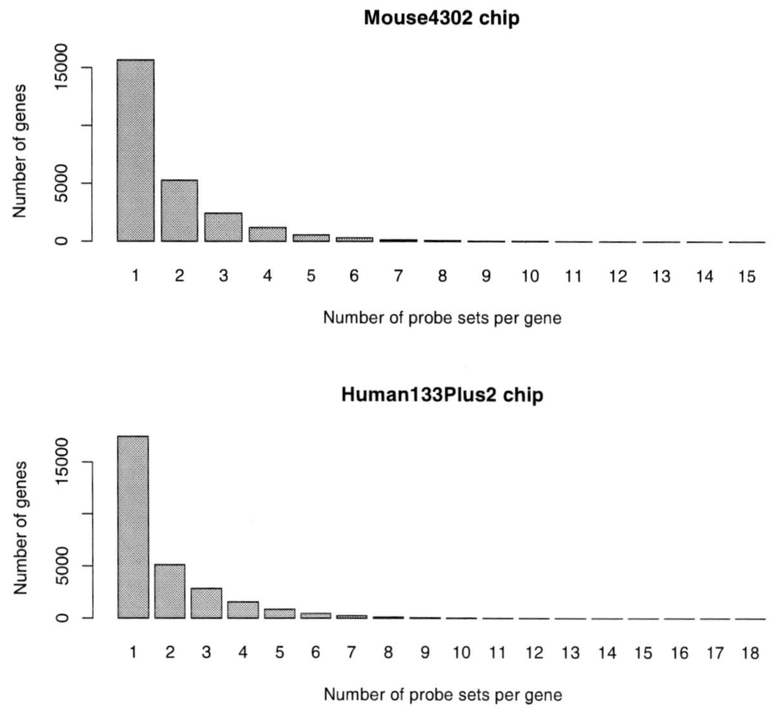

Figure 1. The distribution of sibling probe set numbers per gene in Affymetrix Human and Mouse chips. The figure describes the distribution of the probe set numbers per gene on the Affymetrix mouse moe4302 chip and human hgu133plus2. About 40% of genes are presented by multiple probe sets, half of which are represented by three or more sibling probe sets, and some genes have even more than ten sibling probe sets.

If there is a one-to-one relationship between a probe set and a gene, then the many-to-one situation reduces to one probe set to many probe pairs. One widely adopted approach is to transform a set of probe pair's intensities into a single probe-set measurement using robust average or robust modeling fitting. The representative algorithms for this summarization procedure are MAS5 [5], MEBI [6], RMA [7], GCRMA [8], etc. Averaging or summarizing probe level data could reduce the complexity of the data and subsequent analysis. However, large probe level variation carries important information that is useful for statistical analysis and therefore should be considered. Another approach is to use probe level probe intensities directly in the analysis for identifying differentially expressed genes, which are referred to probe level linear model analysis by Chu et al. [4]. This model does not include the summarization step, but rather uses normalized probe level intensities in the model for detecting differential expression.

If there are multiple probe sets mapping to the same specific gene, the many-to-one situation exists in both the probe set level and the probe level, which is often ignored in the microarray data analysis. Almost all the analysis in the literature treat sibling probe sets as distinct genes without considering the possible correlation among them. Until recently, several groups independently studied how well sibling probe sets measure the same gene expression [9], [11], [10] and showed that expression values from sibling probe sets are not always highly correlated as one would expect.

In this chapter, we review the basics of using ANOVA models to analysis Affymetrix GeneChip data. We discuss issues associated with probe level analysis for detection of differentially expressed genes. We introduce a generalized ANOVA framework to consolidate sibling probe sets whenever appropriate. Although we use Affymetrix as the example in this chapter, models and methods discussed in this chapter apply to high-density oligonucleotide platforms in general.

Generalized Analysis of Variance: Fixed and Mixed Models

ANOVA is one of the most widely used statistical techniques. It is often used to model the effect of a linear combination of different factors and to decompose the total variation of a data set into meaningful component parts associated with specific sources of variation. In general, we could write a linear model in a simple form as follows:

$$\mathbf{Y} = \mathbf{X}\beta + \mathbf{Z}\mathbf{u} + \epsilon \tag{1}$$

This model (1) is the simplified version for the format using vectors and matrix as follows:

$$\begin{bmatrix} y_1 \\ y_2 \\ \vdots \\ y_n \end{bmatrix} = \begin{bmatrix} x_{11} & x_{12} & \cdots & x_{1p} \\ x_{21} & x_{22} & \cdots & x_{2p} \\ \vdots & \vdots & & \vdots \\ x_{n1} & x_{n2} & \cdots & x_{np} \end{bmatrix} \begin{bmatrix} \beta_1 \\ \beta_2 \\ \vdots \\ \beta_p \end{bmatrix} + \begin{bmatrix} z_{11} & z_{12} & \cdots & z_{1q} \\ z_{21} & z_{22} & \cdots & z_{2q} \\ \vdots & \vdots & & \vdots \\ z_{n1} & z_{n2} & \cdots & z_{nq} \end{bmatrix} \begin{bmatrix} u_1 \\ u_2 \\ \vdots \\ u_q \end{bmatrix} + \begin{bmatrix} \epsilon_1 \\ \epsilon_2 \\ \vdots \\ \epsilon_n \end{bmatrix}$$

$\mathbf{Y}$: is an $n \times 1$ vector, n is the number of observed or measured values in an experiment.

β and $\mathbf{u}$: β is a $p \times 1$ vector, containng p fixed effect(s) of a single independent variable or several independent variables and their interactions. $\mathbf{u}$ is a $q \times 1$ vector, containing q random effect(s) of independent variables. We usually call these independent variables as factors as these factors are controlled and varied by the experimenter. In principle, a factor is considered fixed if the levels of the factor constitute the entire population. Examples include male and female sexes, different cell lines, probes of a probe set. On the other hand, a factor is considered as random when one wants to make inferences on an entire population, and the levels in the experiment only represent samples from that population. For example, array effects are random, as each array in a given microarray experiment is just a sample from all arrays that could have been used. A different array would be used in a different experiment.

X and **Z** are design matrices for the corresponding factors, which usually contain indicator variables (ones and zeros) that indicate group membership of β and **u** in ANOVA.

ϵ is the residual, modeling the random variation in the experiment.

If all the effects in a model are fixed and no random effects of **u** are included, then the model is called a fixed effect model in the format of $\mathbf{Y} = \mathbf{X}\beta + \epsilon$. Very often, some factors are fixed and others are random, and then it's called a mixed model (model (1)). A key assumption in the mixed model is that **u** and ϵ are normally distributed with

$$\mathbf{E}\begin{bmatrix} u \\ \epsilon \end{bmatrix} = \begin{bmatrix} 0 \\ 0 \end{bmatrix}$$

$$\mathbf{Var}\begin{bmatrix} u \\ \epsilon \end{bmatrix} = \begin{bmatrix} G & 0 \\ 0 & R \end{bmatrix}$$

The variance of Y, V, is then $V = Z'GZ + R$. G contains variance components in a diagonal structure, i.e. $\sigma_a^2 I_n$ for the array random factor where I_n denotes the $n \times n$ identity matrix. R models the random variation, i.e. $R = \sigma_e^2 I_n$ for the unrelated individual measurements.

ANOVA is used to test whether the means among different groups are equal. In the fixed effect model, one could compute F ratios of the variance due to the factor of interest to an appropriate measure of the residual variances in the closed form. In the mixed model, computation of F ratio is more complicated. One needs to estimate the residual variances and other variance components (σ_e^2, and σ_a^2 in model (1)) iteratively using Expectation-Maximum or Newton-Raphson algorithms by maximizing the likelihood or the residual likelihood, and then to calculate F ratios similarly as in the fixed model by substituting the estimated residual and other variances.

Identification of Differentially Expressed Genes Using Probe Level Linear Models

PLLMM: Probe Level Linear Mixed Model

Identification of differentially expressed genes is one of the most fundamental tasks in microarray experiments. Since genes are represented by probe sets, a widely used approach is to analyze each probe set individually with the same statistical model (also called gene-specific or probe set specific model). For any given probe set, the generic form of the probe level linear mixed model (PLLMM) involving a single treatment factor with two or more levels could be written in the following form:

$$Y_{ijk} = \mu + T_i + P_j + TP_{ij} + A_k + \epsilon_{ijk} \tag{2}$$

Y_{ijk} is the normalized probe intensity of the ith level of treatment at the jth probe in the kth replicate. Normalization is necessary to remove any systematic biases that do not reflect true biological variations between arrays in a multiple array experiment. Several normalization methods have been developed and reviewed in the literature [12], and will not be discussed in this chapter. The overall mean μ and the residual error ϵ are different for each probe set. T is the factor of interest, the treatment effect. P, and TP represent probe and probe-by-treatment interaction effects. T, P, and TP are considered as fixed effect, mapping to the fixed effect β in model (1). A denotes array effect, which is a random factor and assumed to be normally distributed with mean 0 and variance σ_a^2. σ_a^2 models the common correlation across all observations on the same chip.

More complex models can be fit to data by adding terms and interactions representing different factors in a complex experiment. This is particularly useful for simultaneously estimating the relative contribution of two or more treatments to expression and for determining whether the effect of one treatment affects the others. For example, one wants to compare transcriptome changes of eight soybean cultivars with various levels of quantitative resistance in response to *P. sojae* infection at two time points (three and five days after infection). In this experiment, we have two factors: cultivar and time. Plus we still need to consider probe and array effects. Model (2) becomes:

$$Y_{ijkl} = \mu + T_i + C_j + TC_{ij} + P_k + TP_{ik} + CP_{jk} + A_{l(ij)} + \epsilon_{ijkl} \tag{3}$$

Where T, C, P, TC, TP, CP are effects for time, cultivar, probe and all their possible interaction effects. This model allows us not only to compare expression changes among the eight cultivars and between two time points, but also to monitor whether the gene expression difference between two time points depends on cultivars by testing the TC interaction effect.

One advantage of the PLLMM model over the summarization methods is that it includes probe and probe-by-treatment interaction effects. All probes within a probe set are expected to estimate expression of the same gene, therefore ideally to have similar expression values. However, high levels of variation among probes, consistent across arrays (Fig.2), are often observed [6]. It is still not clear what causes such a large variation in probe intensities. Wu et al [8] showed that it is partially related to affinities, or position dependent base effects. Other researchers [13] [14] [15] demonstrated that it is also related to cross-hybridization. Individual probes have different levels of cross-hybridization, resulting in large probe intensity variation. Moreover, the probe expression profiles often vary between treatments, which is modeled as probe-by-treatment interaction(Fig.2). The cause of probe-by-treatment interaction could also be due to cross-hybridization [4]. The rational underlying argument for this is that: if the signal due to the cross-hybridization is larger relative to the signal due to the target sequence, then probe intensities for treatment would be different (Fig.2). For example, in the comparison of expression between two species, polymorphism causes one species to hybridize less intensively to a particular oligonumcleoitide [14], which appear as a probe-by-species interaction effect. The cause of cross-hybridization is complicated. Probes that do not match the target sequence could account for part of it, but there is no obvious connection between the probe intensity and how good a probe matches its target sequence [15].

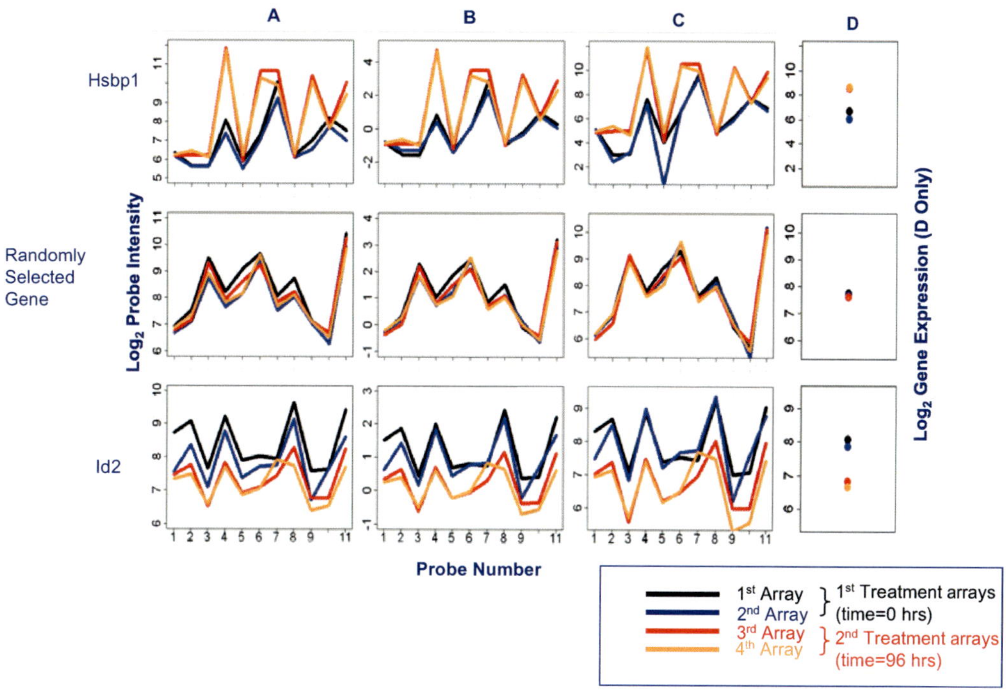

Figure 2. Comparisons of probe level patterns across replicates and treatments for selected genes. Each line color (in Panels A, B, and C) represents a probe level pattern on a specific array. Blue and black lines show probe level patterns on arrays from time = 0. Red and orange lines show probe level patterns on arrays for time = 96 hours. The first row shows probe level patterns for the highest up-regulated gene by fold change (Hspb1). The second row is a randomly selected gene. The third row is the most down-regulated gene by fold change. The x-axis for each plot in (Panels A, B, and C) is probe number (1 through 11). (Panel A) shows log2 of raw probe intensities by probe number. (Panel B) shows log2 array-centered intensities by probe number. The log2 quantile normalized mas-background-corrected probe intensities are shown in (Panel C). (Panel D) shows summary gene expression estimates using median polish. The summarization method used is exactly like the RMA method except that mas-background correction is used instead of rma background correction. (Figure courtesy of Alexander Cambon [15])

Choosing which value to denote the probe expression values Y is critical. As we discussed before, Affymetrix arrays measure both PM and MM for each probe. The MM probe is designed to serve as an internal control of hybridization specificity , therefore using the difference between PM and MM as the expression value for a specific probe [16] sounds reasonable. However, empirical studies show that that PM and MM data typically have a correlation of around 0.8 within one array [17]. Around 30% of MM signals are greater than PM signals and the number of probe pairs within probe sets for which MM $>$ PM varies from 0 to 14. These observations suggest that MM may detect specific bind signals except non-specific binding signals. However, there seems to be a lack of appropriate models to estimate signals given by MM probes. Since MM data tend to increase the noise [4] and differences of PM-MM tend to attenuate signals [7], most researchers tend to use only PM intensities as expression measures. Hsieh et al [17] propose to use the correlation between MM and PM and treat them as a vector of dependent variables in the bivariate mixed model analysis. The bivariate model integrates more data points and provides dynamic estimation of covariance structure between PM and MM probes. Simulation studies showed that it has more statistical power to detect small differences betwen treatments compared to the univariate model (PM only), but needs 10-20 times more computational time. For an experiment of a large sample size, the improvement becomes insignificant.

Two-Way ANOVA of Probe Level Data Analysis

Barrera et al [18] also advocate using probe level data analysis, but simplify the model (2) by dropping the random array effect. Model (2) is reduced to a two-way ANOVA model:

$$Y_{ijk} = \mu + T_i + P_j + TP_{ij} + \epsilon_{ijk} \tag{4}$$

This model is a reduced model of the model (2) with no array effect. When array correlation is very high and array variance σ_a^2 is small, the results from this reduced model are similar to that from model (2). In the cases that σ_a^2 is relatively large compared to the error variance σ_e^2, the statistical power of this reduced model decreases because of the over-estimation of σ_e^2 which is confounded with the array variance σ_a^2.

Global Model vs. Gene-specific Model

PLLMM applies to one gene at a time. This means that the inference of whether a particular gene is differentially expressed depends only on the data from that gene. It's computationally efficient, but the problem is that there is usually not enough data for individual genes to get accurate estimates of expression value and error variance. It's possible that the estimated small error variance for some genes just happens by chance over thousands of genes, which in fact leads to many "false positives" in detecting differential expression [19] [1] . The global model, in the sense that it applies to the expression data for all genes at once, enables one to combine data across genes and estimate error variance distributions. It is first introduced by Kerr [2] in the two-color cDNA microarray data analysis and has been shown to give more robust inference than the gene-specific model. Hoeschele and Li [20] extended the global model to the Affymetrix data analysis and proposed a gene-specific algorithm to address the computational related issues. Joint analysis of expression data on all genes provides flexibility on how the correlation between factors are modeled and permits model evaluation and residual analysis. When the data has been properly normalized, the overall treatment effect across genes is essentially zero, and the point estimates of gene expression using the global ANOVA model is essentially equivalent to that of a single gene ANOVA model on the data that has been "centered" [21] [20].

Consolidate Multiple Probe Sets to Improve the Power of Differentially Expressed Gene Detection

Probe-set specific models analyze each probe set respectively. For genes that are represented by sibling probe sets (multiple probe sets mapping to the same gene), inference from one probe set should not be simply extended to the corresponding gene since these multiple probe sets are not always highly correlated. It's also inappropriate to arbitrarily choose one sibling probe set as the representative of the gene and ignore the others [22] [23] [24].

For Affymetrix arrays, genes that are represented by multiple probe sets are designed for historical reasons. According to Affymetrix [http://www.affymetrix.com/support/help /faqs/ge_assays/faq_16.jsp], there are three primary reasons for designing sibling probe sets for the same gene: first, some cDNAs may be thought to come from different loci at the time of chip design, but later genome annotation maps them to the same gene; second, some probe sets turn out to cross-hybridize in an unpredictable manner, and additional probe sets with better specificity have to be designed for the same gene; third, probe sets specific to RNA variants, such as products of alternative splicing, or highly similar gene family or transcripts with different poly(A) sites, may be designed on purpose. Correspondingly, Affymetrix probe set name suffixes try to indicate these design purposes, e.g. probe sets with "s" and "x" suffixes are prone to cross-hybridization, and probe sets with an "a" suffix are more likely to represent alternative splicing variants. However, two independent studies showed that different expression scores of sibling probe sets are not due to the inclusion of these suboptimal probe sets, and there is lack of evidence showing that these suboptimal probe sets performed worse than "better designed" probe sets [24] and [9].

Ideally, one wants to separate sibling probe sets into two types: those that behave similarly across treatments and those that behave differently across treatments for a given experiment. We cast the problem of automatic determination of the sibling probe set type in the ANOVA framework, in which the differential expression between sibling probe sets, treatments and their mutual influence are simultaneously inspected in a two-way ANOVA model or it's extension with block effect and test whether their interaction is significant. Insignificant interaction effect indicates that sibling probe sets are more likely to behave similarly and provide evidence for consolidation.

We choose to use the properly summarized probe set expression to reduce the complexity of the analysis. Using three publicly available Affymetrix data sets [25] [26] [27], we show that our approach (defined as "per-gene" approach) is able to call more biologically relevant genes than the the existing approach: treat all sibling probe sets as distinct genes (defined as "per-probeset" approach throughout this chapter).

The Statistical Framework for Consolidating Sibling Probe Sets

The outline of automatic identification and consolidation of qualified sibling probe sets based on statistically supported evidence is shown in Figure 3. We start our analysis from properly normalized and summarized expression scores for each probe set, e.g. RMA score [7], GCRMA score [8] or Model-Based Expression Index (MBEI) [6]. We ask whether the differential expression over treatments among sibling probe sets follows the same trend or not in a two-way ANOVA model, which includes treatment (T), probe set (Ps), as well as their interaction effect (TPs). Non-significant interaction effect indicates that the sibling probe sets have the same trend of differential expression over treatments. As shown in the top row of Fig.4, several probe sets show similar expression profile (slopes) between wild type and treatment (knock-out). These kinds of sibling sets are to be consolidated, as they essentially are hidden replicated measures of the target gene expression. Consequently, the p-value of the treatment effect should be reported based on the two-way ANOVA model (model(5)) since it accounts for all measures from sibling probe sets for the same gene. Significant interaction effect indicates that the expression profiles from the probe sets are different in slopes shown in the middle and bottom rows of Fig.4 These sibling probe sets are more appropriately treated as independent probe sets although they share the same gene symbol. For independent probe sets or single probe sets, we compared the differential expression over treatments using only one-way ANOVA model (model (7)). In this case, p-values of the treatment effect were reported from the one-way ANOVA model.

It is often seen that the microarray experiment involves paired samples, for example, a pair of treatment and control samples are from the same individual. We need to add a block factor to the existing one-way (model(8)) or two-way ANOVA model (model(6)) to take into account the corresponding relationship between each pair.

The ANOVA Models

For genes with sibling probe sets, we fit the full two-way ANOVA model (model(5)) with a probe set by treatment interaction to the pooled data as follows:

$$Y_{ijk} = \mu + T_i + Ps_j + TPs_{ij} + \epsilon_{ijk} \tag{5}$$

Let y_{ijk} be the normalized and summarized probe set intensity score for the i_{th} gene, j_{th} probe set and k_{th} replicate of this probe set. We model treatment effect (T), probe set effect Ps, and their interaction effect TPs. Let B represents the block factor, where k presents block size, $k = 1, 2$, the two-way ANOVA model with block effect is:

$$Y_{ijk} = \mu + T_i + Ps_j + TPs_{ij} + B_k + \epsilon_{ijk} \tag{6}$$

If the interaction effect TPs is insignificant after multiple-test correction (as we used False Discovery Rate (FDR)$\leq$ 0.01,Benjamini-Hochberg Procedure [28], we then report P-values of the treatment effect T; otherwise we consider sibling probe sets as independent probe sets. For the gene corresponding to a single probe set and these independent probe sets, we fit the one-way ANOVA model (7) where only the treatment effect is included.

$$y_{jk} = \mu + T_j + \epsilon_{jk} \tag{7}$$

Similarly, the one-way ANOVA model with block effect is:

$$y_{jk} = \mu + T_j + B_k + \epsilon_{jk}, \tag{8}$$

where $k = 1, 2$.

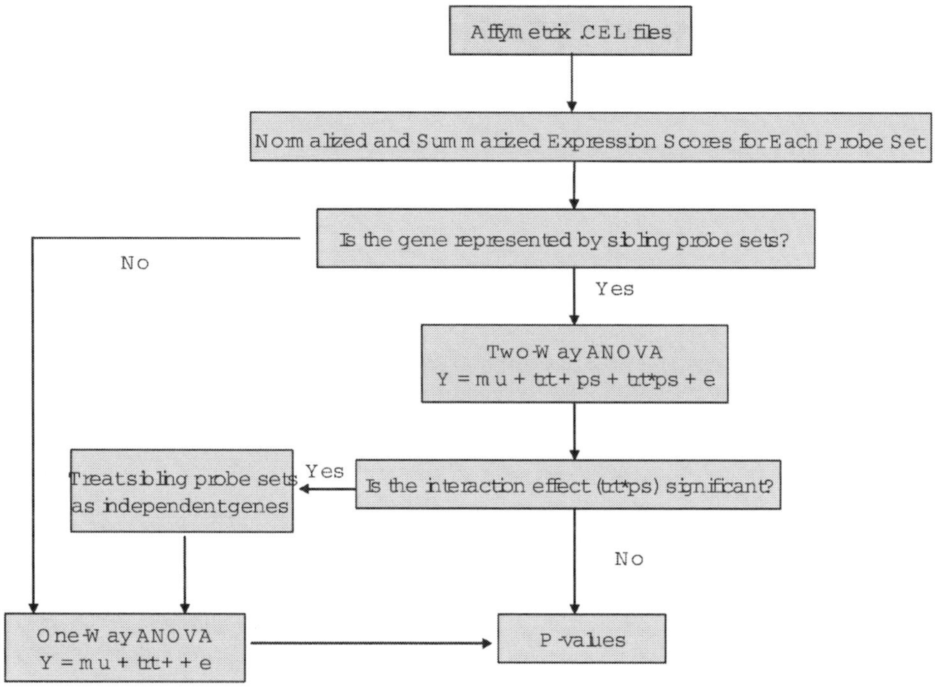

Figure 3. The algorithm flowchart. The figure demonstrates the outline of identification and consolidation of qualified sibling probe sets based on statistical tests. We are interested in studying the differentially expressed genes across treatments. The analysis starts from properly normalized and summarized expression scores for each probe set. For genes that are represented by multiple probe sets (sibling probe sets), insignificant interaction effect (trt*ps) between treatment (trt) and probesets (ps) suggests consolidating sibling probe sets and p-values of the treatment effect are obtained from the two-way ANOVA model. For the gene corresponding to a single probe set and those probe sets that are not eligible for consolidating, i.e. significant interaction effect (trt*ps), p-values of the treatment effect are reported from the one-way ANOVA model. Then p-values are combined as a final result for screening differentially expressed genes across treatments.

GO Enrichment Analysis

For a gene list generated by the per-gene or the per-probeset approach, we used Bioconductor package "GOstats" [http://www.bioconductor.org/] to perform GO enrichment analysis [36].

Results

We compared the proposed the per-gene approach with the existing per-probeset approach on Affymetrix platforms moe4302 and hgu133plus2. Under the same FDR cut-off as well as p-value cut-off, we say that approach A **dominates** approach B if the gene list generated by A is much longer than that generated by B, and the vast majority of list B falls into list A (Fig.5). The approach that identifies the gene list enriched with experimentally

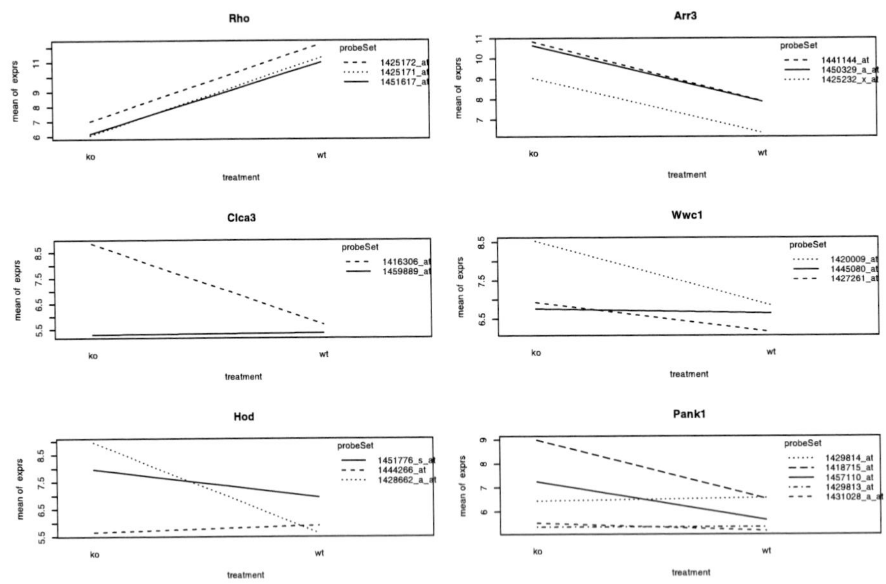

Figure 4. Examples of sibling probe sets need to be consolidated (top row) and otherwise (middle and bottom rows). In the top row, the differential expression of genes "Rho" and "Arr3" between wild type and treatment (Nrl knockout) does not change over sibling probe sets (raw p-values for interaction effect are 0.68 and 0.89 respectively). In the middle and bottom rows, the differential expression between wild type and treatment for sibling probe sets have either different magnitude or reversed trend. The former is translated into a need for consolidation but not the latter.

relevant GO terms indicates better performance.

Example1: Discriminative Analysis over Treatment and Control

The first Affymetrix data set we used was generated by Akimoto et al [25] using Affymetrix moe4302 chip. The data was downloaded from the Gene Expression Omnibus (GEO) database using accession number GSE4051. We focused on identifying differentially expressed genes at developmental maturity stage P10 with four replicates in both wild type and Nrl-knockout conditions.

For $15,632$ genes that Affymetrix designed single probe set for mouse $4,302$ GeneChip, we performed one-way ANOVA with both equal variance and unequal variance assumptions. Correspondingly for the $10,049$ genes that Affymetrix designed sibling probe sets for, we performed two-way ANOVA analysis with interaction between the two fixed effects T and Ps (model(5)). Specifically, we model probe set (Ps) and treatment (T) (Wild Type vs. Nrl-ko) as two factors as well as their interaction (whether differential expression changes over probe sets or *vice versa*). There are 62 sibling probe set genes whose interaction terms were called significant at FDR [28] no larger than 1%. It means that the differential expression over wild type and Nrl-ko conditions is dependent on the sibling probe sets or *vice versa*. For this reason we treated the 255 probe sets mapping to these 62 genes as individual probe sets, followed by fitting the one-way ANOVA model with treat-

Table 1. Performance comparisons in terms of numbers of differentially expressed genes. Comparison of the per-gene approach and per-probeset approach in terms of screening differentially expressed genes between wild type and Nrl knockout.

	Mult-test Algorithm	Unequal Variance	equal variance
Per-Gene	Bonferroni (.05)	39	45
Per-Gene	FDR-BH (.05)	313	434
Per-Gene	FDR-BY (.05)	63	84
Per-Gene	RawP cut-off (6.5e-05)	124	151
Per-PS	Bonferroni (.05)	2	6
Per-PS	FDR-BH (.05)	6	59
Per-PS	FDR-BY (.05)	1	4
Per-PS	RawP cut-off (6.5e-05)	6	59

ment effect only (model(7)). Finally, raw p-values of the treatment effect were combined from the full two-way ANOVA and the one-way ANOVA. The number of hypotheses tests reduced from $45,101$ in the per-probeset approach to $25,917 = 15,675+10,049-62+255$ in the per-gene approach.

Given an FDR cutoff (e.g.1%), the per-gene approach calls more true differentially expressed genes between wild type and Nrl-knockout under different statistical assumptions and dominates over the per-probeset approach. Assuming the wild type and Nrl-ko populations have unequal variance, thr per-probeset approach called only six probe sets to be significantly differentially expressed with an FDR smaller or equal to 5% while the per-gene procedure called 313 genes at the same FDR cut-off (Table 1). Five out of six probe sets determined to be differentially expressed in the per-probeset approach (Nrl, Rho, 4921511K06Rik, Gnb1, Lrp4) are included in the 313 probe set list (Fig. 5a, overlap 83.3%). For example, Nrl acts synergistically with Crx to regulate the transcription of Rho [30], and Gnb1 serves as a membrane receptor for the signal transduction cascade regulating Rho [31]. We also examined 308 additional probe sets discovered at the same FDR by the per gene approach and found many well-characterized genes that are directly or indirectly regulated by Nrl. For example, Pde6b and Pde6h are directly regulated by Nrl [32]. Rp1h and Arr3 are also responsive to Nrl knockout [30] [33]. Analysis under the more stringent assumption of equal variance and using the other multi-test correction methods such as Bonferroni, raw p-values cut-off and FDR under general dependency (Benjamini-Yeuketieli [29]) follow the same trend (Table 1).

Example2: Cancer Gene Marker Identification Using Paired Samples

The second Affymetrix data set we used was generated by Reyes et al [26] using Affymetrix $hgu133plus2$ chip. The data was downloaded from the GEO database using accession number GSE3678. The experiment profiles gene expression in seven paired PTC patient samples and normal samples. We compare the two approaches by inspecting differentially expressed gene lists between normal and PTC tissues.

Table 2. Performance comparisons in terms of numbers of differentially expressed genes. Comparison of the per-gene approach and per-probeset approach in terms of screening differentially expressed genes between seven paired PTC patient samples and normal samples.

	Per-Gene	Per-PS
Bonferroni (.01)	68	1
FDR-BH (.01)	402	32
FDR-BY (.01)	141	0
RawP cut-off (4.83e-06)	148	32

GCRMA [8] was used to normalize and summarize expression scores for each probe set in each tissue sample. Since this data set is different from the mouse chip data analysis because of paired data, we use the extended two-way and one-way ANOVA model with the patient as a block effect (model(6) and (8)). Controlling FDR at the level of 0.01 using the "BH" procedure, the per-gene approach and per-probeset approaches call 402 and 32 differentially expressed genes between normal and PTC samples. It consistently shows that the per-gene approach dominates the per-probeset approach in that 31 out of 32 probe sets (Fig.5b) called by the per-probeset approach were also called by the per-gene approach. Using other multiple test correction procedures follow the same trend (Table 2).

We compare our approach with the per-probeset using two strata of biological knowledge: cancer related gene ontology terms and true positive genes that are individually validated using traditional biochemical and genetic approaches. According to five cancer related GO terms, the per-gene approach outperforms the per-probeset approach in light of enriched p-values (Table 3). Huang et al [34] collected seven well-studied genes in their publication as true positive PTC marker. DPP6, DPP4 (liver dipeptidyl peptidase IV), SFTPB, CHI3L1, and MUC1 are known over-expressed genes in PTC samples. TPO and DIO2 are genes involved in thyroid metabolism. TPO plays central role in thyroid gland function, and DIO2 activates thyroid hormone by converting the prohormone thyroxine (T4) by outer ring deiodination (ORD) to bioactive 3,3',5-triiodothyronine (T3). At FDR cutoff 1%, our approach is able to pick out all seven genes while none was picked out by the per-probeset approach. It provides compelling evidence that our approach dominates over the competitor and is capable of identifying more biologically relevant genes.

Example3: Spermatogonial Stem Cell Self-Renewal Gene Marker Identification

In order to determine whether the per-gene approach consistently outperforms per-probeset under varied experiment conditions such as multiple treatment, normalization and summarization methods, we further compare two approaches on a third data set. The third data set (GES4799) we used was generated by Oatley [27] who studied GDNF-regulated gene expression over five time points before and after GDNF/GFRα1 replacement with a total of 15 samples.

Table 3. Comparison in terms of cancer functional categories. Comparison of per-gene approach, per-probeset approach in terms of enrichment of informative GO terms.

GO term	PerGene	PerPS
Apoptosis	0.0757	0.6480
Cell Growth	0.0912	0.2194
Cell Differentiation	0.0007	0.9184
Cell Adhesion	0.0006	0.0787
Blood Vessel Development	0.0041	NA

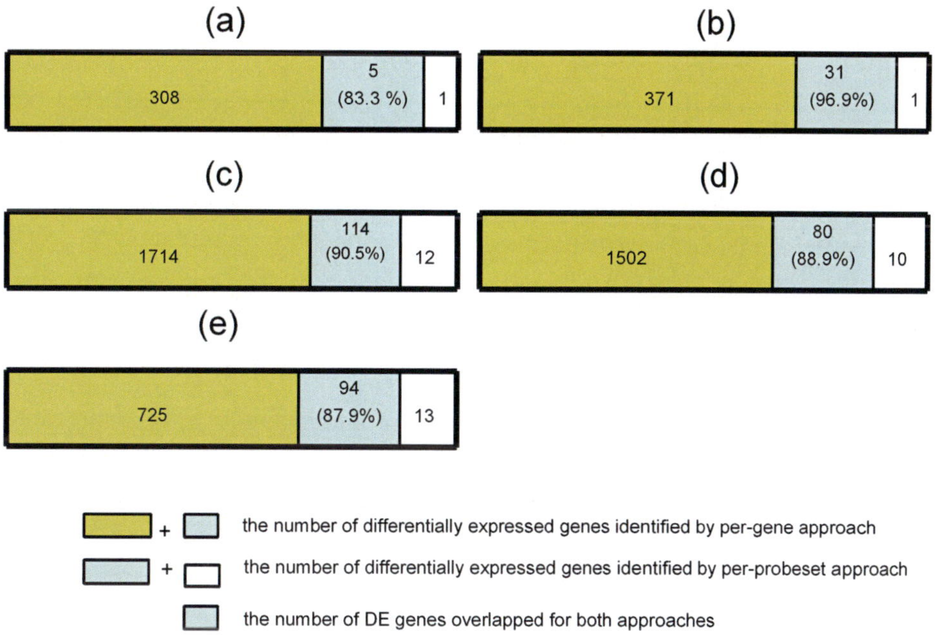

Figure 5. Per-gene approach dominates per-probeset approach. The percentages of overlapped genes detected by both approaches are shown in the overlapped areas. The number of differentially expressed genes identified by the per-gene approach and the per-probeset approach are: a. Data Set1, 313 vs. 6, overlap: 83.3% b. Data Set2, 402 vs. 32, overlap: 96.9% c. Data Set3, MBEI pre-processing, 1828 vs. 126, overlap: 90.5% d. Data Set3, RMA pre-processing, 1592 vs. 90, overlap: 88.9% e. Data Set3, GCRMA pre-processing, 832 vs. 107, overlap: 87.9%

For this data set, we use GCRMA [8], RMA [7] and MBEI [6] pre-processing methods to see if the normalization methods have an effect on the per-gene approach. Similar to our previous analysis, we reported p-values from (model(7) or (5)) depending on whether the interaction effect is significant. Compared to the per probeset approach, we once again found the per-gene approach dominates the per-probeset approach using the FDR cutoff of 0.01 (BH procedure) for all three pre-processing methods (Fig.5c-e).

Table 4. Comparison of the per-gene approach and the per-probeset approach in terms of enrichment of informative GO terms. When the exact GO term was not found P-value of a similar GO term was returned.

		Chromatin Remodeling	Cell Differentiation	Regulation of Cell Growth
MBEI	Per-gene	0.0575	0.0013	0.0512
	Per-ps	0.0651	0.0300	0.0345
RMA	Per-gene	NA	0.0021	0.0013
	Per-ps	NA	0.0161	NA
GCRMA	Per-gene	0.0297	0.0060	2.52E-05
	Per-ps	NA	0.1064	0.01738

We compare the two approaches in terms of associated important GO terms such as "Chromatin Remodeling", "Cell Differentiation" and "Regulation of Cell Growth" (Table 4). Although the results using different normalization methods are slightly different, the per-gene approach consistently shows significant enrichment for all three GO terms (p-value < 0.05), suggesting it is the best approach to identify genes that are associated with the stem cell self-renewal process.

Linear Mixed Model Including Probe and Probe Set Level Expression

We have reviewed linear models (2) and (5) that could account for the variation of probe intensities, probe set intensities and their interaction with treatment effects, respectively. In both cases, the statistical power for detecting differentially expressed genes is substantially improved because of incorporation of replicated data at different levels that aims to improve the specificity of the gene expression measure. It is possible to include both probe and probe set effects into the same model to account for variations at both the probe level and the probe set level. For a simple experiment that only have one experimental factor, the model could be written in the following form:

$$Y_{ijlk} = \mu + T_i + Ps_j + TPs_{ij} + P(Ps)_{kj} + PT(Ps)_{ik(j)} + A_l + \epsilon_{ijkl} \quad (9)$$

Where T, Ps and TPs are effects for probe set, treatment and their interaction. $P(Ps)$ is the probe effect for different probes. Because probes for every sibling probe set are surely different, the probe effect is treated as a nested factor within the probe set. Probe by treatment interaction is still included in the model, but nested within the probe set. In theory, this model combines models (2) and (5), and is expected to result in high statistical power for detecting differentially expressed genes.

Conclusion

In this chapter, we review basics of the generalized ANOVA model and it's application to oligobucleotide based microarray data analysis for differentially expressed gene identification. Using Affymetric genechip arrays as representative array, we overview the designed many-to-one feature for gene expression measurement and discuss how to effectively use the abundant data in analysis to improve the statistical power.

Using probe level intensities directly in the ANOVA model is similar to treating the probes as replicated measures for the same probe set, but with large probe specific effects. It includes probes and probe-by-treatment effects to model the large but unknown variation among probes and among probe-by-treatment, providing an ability to fully utilize all related information in the statistical analysis. Recently, Zhu [35] proposed to treat the multiple probes within a probe set as replicates when estimating the correlation coefficient between two probe sets. Rather than calculating Pearson correlation coefficient based on the summarized probe sets scores, the correlation estimator is estimated through the EM algorithm assuming multivariate normal distribution of probes of two probe sets. Authors are able to show that the estimated correlation matrix is closer to the nominal correlation matrix using two Affymetrix Spike-in data sets. It indicates that probe level data analysis could provide more accurate inferences if variations among probes have been properly modeled.

We present an ANOVA-based approach to consolidate sibling probe sets whenever possible in the context of detecting differential expression using popular Affymetrix mouse moe4302 and human hgu133plus2 platforms. Consolidating sibling probe sets into a gene-specific probe set is determined automatically through a statistical test of probe set by treatment interaction effect in the two-way ANOVA model. It improves the analysis in two ways. First, pooling data from sibling probe sets improves the estimation of mean and variance of the observed gene expression level so that the significance of differential expression (p-value) is more accurately estimated. Second, pooling enhances the power of statistical tests, because it consolidates the redundant sibling probe sets into one probe set and reduces the number of hypothesis tests. The ANOVA model allows us to separate the sibling probe sets into two types: those that behave similarly across treatments and those that behave differently across treatments. Consolidation of sibling probe sets of the former type results in a large increase in the number of differentially expressed genes under various statistical criteria.

References

[1] Tusher, V. G., Tibshirani, R., & Chu, G. (2001). Significance analysis of microarrays applied to the ionizing radiation response. *Proc. Natl Acad. Sci. USA*, **98**, 5116-5121.

[2] Kerr, M. K., Martin, M., & Churchill, G. A. (2000). Analysis of variance for gene expression microarray data. *J. Comput. Biol.*, **7**, 819837.

[3] Wolfinger, R. D., Gibson, G., Wolfinger, E. D., Bennett, L., Hamadeh, H., Bushel, P., Afshari, C., & Paules, R. S. (2001). Assessing gene significance from cDNA microarray expression data via mixed models. *J. Comp. Biol.*, **8**, 625637.

[4] Chu, T., M., Weir, B., & Wolfinger, R. (2002). A systematic statistical linear modeling approach to oligonucleotide array experiments. *Math. Biosci.*, **176**, 35 - 51

[5] *Affymetrix Inc. Statistical Algorithms Dscription Document*. (2002).

[6] Li,C and Wong,W. (2001) Model-based analysis of oligonucleotide arrays: expression index computation and outlier detection. *Proc Natl Acad Sci USA*, *98*, 31-36.

[7] Irizarry, R.A., Hobbs, B., Collin, F., Beazer-Barclay, Y.D., Antonellis, K.J., Scherf, U., & Speed, T.P. (2003). Exploration, Normalization, and Summaries of High Density Oligonucleotide Array Probe Level Data. *Biostatistics*, **4(2)**, 249-264.

[8] Wu, Z., Irizarry, R.A., Gentleman, R., Martinez-Murillo, F., & Spencer, F. (2004). A model-based background adjustment for oligonucleotide expression arrays. *J. Am. Stat. Assoc.*, **99**, 909-917.

[9] Elbez Y., Farkash-Amar, S., & Simon, I. (2006). An analysis of intra array repeat: the good, the bad and the noninformative. *BMC Genomics*, **7**, 136.

[10] Stalteri,M.A., & Harrison, A.P. (2007). Interpretation of multiple probe sets mapping to the same gene in Affymetrix GeneChips, *BMC Bioinformatics*, **8**, 48-55.

[11] Cui, XQ., & Loraine, A. E. (2006). Global Correlation Analysis Between Redundant Probe Sets Using A Large Collection Of Arabidopsis Ath1 Expression Profiling Data. *CSB2006 Proceedings of the LSS Computational Systems Bioinformatics Conference*. Stanford, CA, August, p 223-226.

[12] Bolstad, BM. Irizarry, RA, Astrand, M., & Speed, TP. (2003) A comparison of normalization methods for high density oligonucleotide array data based on variance and bias. *Bioinformatics*, **19**, 185-193.

[13] Zhang, L. Miles, M.F., & Aldape, K.D. (2003). A model of molecular interactions on short oligonucleotide microarrays. *Nature Biotechnology*, **21(7)**, 818-821.

[14] Hsieh, W.P., Chu, T. M., Wolfinger, R. D., & Gibson, G. (2003). Mixed model reanalysis of primate data suggests tissue and species biases in oligonucleotide based gene expression profiles. *Genetics*, **165**, 747 - 757.

[15] Cambon, A.C., Khalyfa, A., Cooper, N.G.F., & Thompson, C.M. (2007). Analysis of probe level patterns in Affymetrix microarray data. *BMC Bioinformatics*, **8**, 146.

[16] Lockhart DJ, Dong H, Byrne MC, Follettie MT, Gallo MV, Chee MS, Mittmann M, Wang C, Kobayashi M, Horton H, & Brown EL. (1996) Expression monitoring by hybridization to high-density oligonucleotide arrays. *Nat Biotechnol*, **14(13)**, 1675-1680.

[17] Hsieh, WP, Chu, TM, & Wolfinger, R. (2007). Comparison of statistical performance of univariate and bivariate mixed models for Affymetrix(r) probe level data. *Journal of Statistical Computation and Simulation*, **77(3)**, 251 - 264

[18] Barrera,L., Benner, C., Tao, YC, Barrera, E., & Zhou, Y. (2007). Leveraging two-way probe-level block design for identifying differential gene expression with high-density oligonucleotide arrays. *BMC Bioinformatics*, **5**, 42.

[19] Efron, B., Tibshirani, R., Storey, J.D., & Tusher, V. 2001. Empirical Bayes analysis of a microarray experiment. *J. Am. Stat. Assoc.*, **96**, 1151-1160.

[20] Hoeschele, I, & Li, H. (2005). A note on joint versus gene-specific mixed model analysis of microarray gene expression data. *Biostatistics*, **6(2)**, 183-186.

[21] Kerr, M.K. (2003). Linear Models for Microarray Data Analysis: Hidden Similarities and Differences. *Journal of Computational Biology*, **10(6)**, 891-901.

[22] Yanai I., Graur D., & Ophir R. (2004). Incongruent expression profiles between human and mouse orthologous genes suggest widespread neutral evolution of transcription control. *OMICS*, **8**, 15-24.

[23] Jordan I, Marino-Ramirez L, & Koonin E. (2005). Evolutionary significance of gene expression divergence. *Gene*, **345**, 119-126.

[24] Liao,BY, & Zhang,J. (2006). Evolutionary conservation of expression profiles between human and mouse orthologous genes. *Mol Biol Evol*, **23**, 530-540.

[25] Akimoto M., Cheng H., Zhu D., Brzezinski J., Khanna R., Filippova E., Oh E, Jing Y.,.Linares J., Brooks M., Zareparsi S., Mears A., Hero A., Glaser T., & Swaroop A. (2006). Targeting of GFP to newborn rods by Nrl promoter and temporal expression profiling of flow-sorted photoreceptors. *Proc Natl Acad Sci USA*, **103**, 3890-3895.

[26] Reyes,I, & Geliebter,J. (2005). http://www.ncbi.nlm.nih.gov/geo.

[27] Oatley J., Avarbock M., Telaranta A., Fearon D., & rinster R. (2006). Identifying genes important for spermatogonial stem cell self-renewal and survival. *Proc Natl Acad Sci U S A*, **103**, 9524-9529.

[28] Benjamini,Y , & Hochberg,Y. (1995). Controlling the false discovery rate-A practical and powerful approach to multiple testing. *J Roy Stat Soc B Met*, **57**, 289-300.

[29] Benjamini,Y , & Yekutieli,D. (2001). The control of the false discovery rate under dependency. *Ann. Stat.*, **29**, 1165-1188.

[30] Mears A., Kondo M., Swain P., Takada Y., Bush R., Saunders T., Sieving P., & Swaroop A. (2001). Evolutionary conservation of expression profiles between human and mouse orthologous genes. *Mol Biol Evol*, **29**, 447-452.

[31] Lee R., Ting T., Lieberman B., Tobias D., & Ho Y. (1992). Regulation of retinal cGMP cascade by phosducin in bovine rod photoreceptor cells. Interaction of phosducin and transducin. *J Biol Chem*, **267**, 25104-25112.

[32] Pittler S., Zhang Y., Chen S., Mears A., Zack D., Ren Z., Swain P., Yao S., Swaroop A., & White J. (2004). Functional analysis of the rod photoreceptor cGMP phosphodiesterase alpha-subunit gene promoter: Nrl and Crx are required for full transcriptional activity. *J Biol Chem*, **179**, 19800-19807.

[33] Cheng H., Aleman T., Cideciyan A., Khanna R., Jacobson S., & waroop A. (2006). In vivo function of the orphan nuclear receptor NR2E3 in establishing photoreceptor identity during mammalian retinal development. *Human Molecular Genetics*, **15**, 2588-2602.

[34] Huang Y., Prasad M., Lemon W., Hampel H., Wright F., Kornacker K., LiVolsi K., Frankel W., Kloos R., Eng C., Pellegata N., & Chapelle A. (2001). Gene expression in papillary thyroid carcinoma reveals highly consistent profiles. *Proc Natl Acad Sci U S A*, **98**, 15044-15049.

[35] Zhu, D., Li, Y. , &Li, H. 2007. Multivariate correlation estimator for inferring functional relationships from replicated genome-wide data. *Bioinformatics*, **23(17)**, 2298-2305.

[36] Jiang,Z and Gentleman,R. (2007). Extensions to gene set enrichment, *Bioinformatics*, **23**, 306-313.

In: Oligonucleotide Array Sequence Analysis
Editors: M.K. Moretti and L.J. Rizzo, pp. 265-295

ISBN: 978-1-60456-542-3

Chapter 9

ONLINE TOOLS FOR PRESENTATION AND ANALYSIS OF PLANT MICROARRAY DATA

David Honys[1,2,*], ***Nikoleta Dupľáková***[1,2] ***and David Reňák***[1,2,3]

[1]Laboratory of Pollen Biology, Institute of Experimental Botany ASCR, v.v.i., Rozvojová 263, 165 02 Prague 6, Czech Republic

[2]Department of Plant Physiology, Faculty of Science, Charles University, Viničná 5, 128 44, Prague 2, Czech Republic

[3]Department of Plant Physiology and Anatomy, Faculty of Science, University of South Bohemia, Branišovská 31, 370 05 České Budějovice, Czech Republic

ABSTRACT

The complete sequencing and annotation of the *Arabidopsis thaliana* genome represented a major step in biological research. This knowledge enabled gene prediction, assignment of functional categories and gave an opportunity to characterize global gene expression patterns at the transcriptome level at different developmental stages and under various physiological and stress conditions. For the discovery of partial or whole regulatory or functional networks, the development of high-throughput technologies was inevitable with genome-wide transcriptomic studies providing an essential input. DNA microarray technologies are among the most frequently used methods for parallel global analysis of gene expression. Microarray technologies now belong to the standard functional genomics toolbox and have undergone massive development leading to increased genome coverage, accuracy and reliability. Whole genome microarrays developed by Affymetrix in collaboration with Syngenta represented the first standard in genome-wide transcriptomic studies in plants. Whole genome Affymetrix ATH1 GeneChips covers about 76% of the *Arabidopsis thaliana* genome. Moreover, the introduction of the Minimum Information About Microarray experiments (MIAME) has increased the value and reproducibility of microarray experiments and has become a standard in documentation of array experiments and in the creation of databases of comparable transcriptomic experiments. The number of experiments exploiting microarray technology has markedly increased in recent years. Not surprisingly, there are potential difficulties in the orientation in available data sets. Microarray expression data are deposited on servers, many of which are publicly accessible. Currently, these databases store several thousands of individual datasets and some of these offer online tools for data normalization, filtering,

* Author for correspondence

statistical testing and pattern discovery. In parallel with the rapid accumulation of transcriptomic data, on-line analysis tools are being introduced to simplify their use. Global statistical data analysis methods contribute to the development of overall concepts about gene expression patterns and to query and compose working hypotheses. More recently, these applications are being supplemented with more specialized products offering visualization and specific data mining tools. In this chapter, an overview of available on-line databases and web-based applications using microarray data is presented together with the information about their structure and elementary principles.

INTRODUCTION

Microarray technologies now belong to the standard functional genomics toolbox and have undergone massive development leading to increased genome coverage, accuracy and reliability. The number of experiments exploiting microarray technology has markedly increased in recent years. In parallel with the rapid accumulation of transcriptomic data, on-line analysis tools are being introduced to simplify their use. Global statistical data analysis methods contribute to the development of overall concepts about gene expression patterns and to query and compose working hypotheses. More recently, these applications are being supplemented with more specialized products offering visualization and specific data mining tools.

The knowledge of whole sequenced and annotated plant genomes that started with Arabidopsis thaliana [AGI 2000] enabled gene prediction, assignment of functional categories and gave an opportunity to study gene and chromosome organization including the distribution of transposable elements. Finally it has enabled the characterization of global gene expression patterns at the transcriptome level at different developmental stages and under various physiological and stress conditions. Efforts to reveal the biological functions of thousand of genes and their integration into proteome, metabolome and interactome networks has become the principal focus of many studies and represents a key objective of the 2010 Project (http://www.nsf.gov/pubs/2006/nsf06612/nsf06612.htm).

A number of efficient and accurate gene expression analysis technologies to determine the expression levels of individual genes have been widely exploited in recent decades (Northern blot analysis, quantitative reverse transcription-PCR, cDNA library screening). Most of these methods enable analysis of the expression of single or relatively few selected genes. For the discovery of partial or whole gene functional or regulatory networks, the development of high-throughput technologies is essential with genome-wide transcriptomic studies providing a major input [Donson et al. 2002]. Several such methods have been developed including, cDNA fingerprinting [Money et al. 1996], serial analysis of gene expression - SAGE [Velculescu et al. 1995], massively parallel signature sequencing - MPSS [Brenner et al. 2000], high-density DNA oligonucleotide probe microarrays [Lockhart et al. 1996, Lipshutz et al. 1999] or cDNA arrays [Schena et al. 1995]. DNA microarray technologies are among the most frequently used methods for parallel global analysis of gene expression. These methods are based on the principle of selective and differential hybridization between sample target molecules and immobilized DNA probes. Hybridisation to probes arrayed on a solid surface report the relative abundance of DNA or RNA target molecules by fluorescent signal detection [van Hall et al. 2000, Clarke and Zhu 2006]. Microarray technologies now belong to the standard functional genomics toolbox [Zhu 2003,

Aharoni and Vorst 2002] and have undergone massive development leading to increased genome coverage, accuracy and reliability. Whole Genome microarrays developed by Affymetrix (Santa Clara, CA, USA) in collaboration with Syngenta represented the first standard in genome wide transcriptomic studies in plants. Whole genome Affymetrix ATH1 GeneChips cover about 76% of the *Arabidopsis thaliana* genes [Hennig et al. 2003]. Moreover, the introduction of the Minimum Information About Microarray experiments (MIAME) as standard documentation for array experiments and in transcriptomic databases, increasing the value and comparability of microarray data [Brazma et al. 2001].

As microarray technology has become standard technique in large scale gene expression studies, the number of transcriptomic experiments has markedly increased in recent years. Not surprisingly, there are potential difficulties in navigating between different available data sets. Microarray expression data are deposited on servers, many of which are publicly accessible. Public plant microarray data are deposited in several databases including ArrayExpress, GEO, NASCArrays and the Stanford Microarray Database. Currently these databases store several thousands of individual datasets and some of these offer on-line tools for data normalization, filtering, statistical testing and pattern discovery [Parkinson et al. 2005, Barrett et al. 2005, Edgar et al. 2002, Gollub et al. 2003, Craigon et al. 2004]. In parallel with the rapid accumulation of transcriptomic data, on-line analysis tools are being introduced to simplify their use. Global statistical data analysis methods contribute to the identification of overall gene expression patterns and to query and compose new working hypotheses based on these findings [Clarke and Zhu 2006, Zhu 2003, Hughes et al. 2000, Mandaokar et al. 2006]. More recently, these applications are being supplemented with more specialized products offering visualization and specific data mining tools. Genevestigator, Botany Array Resource, Arabidopsis co-expression tool, and ArrayExpress Expression Profiler offer Web-based tools to analyse large microarray datasets.

In this chapter we describe and collate several available on-line portals devoted to storage and analyses of plant microarray data. Summary of individual portals is provided in Table 1. For each database, basic information is available like URL, annotation of Arabidopsis genome, microarray data resource, used platforms, data normalisation, number of microarray datasets covered and access options. It is also distinguished between data repositories and/or data analysis tools. Moreover, gene- or genome-centric database concepts are compared. Gene-centric approach means that the particular gene is in the centre of interest and relevant data can be acquired from individual experiments. On the contrary, genome-centric approach enables the identification of genes fulfilling given criteria. Most portals enable direct data download. However, they offer different formats that are listed in Table 1. In the following section we provide more detailed description of all portals. We always shortly introduce the overall database concept and general features with special attention paid to specific tools for data analysis.

Table 1. Overview of plant microarray databases

Portal	ACT	aGFP	ArrayExpress	BAR	GEO	Genevestigator V3	NASCArrays	SMD
Database type	Data repository and simple toolbox	Data visualisation toolbox	Data repository and toolbox	Data analysis and visualisation toolbox	Data repository	Data repository and toolbox	Data repository and simple toolbox	Data repository and toolbox
Concept	Genome-centric	Gene-centric	Gene-centric and genome-centric	Gene-centric and genome-centric	Gene-centric and genome-centric	Gene-centric and genome-centric	Gene-centric	Gene-centric and genome-centric
Arabidopsis genome annotation	TAIR7	TAIR7	Various	TAIR7	n/a	TAIR7	TIGR5	n/a
Microarray data resource	NASCArrays/GARNet	AtGenExpress, NASCArrays	EBI Array express or user-uploaded data	Bio-Array Resource Database, AtGenExpress, NASCArrays	User-uploaded data compatible with GEO structure	NASCArrays, FGCZ, GEO, ArrayExpress, AtGenExpress, TAIR	NASCArrays	Various including Affymetrix, Agilent, and spotted arrays
Platform(s)	Affymetrix ATH1 Whole Genome Array	Affymetrix ATH1 Whole Genome Array	Various	Affymetrix ATH1 Whole Genome Array	Various high-throughput gene expression data	Affymetrix AG and ATH1 Arrays	Affymetrix AG and ATH1 Arrays	Various
Data normalisation	MAS5.0	MAS4.0 and MAS5.0	Various	MAS5.0	Various	MAS5.0	MAS5.0	Various
No. of experiments	73 experiments, 544 arrays (Arabidopsis)	120 experiments and 350 arrays	3,051 experiments and over 50,000 arrays from over 200 organisms	BAR 150 arrays, NASCArrays incl. AtGenExpress 330/2954 (Arabidopsis)	4,134 platforms, 181,922 samples and 7,202 series from over 100 organisms	201 experiments and 3132 arrays (Arabidopsis)	330 experiments and 2954 arrays	over 70,000 arrays and 15,451 experiments from 53 organisms

Table 1. Continued

Portal	ACT	aGFP	ArrayExpress	BAR	GEO	Genevestigator V3	NASCArrays	SMD
Access	Free	Free	Free with optional password-protected account	Free	Free	Open (free), Classic (free for academic users) and Advanced (paid)	Free	Free with restricted access to non-public data
Data download	Yes, via NASCArrays	Yes	Yes	Yes	Yes	Yes	Yes	Yes
Downloaded data format	Tab delimited, XLS files, Gnumeric, CSV files	TAB-delimited text	TXT or XLS format, MEGA-ML format	Depends on selected tool	Web-based interactive forms, XLS files, MINiML, SOFT text files	Depends on selected tool	CSV or TAB-delimited files	XLS and TXT files
URL	http://www.arabidopsis.leeds.ac.uk/act/index.php	http://aGFP.ueb.cas.cz	http://www.ebi.ac.uk/arrayexpress	http://bbc.botany.utoronto.ca	http://www.ncbi.nlm.nih.gov/geo/	https://www.genevestigator.ethz.ch/	http://affy.arabidopsis.info/narrays/experimentbrowse.pl	http://genome-www5.stanford.edu/
Key references	*Jen et al. 2006, Manfield et al. 2006*	*Duplakova et al. 2007*	*Torrente et al. 2005, Parkinson et al. 2006*	*Toufighi et al. 2005, Winter et al. 2007*	*Edgar et sl. 2002, Barrett et al. 2007*	*Zimmermann et al. 2004, Laule et al. 2006,*	*Craigon et al. 2004*	*Ball et al. 2005, Demeter et al. 2007*

Portals

Arabidopsis Co-expression Tool (ACT)

ACT is a web-based tool for gene expression analysis using large microarray data set from the Nottingham Arabidopsis Stock Centre. The database stores pre-calculated co-expression results for cca 21 800 genes based on data from over 400 arrays (Figure 1). ACT enables identification of gene co-expression patterns across user-selected single or multiple arrays. An additional Clique Finder tool provides quantitative method for the determination of correlation cut-offs leading to the generation of groups of genes that may share a common purpose [Manfield et al. 2006]. For further analysis of co-regulated genes, promoter element detection software is included for the identification of potential DNA upstream elements [Jen et al. 2006]. The bioinformatic analyses consist of three steps: (i) generation of a list of

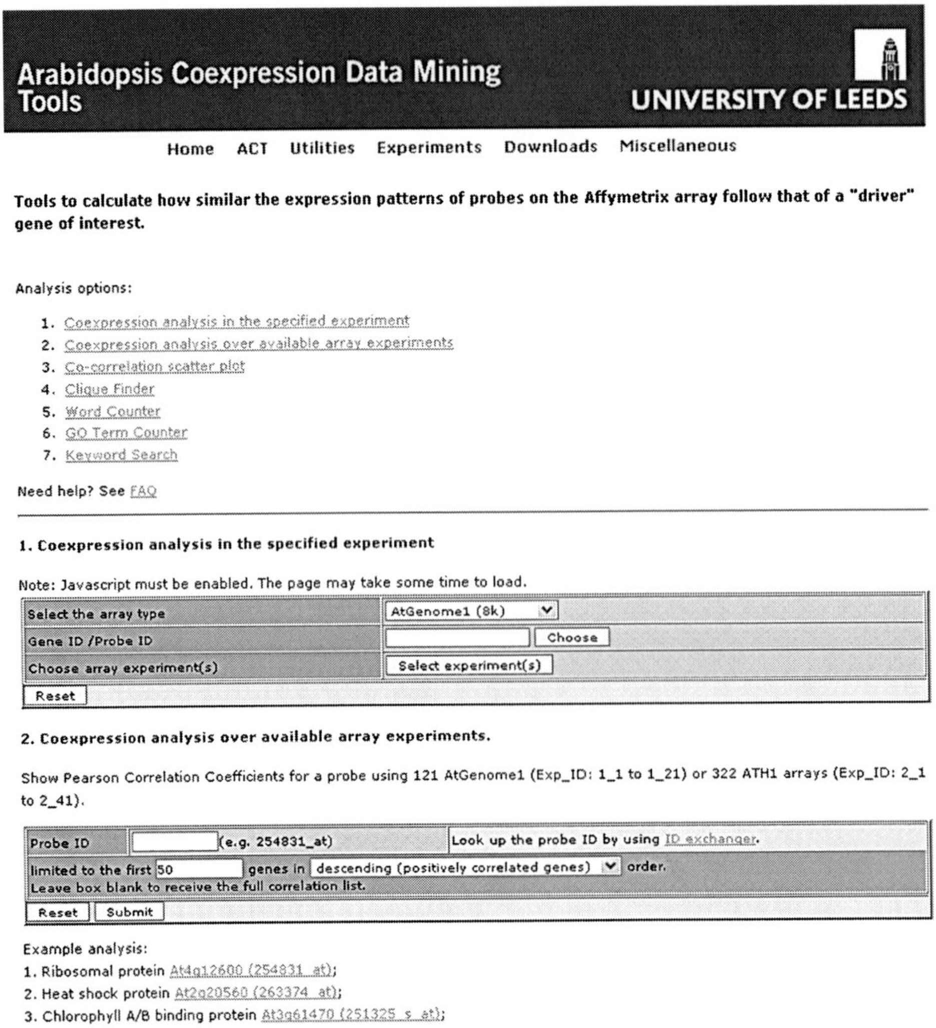

Figure 1. Homepage of the Arabidopsis Co-expression Tool available at http://www.arabidopsis.leeds.ac.uk/act/index.php.

negatively/positively co-expressed genes for a given gene of interest (driver); (ii) calculation of a co-correlation scatter plot of two selected genes; and (iii) identification of groups of genes that share statistical significant co-expression patterns. Co-correlation scatter plot analyses showing positively and negatively correlated genes may identify similar transcription regulatory mechanisms and can help to reveal the related biological functions of unannotated genes. Moreover, groups of genes with significant co-expression patterns often share the same biological theme and their members may overlap with other groups. It may then suggest their involvement in different biological processes. ACT provides a genome-centric approach allowing generation of a list of genes co-expressed within the driver gene throughout all experiments available. As an output, ACT offers a graphical view of gene expression patterns, list view of selected genes, co-correlation scatter plots; clusters of closely-associated probes and annotations of the most correlated genes to a given probe. Data can be downloaded as simple Tab-delimited TXT files, CSV, XLS files or in a Gnumeric format.

Specific Tools

Expression Pattern Displayer allows an exploration of gene expression patterns in a specified experiment or over all arrays available in the database.

Coexpression analysis was designed for the generation of lists of genes co-expressed with the probe (driver gene) according to their r-, p- and E-values.

Co-correlation scatter plot results in 2-D visualization of co-correlated data within two probes.

Clique Finder identifies genesets that are consistently co-expressed with each other throughout all microarray data.

Word count highlights annotations of the most correlated genes to a probe.

GO Terms offers Gene Ontology terms of the most correlated probes.

Sequence extractor extracts the specific region of cis-upstream or downstream sequences of single or multiple genes.

cis-Element Analyser analyses of cis-elements in the given sequence while **cis-Element_Locator** can be used for the localization of user-defined DNA motifs in promoters of single or multiple genes.

ID/Function Linker converts multiple Affymetrix probe IDs and AGI codes.

Distinguishable Features

Even though ACT uses the simplest measuring algorithm using the Pearson correlation coefficient, it is suggested to be an effective tool for the calculation of statistical significance [Manfield et al. 2006]. Researchers looking for positively and negatively correlated genes may (not) consider similar transcription regulatory mechanisms involved in related biological processes when revealing possible functions of unannotated genes. Moreover, the result of overlapping genes in each group may lead to a conclusion of being involved in different biological processes. In addition, the 2-D co-correlation plots can help in identifying sets of genes acting in particular metabolic, signaling or developmental processes. Finally, ACT tools for promoter analysis can be used to identify motifs which are often seen to underpin the observed expression correlation pattern. Other tools for identifying over-represented (but not necessarily characterized) sequence motifs such as the Inclusive Motif Sampler are also useful for analyzing promoters of correlated genes.

Arabidopsis Gene Family Profiler (aGFP)

Currently, on-line analysis tools have been developed to simplify particular analyses of transcriptomic data from wide-ranging microarray analysis. Database Arabidopsis Gene Family Profiler (aGFP; http://aGFP.ueb.cas.cz) [Dupľáková et al. 2007] is a web-based tool offering visualization of microarray expression data of single genes, pre-defined gene families or even custom gene sets (Figure 2). aGFP gives users the possibility to analyze current *Arabidopsis* developmental transcriptomic data starting with simple global queries that can be expanded and further refined to visualize comparative and highly selective gene expression profiles. One of new features of aGFP is the possibility to display gene expression of pre-defined gene families and to easily adjust their composition by addition or reduction their members.

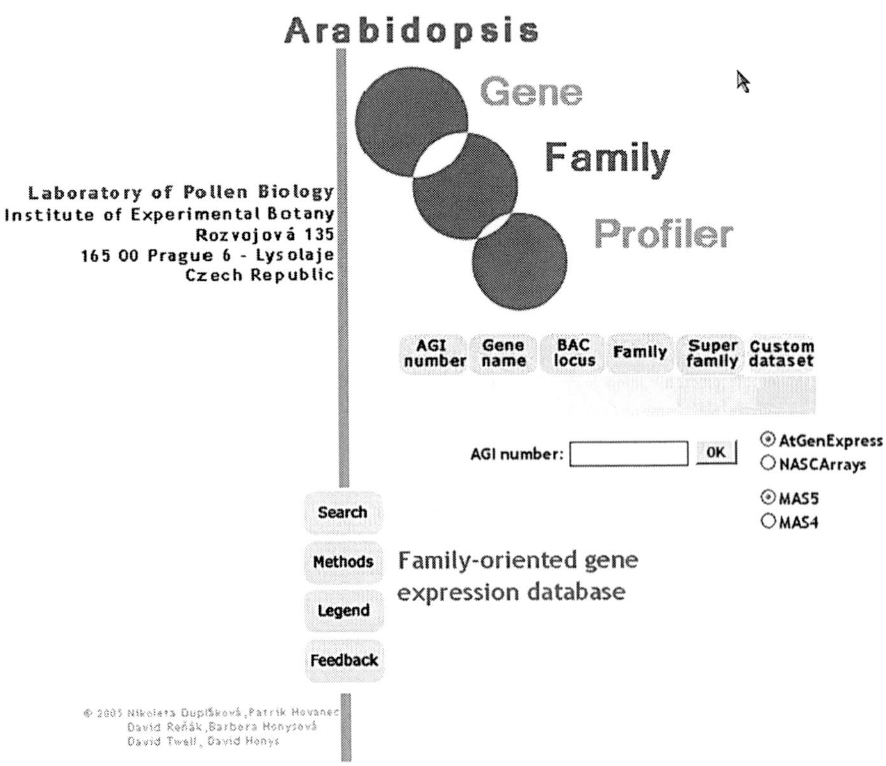

Figure 2. Homepage of the ArabidopsisGFP database available at http://aGFP.ueb.cas.cz.

aGFP database is based on a "gene-centric approach". Authors adopted the general concept "from simple to complex". In the first approximation, an arithmetical mean expression signal from multiple experiments is displayed. In subsequent steps the user can choose to display expression data for individual plant organs or tissues at particular growth stages. This is accompanied by the option of progressive replacement of arithmetical means by individual expression values. So users have the option to choose different levels of visualization to suit their needs.

aGFP offers several data search options together with two other input parameters; the gene detection algorithm (MAS4.0 or MAS5.0) and the source of expression data (AtGenExpress or NASCArrays). At any stage of the query the user has the possibility to directly switch options between these pairs of parameters.

aGFP uses two large freely available transcriptomic databases for various tissues at different developmental stages of wild type plants *Arabidopsis thaliana* grown at physiological conditions. The first subset contains data obtained within a scope of the AtGenExpress project [Schmid et al. 2005], the second comprises all other datasets deposited at NASC and was labeled NASCArrays [Craigon et al. 2004]. Data in each subset are presented using several different graphical displays and, the user has an option to instantly switch between subsets in each environment. Only experiments using Affymetrix ATH1 whole genome arrays with at least two biological replicates were included.

The complete datasets used in the aGFP database are described in the Legend available from the homepage and it is possible to trace the origin of all datasets. Gene expression data can be downloaded for individual and selected gene sets as a TAB-delimited TXT files. This enables the direct import of downloaded data into spreadsheet editors such as Excel and database software like Microsoft Access and FileMaker.

Moreover, aGFP allows a direct comparison of the influence of the detection algorithm or data resource on expression profiles. The database presents data normalised using two different algorithms, empirical Affymetrix MAS 4.0 and statistical MAS 5.0 with the possibility to switch between them at any point.

Genes in arabidopsisGFP are organized into two hierarchical levels consisting of gene families and superfamilies. All data were assembled from various relevant resources, the majority from TAIR – Arabidopsis Gene Family Information (http://www.arabidopsis.org/browse/genefamily/index.jsp) and AGRIS (http://arabidopsis.med.ohio-state.edu/) [Palaniswamy et al. 2006].

An interactive "virtual plant" represents the main tool for the visualization of the spatial and developmental gene expression profiles. Virtual plant uses a white-yellow-green scale to show relative expression signals of individual genes, or gene families throughout the *Arabidopsis* life cycle. In first step, the „virtual plant“ is displayed an mean expression signal from multiple experiments. In subsequent steps, the user can choose to display expression data for individual plant organs or tissues at particular growth stages. This is accompanied by the option of progressive replacement of arithmetical means by individual expression values. Finally, the user can switch from "virtual plant" visualization to a simple bar chart (standard or log-scaled) or tabulated display. The user can also browse through individual experiments down to normalized or even raw data extracted from individual gene chips. Gene family expression data can also be visualized as a colorized spot chart (heat map) with colour scale identical to that of the virtual plant. The interactivity of modeling is based on the possibility of ad hoc addition and removal of genes to and from currently active set. Moreover, each spot contains mouse-over-activated information about expression signal value and experiment. Annotation of individual experiments meets MIAME standard [Brazma et al. 2001].

Distinguishable Features

A novel feature of aGFP is that it enables the evaluation of the impact of normalization procedures on microarray gene expression data as well the possibility of rapid definition of user-defined gene families or other groups. Simultaneously, aGFP serves as a facile and synoptic developmental reference guide for expression profiles of individual genes or gene families in wild-type *Arabidopsis thaliana* plants. arabidopsisGFP contains lists of pre-defined gene families and superfamilies enabling the rapid comparative visualization of expression profiles of their members.

ArrayExpress

ArrayExpress (Figure 3) was developed as a public repository for gene expression data at the centre of a wider microarray informatics system at European Bioinformatics Institute (EBI) [Brazma et al. 2003]. It comprises well-annotated raw and processed microarray data originated from several platforms supporting MIAME requirements [Parkinson et al. 2005]. Currently, the database stores data of 1,500,000 gene expresion profiles from more than 50 000 hybridizations representing more than 200 organisms [Parkinson et al. 2007]. Microarray data are either part of EBI ArrayExpress or they can be user-uploaded from various resources including Affymetrix, Agilent, Illumina, Nimblegen, etc. [Brazma et al. 2003]. The majority of experimental data relates to gene expresion profiling studies, the remainder are array-based chromatin immunoprecipitation or comparative genomics studies [Parkinson et al. 2005].

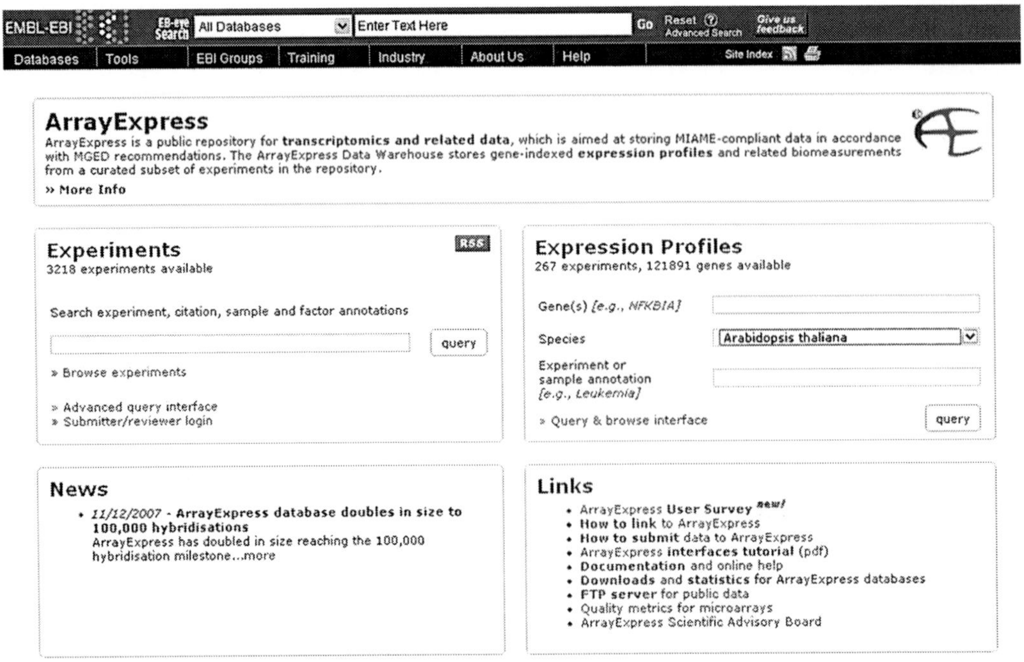

Figure 3. Homepage of the ArrayExpress database available at http://www.ebi.ac.uk/arrayexpress.

ArrayExpress was created to achieve three major goals, first, to serve as an archive for microarray data, second, to provide access to genes expression profiles, and third, to facilitate the sharing of microarry design and experimental protocols [Sarkans et al. 2005; Brazma et al. 2006].

ArrayExpress infrastructure consists of database itself and individual modules enabling user data submission, database query and data analyses [Brazma et al. 2003]. For data submission, three types of data are accepted, arrays, experiments and protocols. Altogether, they provide full information about the respective dataset but they can be uploaded separately under individual accession numbers. It simplifies the process of microarray data submission using standard platforms like Affymetrix. Data are submitted directly using three major submission routes: (i) online via the MIAMExpress data submittion tool [Brazma et al. 2003], (ii) via a spreadsheet suitable for all types and sizes based on technology type, species and experimental type, and (iii) via a MAGE-ML or MAGE-TAB (Microarray gene Expression Markup Language) [Spellman et al. 2002] pipelines set-up with external databases [Kapushensky et al. 2004]. As plant-based data represent about 25% of ArrayExpress datasets and vast majority of those based on Arabidopsis, specific open-source software application for the submission of Arabidopsis-based microarray data is available [Mukherjee et al. 2005]. MIAMExpress itself enables download of limited number of hybridisations with the requirement for filling out a web-form for each individual dataset. For this reason, BLoader software application was developed and implemented to generate the annotation information rapidly and submit the entire set in one batch [Schwager et al. 2005].

Since its introduction, ArrayExpress has undergone significant improvement as it has two major components. Original “ArrayExpress experiment repository” represents the main database comprising complete data while “ArrayExpress gene expression profile data warehouse” contains gene-indexed expression profiles from a curated subset of experiments from the repository [Brazma et al. 2006]. Both ArrayExpress and user-uploaded microarray data can be analysed and visualized exploiting integrated ExpressionProfiler toolbox that is described in detail below [Kapushesky et al. 2004].

Specific Tools

General tools:

Browse Experiments enables filters on species, array, experiment type, experimental factors, author, laboratory, publications, array design name and protocol type.

Browse Expression Profiles selects gene name or AGI, species, experiment or sample annotation.

Similarity Search is designed for arrangement of additional genes with similar expression profiles.

Expression Profiler — next generation:

Expression Profiler (EP) [Kapushesky et al. 2004] represents microarray data analysis and visualization toolbox implemented in the ArrayExpress database [Brazma et al. 2003]. EP was originally designed as a web-based platform for microarray data and other functional genomics-related data analysis [Kapushesky et al. 2004]. After the implementation of new architecture enabling the modularization of the original design and de-centralization of database development, the toolbox was renamed to Expression profiler: next generation

(EP:NG). Individual tools are described below. EP:NG can be accessed from ArrayExpress page or directly at http://www.ebi.ac.uk/expressionprofiler.

Individual EP:NG tools:

User-defined **Data Upload** of microarray data following specific data-organization and data description matrices including optional various metadata.

Data Selection provides brief basic statistical overview of microarray data and user-defined data selection and sub-selection.

Data Transformation represents optional procedures preceding or following data subselection. They include K-nearest neighbour imputation [Troyanskaya et al. 2001] to fill in missing values, LOWESS normalization [Yang et al. 2002], data conversion in two-channel experiments.

Similarity Search extracts and visualizes of a group of co-expressed genes in relation to given one(s).

Clustering is application of hierarchical and partitioning-based K-groups clustering algorithms [Torrente et al. 2005].

Clustering Comparison is a tool for the comparison and matching of two independent K-groups clustering results.

Signature Algorithm identifies co-expressed subset of genes in user-submitted geneset.

Ordination offers several tools for multivariate analysis methods; principal component analysis and correspondence analysis.

Between Group Analysis allows multiple discriminant approach used with expression data matrices.

Distinguishable Features

ArrayExpress database provides a free public repository of microarray gene expression data of various origins. In its modular architecture it links together gene expression data and number of data processing and analysis tools. Moreover, data submission in three separate steps resembling GEO enables future comparison and analyses of microarray data of different origin. Extra value was added by the subselection of specific gene expression data into growing curated ArrayExpress data warehouse. Data analysis and visualization tools are implemented in the integrated ExpressionProfiler package. This, together with unified data submission format, enables direct link to downloaded gene expression data, their efficient processing and analyses. Among most distinguishable features of EP:NG belong its gene clustering and statistical tools.

Bio-Array Resource (BAR)

The Bio-Array Resource (BAR) is an extensive microarray data repository accompanied by number of very helpful tools designed for data visualization and detailed analyses (Figure 4) [Toufighi et al. 2005]. BAR is a completely web-based database, exploiting MySQL with interfaces implemented in Perl and C. The site was originally named Botany Array Resource as it was designed for Arabidopsis-based data. Recently, a toolbox for the visualisation of gene expression in mouse tissues have been recently added [Winter et al.

2007] so further development and extension of the coverage is likely to be expected. Currently, BAR database comprises microarray data collected from University of Toronto microarray facility as well as data loaded from NASCArrays and AtGenExpress Consortium. The database has been designed as MIAME-compliant [Brazma et al. 2001] and all provided data and tools are publicly available through a web interface. BAR offers large number of various data analysis tools. They include not only gene expression visualization and analysis tools but also set of assorted on-line tools especially useful for genetic mapping and data formatting, visualization and clean up. An overview and brief characterization of individual tools is provided in the specific section.

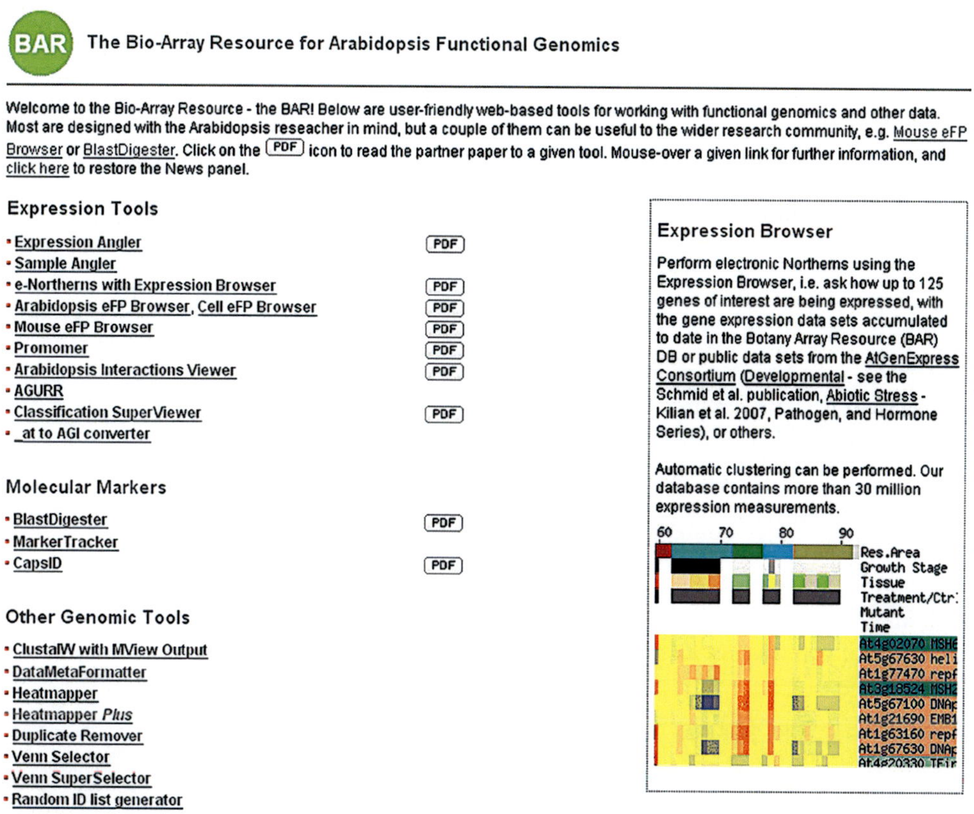

Figure 4. Homepage of the Bio-Array Resource database and toolbox available at http://bbc.botany.utoronto.ca.

Specific Tools

Gene expression tools:

Expression Angler identifies sets of co-regulated, anti-correlated, or condition/tissue-specific genes in various data sets.

Sample Angler identifies samples exhibiting similar expression profiles.

e-Northerns with Expression Browser is an electronic Northerns tool providing a graphical display of expression data of particular set of genes across all selected datasets.

Arabidopsis eFP Browser provides intuitive visualization of particular gene expression patterns.

Cell eFP Browser enables intuitive visualization of particular protein subcellular localization patterns based on documented and predicted subcellular localizations according to the SUBA database [Heazlewood et al. 2005, 2007].

Mouse eFP Browser offers intuitive visualization of particular gene expression patterns based on the Zhang et al. [2004] Functional Landscape of Mouse Gene Expression.

Promomer was designed for the identification of over-represented motifs in the promoter of a gene of interest, or in promoters of co-regulated genes.

Arabidopsis Interactions Viewer returns predicted or documented interaction partners with a gene of interest.

AGURR (Arabidopsis Genetic Uniqueness and Redundancy Revealer) is a program for the identification of samples in which one gene is uniquely expressed or in which several genes are potentially redundantly expressed.

Classification SuperViewer returns functional classification a of AGI IDs list based on the MIPS database.

at to AGI converter converts 22k Affymetrix GeneChip IDs into AGI IDs and vice versa.

Molecular Markers:

BlastDigester analyses a nucleotide BLAST output for restriction enzymes differentially cutting the aligned sequences [Ilic et al. 2005].

MarkerTracker is a repository for genetic markers.

CapsID serves as automated CAPS marker-based genotyping system [Taylor and Provart 2006].

Other general tools for basic genomic analyses:

ClustalW with MView Output is a web-based version of ClustalW.

DataMetaFormatter can be used for Arabidopsis gene expression data reformatting and adding pieces of meta-information (protein-protein interactions, functional classification).

Heatmapper and Heatmapper Plus applies a third dimension of information to a 2-D table via colour-coding.

Duplicate Remover removes duplicates from data lists.

Venn Selector shows identifiers in common and unique to two sets of sequences and fetching annotation.

Venn SuperSelector organizes multiple lists of genes or terms with associated values with an option of data export to Excel.

Random ID list generator is useful for the generation of *n* sets of *y* genes containing *z* number of randomly generated Arabidopsis AGI IDs.

Distinguishable Features

The BAR provides number of sophisticated and easy-to-use web-based tools for microarray data visualization and analyses. Moreover, these tools are accompanied by other "simple" but extremely helpful tools for data formatting, presentation and clean-up that certainly may be time-consuming. Microarray data are divided into several databases according to their origin and experiment type. In most tools, users can choose from Bio-Array resource database, several AtGenExpress Consortium databases, NASC Arrays and sometimes even their own in-house databases.

Gene Expression Omnibus (GEO)

The Gene Expression Omnibus (GEO) was originally designed as a public data repository and retrieval system for high throughput gene expression data with an ambition to complement number of existing in house gene expression databases and to act as the central data distribution hub (Figure 5) [Edgar et al. 2002]. It represents one of large number of computational resources for the analysis of data stored in GenBank and other biological data provided by NCBI and available through NCBI's website http://www.ncbi.nlm.nih.gov [Wheeler et al. 2005, 2007]. Currently, the database has a MIAME-compliant infrastructure and contains both raw and processed data derived from over 100 organisms submitted by over

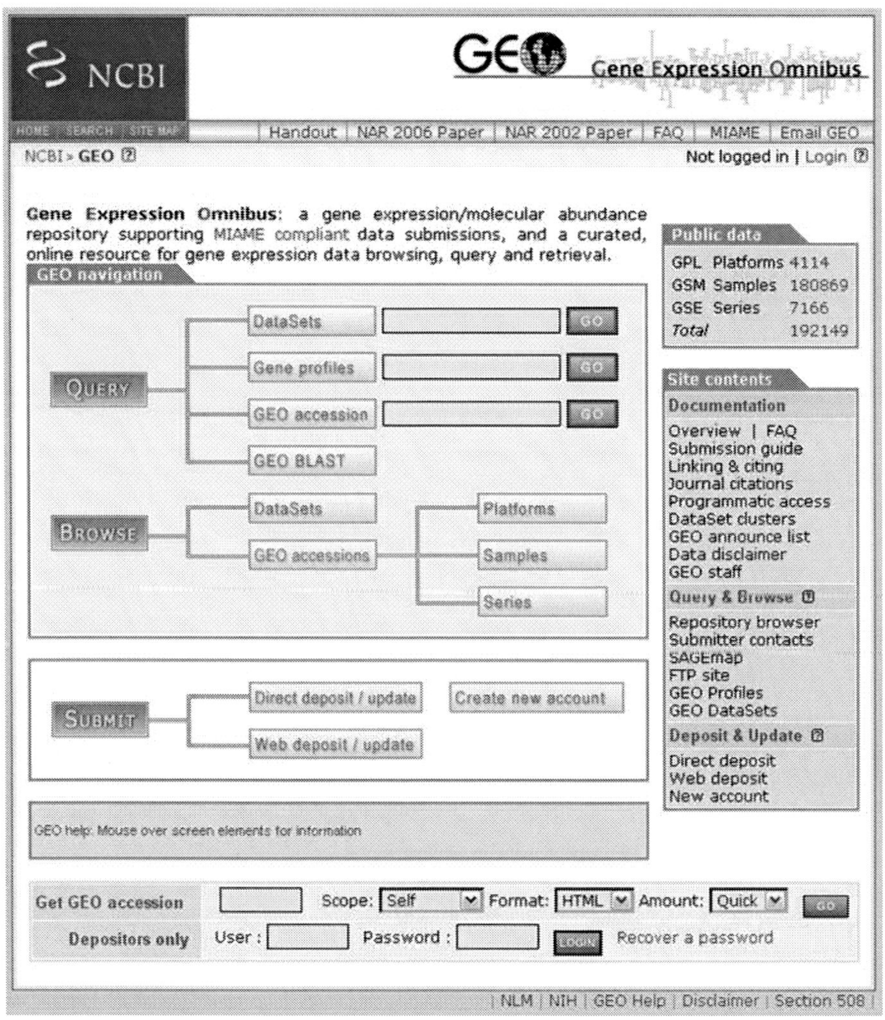

Figure 5. Homepage of the GEO database available at http://www.ncbi.nlm.nih.gov/geo/.

1,500 laboratories [Barrett and Edgar 2006a, Barrett et al. 2007]. GEO is not devoted to one technology but it contains microarray-based experiments measuring the abundance of mRNA, genomic DNA and protein molecules, as well as non-array-based technologies such as SAGE and mass spectrometry peptide profiling [Barrett et al. 2005].

The unifying feature of GEO database is that all data are segregated and stored as three principle components, platforms, samples and series. Platform provides a summary description of the array and data table by defining the array template. Sample describes the biological material, the experimental conditions under which the sample was handled and a respective data table containing hybridization signal levels for each feature on the corresponding previously defined platform. Finally, series defines a set of related samples that are considered as a part of an individual experiment [Barrett et al. 2007].

Specific Tools

There are two key options to query GEO using two NCBI Entrez databases [Barrett and Edgar 2006b]. **Entrez GEO DataSets** allows a genome-centric view on gene expression data in GEO. It returns a keyword-based identification of experiments of interest. **Entrez GEO Profiles** provides a gene-centric view on gene expression data in GEO. It returns expression profiles of genes of interest. Particular genes are identified using various search options. Both NCBI Entrez databases are supplemented with several additional tools. Alternatively, GEO database can be queried using GEO BLAST tool. It retrieves sequences and corresponding expression profiles similar to user-defined sequence.

Entrez GEO DataSets supplementary tools:

DataSet clusters is a hierarchical and K-means clustering tool for data visualization, selection and download.

Query group A versus B identifies genes with expression profiles meeting user-defined statistical criteria.

Subset effect flags was designed for the identification of potentially interesting genes by showing those with significantly different expression levels among experimental variables.

Entrez GEO Profiles supplementary tools:

Profile neighbors link allows connection and display of genes sharing similar expression profiles.

Sequence neighbors link enables connection and display of groups of genes related by sequence similarity.

Homolog neighbors link connects and displays groups of genes related by Homologene groups.

Links link links GEO data to related data in other NCBI resources (ie. PubMed, GenBank, Gene, UniGene, OMIM).

Distinguishable Features

GEO is a fast-growing data repository representing large compendium of various gene expression data. Its main impact lies in the complexity of microarray data and in their integration with other resources like sequence information, mapping and bibliographic data.

Moreover, GEO offers the possibility to download gene expression data to enable researchers the use of their own software. One example of such software package can be downloaded as a part of R/Bio/Conductor at http://bioconductor.org/packages/release/bioc/html/GEOquery.html. On the contrary, the diversity of area covered makes the development of robust and user-friendly statistical and data analysis tools difficult. Authors also plan further development of GEO and its extension in order to include non-gene-expression data types like chromatin-immunoprecipitation on arrays (ChIP-chip) studies, array comparative genomic hybridization (aCGH), SNP arrays and even proteomic data [Barrett et al. 2007].

Genevestigator V3 (GV3)

In 2007, new version of Genevestigator (V3 beta) was released (Figure 6). It serves as a bioinformatic service for microarray gene expression data from five organisms (*Arabidopsis thaliana, Mus musculus, Rattus norvegicus, Hordeum vulgare, Homo sapiens*) with total number 16,319 microarrays [Zimmermann et al. 2004, 2005, Laule et al. 2006]. For Arabidopsis, Genevestigator V3 covers 3222 experiments. Initial Genevestigator database was launched in 2004 as a web-based MySQL/PHP software application to query Arabidopsis microarray database with several analysis tools – Digital Northern, Gene Correlator, The, Gene Atlas, Gene Chronologer, Response Viewer and Meta-Analyser. Currently Genevestigator V3 operates as Java applet at a client's side. This client-server structure offers more sophisticated functions and allows more comfortable interface. Obviously, it is necessary to have Java runtime environment (JRE) version 1.4.2 or higher installed. The database is supplemented with several tools - Meta-Profile Analysis, Biomarker Search, Clustering Analysis, Pathway Projector. Genevestigator belongs among the most professional microarray analysis applications and provides many data mining tools allowing multilevel analyses. However, it does not serve as a data repository thus the user-controlled direct data upload is not possible. Currently, it hosts only expression data from Affymetrix GeneChips; for Arabidopsis, AG and ATH1 arrays are covered.

Transcriptomic data and their annotations were retrieved from many sources and repositories. Of these, NASC (http://affymetrix.arabidopsis.info/) [Craigon et al. 2004], FGCZ (http://www.fgcz.ethz.ch/), GEO (http://www.ncbi.nlm.nih.gov/geo/), [Edgar et al. 2002, 2006, Barrett et al. 2005, 2007, Barrett and Edgar 2006a], ArrayExpress (http://www.ebi.ac.uk/microarray-as/aer/?#ae-main[0]) [Parkinson et al. 2005, 2007, Brazma et al. 2003, 2006, Sarkans et al. 2005, Rocca-Serra et al. 2003], AtGenExpress [Schmid et al. 2005] and TAIR (http://www.arabidopsis.org/) [Rhee et al. 2003] belong among the public repositories. Genevestigator enables two main query approaches – a gene-centric and genome-centric. It also offers an extensive selection of data-searching options that are well described in the accompanying documentation (https://www.genevestigator.ethz.ch/index.php?option=com_content&task=view&id=34&Itemid=95).

Genevestigator V3 is generally publicly available. However, it offers three levels of access – Open, Classic and Advanced. Open access does not require the registration of user account but it offers only two tools (Northern and Selection tool). Classic access enables also the Meta-Profile Analysis toolbox and is free for academic institutions. Advanced access offers all tools available in GV3 (also Biomarker search, Clustering analysis and Pathway projector) but this access option is paid.

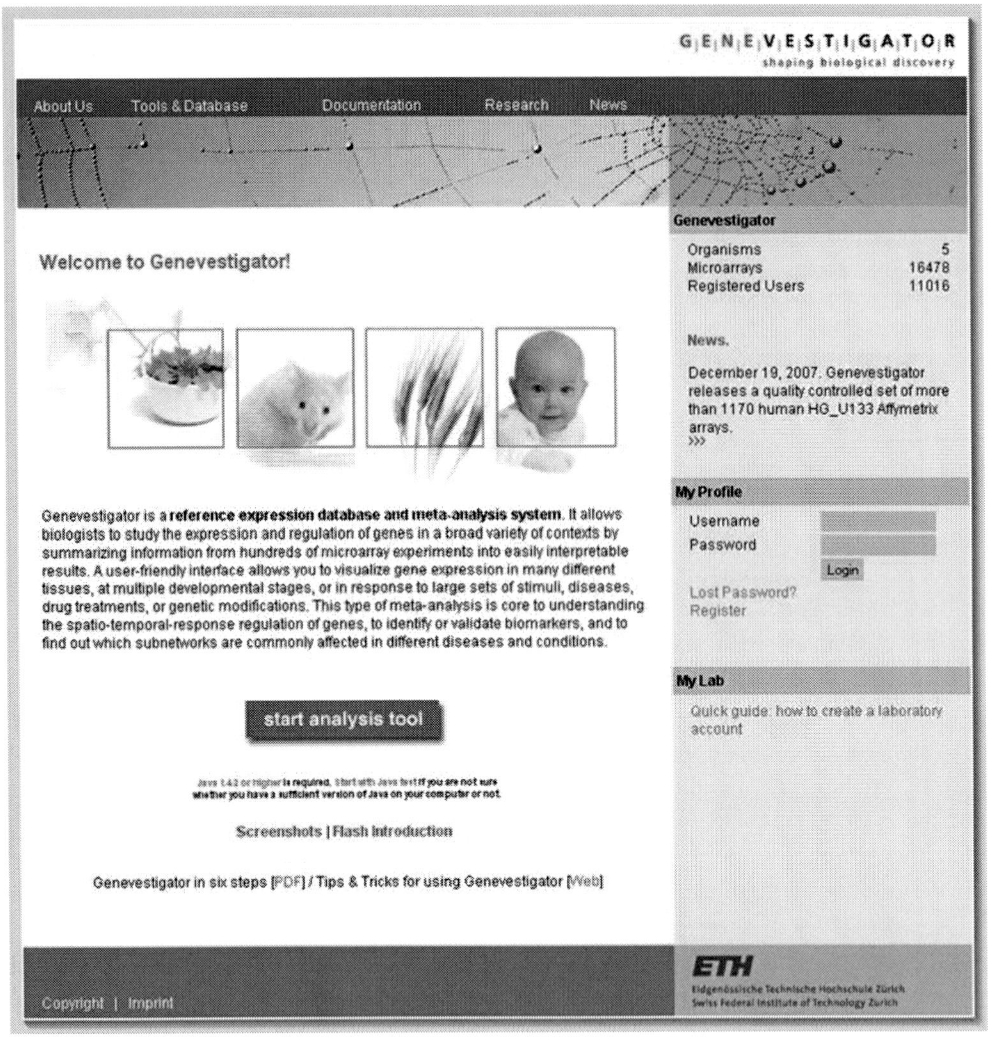

Figure 6. Homepage of the Genevestigator V3 database and toolbox available at https://www.gene vestigator.ethz.ch/.

Data Mining and Analysis Tools

Meta-Profile Analysis comprise six tools - Selection, Northern, Anatomy, Development, Stimulus and Mutation that allow the discovery of gene expression pattern across various arrays (Selection, Northern) or categories (Anatomy, Development, Stimulus and Station). All Meta-profile Analysis tools are based on „gene-centric approach“ so they return signal intensity values of pre-defined genes and genesets. Further, these tools can be divided into two categories according to conditions and gene-array approach used. The gene-condition approach groups all arrays with same conditions for instance identical developmental or anatomical category („meta-profiles“).

Selection tool displays how selected gene or genes are expressed across all arrays in the database. The user can choose arrays and genes for display. Moreover, each experiment can be selected and highlighted in context of all experiments. Results are visualised as a scatter plot with magnifying rectangle.

Northern displays how selected gene or genes are expressed across chosen arrays in the database. Selection and Northern tools may depict expression profiles of many genes through many arrays simultaneously. Results are shown as an editable scatter plot with possibility to switch the axes (horizontal or vertical).

Anatomy depicts how gene(s) of interest are expressed in different plant tissues, organs and cell cultures organised in a tree of anatomical categories. In each category, the average signal intensity value (mean value) from all selected arrays belonging to this category is plotted. Expression values from children nodes are always involved in the mean expression value in a parent node. Results are visualised as a scatter plot or heat map. Users shall be aware that the scatter plot displays only up to ten genes.

The **Development** tool is very similar to the Anatomy tool (identical output formats and expression data types) but it shows how strongly the selected gene(s) are expressed in different developmental stages.

The **Stimulus** tool visualises gene expression responses to external stimuli and reveals up- or down-regulated genes. The same stimuli from different experiments are processed separately. The user then can see one stimulus in one plot repeatedly. Results are shown as scatter plot or heath map with the indication of number of arrays under treatment (T) and control (C) conditions used for the calculation of ratio and log ratio. Again, the scatter plot displays up to ten genes.

Mutation displays the gene expression response to the mutations or other genetic modifications in the genome as a ratio or log ratio value. The Mutation tool features are identical as in the Stimulus tool._Both tools also have identical layout, output and options.

Biomarker Search helps user to identify or validate biomarker genes in given category. It consists of the same toolbox as Meta-Profile Analysis (Anatomy, Development, Stimulus, Mutation). Specific biomarker identification is based on target-base search. The score defined to identify specifically expressed genes is derived from the sum of averages from the selected target and base categories. It is possible to display up to 400 genes per search. All available tools (**Anatomy, Development Stimulus and Mutation**) return the list of probe sets with the highest scoring genes in given target group over against a base. Whereas Anatomy and Development work with absolute signals, Stimulus and Mutation use relative expression values. The Anatomy tool is visualized as the expandable tree menu of individual categories. The Development tool is displayed as a pictorial list of development categories. The Mutation tool simply shows a list of mutation categories. Results are shown as a heat map with blue-white (absolute expression values) or green-red color coding (relative expression values)

Clustering Analysis is used to identify local patterns in gene expression data in given categories (Arrays, Anatomy, Development, Stimulus, Mutation) and to discover the possible relationship among gene expression patterns. Huge advantage of GV3 clustering tool is that it enables clustering of genes within their biological context using meta-profiles in the analysis (Anatomy, Development, Stimulus, Mutation). After selection of particular experiments and genes the user chooses the clustering method. There are two different algorithms available – Hierarchical Clustering and Biclustering. Whereas the first results in a two-dimensional matrix consisting of genes and conditions by selected gene cluster and profile cluster, the other uses exact BiMax algorithm that identifies all biclusters in a matrix.

Pathway Projector represents a reaction network analysis system. It combines expression data from transcriptomic studies and literature knowledge about signaling and metabolic pathways. Currently, the Pathway Projector analysis tool is available only for two organisms – *Arabidopsis thaliana* and *Mus musculus*. In few subsequent steps, the user defines comparison set (array set) that contains treatment and control sets and then selects appropriate subnetworks. The pre-defined or user-defined datasets for four ontologies (anatomy, development, stimulus, mutation) represent great advantage of this tool. In addition to display a network of interest, Pathway Projector also enables the user to build a new network from predefined pathways. Moreover, it is possible to import and export pathway maps as a GVP files.

Distinguishable Features

The possibility to download graphical output in format and resolution ready for publication or web presentation represents very useful feature of newly released Genevestigator V3. The user can select between EPS, PNG or JPEG formats and set the figure parameters for export (width, height). Manual curation of microarray experiments represents great advantages of GV3. It results in Quality control (QC) that allows the detection of arrays with lower or poor quality and subsequently elimination of these arrays or experiment excluded from the microarray database. Identification of problematic arrays increases the output reliability of given data analysis. Genevestigator V3 categorizes experimental data that passed the QC check into separated subset called "High quality arrays only".

NASCArrays

NASCArrays [Craigon et al. 2004] is the Nottingham Arabidopsis Stock Centre's microarray database (http://affy.arabidopsis.info/narrays/experimentbrowse.pl) and it serves primarily as a repository for microarray data produced by NASC's transcriptomics service (Fig. 7). NASCArrays database harbours data from two types of Affymetrix GeneChips, AG Arrays and ATH1 Whole GenomeArrays. Microarray data are shared with other databases worldwide. In this database, individual arrays are referred to as "slides". Index page of NASCArrays allows obtaining expression data by Search tools, Treeview and Data mining tools. The easiest way of data selection is the keyword search in Experiment Search, NASCArrays Reference Number and Slide search with using many selection criteria e.g. growth conditions, development stage, etc. Alternatively, individual experiments can be selected in a Treeview (list of experiments) containing all experiments categorized according to various criteria [Whetzel et al. 2006]. Moreover, the Slide selection tool allows researcher to perform desired data analysis by choosing individual slides. Immediately after selection of appropriate slides, set of data mining tools can be applied with the possibility of data download.

The experiment page shows details for each experiment (abstract, contact details of the author, information about the experiment and list of all slides belonging to the respective experiment with precise experimental information.

For download, one or several slides can be selected using "Slide selection" function for the whole experiment or for up to 300 genes selected by bulk gene download. Data are supplied in one of two common data formats, CSV files or TAB-delimited text files. All data are well annotated with sample preparation details. Moreover, it is possible to download several data categories - Signal, StatPairsUsed, PresentCall, Detection P-value. For clustering purposes, only signal values may be downloaded.

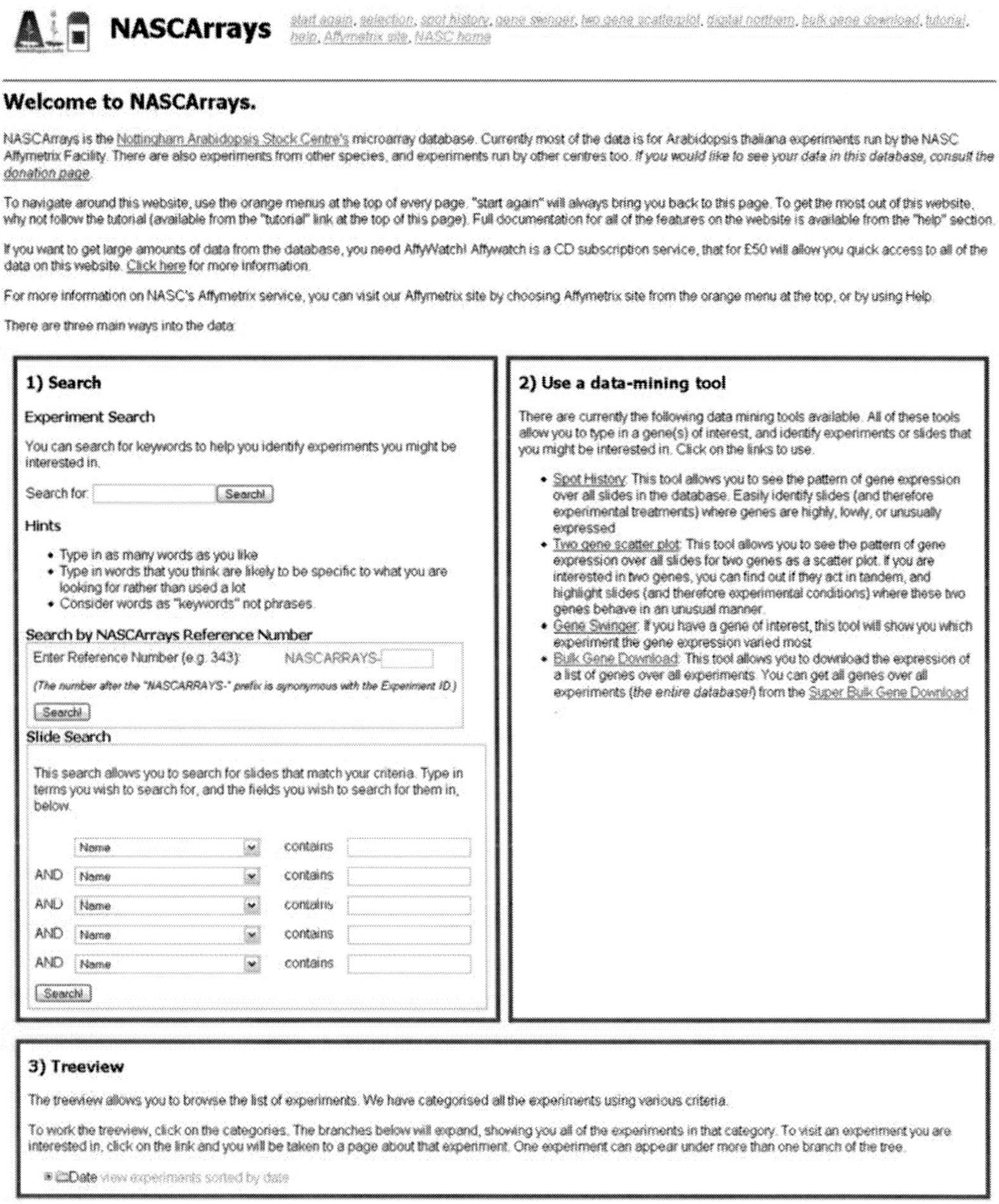

Figure 7. Homepage of the NASCArrays database available at http://affy.arabidopsis.info/narrays/experimentbrowse.pl.

Arabidopsis genome annotation supplied from Arabidopsis Ensembl (http://atensembl.arabidopsis.info/index.html) is based on TIGR3 version [Childs et al. 2007], (http://www.tigr.org/). For this reason, data must be interpreted carefully. However, user-supplied annotation for each sample is MIAME-compliant [Brazma et al. 2001] including information on how the original RNA sample preparation and treatment. The information about the each experiment and all used microarrays is very comprehensive. It can be accessed on the experiment page where each slide is represented by a slide box. Moreover, specially formatted data for clustering analysis using EBI-EPCLUST software [Kapushesky et al. 2004; http://ep.ebi.ac.uk/EP/EPCLUST/) can be acquired and all microarray data produced by NASC's Affymetrix service can be also mailed to users on CDs (AffyWatch, paid service).

Specific Tools

Spot history displays the distribution of expression of a gene of interest over all experiments in the database and gives the user the chance to find experiments or individual slides with distant expression values for given gene and therefore to identify "unusual experiments".

Two-gene scatter plot visualises expression profiles of two genes and reveals the relationships between these genes (co-expression) over all slides in a database.

Gene swinger is a web-based tool for experiment ranking according to the Coefficient of Variance value. It identifies experiments where the gene of interest shows the highest variability.

Digital Northern is an elementary tool for the visualization of relative expression signal and detection call (present, absent) of up to ten genes over all slides or over selected experiment or slides.

Simple Pairwise Analysis is a simple mining tool for identification of upregulated genes. After selection of just two slides it shows top 200 upregulated genes at each slide.

Bulk gene download enables download of up to 300 genes. After submitting list of AGI codes, related data from all the experiments in the database are downloaded.

Distinguishable Features

Extensive user-supplied annotation for each sample meets MIAME requirements [Brazma et al. 2001]. Data describing each experiment are very comprehensive including information on how the original RNA sample was prepared and how the RNA sample was subsequently hybridised. This information can be accessed on the experiment page. Among advantageous features belongs the availability of expression data specially formatted for clustering analysis by EBI-EPCLUST software [Kapushesky et al. 2004; http://ep.ebi.ac.uk/EP/EPCLUST/) and the availability of paid AffyWatch mailing service.

The Stanford Microarray Databases (SMD)

Originally, the database was established in 1999 for Stanford investigators but it is no longer restricted to local Stanford users. Now SMD serves as a data repository and microarray research database that is publicly available for all researches worldwide (Figure 8) [Sherlock et al. 2001, Gollub et al. 2003]. Currently, the database contains the data for 53 organisms that represents almost 70,000 experiments (http://smd.stanford.edu/statictics.html). SMD supports wide collection of microarray platforms and software packages such as Affymetrix, Agilent, Combimatrix, GenePix, ScanAlyze, etc. Most data are public; however access to non-public data is limited to registered Stanford researchers and their collaborators. In principle, users can upload the data to the repository, subsequently retrieve and analyze their data using variety of data mining, analysing and displaying tools, and then, after publishing, made them available for the scientific community [Ball et al. 2005].

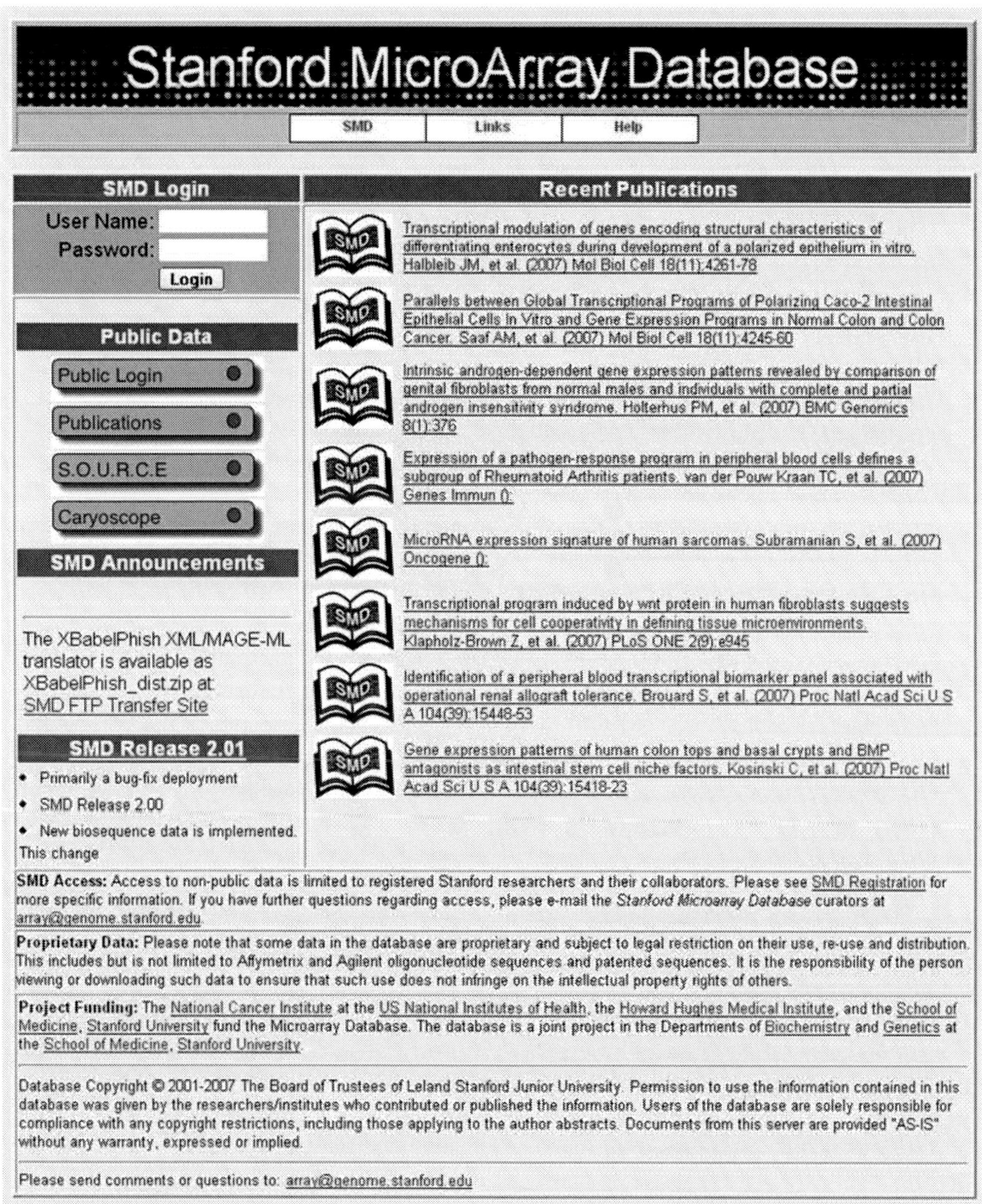

Figure 8. Homepage of the Stanford Microarray Database available at http://genome-www5.stanford.edu/.

After data retrieval, SMD provides various data processing and analysing tools. First, all data undergoes normalization and centering enabling cross-array comparison. Second, clustering algorithms reveal specific patterns within data using hierarchical clustering, self organising maps, singular value decomposition to identify missing data, alternatively, KNNImpute to estimate missing values in data matrices. Third, quality assessment tools offer other options including ANOVA analysis to detect spatial and print-tip bias on the array, HEEBO/MEEBO plots to view diagnostic and doping control, and graphing tools to produce histograms or scatter plots of user-selected fields [Demeter et al. 2007].

When searching for particular data, users can access variety of data mining tools in two different search forms – "Basic Search" and "Advanced Search". Options available in these forms give users great flexibility in their searches to specify the organism of interest or authors, and to further refine categories and subcategories by application of cellular, biochemical or physiological limits. Moreover, researchers can select from particular experiments, specify genes and follows other filtering criteria and are allowed to edit or delete the annotations associated with the data. Finally, obtained results can be displayed in different formats and links are provided to explore more details about genes, arrays and experiments of interest.

Specific Tools

Quality Assessment is an exploratory graphic tool producing histograms or scatter plots of user-selected fields to look at data points distribution, an ANOVA analysis to detect spatial and print tip bias on the array and a tool from BioConductor's ArrayQuality package to view diagnostic and doping control plots for HEEBO/MEEBO arrays.

Hierarchical Clustering and other Data Analysis Tools are intended for data clustering and visualizating of co-expressed genes.

Singular Value Decomposition provides a mathematical algorithm for the determination of unique orthogonal gene and corresponding array expression patterns.

KNNImpute enables computation of missing values in data matrices before applying singular value decomposition (SVD) displayed as raster image or bar graph; SVD is a mathematical approach allowing determination of unique orthogonal genes and corresponding array expression patterns.

GO TermFinder determines whether a list of genes produced by any number of microarray analysis software tools has a significant enrichement of GO terms using Gene Ontology annotation.

Array Color Tool provides a simplified view of the data ratio for a given microarray allowing the user to quickly examine the microarray for evidence of global effects such as printing biases using ANOVA calculation to measure the dependence of per-sot ratio on printing plate and on sector.

Experiment Selection Tool selects from variety of experiments by slide name, organism, category and subcategory of experiments.

Expresion History Tool allows users to find microarray expression data for a gene throughout all experiments available.

Data Filtering Options offers variety of statictical measurements to filter and select data.

Clustering and Image Generation provides choice of self organizing maps or clustering tools using Pearson Correlation or Euclidean Distance.

Gene Selection and Annotation Tool enables selection of one or all genes and their annotation by description, gene model, GenBank accession, clone ID, or BAC locus.

Distinguishable Features

SMD is a fast expanding database using extensive data mining tools for microarray data processing, analyzing, visualizing and sharing. Since employing wide spectra of microarray platforms, software, organisms and experimental conditions, the database is able to meet broad users' expectancy. SMD is currently the largest public repository for Arabidopsis Affymetrix microarray data. However, using complete data repository and all associated tools is possible only after paid registration for long-term data storage. SMD users can store dataset before and after filtering and centering in preclustering files. One can find useful to be able to precisely set up criteria for data mining and use the Expression History tool to explore the behavior of a gene of interest across all microarrays. SMD has also implemented pipe-line that converts sets of microarray data into MEGA-ML files and deposits them directly to ArrayExpress and GEO database.

CONCLUSION

Enormous increase of gene expression profiling studies in recent years have supported rapid formation of extensive data repositories and development of number of on-line software applications allowing researchers to exploit available gene expression data. This chapter provided brief description and comparison of microarray-based gene expression databases and web-based tools. Special attention was paid to Affymetrix GeneChip platform that is the most widely used and serves as a standard for the model species *Arabidopsis thaliana*. However, several included databases comprise data acquired from various platforms of microarray-based experiments (ArrayExpress, SMD) or even non-array based technologies (GEO). Moreover, these databases contain expression data from different organisms such as SMD storing data from 53 organisms, GEO from over 100 individual organisms and ArrayExpress carrying data from even more than 200 organisms. From Affymetrix-focused portals, number of databases contains data from both array types – the first generation GeneChip® 8K AG Genome Arrays covering approximately one third of Arabidopsis genome and later and more complete ATH1 Whole Genome Array (22K). Fewer databases are focused solely on the later type of Affymetrix ATH1 GeneChip (BAR, aGFP).

The largest data repositories and databases with associated analysis tools often stand as integrated components of extensive bioinformatic projects run by large institutions like European Bioinformatics Institute (EMBL-EBI) or National Center for Biotechnology Information (NCBI). Other databases and web-based tools represent extra value added to microarray service (NASCArrays) or were created by established laboratories working in transcriptomic or bioinformatic field (BAR, aGFP). A subset of microarray databases were originally designed as public data repositories and retrieval systems for high throughput expression data. They serve mainly as data storage and offer fewer analysis tools (GEO, ArrayExpress, SMD). On the other hand, other software applications were designed primarily for data analysis. They exploit microarray data from many gene expression databases and offer large number of various analysis tools (BAR, Genevestigator V3).

Microarray data analysis tools can be divided into two main categories. The first category comprises gene expression visualization tools, whereas tools forming the second one offer more thorough analyses of gene expression data, for example gene grouping and display of co-expressed genes. The most advanced software applications (Genevestigator V3 or BAR) offer both types of analyses. Other tools were designed specifically for intuitive visualization of gene expression at various morphological and developmental stages (aGFP).

Almost all databases and web-based tools presented in this chapter are freely publicly available. Moreover, some databases offer the possibility of password-protected account to increase the work comfort (ArrayExpress). As an exception, Genevestigator V3 offers three levels of access – Open, Classic and Advanced. Classic and Advanced access options supplement free Open access, they are paid (with the exception of Classic access for academic users) and offer more data analysis tools as described previously. In addition, paying registration is required in the Stanford MicroArray Database for long-term storage.

Individual databases exploit various microarray data resources. Most databases process data obtained from publicly available data resources and repositories (NASC, FGCZ, GEO, TAIR, etc). In specialized cases like NASCArrays, they represent a functional overlay of existing microarray service. As a supplement, several projects allow user-controlled data submission (GEO, ArrayExpress),

And how to recognize the quality of experimental data in the database? First, the use of curated experiments is important prerequisite for correct interpretation of expression data. Currently, the large number of databases is strictly MIAME-compliant. The minimum information about microarray experiment (MIAME) contributes to increased comparability and of microarray data and so increases the value of interpretation of obtained results [Brazma et al. 2001]. In some cases, the use of alternative normalization algorithms may prove useful (aGFP). So, the user can identify the quality database by means of well-annotated raw and processed microarray data too. Further criteria comprise the assessment of expression data quality by measurement and evaluation of several quality control metrics. Quality control (QC) allows the detection of arrays with lower or poor quality and subsequent elimination of these arrays or whole experiments from the microarray database. Identification of problematic arrays increase the output reliability of given data analysis. Genevestigator V3 categorizes experimental data that passed the QC check into separated subset called "High quality arrays only". Moreover, several databases allow users not only to extract gene expression data, but also facilitate the sharing of microarray experiment design and experimental protocols. This possibility is very helpful as it markedly improves the design of new experiments. In addition to gene expression data characterized by expression signal (sometimes along with detection call, p-value) and data obtained as a result of database query and data analysis, selected databases also include other data labels (gene annotation, gene ontology). On the other hand, some databases do not clearly mention the version of Arabidopsis genome annotation and this imperfection can contribute to data misinterpretation.

In addition to the database query, data analyses and data submission, most databases offer also data download. Database commonly support various data format options, from well-known XLS, CSV or Tab-delimited TXT files to specialised formats like SOFT text file (GEO). When downloading defined datasets, the user can often acquire additional information e.g. StatPairsUsed, PresentCall, Detection P-value (Genevestigator V3).

To date, the vast amount of transcriptomic data is publicly available in number of data repositories together with various data analysis tools providing instruments for their exploitation as well as assigning putative functional information to analysed genes. The large number of software applications has a user-friendly web-interface and broad scale of analysis tool, which may help researchers to open up new perspective of gene expression data and to exploit them for the planning of specifically targeted experiments and verification of their hypotheses.

Acknowledgments

Authors gratefully acknowledge the financial support from GACR (522/06/0896), GAAVCR (KJB600380701) and MSMT CR (LC06004, OC08011).

References

Aharoni, A; Vorst, O. DNA microarrays for functional plant genomics. *Plant Mol Biol*, 2002, 48, 99-118.

Arabidopsis Genome Initiative. Analysis of the genome sequence of the flowering plant Arabidopsis thaliana. *Nature*, 2000, 408, 796-815.

Ball, CA; Awad, IAB; Demeter, J; Gollub, J; Hebert, JM; Hernandez-Boussard, T; Jin, H; Matese, JC; Nitzberg, M; Wymore, F; Zachariah, ZK; Brown, PO; Sherlock, G. *Nucleic Acids Research,* 2005, 33 D580-D582

Barrett, T; Suzek, TO; Troup, DB; Wilhite, SE; Ngau, WC; Ledoux, P; Rudnev, D; Lash, AE; Fujibuchi, W; Edgar, R. NCBI GEO: mining millions of expression profiles - database and tools. *Nucleic Acids Res,* 2005, 33 Database Issue, D562-566.

Barrett, T; Edgar, R. Gene expression omnibus: microarray data storage, submission, retrieval, and analysis. *Methods Enzymol,* 2006a, 411, 352-369.

Barrett, T; Edgar, R. Mining microarray data at NCBI's Gene Expression Omnibus (GEO). *Methods Mol Biol*, 2006b, 338, 175-190.

Barrett, T; Troup, DB; Wilhite, SE; Ledoux, P; Rudnev, D; Evangelista, C; Kim, IF; Soboleva, A; Tomashevsky, M; Edgar, R. NCBI GEO: mining tens of millions of expression profiles--database and tools update. *Nucleic Acids Res,* 2007, 35 Database issue, D760-765.

Brazma, A; Hingamp, P; Quackenbush, J; Sherlock, G; Spellman, P; Stoeckert, C; Aach, J; Ansorge, W; Ball, CA; Causton, HC; Gaasterland, T; Glenisson, P; Holstege, FC; Kim, IF; Markowitz, V; Matese, JC; Parkinson, H; Robinson, A; Sarkans, U; Schulze-Kremer, S; Stewart, J; Taylor, R; Vilo, J; Vingron, M. Minimum information about a microarray experiment (MIAME)-toward standards for microarray data. *Nat Genet,* 2001, 29(4), 365-371.

Brazma, A; Parkinson, H; Sarkans, U; Shojatalab, M; Vilo, J; Abeygunawardena, N; Holloway, E; Kapushesky, M; Kemmeren, P; Lara, GG; Oezcimen, A; Rocca-Serra, P; Sansone, SA. ArrayExpress - a public repository for microarray gene expression data at the EBI. *Nucleic Acids Res,* 2003, 31(1), 68-71.

Brazma, A; Kapushesky, M; Parkinson, H; Sarkans, U; Shojatalab, M. Data storage and analysis in ArrayExpress. *Methods Enzymol,* 2006, 411, 370-386.

Brenner, S; Johnson, M; Bridgham, J; Golda, G; Lloyd, DH; Johnson, D; Luo, S; McCurdy, S; Foy, M; Ewan, M; Roth, R; George, D; Eletr, S; Albrecht, G; Vermaas, E; Williams, SR; Moon, K; Burcham, T; Pallas, M; DuBridge, RB; Kirchner, J; Fearon, K; Mao, J; Corcoran, K. Gene expression analysis by massively parallel signature sequencing (MPSS) on microbead arrays. *Nat Biotechnol*, 2000, 18, 630-634.

Childs, KL; Hamilton, JP; Zhu, W; Ly, E; Cheung, F; Wu, H; Rabinowicz, PD; Town, CD; Buell, CR; Chan, AP. The TIGR Plant Transcript Assemblies database. *Nucleic Acids Res,* 2007, 35(Database issue):D846-851.

Clarke, JD; Zhu, T. Microarray analysis of the transcriptome as a stepping stone towards understanding biological systems. practical considerations and perspectives. *Plant J*, 2006, 45, 630-650.

Craigon, DJ; James, N; Okyere, J; Higgins, J; Jotham, J; May, S. NASCArrays: a repository for microarray data generated by NASC's transcriptomics service. *Nucleic Acids Res*, 2004, 32(Database issue), D575-577.

Demeter, J; Beauheim, C; Gollub, J; Henandez- Bouccard, T; Jin, H; Maier, D; Matese, JC; Nitzberg, M; Wymore, F; Zachariah, ZK; Brown, PO; Sherlock, G; Ball, CA. *Nucleic Acids Research,* 2007, 35, D766-D770

Donson, J; Fang, Y; Espiritu-Santo, G; Xing, W; Salazar, A; Miyamoto, S; Armendarez, V; Volkmuth, W. Comprehensive gene expression analysis by transcript profiling. *Plant Mol Biol*, 2002, 48, 75-97.

Dupľáková, N; Reňák, D; Hovanec, P; Honysová, B; Twell, D; Honys D. Arabidopsis Gene Family Profiler (aGFP)--user-oriented transcriptomic database with easy-to-use graphic interface. BMC Plant Biol, 2007, 23, 7:39.

Edgar, R; Domrachev, M; Lash, AE. Gene Expression Omnibus: NCBI gene expression and hybridization array data repository. *Nucleic Acids Res,* 2002, 30(1), 207-210.

Edgar, R; Barrett, T. NCBI GEO standards and services for microarray data. *Nat Biotechnol,* 2006, 24(12), 1471-1472.

Gollub, J; Ball, CA; Binklay, G; Demeter, J; Pinknkelstein, DB; Hebert, JM; Hernandez-Boussard, T; Heng, J; Kaloper, M; Matese, JC; Schroeder, M; Brown, PO; Botstein, D; Sherlock, D. *Nucleic Acids Research,* 2003, 31 94-96

Heazlewood, JL; Tonti-Filippini, J; Verboom, RE; Millar, AH. Combining experimental and predicted datasets for determination of the subcellular location of proteins in Arabidopsis. *Plant Physiol*, 2005, 139(2), 598-609.

Heazlewood, JL; Verboom, RE; Tonti-Filippini, J; Small, I; Millar, AH. SUBA: the Arabidopsis Subcellular Database. *Nucleic Acids Res*, 2007, 35(Database issue), D213-218.

Hennig, L; Menges, M; Murray, JA; Gruissem, W. Arabidopsis transcript profiling on Affymetrix GeneChip arrays. *Plant Mol Biol*, 2003, 53, 457-465.

Hughes, TR; Marton, MJ; Jones, AR; Roberts, CJ; Stoughton, R; Armour, CD; Bennett, HA; Coffey, E; Dai, H; He, YD; Kidd, MJ; King, AM; Meyer, MR; Slade, D; Lum, PY; Stepaniants, SB; Shoemaker, DD; Gachotte, D; Chakraburtty, K; Simon, J; Bard, M; Friend, SH. Functional discovery via a compendium of expression profiles. *Cell*, 2000, 102, 109-126.

Ilic, K; Berleth, T; Provart, NJ. BlastDigester - a web-based program for efficient CAPS marker design. *Trends Genet*, 2005, 21(1), 36.

Jen, CH; Manfield, IW; Michalopoulos, I; Piney, JW; Willats, WGT; Gilmartin, PM; Westhead, DR. *Plant Journal*, 2006, 46, 336-348.

Kapushesky, M; Kemmeren, P; Culhane, AC; Durinck, S; Ihmels, J; Körner, C; Kull, M; Torrente, A; Sarkans, U; Vilo, J; Brazma, A. Expression Profiler: next generation - an online platform for analysis of microarray data. *Nucleic Acids Res*, 2004, 32, W465-470.

Laule, O; Hirsch-Hoffmann, M; Hruz, T; Gruissem, W; Zimmermann, P. Web-based analysis of the mouse transcriptome using Genevestigator. *BMC Bioinformatics,* 2006, 7:311.

Lipshutz RJ, Fodor SP, Gingeras TR, Lockhart DJ. High density synthetic oligonucleotide arrays. *Nat Genet*, 1999, 21, 20-24.

Lockhart DJ, Dong H, Byrne MC, Follettie MT, Gallo MV, Chee MS, Mittmann M, Wang C, Kobayashi M, Horton H, Brown EL. Expression monitoring by hybridization to high-density oligonucleotide arrays. *Nat Biotechnol*, 1996, 14, 1675-1680.

Mandaokar A, Thines B, Shin B, Markus Lange B, Choi G, Koo YJ, Yoo YJ, Choi YD, Browse J. Transcriptional regulators of stamen development in Arabidopsis identified by transcriptional profiling. *Plant J*, 2006, 46, 984-1008.

Manfield, WI; Jen, CH; Pinney, JW; Michalopoulos, I; Bradford, JR; Gilmartin, PM; Westhead, DR. *Nucleic Acids Research,* 2006, 34, 504-509.

Money, T; Reader, S; Qu, LJ; Dunford, RP; Moore, G. AFLP-based mRNA fingerprinting. *Nucleic Acids Res*, 1996, 24, 2616-2617.

Mukherjee, G; Abeygunawardena, N; Parkinson, H; Contrino, S; Durinck, S; Farne, A; Holloway, E; Lilja, P; Moreau, Y; Oezcimen, A; Rayner, T; Sharma, A; Brazma, A; Sarkans, U; Shojatalab, M. Plant-based microarray data at the European Bioinformatics Institute. Introducing AtMIAMExpress, a submission tool for Arabidopsis gene expression data to ArrayExpress. *Plant Physiol,* 2005, 139(2), 632-636.

Palaniswamy, SK; James, S; Sun, H; Lamb, RS; Davuluri, RV; Grotewold, E. AGRIS and AtRegNet. a platform to link cis-regulatory elements and transcription factors into regulatory networks. *Plant Physiol,* 2006, 140(3), 818-829.

Parkinson, H; Sarkans, U; Shojatalab, M; Abeygunawardena, N; Contrino, S; Coulson, R; Farne, A; Lara, GG; Holloway, E; Kapushesky, M; Lilja, P; Mukherjee, G; Oezcimen, A; Rayner, T; Rocca-Serra, P; Sharma, A; Sansone, S; Brazma, A. ArrayExpress - a public repository for microarray gene expression data at the EBI. *Nucleic Acids Res,* 2005, 33 Database Issue, D553-555.

Parkinson, H; Kapushesky, M; Shojatalab, M; Abeygunawardena, N; Coulson, R; Farne, A; Holloway, E; Kolesnykov, N; Lilja, P; Lukk, M; Mani, R; Rayner, T; Sharma, A; William, E; Sarkans, U; Brazma, A. ArrayExpress - a public database of microarray experiments and gene expression profiles. *Nucleic Acids Res,* 2007, 35 Database issue, D747-750.

Rhee, SY; Beavis, W; Berardini, TZ; Chen, G; Dixon, D; Doyle, A; Garcia-Hernandez, M; Huala, E; Lander, G; Montoya, M; Miller, N; Mueller, LA; Mundodi, S; Reiser, L; Tacklind, J; Weems, DC; Wu, Y; Xu, I; Yoo, D; Yoon, J; Zhang, P. The Arabidopsis Information Resource (TAIR): a model organism database providing a centralized, curated gateway to Arabidopsis biology, research materials and community. Nucleic Acids Res, 2003, 31(1), 224-228.

Rocca-Serra, P; Brazma, A; Parkinson, H; Sarkans, U; Shojatalab, M; Contrino, S; Vilo, J; Abeygunawardena, N; Mukherjee, G; Holloway, E; Kapushesky, M; Kemmeren, P; Lara, GG; Oezcimen, A; Sansone, SA. ArrayExpress: a public database of gene expression data at EBI. *C R Biol,* 2003, 326(10-11), 1075-1078.

Sarkans, U; Parkinson, H; Lara, GG; Oezcimen, A; Sharma, A; Abeygunawardena, N; Contrino, S; Holloway, E; Rocca-Serra, P; Mukherjee, G; Shojatalab, M; Kapushesky, M; Sansone, SA; Farne, A; Rayner, T; Brazma, A. The ArrayExpress gene expression database: a software engineering and implementation perspective. *Bioinformatics,* 2005, 21(8), 1495-1501.

Schena, M; Shalon, D; Davis, RW; Brown, PO. Quantitative monitoring of gene expression patterns with a complementary DNA microarray. *Science*, 1995, 270, 467-470.

Schmid, M; Davison, TS; Henz, SR; Pape, UJ; Demar, M; Vingron, M; Scholkopf, B; Weigel, D; Lohmann, JU. A gene expression map of Arabidopsis thaliana development. *Nat Genet*, 2005, 37(5), 501-506.

Schwager, C; Blake, J. Bloader - a batch loader application for MIAMExpress. *Bioinformatics*. 2005, 21(8):1727-1729.

Sherlock, G; Hernandez-Boussard, T; Kasarskis, A; Binkley, G; Matese, JC; Dwight, SS; Kaloper, M; Weng, S; Jin, H; Ball, CA; Eisen, MB; Spellman, PT; Brown, PO; Botstein, D; Cherry, JM. Nucleic Acids Research, 2001, 152-155.

Spellman, PT; Miller, M; Stewart, J; Troup, C; Sarkans, U; Chervitz, S; Bernhart, D; Sherlock, G; Ball, C; Lepage, M; Swiatek, M; Marks, WL; Goncalves, J; Markel, S; Iordan, D; Shojatalab, M; Pizarro, A; White, J; Hubley, R; Deutsch, E; Senger, M; Aronow, BJ; Robinson, A; Bassett, D; Stoeckert, CJ Jr; Brazma, A. Design and implementation of microarray gene expression markup langure (MAGE-ML). *Genome Biol*, 2002, 3, RESEARCH0046.

Taylor, J; Provart, NJ. CapsID: a web-based tool for developing parsimonious sets of CAPS molecular markers for genotyping. *BMC Genet*, 2006, 10, 7:27.

Torrente, A; Kapushesky, M; Brazma, A. A new algorithm for comparing and visualizing relationships between hierarchical and flat gene expression data clusterings. *Bioinformatics*, 2005, 21(21), 3993-3999.

Toufighi, K; Brady, SM; Austin, R; Ly, E; Provart, NJ. The Botany Array Resource: e-Northerns; Expression Angling; and promoter analyses. *Plant J*, 2005, 43(1), 153-63.

Troyanskaya, O; Cantor, M; Sherlock, G; Brown, P; Hastie, T; Tibshirani, R; Botstein, D; Altman, RB. Missing value estimation methods for DNA microarrays. *Bioinformatics*, 2001, 17(6), 520-525.

van Hal, NL; Vorst, O; van Houwelingen, AM; Kok, EJ; Peijnenburg, A; Aharoni, A; van Tunen, AJ; Keijer, J. The application of DNA microarrays in gene expression analysis. *J Biotechnol*, 2000, 78, 271-280.

Velculescu, VE; Zhang, L; Vogelstein, B; Kinzler, KW. Serial analysis of gene expression. *Science*, 1995, 270, 484-487.

Wheeler, DL; Barrett, T; Benson, DA; Bryant, SH; Canese, K; Church, DM; DiCuccio, M; Edgar, R; Federhen, S; Helmberg, W; Kenton, DL; Khovayko, O; Lipman, DJ; Madden, TL; Maglott, DR; Ostell, J; Pontius, JU; Pruitt, KD; Schuler, GD; Schriml, LM; Sequeira, E; Sherry, ST; Sirotkin, K; Starchenko, G; Suzek, TO; Tatusov, R; Tatusova, TA; Wagner, L; Yaschenko, E. Database resources of the National Center for Biotechnology Information. *Nucleic Acids Res*, 2005, 33(Database issue), D39-45.

Wheeler, DL; Barrett, T; Benson, DA; Bryant, SH; Canese, K; Chetvernin, V; Church, DM; DiCuccio, M; Edgar, R; Federhen, S; Geer, LY; Kapustin, Y; Khovayko, O; Landsman, D; Lipman, DJ; Madden, TL; Maglott, DR; Ostell, J; Miller, V; Pruitt, KD; Schuler, GD; Sequeira, E; Sherry, ST; Sirotkin, K; Souvorov, A; Starchenko, G; Tatusov, RL; Tatusova, TA; Wagner ,L; Yaschenko, E. Database resources of the National Center for Biotechnology Information. *Nucleic Acids Res*, 2007, 35(Database issue), D5-12.

Whetzel, PL; Parkinson, H; Causton, HC; Fan, L; Fostel, J; Fragoso, G; Game, L; Heiskanen, M; Morrison, N; Rocca-Serra, P; Sansone, SA; Taylor, C; White, J; Stoeckert, CJ Jr. The MGED Ontology: a resource for semantics-based description of microarray experiments. *Bioinformatics*, 2006, 22(7), 866-73.

Winter, D; Vinegar, B; Nahal, H; Ammar, R; Wilson, GV; Provart, NJ. An "electronic fluorescent pictograph" browser for exploring and analyzing large-scale biological data sets. *PLoS ONE*, 2007, 2(1), e718.

Yang, YH; Dudoit, S; Luu, P; Lin, DM; Peng, V; Ngai, J; Speed, TP. Normalization for cDNA microarray data: a robust composite method addressing single and multiple slide systematic variation. *Nucleic Acids Res*. 2002, 30(4), e15.

Zhang, W; Morris, QD; Chang, R; Shai, O; Bakowski, MA; Mitsakakis, N; Mohammad, N; Robinson, MD; Zirngibl, R; Somogyi, E; Laurin, N; Eftekharpour, E; Sat, E; Grigull, J; Pan, Q; Peng, WT; Krogan, N; Greenblatt, J; Fehlings, M; van der Kooy, D; Aubin, J; Bruneau, BG; Rossant, J; Blencowe, BJ; Frey, BJ; Hughes, TR. The functional landscape of mouse gene expression. *J Biol*. 2004, 3, 21.

Zhu, T. Global analysis of gene expression using GeneChip microarrays. *Curr Opin Plant Biol*, 2003, 6, 418-425.

Zimmermann, P; Hirsch-Hoffmann, M; Hennig, L; Gruissem, W. GENEVESTIGATOR. Arabidopsis Microarray Database and Analysis Toolbox. *Plant Physiol,* 2004, 136(1), 2621-2632.

Zimmermann, P; Hennig, L; Gruissem, W. Gene-expression analysis and network discovery using Genevestigator. Trends Plant Sci, 2005, 10(9), 407-409.

In: Oligonucleotide Array Sequence Analysis
Editors: M.K. Moretti and L.J. Rizzo, pp. 297-321
ISBN: 978-1-60456-542-3

Chapter 10

METHODS, STRATEGIES AND FACTS FOR IMPROVING THE ANALYSIS OF GENETIC VARIATIONS USING DNA ARRAYS

Angélica Rangel-López[1,a], Alfonso Méndez-Tenorio[2,b], Rogelio Maldonado-Rodríguez[2,c] and Kenneth L. Beattie[3,d]

[1]Unidad de Investigación Médica en Enfermedades Nefrológicas, Hospital de Especialidades, Centro Médico Nacional Siglo XXI, IMSS, México 06725, D. F.
[2]Laboratorio de Biotecnología y Bioinformática Genómica.
Escuela Nacional de Ciencias Biológicas. IPN. México 11340, D. F.
[3]Amerigenics, Inc. 1326 Open Range Rd, Crossville, TN 38555, USA.

ABSTRACT

Cancer is a disease resulting from the breakdown of several checkpoints and tumor-suppressing mechanisms. In cancer research, the development of new technologies which have produced genomic tools indispensable for understanding how gene products are regulated in normal and diseased conditions on a global genome scale; one of these technologies are the DNA arrays. Although the most common use of DNA arrays is gene expression profiling and mutation detection, scientists have successfully used them for multiple applications, including genotyping, re-sequencing, DNA copy number analysis and DNA-protein interactions.

This section then will be dedicated to the use of public sequence database that can be accessed, and the design of DNA oligonucleotide probes for oligoarrays derived from sequences of special interest. Single probe and stacking hybridization are explored as possible microarray designs. Then the use of thermodynamic models and *in silico* hybridization are explored in order to access the sensitivity and specificity of the oligoarray and how its design can be improved. However, many commercial and public applications do not consider that the

[a] E-mail address: ragn62@prodigy.net.mx
[b] E-mail address: amendezt@ipn.mx
[c] E-mail address: romaldodr@hotmail.com
[d] E-mail address: kbeattie@frontiernet.net
Address for correspondence: Nephological Research Unit, Speciality Hospital, National Medical Center, SXXI, IMSS. Av. Cuauhtémoc 330, 06725 México, D. F., México; tel/fax.: 525-627-69-00 ext.:21371.

hybridization between target DNAs and microarray probes is a chemical reaction which is influenced by several thermodynamic parameters. One of the most important of such parameters is the thermal stability of the nucleic acid duplexes, which are formed as a result of the hybridization. According to the reaction conditions, these duplexes can be perfectly or imperfectly paired, which is of critical importance when a diagnostic kit is developed in order to asses its sensitivity and specificity.

In silico hybridization can be used to avoid undesired hybridization events (such as multiple target-probe interactions and stable mismatched hybrids). Also, the strategy can facilitate (through use of different length probes) selection of a probe set that has a narrow duplex stability, allowing maximal specificity under a single hybridization condition.

Introduction

Cancer is a genetic disorder arising from the breakdown of several checkpoints and tumor-suppressing mechanisms or by accumulation of somatic mutations in genes involved in critical cellular pathways. These mutations typically result in proteins which exhibit their oncogenic effect by altering signaling through vital transduction networks, or in haploinsufficiency of critical tumor suppressor proteins. The detection of these variations in DNA is very important for the early prediction of cancer, diagnostics of disease and clinical prognosis.

In cancer, the development of new technologies has changed the way we do research in biology and medicine, one of these technologies is the DNA arrays. Array-based technologies are revolutionizing genomics, especially the analysis of DNA variation. Array technologies are not without limitations, however, and one major drawback is the poor flexibility of typical array formats. It is cumbersome to create one's own tailored arrays by spotting DNA. Commercially available microarrays, on the other hand, either contain a fixed and usually broadly applicable content or are expensive to purchase with customized features.

This review gives a brief perspective of some of the landmark discoveries in array-based technology related to analysis of DNA variation. The fundamental concepts behind the technology and some characteristics of DNA oligoarrays are then covered, followed by an overview of how those can be studied in the laboratory. Finally, the use of public sequence database that can be accessed, and the design of DNA oligonucleotide probes for oligoarrays arising from sequences of special interest as possible microarray designs are discussed.

Microarray Definition

The term microarray is both descriptive and somewhat ambiguous as it is commonly used to describe a collection of microscopic spots arranged in an array or grid-like format and attached to a solid support, typically a glass, filter, or silicon wafer, hence the term. These spots are typically referred as "probes" and are designed such that each probe binds a specific nucleic acid sequence ("target") corresponding to a particular gene or other sequence through a process termed hybridization. The sequence bound to a probe, often referred to as the target, is labeled with some kind of detectable molecule or dye such as a fluorophore. The level of binding between a probe and its target is quantified by measuring the fluorescence or signal emitted by the labeling dye when scanned. This signal, in turn, provides a measure of the expression of the specific gene containing the target sequence [1]. This term is commonly used to describe also a variety of platforms including protein microarrays and tissue microarrays [2].

Microarray Nomenclature

Oligonucleotide arrays, also known as "DNA chips", "gene chips", or "genesensor chips" [3]. Currently there are in the literature at least two nomenclature systems for referring to hybridization partners. Both use common terms: "probes" and "targets". With symmetry akin to the hybridization reaction itself, each system mirrors the other. What one describes as "probes", the other describes as "targets"; what one describes as "probe". With respect to the nucleic acids whose entwining represents the hybridization reactions, the identity of one is defined—it tends to be tethered to the solid phase, making up the microarray itself. The identity of the other is revealed by hybridization. The strategy of the "standard" microarray therefore parallels that of a reverse dot-blot, in which the probe is immobilized. For this reason, here we described the tethered nucleic acid as "probe", and the free nucleic acid as "target" [3].

Microarray Technology

The latest microarray technology is a result of rapid development in the 1990s and early 2000s. Although the foundation for microarray technology was established much earlier in 1975, when the first hybridization between nucleic acids was performed [4], it took researchers another 20 years to develop microarrays, due to underdevelopment computer, robotic and DNA technologies. DNA microarray technologies initially were designed to measure the transcriptional levels of RNA transcripts derived from thousands of genes within a genome in a single experiment. Today's arrays are vastly superior to their predecessors in terms of quality, probe density, and structural layout [1-3].

This technology has made it possible to relate physiological cell states to gene expression patterns for studying tumors, diseases progression, cellular response to stimuli, and drug target identification. At present, DNA microarrays have become a vital component in genomic research for high-throughput gene expression analysis [5, 6], mutation detection [7, 8], gene discovery and genetic mapping studies [9] and protein–DNA interaction analysis [10-13].

More recent uses of DNA microarrays in biomedical research are not limited to gene expression. DNA microarrays are being used to detect single nucleotide polymorphisms (SNPs) of our genome (Hap Map project) [14], aberrations in methylation patterns [15], alterations in gene copy number [16], alternative RNA splicing [17], and pathogen detection [18, 19].

Several platforms for fabricating DNA arrays have been described [20]. All platforms use hybridization between probe nucleic acids bound to solid support (microarray) and labeled target nucleic acids. It is well known that complementary single-stranded sequences of nucleic acids form double stranded hybrids. This property is the basis of the very powerful molecular biology tools such as Southern and Northern blots, in situ hybridization, and Polymerase Chain Reaction (PCR). In these, specific single-stranded DNA sequences are used to probe for its complementary sequence (DNA or RNA) forming hybrids. This same idea also is used in DNA microarray technologies.

The aim, however, is not only to detect but also to measure the expression levels of not a few but rather thousands of genes in the same experiment. For this purpose, thousands of single-stranded sequences that are complementary to target sequences are bound, synthesized, or spotted to a glass support whose size is similar to a typical microscope slide.

Platforms can be grouped according to the type of fabrication and type of probe. There are mainly two types of DNA arrays, depending on the type of spotted probes. One uses small single-stranded oligonucleotides (~22 nt) synthesized in situ whose leading provider is Affymetrix (Santa Clara, CA, USA) [21]. The other type of arrays uses complementary DNA (cDNA) obtained by reverse transcription of the genes' messenger RNAs (mRNA), completion of the second strand, cloning of the double-stranded DNAs, and typically PCR amplification of their open reading frames (ORF), which become the bound probes. One of the limitations of using large ORF or cDNA sequences is an uneven optimal melting temperature caused by differences in their sizes and the content of GC-paired nucleotides. A second problem is cross-hybridization of closely related sequences, overlapped genes, and splicing variants. In oligo-based DNA arrays, the targeted nucleic acid specie is redundantly detected by designing several complementary oligonucleotides spanning each entire target sequence by segments. The oligonucleotides are designed in such a way to avoid the cDNA probe drawbacks and to maximize the specificity for the target gene [22, 23].

See some links for more detailed information about the microarray technology [24-26].

Some Limitations of Oligonucleotide Array Technology

Several technical challenges remain to be solved, however, before oligonucleotide arrays can reach their full potential and be implemented in a robust fashion. Hairpin structures are a common feature of single-stranded DNA and RNA molecules. In relation to arrays oligonucleotides, one limitation of the method is that stable secondary structures in the target nucleic acid can make the target sequence inaccessible to intermolecular Watson-Crick base pairing [27-29]. Several techniques have been explored to alleviate this problem, but none appears to be totally successful. Fragmenting the nucleic acid sequence, preferably to a size close to that of the oligonucleotides on the array, by heating the DNA or RNA in the presence of Mg^{2+} [30] or by the use of uracil-*N-glycosylase* [31] can reduce the secondary structure effect. However, the extent of fragmentation is difficult to control. Oligonucleotide analogues such as peptide nucleic acid (PNA) exhibit PNA-DNA heteroduplexes with a high thermal stability at low salt concentration. The authors suggest that under these conditions the DNA strands are less likely to fold to form stable secondary structure and therefore would be accessible to the oligonucleotide probes [32]. However, there is an increasing tendency for non specific DNA binding [33]. It's important to point that these stable structures probably explain why some regions of targets are not accessible for heteroduplex formation with complementary oligonucleotides. Heteroduplex formation between a hairpin structure and a complementary oligonucleotide probe implies two main steps: opening of the structure to allow formation of one or a few base pairs in a transient nucleation complex followed by further base pairing between the two complementary strands to form a heteroduplex which is more stable than the starting structures. The secondary structure of nucleic acid targets can be minimized by using a hybridization temperature above the melting temperature of intramolecular structure or low salt concentration solutions. However, these conditions are not appropriated for hybridization of nucleic acid molecules with short oligodeoxynucleotides.

The Mirzabekov laboratory previously described a contiguous stacking hybridization strategy for potential use in *de novo* sequencing by hybridization [34, 35], and more lately reported the successful use of contiguous stacking hybridization to investigate point mutations [36]. In the latter study, the hybridization was carried out in two steps: The target DNA was first hybridized to decanucleotide probes (10 mer) covalently attached within a thin polyacrylamide gel matrix, to place the mutant site adjacent to the 10-mer duplex, then fluorescently labeled contiguous stacking pentanucleotide probes (5 mers) were applied to gel matrix to detect point mutations within the target sequence immediately adjacent to 10-mer duplex.

Tandem Hybridization on Oligonucleotide Microarrays

Tandem hybridization (TH) or coaxial hybridization as is known also is a novel technology with high sensitivity and specificity which is especially useful for the detection of point mutation.

This strategy is designed to minimize or eliminate some difficulties in simultaneous hybridization of DNA sample to numerous oligonucleotide probes attached to a solid support material. Some of these challenges are as follows:

1. The need to generate single-stranded target nucleic acids in order to achieve optimal hybridization signals.
2. Spontaneous formation of secondary structure in the single-stranded target nucleic acid, causing certain stretches of target sequences to be poorly accessible to hybridization.
3. Imperfect specificity of hybridization, making it difficult or impossible to distinguish between certain sequences variations.
4. The strong influence of base composition on the stability of short duplex structures, making it difficult to use an extensive array of oligonucleotides (differing in base composition) to analyze a nucleic acid sample under a single hybridization condition.
5. Multiple occurrences of sequences complementary to other oligonucleotide probes with the nucleic acid sample, limiting the genetic complexity of a nucleic acid sample than can be analyzed by arrays of short oligonucleotide probes.
6. The need to label each nucleic acid analyte prior to hybridization to DNA probe array, a significant factor in the overall time and cost of analysis.

The principle of TH is: unlabeled, single-stranded target DNA is preannealed to one or more labeled "auxiliary" oligonucleotides which can be designed to disrupt interfering secondary or higher-order structures, or to cover up alternative hybridization sites that any of surface tethered probes may have within the target. TH method archives sequence specificity by using short, glass-tethered capture probe along with one or two longer, contiguously stacking oligonucleotides that are preannealed to the target. Good discrimination of point mutations is achieved within the short capture probe, while the longer stacking probe(s) contribute the label and stabilize the binding of the target via base-stacking interactions. The stacking probe enables hybridization to be conducted at an elevated temperature at which the capture probe alone cannot bind to the target, and directs the analysis to a single site within the target, even within a complex nucleic acid analyte (see Figure 1). This preannealing strategy may facilitate more reliable applications of genesensor technology, and avoids the inconvenient and costly labeling of numerous DNA samples [37].

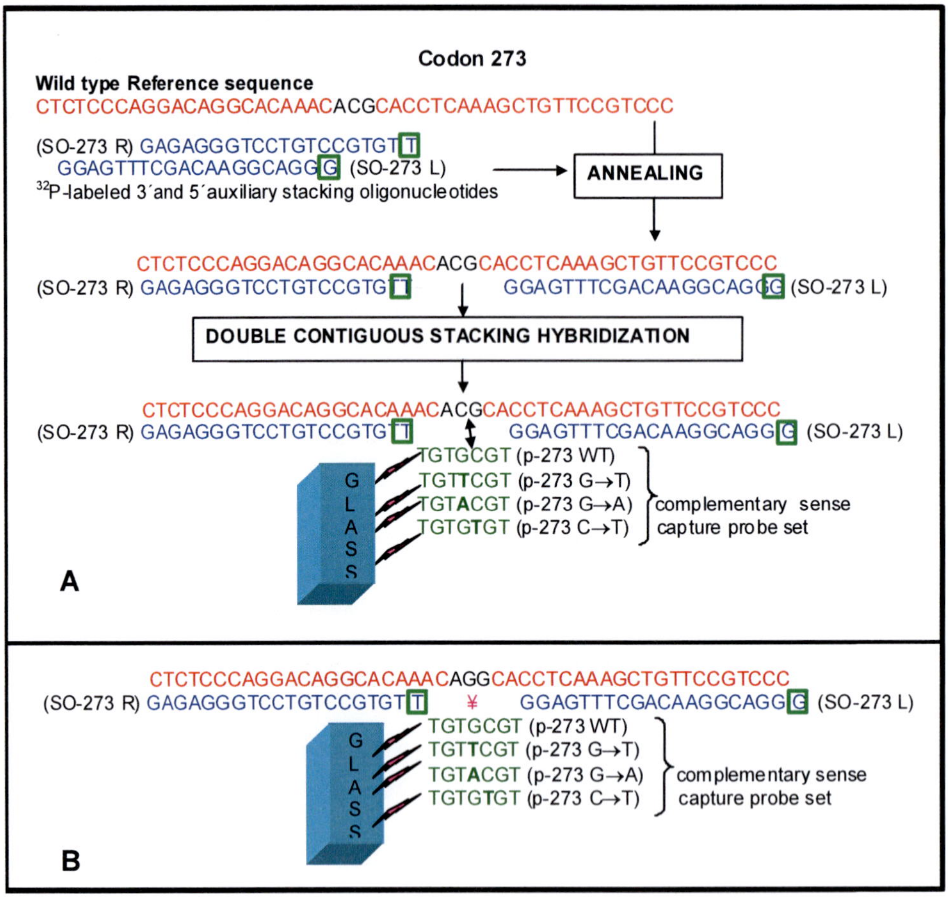

Figure 1. Tandem hybridization strategy on design of *TP53* Low density DNA microarray. (A) An example of the alignment of stacking oligonucleotides and probes to their respective synthetic wild type target sequence in codon 273 of the *TP53* gene. As shown, a nucleic acid target is first annealed with a molar excess of labeled auxiliary oligonucleotide (the green square marks the 32P-labeled end of the oligonucleotide). The labeled auxiliary oligonucleotide serves as a reagent to introduce the detection tag into a single specific site within the nucleic acid target. The partially duplex labeled target is then applied to the array of 7-mer capture probes, end-tethered to a glas slide or chip. Each capture probe is designed to hybridize in the gap formed on the target strand, in tandem with both labeled stacking probes preannealed to the target. Base stacking interactions between the short capture probe and the long stacking probe stabilize the binding of partially duplex labeled target to the glass-tethered capture probe. Hybridization is carried out at an (elevated) temperature at which the short capture probe by itself does not form a stable duplex structure with the target; capture of label to the glass occurs only when the capture probe hybridizes in tandem with the labeled stacking probes. As depicted in (B), a single base mismatch between the short capture probe and the target sequence will disrupt the short duplex, preventing binding of label to the glass.

The TH approach and double tandem hybridization strategy (DTH), wherein 7-mer capture probes are hybridized with duplex polymerase chain reaction products that have been heat-denatured and preannealed with two longer stacking probes to form a 7-nucleotide gap at the site of capture-probe binding, were successfully tested in several experimental models. Between the experimental models tested are: the ΔF508 mutation and other sequence changes responsible for cystic fibrosis (a genetical disease) [38], the identification of mutations in the *RET* gene associated to medullary thyroid carcinoma [39], the identification of *TP53* mutations hotspots in lung cancer samples [40] and, the identification of the Asiatico-American (AA) and European (E) HPV 16 variants [41].

Some Relevant Guidelines for the Design of Oligonucleotide Probes

As it has been previously mentioned the detection of genetic variation with microarray approaches can in theory be capable of diagnoses thousands of genetic variations due to the parallel nature of this technology. However this technology still suffers of important limitations in detecting known single polymorphism. In particular there is a high rate of false positives when specificity in determined by the hybridization, is difficult to distinguish all the polymorphisms, is difficult to detect low level of polymorphism [42].

An important and key issue for the multiple possible applications of microarrays for the study of genetic variations is the design of probes. Typically the design of these probes is a process involving three stages:

1. Selecting the appropriated (desired) targets.
2. Calculating the sequence of the probe.
3. Validating the specificity of the probe

Therefore the process of selecting probes looks similar to that commonly used for genetic expression. However some important precautions are necessary for this type the design. We will explain some guidelines which can be applied for selection of probes for the analysis of genetic variations.

Selecting the Appropriated (Desired) Targets

There are two different models for exploring for genetic variations on the genes of interest:

1. Exploring the whole sequence of the gene of interest or
2. Designing a set of probes for detection of the most common or important genetic variations on a gene of interest.

In the first case the microarray must be integrated by a high number of probes in order that the whole gene can be scanned for searching potential mutations. This approach is then similar to the concept of Sequencing by Hybridization [42, 43]. However this approach has some important limitations:

If the array includes only the probes that allows to screen the "reference" sequence of the target, then this microarray would be able to detect only the presence of the mutation but not its identity.

If additional probes are included for detection of the all possible variations (in the case of single base variations), then the number of require probes could be excessive.

Alternatively the microarray can include the whole set of combinations of probes of a defined length as the microarrays for sequencing by hybridization or resequencing. However, this design can be problematic as the variation of stability is very broad and then it will be difficult for distinguishing ambiguous hybridizations.

Most part of the potential sequence variations will no be observed in any sequence, then most part of the data provided by the microarray will not be informative.

A more efficient microarray can be designed for the second case, as the number of required probes will be significantly reduced. In such case the sets of probes can be created for detection of all the potential variations and approaches as the tandem hybridization can be used in order to allow a more sensitive detection. However the main limitation of this technique is that some genetic variations will be skipped. Fortunately most part of the genetic variations observed in diseases show important "hot spots" or variations which are especially frequent on human populations. For example the *TP53* database of the International Agency for Research on Cancer (see Table 1) shows six especially abundant somatic mutations at codons 175, 245, 248, 249, 273 and 282. These mutations represent almost the 30% of the genetic variations observed for this gene [44]. Therefore, although some information is missed this type of microarray still can be important for the diagnosis of variations commonly observed or associated with some diseases.

Table 1. Some databases on genetic variations related with cancer

Database name	URL	Brief description
IARC TP53 Database	http://www-p53.iarc.fr/index.html	Most frequent somatic mutations in the tumor suppressor gene TP53 associated with most human cancers and germline TP53 mutations.
Tumor Associated Gene database	http://www.binfo.ncku.edu.tw/TAG/GeneDoc.php	Provides information as gene content and mutation, from well-characterized oncogenes and tumor suppressor genes to facilitate cancer research.
SNP500Cancer	http://snp500cancer.nci.nih.gov/	Re-sequenced SNPs from 102 reference samples
Cosmic	http://www.sanger.ac.uk/genetics/CGP/cosmic/	Catalogue of somatic mutations in cancer
HGMD	http://www.hgmd.cf.ac.uk/ac/index.php	Human Gene Mutation Database at the Institute of Medical Genetics
Human Genome Variation Society HGVS	http://www.hgvs.org/dblist/dblist.html	Formerly the HUGO Mutation Database Initiative. Includes Variation Databases and Related Sites
OMIM	http://www.ncbi.nlm.nih.gov/sites/entrez?db=omim	A catalog of human genes and genetic disorders for the World Wide Web by NCBI
dbSNP	http://www.ncbi.nlm.nih.gov/SNP/	Database of genome variation of the NCBI which provides a comprehensive catalogs of common genomic variations in humans and other organisms
OncoDB.HCC	http://www.oxfordjournals.org/nar/database/summary/1026	Oncogenomic database of hepatocellular carcinoma (HCC).
Tumor Gene Family Databases (TGDBs)	http://www.tumor-gene.org/tgdf.html	Contain a broad range of data about the genes involved in cancer (e.g. proto-oncogenes and tumor supressor genes).

For a convenient selection of the targets for designing of the probes probe there are several databases available up to date [45]. Table 1 resumes some important databases that can be used for such purpose. It is important to mention that some of this databases show genetic variations observed in some important genes, but sometimes there is a lack of statistical evidence that can definitively associate the variation with the disease as in the case of the SNPs database. For such cases the information should be carefully interpreted. Sometimes, additional annotation of the genes can provides information about the relatedness of the observed variations with the disease of interest.

Deriving the Sequences of the Probes

When probes are directed against a sequence variation, there are few choices about the sequence zone on the targets that can be used to derive the sequence of the probe. Most important is to define the length of the probe and the possible placement of the recognized area when the duplex is formed. However as the probes are usually required for detection of small sequence variations, the thermal stability of the duplex formed during the hybridization reaction becomes of critical importance. In this context it is important to mention that the stability of duplex formation between probes and targets, in the presence of small sequence variations, is strongly dependent on the sequential context [46].

Table 2. Unified set of DNA parameters for the nearest-neighbor model for predicting stability of DNA/DNA duplexes. [1]

NN	ΔH° (kcal/mol) [2]	ΔS° (cal/mol K) [2]	ΔG° (kcal/mol) [2]
AA/TT	-7.9	-22.2	-1.00
AT/TA	-7.2	-20.4	-0.58
TA/AT	-7.2	-21.3	-0.58
CA/GT	-8.5	-22.7	-1.45
GT/CA	-8.4	-22.4	-1.44
CT/GA	-7.8	-21.0	-1.28
GA/CT	-8.2	-22.2	-1.30
CG/GC	-10.6	-27.2	-2.17
GC/CG	-9.8	-24.4	-2.24
GG/CC	-8.0	-19.9	-1.84
Initiation with GC	0.1	-2.8	0.98
Initiation with AT	2.3	4.1	1.03
Symmetry correction	0.0	-1.4	0.40

Source data from SantaLucia, 1998.
ΔH^0 enthalpy, ΔS^0 entropy, ΔG^0 free energy

Typically thermodynamic nearest-neighbor models, based on experimental data, have been developed for predicting the thermal stability of short DNA duplexes [46-51]. According with these models, the main contribution of duplex stability are the stacking interactions between neighboring bases. For perfect matches there are 10 different types of such interactions (see Table 2). Moreover, the nearest-neighbor model is appropriated for calculating the stability of single mismatched duplexes and it can be expanded to include the contribution of other components of the secondary structure [46, 52, 53]. Some stability values are affected by nucleic acid concentrations and by the ionic strength. But the relative values between them are conserved on a wide range of ionic conditions [54]. Most part of the thermodynamic parameters used for this model have been derived from experiments in solution. It can be possible to find significant differences when the model is applied to solid-liquid system as that used in the microarrays, where one of the strands is fixed to a solid support. However, it has been observed that the relative stabilities are conserved also in these systems [55].

```
5'-TACGTAGTCGTACGTGATCGATCGTAGCTAGCTAGCTAGTCGATCATACA-3'
   ||||||||||||||||||||||||||||||||||||||||||||||||||
3'-ATGCATCAGCATGCACTAGCTAGCATCGATCGATCGATCAGCTAGTATGT-5'
```

ΔG° = -65.85 kcal/mol Tm = 77.5°C

```
5'-TACGTAGTCTTACGTGATCGATCGTAGCTAGCTAGCTAGTCGATCATACA-3
   ||||||||| ||||||||||||||||||||||||||||||||||||||||
3'-ATGCATCAGCATGCACTAGCTAGCATCGATCGATCGATCAGCTAGTATGT-5
```

ΔG° = -61.24 kcal/mol Tm = 76.6°C

Stability decreases about 7% (ΔTm = 0.9°C)

```
5'-TACGTAGTCGTACGTGATC-3'
   |||||||||||||||||||
3'-ATGCATCAGCATGCACTAG-5'
```

ΔG° = -24.40 kcal/mol Tm = 66.0°C

```
5'-TACGTAGTCTTACGTGATC-3'
   ||||||||| |||||||||
3'-ATGCATCAGCATGCACTAG-5'
```

ΔG° = -19.79 kcal/mol Tm = 63.3oC

Stability decreases about 19%

(ΔTm = 2.7°C)

Figure 2. Effect of the probe length on the discrimination power for single mismatches with microarray DNA probes. Thermal stability of the probes was calculated with the Nearest-Neighbor (NN) model which considers that the main source of duplex stability is the stacking between neighboring bases. It can be seen that the discrimination power (measures as the difference of thermal stabilities between the perfect and the mismatched target-probe duplex) increases as the probe length reduces. Thermodynamic values calculated with NN data from SantaLucia and Allawi et al. setting [Na+] = 50 mM and [oligonucleotide] = 100 μM. The Tm estimates are intended only to describe the relative stability of different oligonucleotides in a given hybridization reaction, since the absolute values will depend on variables of DNA and ionic concentrations.

Probe length can be adjusted in order to have a similar stability for the probes in the microarray. Additionally, the probe length is an important factor which determines the ability of the probes for the detection of small sequence variations. The use of the nearest-neihgbor model allows the prediction of the effect of the probe length on the discrimination power of single sequence variations. Figure 2 illustrates that the presence of a CT mismatch produces a higher variation of Tm in the hybridization with a short probe (about 3°C for a 19mer probe) than a longer probe (about 1°C for a 50mer probe). However the identity of the mismatch is very important as is showed in Table 3, where it can be seen than AC and CT mismatches are highly destabilizing, whereas GA, GT and GG mismatches are highly stable in some sequential contexts. Therefore short probes are better suitable for the detection of single variations as SNPs. However the special stability of some single mismatches represents an important limitation for the application of the "classic" microarray technology for detecting single variation where single probes are used for detection of individual mutations. In such cases even the use of short probes will produce intense signals (ambiguous hybridization) in the presence of highly stable mismatches. Moreover, if the hybridization conditions are not rigorously controlled for detection of these mutations, ambiguous hybridization is produced. This issue becomes critical when a high number of probes are hybridized simultaneously where is not possible to control rigorously the hybridization conditions for all them. A possible solution for such problem could the use of probes of different length which have similar stability.

Table 3. Tm Differences for single internal mismatches

Mismatch = G-T				
Mismatch	3' match	^{0}C	5' match	^{0}C
AGA/TTT	AGA/TCT	9.4	AAA/TTT	7.3
AGC/GTT	AGC/GCT	9.2	AAC/GTT	5.7
AGG/CTT	AGG/CCT	8.7	AAG/CTT	5.6
AGT/ATT	AGT/ACT	9.1	AAT/ATT	6.3
CGA/TTG	CGA/TCG	7.9	CAA/TTG	5.1
CGC/GTG	CGC/GCG	7.9	CAC/GTG	3.8
CGG/CTG	CGG/CCG	7.5	CAG/CTG	3.8
CGT/ATG	CGT/ACG	7.7	CAT/ATG	4.2
GGA/TTC	GGA/TCC	7.5	GAA/TTC	4.4
GGC/GTC	GGC/GCC	7.2	GAC/GTC	2.8
GGG/CTC	GGG/CCC	6.7	GAG/CTC	2.6
GGT/ATC	GGT/ACC	7.3	GAT/ATC	3.5
TGA/TTA	TGA/TCA	8.8	TAA/TTA	5.0
TGC/GTA	TGC/GCA	8.7	TAC/GTA	3.5
TGG/CTA	TGG/CCA	8.2	TAG/CTA	3.4
TGT/ATA	TGT/ACA	8.6	TAT/ATA	4.0
G:T Median discriminatory values		8.15		4.40
Mismatch = C-T				
Mismatch	3' match	^{0}C	5' match	^{0}C
ACA/TTT	ACA/TGT	11.1	AAA/TTT	8.2
ACC/GTT	ACC/GGT	12.2	AAC/GTT	9.2
ACG/CTT	ACG/CGT	11.6	AAG/CTT	7.9
ACT/ATT	ACT/AGT	11.2	AAT/ATT	8.4
CCA/TTG	CCA/TGG	12.4	CAA/TTG	9.4
CCC/GTG	CCC/GGG	13.4	CAC/GTG	10.2
CCG/CTG	CCG/CGG	13.0	CAG/CTG	9.3
CCT/ATG	CCT/AGG	12.6	CAT/ATG	9.6
GCA/TTC	GCA/TGC	12.7	GAA/TTC	8.4
GCC/GTC	GCC/GGC	13.6	GAC/GTC	9.1
GCG/CTC	GCG/CGC	13.0	GAG/CTC	8.1
GCT/ATC	GCT/AGC	12.9	GAT/ATC	8.6
TCA/TTA	TCA/TGA	11.9	TAA/TTA	8.1
TCC/GTA	TCC/GGA	13.1	TAC/GTA	9.1
TCG/CTA	TCG/CGA	12.3	TAG/CTA	7.8
TCT/ATA	TCT/AGA	12.0	TAT/ATA	8.3
C:T Median discriminatory values		12.43		8.73

Typically the presence of mutations with classic microarrays is carried up by comparing the signal intensity of the mismatched probe against the perfectly complementary duplex. Moreover several probes are designed to screen the entire sequential context around the mismatch. If the loss of the signal intensity is the observed within several probes this will constitute a confirmation of the mutation presence. However, even with this approach, the detection of mutations that produce especially stable mismatches would be difficult when the Tm variation is less of 0.5°C. Even with the use of several probes for screening the same mutation will produce similar sequential contexts, and then the detection of such special mutations remains difficult. Some developing technologies are promising for solving several of these problems. The method termed DASH (dynamic allele-specific hybridization), which entails dynamic tracking of probe (oligonucleotide) to target (PCR product) hybridization as reaction temperature is steadily increased allows for an more accurate genotype assignment [56]. Current failures on this methodology are assumed to be caused by secondary structures in the target molecule, which could be reliably predicted from thermodynamic theory. Electric chips use probes coupled to electric contacts sending a current to denature target-probe hybrids, allow a more strict control of the hybridization-denaturing conditions [57].

Tandem mismatches are other kind of sequence variations that can be present in some special cases. From the thermodynamic point of view, internal tandem mismatches can not be considered as the combination of two consecutive single mismatches [58]. Tandem mismatches follow a special behavior, some tandem mismatches are exceptionally stable with a stability even similar to the AT base pairs [59]. Therefore tandem mismatches are especially problematic for their detection with this technology. The deletion of one or few consecutive bases will produce bulges in the bimolecular secondary structure of the target-probe duplexes. Some one base bulges are also relatively stable [60]. Moreover, when short probes are used, one-base bulges can be found easier by chance, allowing the formation of non-desired cross reactions, but it has been reported an efficient detection of several one-deletions with microarrays [61].

The placement of the mismatch in the duplex is also important. In general the higher discrimination power for mismatches is reached when they are located at the middle of the probe. However mismatches located at the ends or close to them (especially at the penultimate positions) are highly stable in general [33, 62].

All these facts indicate that the classical microarray technology is not well suitable for the detection of all single base variations with this technology. In particular there is a high rate of false positives for distinguishing all the polymorphisms and it is difficult to detect low level of polymorphism [42]. However an interesting alternative is the use of tandem or coaxial hybridization (TH) as it has been previously described. In general the use of coaxial stacking highly increases the ability for discriminating single mismatches .

The formation of a tandem probe-auxiliary oligonucleotide duplex with the target is accompanied by the appearance of the stacking interaction between the terminal base pairs of the contiguous duplexes and then it contributes to the stabilization of the complex. Some studies have showed that such stabilization reach the maximum when there are not gaps between the duplexes and indicate that such stabilization can produce an increase of 1 to 4°C of the Tm of the probe [63-66]. This discrimination power reaches a maximum when both flanking sequences to binding site of the probe are hybridized with auxiliary (stacking) oligonucleotides [40]. Moreover the presence of additional auxiliary oligonucleotides, which must be pre-aligned with the samples, helps to eliminate the any interfering secondary structure of the targets. They also allow the correct location of the site where the probe must be hybridized.

When the probe produces mismatches with the target, the duplex stability is highly disturbed. Although this stabilization is dependent on the sequence context as in the case of the simple hybridization, the effect is higher with the tandem hybridization approach even for highly stable mismatches. Tandem hybridization has been successfully tested in the laboratory and it has been demonstrated that it can reach highly sensitivity levels for single mutation detection. However the hybridization with this approach requires the presence of additional auxiliary oligonucleotides, which must be pre-aligned with the targets. This pre-annealing must be carried up in a multiplex way and then special precautions must be taken for the design of the auxiliary oligonucleotides to avoid the formation of stable dimers between them.

Validating the Specificity of the Probe

There are several software applications both commercial and free that can be used for microarray probe design [67-71]. Many of them use thermodynamic models to predict the stability of the duplexes formed by the probes when they hybridize against their targets as well as tools for ensuring the specificity of the probes. However most part of these software applications are aimed to the design microarrays for differential expression analysis. In general these applications could not be appropriated for designing of probes for monitoring single genetic variations for the reasons exposed next.

Small sequence variations are not of special interest when monitoring the gene-expression. As long probes are used for such experiments the formation of ambiguous mismatched duplexes is quite possible within highly similar sequences. No validation of such variations is considered for designing probes for gene expression. However this aspect is critical when designing probes for evaluation of genetic variations. The validation of the specificity of the probes is commonly carried up with software tools that evaluate the similarity of the sequence recognized by the probe. If multiple sequences similar to the targeted site are detected then they can produce cross reactions.

However software tools like BLAST or FASTA are not optimized for searching of similarities with short nucleotide sequences. Moreover even when similar sequences the identity of the mismatches is important the correct identification of the binding sites.

Our team has proposed than alternative way for testing microarrays aimed for testing genetic variation is to check their performance *in silico*. The hybridization between the probes in the microarrays and the potential targets can be virtually simulated and this simulation can calculates the hybridization patterns where the differences between spot intensities in the presence of single mismatches can be visualized and therefore it can be possible to anticipate if the microarray could produce the desired results. This approach is called Virtual Hybridization and it has been successfully tested to analyze hybridization results and for improving the microarray design [72].

Search of specificity of the probes in the microarray yields more satisfactory results as in this case all potential sites where ambiguous hybridization can occur are searched, and then thermodynamic models are used to calculate the stability of the potential duplex. Therefore this represents an improvement compared with the tools commonly used for assessing specificity based in the sequence similarity/complementary only.

Moreover the concept of *in silico* hybridization can be used to avoid undesired hybridization events (such as multiple target-probe interactions and stable mismatched hybrids). The strategy can facilitate (through the use of different length probes) selection of a probe set having a narrow duplex stability allowing maximal specificity under a single hybridization condition.

Applications Developed for the Study of Organisms on DNA Arrays

Microarrays have been successfully used for several important purposes. Gene expression, *SNP* analysis, detection of genetic variations, organism identification, sequencing, resequencing and barcode identification. The size of the probes can be critical for some of these applications. In the case of gene expression, organism identification and for detection of some barcode sequences long sizes probes are preferred, since this capture probes are able to detect the gene or the expected sequence even when small sequence variations are present in the sample. However, when single base changes are searched, as happens in the *SNP* analysis or resequencing, the use of relatively smaller probes is obligated to get better discrimination of single base variations. Numerous improvements in the performance of microarrays have been developed through studies for gene expression. The array quality has been estimated by measuring the amount of fluorescence (Cy5-UTP) incorporated by terminal deoxynucleotidyl transferase (TdT) at the probes 3′-end. In another procedure the 3′-end ligation is used. The TdT approach detects the probes 3′-end while the ligation is also detecting the performance of the hybridization. Therefore, both array characteristics can be assessed by the combination of the two assays [73].

The preparation of the target is another important issue. Zhang et al. [74] have successfully tested genomic DNA isolation, for use in microarray hybridization, by a combination of sonication with a heat treatment. This procedure is fast, cheap and easily adaptable for high throughput comparative genomics. In another study, Stangegaard et al. [75] demonstrated that the yield of cDNA is doubled by using 15-mer random primers instead of 6-mer random hexamers. 11-fold more genes were detected in the whole transcriptome DNA microarray tested with this preparative protocol. The homogeneity of the hybridization signal can be critical for quantitative analysis. In some cases the formation of doughnut shaped spots is observed. Doughnut appears as a consequence of evaporation of droplets. However doughnuts can also be formed during the hybridization process by diffusion-limited conditions (the situation in which the actual hybridization rate is slower than that expected on the basis of the intrinsic reaction kinetics). This situation occurs with short hybridization targets which reach the spot from the sides and not from above the spot. It is expected that appropriated microfluidics, such as that obtained in the MAUI hybridization chamber, will solve this problem. Pappaert et al. [76]. Since a 3-fold augment in positive signals was obtained in the MAUI hybridization system while negative controls show lower hybridization. Schaupp et al. [77]. Feature characteristics, such as probe sequence, can cloud the relationship between observed intensity and actual expression. Although this "probe effect" is large, it is also very consistent across hybridizations, which implies that relative measures of expression are substantially more useful than absolute ones. By this reason generally differential expression it is used instead than expressed or unexpressed. However this work uses the consistency in the intensity of each probe target pair as a way to determine if a given gene is or not expressed [78].

One of the common ways to reveal differential gene expression is the use of scatterplot (two-dimensional, 2-D, plot), which is created with the intensity of the hybridization signals. This plots obscure useful information when a large gene count caused saturation across areas of the graph. Park et al. [79] found that topographical plots are useful for quick qualitative analysis that reveals significant information, especially with respect to regions of low differential gene expression not visible in standard analysis.

Two key studies aimed to identify the sources of error and data variability in gene expression analysis were performed by the Toxicogenomics Research Group. In one of them they Izarry et al. [80] demonstrated that standardization of protocols (Sample preparation, microarray processing, hybridization conditions, data acquisition and treatment) good reproducibility was obtained. This work suggests that microarray results, for gene expression, can be comparable across multiple laboratories, especially when a common platform and a set of procedures are used. In this study three microarray platforms (Affymetrix GeneChip, Spotted cDNA arrays and Long oligonucleotide arrays) were compared in ten laboratories using identical RNA samples. In general was found that the lab had a larger effect, on for example, precision than did the platform, and that the results from the best performing lab agreed rather well. In another, well controlled study [81], two platforms were compared (The Affymetrix Mouse Genome 430 2.0 GeneChip and a spotted cDNA array). The results, validated by qRT-PCR, suggest that biological treatment had a far greater impact on measured expression than did platform for more than 90% of genes. In the cases of inconsistency between platforms only one of eleven genes tested by qRT-PCR supported one platform over the other. One of the factors could be the difference in probe design due to splice variants. The qRT-PCR primer was based on the TIGR tentative consensus sequence, which was derived from the assembly of multiple expressed sequence tags (ESTs). Due to imperfections in genome annotation the Affymetrix probe set may target one or more variants, the cDNA probes on the spotted array can test one or more different variants, and the qRT-PCR probe yet others. Another possibility is that some gene family member can cross-hybridize with cDNA or Affymetrix probe sets, testing different combinations of these family members while the qRT-PCR probably tests only a single member. By comparing the Affymetrix probe sequences against the corresponding TIGR EST sequences. For genes that had similar expression patterns across platforms full identity or almost identical sequences were found. However, some of the eleven tentative consensus sequences that disagreed across all three platforms had Affymetrix probes that mapped to the corresponding TIGR EST, supporting the hypothesis that the two platforms were interrogating different sequences for the genes that disagreed across platforms. The alignment of multiple EST sequences to the genome in the region containing the array probes generally suggested multiple splice variants in regions where there was generally a single annotated gene structure in the EnsEMBL database. Therefore the improvement in the annotation of the multiple splice variants will improve the design of oligonucleotide probes for gene expression.

Another important issue in gene expression is the normalization of data. One of the factors is the use of appropriated reference genes. These genes should show minimal expression variability among replicates, tissues and individuals. However, it is impossible to predict how different experimental conditions may affect the expression of putative normalization genes. Therefore it should be imperative to select the best reference genes and use the same normalization strategy in order to reduce the experimental variables [82]. A variation in expression of some housekeeping genes was observed during cancer. Therefore especial quantitation of the expression of these genes should be done previously to use them for normalization [83]. The quantitative expression of genes commonly is validated by real-time PCR and this relies on housekeeping genes. However, no single housekeeping gene can be used for all studies. Abruzzo et al. [84] have developed a statistical approach that automatically identifies those genes whose expression remains unchanged in the samples. Another statistical method is able to show unusual expression patterns or the variability between slides [79]. Both procedures can be used for normalization.

Between the recent advances in differential gene expression arrays there is the hMitChip3, which contains probes to search nuclear and mitochondrial genes [85]. Recently a human exon array has been also developed (AffymetrixGeneChipR Human exon 1.0 ST array). This array has been made to test 1 million known exons. It contains 5.5 million probes. This chip should be able to perform differential exon expression in various tissues under normal or other conditions, revealing more precisely the targets for prevention or the treatment of genetic diseases [86].

Barcode

Another increasingly relevant application of arrays is the detection of barcode sequences. A molecular barcode composed of 384 array features, Affymetrix GeneChip, was tested using defined template quantities of mixed PCR templates. The individual concentration response of each feature to each sample was highly reproducible [87]. Barcode has also been applied to predict tissue type [78]. Besides, plant species can be identified by using a two-locus (The rbcL gene and the non-coding region tmH-psbA spacer region) global DNA barcode [88]. In yet another study a universal barcode array, made with unique 20-base-pair DNA tags, was used in parallel for the analysis of thousands of biological samples [89]. Recently nucleotide barcode was combined with pyrosequencing to reach successful large-scale sample multiplexing analysis [90].

Organism Identification

A set of 60-mer probes to search approximately 100 CNS pathogens: The probes were designed on open reading frames from highly conserved and heterogenic regions within viral families. Additionally viral genes reflecting different stages of pathogen infection were also included to potentially define the stage of the viral infection. This array is aimed to found evidence of viral infection in postmortem tissue from psiquiatric patients [18]. In another study an array of approximately 39,000 probes was done to identify influenza virus subtypes. The array was designed based in the comparison of approximately 5000 viral sequences. Each probe was designed to be exclusive for a given subtype [91].

METAGENOMICS

Metagenomics can also be done with arrays of probes designed to detect specific sequences. Sequence reads can be then assembled to reach gene prediction, functional annotation and metagenomics [92].

RESEQUENCING

Resequencing is an oligonucleotide array based technology aimed to search for sequence changes in a given gene or sequence of interest. The DNA region in study is tested by a set of 25-mer probes covering, base by base, the complete wild type sequence. The chip also contains additional probes looking for the presence of mutations in the sample. Commonly the mutations tested are single base substitutions and single base eliminations or insertions. Other types of mutations, such as double base insertions, can also be included in the probe array. The sequence changes are placed in the middle of the probes and the target is transcribed to RNA with the purpose to increase the sensibility in the detection of point mutations. One of the main advantages of resequencing is that genetic variations, such as point mutations, are more confidently identified by a combination of the loss of signal in the group of probes corresponding to the wild type sequence with a gain of signal in the complementary mutant capture probe. Besides, the analysis is performed in both target DNA strands which increases the discrimination of single base changes, especially in the cases where in one target strand the mutant/wild type, target/probe, forms a relatively stable single mismatched duplex when compared with the perfectly paired wild type target/probe duplex. In these cases the mismatch produced by the base present in the other target strand at the site of the mutation with the probe to capture the wild type sequence produces a big loss in stability in the target/probe duplex formed compared with its respective, wild type target/probe duplex [93].

Successful resequencing of point mutations has been demonstrated since 1996 [94]. Only one, high quality, reference sequence is required to search for variations to that sequence. Resequencing was applied first to search for sequence changes in genes frequently associated to cancer such as *TP53* and *BRCA-1*. It also has been used to search variations in microorganisms between them to detect group A streptococci and their associated antibiotic resistance markers [95], severe acute respiratory syndrome (SARS) virus [96, 97] and smallpox virus genomes [98].

Resequencing can be also be efficiently done by arrayed primer extension (APEX). In this case the 3′-end of each primer used as a capture probe is extended after hybridization to interrogate the template base in the target [99].

Recently, a novel resequencing application has been tested by Affymetrix. 14 125 688 probes were used for whole genome analysis (excluding repetitive sequences but including two plasmids) in *Francisella tularensis*. The main purpose was to perform genotyping and polymorphism discovery. Some of the current limitations in the interpretation of results are the presence of deletions in the sample relative to the reference sequence, since in this case a mixture of no-calls and false positive SNP calls is observed, besides the large population of target DNA fragments in a whole-genome sample may contain sequences capable of high-efficiency hybridization with more than one of the probe pairs, resulting in a false-positive

SNP call (the alternate homology effect) and the local destabilizing effect of genuine SNPs in the sample leads to false-positive calls at adjacent genome locations, which is known as the footprint effect. It is expected that in the near future some improvements, such as the use of smaller-sized probes and hybridization under more astringent conditions will improve the resequencing technology [100].

High throughput resequencing can be done by universal PCR amplification of DNA fragments, with an average of 300 bp, with adaptors ligated to TdT created 3′-adenine overhangs and captured and eluted from a custom high-density microarray consisting of complementary sequences identified from a reference genome sequence SNPs [101].

However in some cases the results were not as expected. This happened when testing SNPs for genotype-calling of individuals and in allele frequency determination of DNA pools from patients with lung cancer. Affymetrix GenChipR Mapping 100K Array-set, which includes 116,204 SNAps with a median and mean intermarker distance of 8.5 and 23.6 kb, respectively. In this study not reliable hybridization data was obtained in a quite large number of SNPs [102].

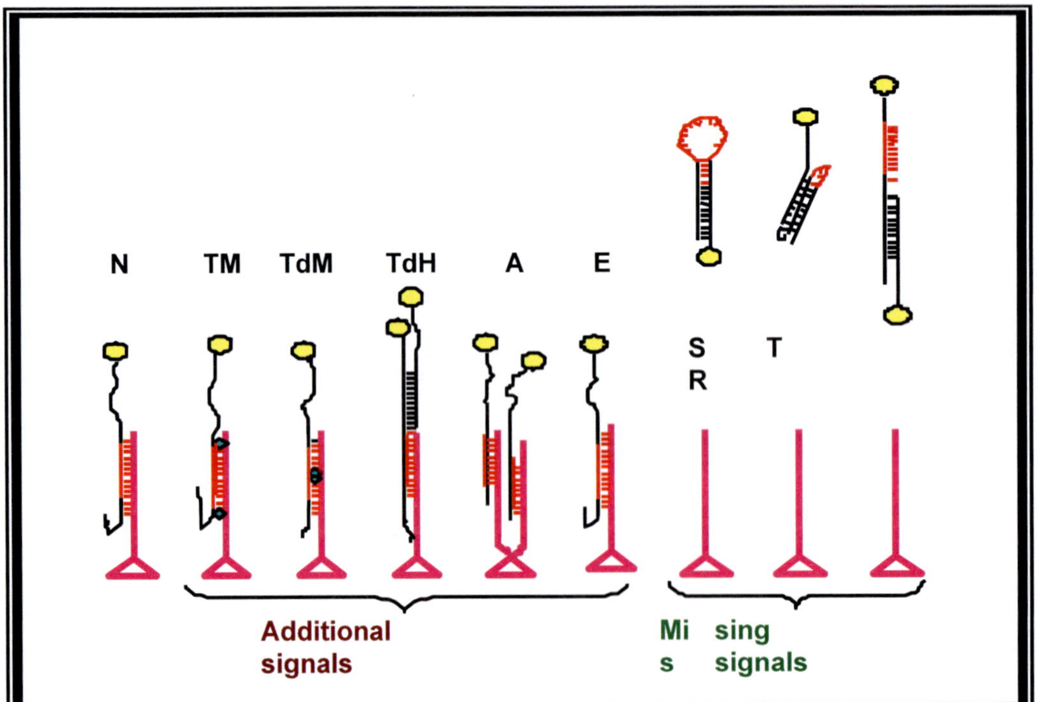

Figure 3. Factors affecting the pattern of hybridization in complex target samples. N, normal hybridization. *Additional hybridization signals* due to: TM, terminal mismatches; TdM, tandem mismatches; TdH, tandem hybridization; A, additive hybridization; E, extrachromosomal DNA. *Missing hybridization signals* due to: S, Secondary structure; T, Tertiary structure and R, repeats. Yellow circles, fluorescent label; blue circles, mismatches; red portions, target sequences; violet lines, probes.

All these works shows the enormous and highly informative applications developed for the study of organisms on DNA arrays. The methodologies in some cases, such as differential expression, are able to detect either relatively long sequences even when some minor and unknown sequence variations can be present. In other cases, such as organism identification or barcoding, predefined sequences are tested with capture probes. In yet another case, such as SNPs or resequencing, point sequence variations are identified. However, there is not an array-based method able to precisely identify the sequence of unknown complex samples. There are many challenges to reach this goal. Between them, the quality of the array, the preparation of the sample for hybridization, the conditions of hybridization and the capture and interpretation of the results. The array of immobilized probes is covered with a solution containing hundreds of thousands of single stranded DNA fragments. Many of the fragments have repeated, inverse repeats or in tandem repetitive sequences, therefore many duplexes can be formed in the solution, besides stable secondary structures can be acquired in the hybridization conditions. Non repeated sequences located in the vicinity of repeated sequences can then become unavailable for hybridization. All these interactions have more chances to occur before interacting with the tethered probes. The remaining exposed target sequences can then produce normal and many unusual interactions with the probes. The following figure (Figure 3) illustrates some of the possible interactions. The effect on the final pattern of hybridization due to these interactions is discussed.

CONCLUSION

Microarrays are a powerful genomic tool that can be applied for biomedical and clinical research. The strength of the technology rests on the several factors including: ease of use, availability of platforms and lower cost relative to other exploratory methods. Biological interpretation however requires the integration of several sources of information. In this context, about guidelines for designing, there some useful databases describing genetic variations which can be accessed in order to obtain sequences for deriving the probes for the microarrays.The data should be carefully examinated because in several cases there is not enough statistical evidence on the degree of relatedness of some variations with the disease. As many microarrays will be directed for the detection of small genetic variations as single nucleotide polymorphisms special attention should be taken about the high stability of some mismatch sequence contexts.

Thermodynamic data based in the nearest-neighbor model are useful for predicting the stability of heteroduplexes as they include data for calculating the stability of duplexes in the presence of single mismatches and some other important components of the secondary structure which can arise when ambiguous hybridization is observed. The length of the probes is critical as short probes show a higher sensitivity for detection of specially stable mismatches. However, short probes are prone to problems of interfering secondary structure and specificity for recognizing the target site. An interesting alternative is the use of coaxial or tandem hybridization (TH), where the use of single or double auxiliary oligonucleotides increases the sensitivity for detecting stable mismatches, reduces interfering secondary structure of the targets and improves the specificity of the probes.

In silico validation of the performance of designed microarray throught virtual hybridization is a useful technique that can be used to avoid undesired hybridization events (such as multiple target-probe interactions and stable mismatched hybrids). Also, this strategy can facilitate (through use of different length probes) selection of a probe set that has a narrow duplex stability allowing maximal specificity under a single hybridization condition.

REFERENCES

[1] Li, X., et al., DNA microarrays: their use and misuse. *Microcirculation*, 2002. 9(1): p. 13-22.

[2] Butte, A., The use and analysis of microarray data. *Nat Rev Drug Discov*, 2002. 1(12): p. 951-60.

[3] Stekel, D., *Microarray bioinformatics*. 2003, USA: Cambridge University Press.

[4] Southern, E.M., Detection of specific sequences among DNA fragments separated by gel electrophoresis. *J Mol Biol*, 1975. 98(3): p. 503-17.

[5] Blaese, R.M., et al., T lymphocyte-directed gene therapy for ADA- SCID: initial trial results after 4 years. *Science*, 1995. 270(5235): p. 475-80.

[6] Nuwaysir, E.F., et al., Gene expression analysis using oligonucleotide arrays produced by maskless photolithography. *Genome Res,* 2002. 12(11): p. 1749-55.

[7] Altshuler, D. and M. Daly, Guilt beyond a reasonable doubt. *Nat Genet,* 2007. 39(7): p. 813-5.

[8] Wang, L., R. Luhm, and M. Lei, SNP and mutation analysis. *Adv Exp Med Biol*, 2007. 593: p. 105-16.

[9] Shai, R.M., Microarray tools for deciphering complex diseases. *Front Biosci*, 2006. 11: p. 1414-24.

[10] Bulyk, M.L., DNA microarray technologies for measuring protein-DNA interactions. *Curr Opin Biotechnol*, 2006. 17(4): p. 422-30.

[11] Bulyk, M.L., et al., Quantifying DNA-protein interactions by double-stranded DNA arrays. *Nat Biotechnol*, 1999. 17(6): p. 573-7.

[12] Mintseris, J. and M.B. Eisen, Design of a combinatorial DNA microarray for protein-DNA interaction studies. *BMC Bioinformatics*, 2006. 7: p. 429.

[13] Warren, C.L., et al., Defining the sequence-recognition profile of DNA-binding molecules. *Proc Natl Acad Sci U S A*, 2006. 103(4): p. 867-72.

[14] Cutler, D.J., et al., High-throughput variation detection and genotyping using microarrays. *Genome Res*, 2001. 11(11): p. 1913-25.

[15] Yan, P.S., et al., Dissecting complex epigenetic alterations in breast cancer using CpG island microarrays. *Cancer Res*, 2001. 61(23): p. 8375-80.

[16] Pollack, J.R., et al., Genome-wide analysis of DNA copy-number changes using cDNA microarrays. *Nat Genet*, 1999. 23(1): p. 41-6.

[17] Relogio, A., et al., Alternative splicing microarrays reveal functional expression of neuron-specific regulators in Hodgkin lymphoma cells. *J Biol Chem*, 2005. 280(6): p. 4779-84.

[18] Conejero-Goldberg, C., et al., Infectious pathogen detection arrays: viral detection in cell lines and postmortem brain tissue. *Biotechniques*, 2005. 39(5): p. 741-51.

[19] Wang, D., et al., Microarray-based detection and genotyping of viral pathogens. *Proc Natl Acad Sci U S A,* 2002. 99(24): p. 15687-92.

[20] Oh, S.J., et al., Surface modification for DNA and protein microarrays. *Omics*, 2006. 10(3): p. 327-43.

[21] Affymetrix. *Affymetrix*. [cited 2007]; Available from: http://www.affymetrix.com.

[22] Jaluria, P., et al., *A perspective on microarrays: current applications, pitfalls, and potential uses.* Microb Cell Fact, 2007. 6: p. 4.

[23] Trevino, V., F. Falciani, and H.A. Barrera-Saldana, DNA microarrays: a powerful genomic tool for biomedical and clinical research. *Mol Med,* 2007. 13(9-10): p. 527-41.

[24] JCVI. *The J. Craig Venter Institute*. [cited 2007]; Available from: http://www.tigr.org/.

[25] MGED. *Microarray Gene Expression Data Society*. [cited 2007]; Available from: http://www.mged.org/.

[26] NCBI. *National Center for Biotechnology Information*. [cited 2007]; Available from: http://www.ncbi.nlm.nih.gov/.

[27] Milner, N., K.U. Mir, and E.M. Southern, Selecting effective antisense reagents on combinatorial oligonucleotide arrays. *Nat Biotechnol*, 1997. 15(6): p. 537-41.

[28] Southern, E., K. Mir, and M. Shchepinov, Molecular interactions on microarrays. *Nat Genet*, 1999. 21(1 Suppl): p. 5-9.

[29] Williams, J.C., et al., Studies of oligonucleotide interactions by hybridisation to arrays: the influence of dangling ends on duplex yield. *Nucleic Acids Res,* 1994. 22(8): p. 1365-7.

[30] Lockhart, D.J., et al., Expression monitoring by hybridization to high-density oligonucleotide arrays. *Nat Biotechnol*, 1996. 14(13): p. 1675-80.

[31] Cronin, M.T., et al., Cystic fibrosis mutation detection by hybridization to light-generated DNA probe arrays. *Hum Mutat*, 1996. 7(3): p. 244-55.

[32] Weiler, J., et al., Hybridisation based DNA screening on peptide nucleic acid (PNA) oligomer arrays. *Nucleic Acids* Res, 1997. 25(14): p. 2792-9.

[33] Doktycz, M. and K.L. Beattie, *Genosensors and model hybridization studies*, in *en Automation technologies for genome characterization*, T.J. Beugelsdijk, Editor. 1997, Wiley: 205-225.

[34] Khrapko, K.R., et al., A method for DNA sequencing by hybridization with oligonucleotide matrix. *DNA Seq,* 1991. 1(6): p. 375-88.

[35] Parinov, S., et al., DNA sequencing by hybridization to microchip octa-and decanucleotides extended by stacked pentanucleotides. *Nucleic Acids Res*, 1996. 24(15): p. 2998-3004.

[36] Yershov, G., et al., DNA analysis and diagnostics on oligonucleotide microchips. *Proc Natl Acad Sci U S A*, 1996. 93(10): p. 4913-8.

[37] Maldonado-Rodriguez, R. and K.L. Beattie; *Nucleic Acid Analysis using sequence-targeted tandem hybridization*; 2001; US Patent #6,268, 147 B1

[38] Maldonado-Rodríguez, R., et al., Hybridization of glass-tethered oligonucleotide probes to target strands preannealed with labeled auxiliary oligonucleotides. *Mol Biotechnol*, 1999. 11(1): p. 1-12.

[39] Maldonado-Rodriguez, R., et al., Detection of mutations in RET proto-oncogene codon 634 through double tandem hybridization. *Mol Biotechnol*, 2003. 25(2): p. 113-29.

[40] Rangel-Lopez, A., et al., Low density DNA microarray for detection of most frequent TP53 missense point mutations. *BMC Biotechnol*, 2005. 5: p. 8.

[41] Mendoza-Lorenzo, P., et al., An oligoarray for the detection of human papillomavirus type 16 variants. *Int J Gynecol Cancer*, 2007. 17(5): p. 1083-91.

[42] Kirk, B.W., et al., Single nucleotide polymorphism seeking long term association with complex disease. *Nucleic Acids Res,* 2002. 30(15): p. 3295-311.

[43] Drmanac, R., et al., DNA sequence determination by hybridization: a strategy for efficient large-scale sequencing. *Science*, 1993. 260(5114): p. 1649-52.

[44] IARC. *IARC TP53 Database*. 2007 [cited 2007]; Available from: http://www-p53.iarc.fr/index.html.

[45] Galperin, M.Y., The Molecular Biology Database Collection: 2008 update. *Nucleic Acids Res*, 2007.

[46] Cantor, C.R. and P.R. Schimmel, *Biophysical Chemistry, Part III: The behavior of biological macromolécules*. 1980, San Francisco USA: W. H. Freeman and Co. 1109-1265.

[47] Borer, P.N., et al., Stability of ribonucleic acid double-stranded helices. *J Mol Biol*, 1974. 86(4): p. 843-53.

[48] Breslauer, K.J., et al., Predicting DNA duplex stability from the base sequence. *Proc Natl Acad Sci U S A,* 1986. 83(11): p. 3746-50.

[49] Wartell, R.M. and A.S. Benight, Thermal-denaturation of DNA-molecules: A comparison of theory with experiment. *Physics Reports*, 1985. 126: p. 67-107.

[50] SantaLucia, J., Jr., A unified view of polymer, dumbbell, and oligonucleotide DNA nearest-neighbor thermodynamics. *Proc Natl Acad Sci U S A,* 1998. 95(4): p. 1460-5.

[51] Delcourt, S.G. and R.D. Blake, Stacking energies in DNA. *J Biol Chem*, 1991. 266(23): p. 15160-9.

[52] SantaLucia, J., Jr. and D.H. Turner, Measuring the thermodynamics of RNA secondary structure formation. *Biopolymers*, 1997. 44(3): p. 309-19.

[53] Vallone, P.M., et al., Melting studies of short DNA hairpins: influence of loop sequence and adjoining base pair identity on hairpin thermodynamic stability. *Biopolymers*, 1999. 50(4): p. 425-42.

[54] Johnson, T., J. Zhu, and R.M. Wartell, Differences between DNA base pair stacking energies are conserved over a wide range of ionic conditions. *Biochemistry*, 1998. 37(35): p. 12343-50.

[55] Fotin, A.V., et al., Parallel thermodynamic analysis of duplexes on oligodeoxyribonucleotide microchips. *Nucleic Acids Res,* 1998. 26(6): p. 1515-21.

[56] Prince, J.A., et al., Robust and accurate single nucleotide polymorphism genotyping by dynamic allele-specific hybridization (DASH): design criteria and assay validation. *Genome Res*, 2001. 11(1): p. 152-62.

[57] Radtkey, R., et al., Rapid, high fidelity analysis of simple sequence repeats on an electronically active DNA microchip. *Nucleic Acids Res*, 2000. 28(7): p. E17.

[58] Allawi, H.T. and J. SantaLucia, Jr., Thermodynamics and NMR of internal G.T mismatches in DNA. *Biochemistry*, 1997. 36(34): p. 10581-94.

[59] Ke, S.H. and R.M. Wartell, The thermal stability of DNA fragments with tandem mismatches at a d(CXYG).d(CY'X'G) site. *Nucleic Acids Res*, 1996. 24(4): p. 707-12.

[60] Zhu, J. and R.M. Wartell, The effect of base sequence on the stability of RNA and DNA single base bulges. *Biochemistry*, 1999. 38(48): p. 15986-93.

[61] Chen, D., et al., Analysis of internal (n-1)mer deletion sequences in synthetic oligodeoxyribonucleotides by hybridization to an immobilized probe array. *Nucleic Acids Res*, 1999. 27(2): p. 389-95.

[62] Urakawa, H., et al., Single-base-pair discrimination of terminal mismatches by using oligonucleotide microarrays and neural network analyses. *Appl Environ Microbiol*, 2002. 68(1): p. 235-44.

[63] Maldonado-Rodríguez, R., et al., Mutation detection by stacking hybridization on genosensor arrays. *Mol Biotechnol*, 1999. 11(1): p. 13-25.

[64] Pyshnyi, D.V., E.L. Goldberg, and E.M. Ivanova, Efficiency of coaxial stacking depends on the DNA duplex structure. *J Biomol Struct Dyn*, 2003. 21(3): p. 459-68.

[65] Pyshnyi, D.V., et al., *Thermodynamic analysis of stacking hybridization of oligonucleotides with DNA template. J Biomol Struct Dyn,* 2001. 19(3): p. 555-70.

[66] Vasiliskov, V.A., D.V. Prokopenko, and A.D. Mirzabekov, Parallel multiplex thermodynamic analysis of coaxial base stacking in DNA duplexes by oligodeoxyribonucleotide microchips. *Nucleic Acids Res*, 2001. 29(11): p. 2303-13.

[67] DNASoftware. *VisualOMP*. [cited 2007]; Available from: http://www.dnasoftware.com/.

[68] Li, F. and G.D. Stormo, Selection of optimal DNA oligos for gene expression arrays. *Bioinformatics*, 2001. 17(11): p. 1067-76.

[69] Rouillard, J.M., C.J. Herbert, and M. Zuker, OligoArray: genome-scale oligonucleotide design for microarrays. *Bioinformatics*, 2002. 18(3): p. 486-7.

[70] Rouillard, J.M., M. Zuker, and E. Gulari, OligoArray 2.0: design of oligonucleotide probes for DNA microarrays using a thermodynamic approach. *Nucleic Acids Res,* 2003. 31(12): p. 3057-62.

[71] Reymond, N., et al., ROSO: optimizing oligonucleotide probes for microarrays. *Bioinformatics*, 2004. 20(2): p. 271-3.

[72] Reyes-López, M.A., et al., Fingerprinting of prokaryotic 16S rRNA genes using oligodeoxyribonucleotide microarrays and virtual hybridization. *Nucleic Acids Res,* 2003. 31(2): p. 779-89.

[73] Guerra, C.E., *Analysis of oligonucleotide microarrays by 3' end labeling using fluorescent nucleotides and terminal transferase.* Biotechniques, 2006. 41(1): p. 53-6.

[74] Zhang, L., et al., *Bacterial genomic* DNA isolation using sonication for microarray analysis. *Biotechniques*, 2005. 39(5): p. 640, 642, 644.

[75] Stangegaard, M., I.H. Dufva, and M. Dufva, Reverse transcription using random pentadecamer primers increases yield and quality of resulting *cDNA. Biotechniques*, 2006. 40(5): p. 649-57.

[76] Pappaert, K., et al., Diffusion limitation: a possible source for the occurrence of doughnut patterns on DNA microarrays. *Biotechniques*, 2006. 41(5): p. 609-16.

[77] Schaupp, C.J., et al., Active mixing during hybridization improves the accuracy and reproducibility of microarray results. *Biotechniques*, 2005. 38(1): p. 117-9.

[78] Zilliox, M.J. and R.A. Irizarry, A gene expression bar code for microarray data. Nat *Methods*, 2007. 4(11): p. 911-3.

[79] Park, T., et al., Diagnostic plots for detecting outlying slides in a cDNA microarray experiment. *Biotechniques*, 2005. 38(3): p. 463-71.

[80] Irizarry, R.A., et al., Multiple-laboratory comparison of microarray platforms. *Nat Methods*, 2005. 2(5): p. 345-50.

[81] Larkin, J.E., et al., Independence and reproducibility across microarray platforms. *Nat Methods*, 2005. 2(5): p. 337-44.

[82] Cai, J.H., et al., Validation of rat reference genes for improved quantitative gene expression analysis using low density arrays. *Biotechniques*, 2007. 42(4): p. 503-12.

[83] Khimani, A.H., et al., Housekeeping genes in cancer: normalization of array data. *Biotechniques*, 2005. 38(5): p. 739-45.

[84] Abruzzo, L.V., et al., Validation of oligonucleotide microarray data using microfluidic low-density arrays: a new statistical method to normalize real-time RT-PCR data. *Biotechniques*, 2005. 38(5): p. 785-92.

[85] Bai, X., et al., Third-generation human mitochondria-focused cDNA microarray and its bioinformatic tools for analysis of gene expression. *Biotechniques*, 2007. 42(3): p. 365-75.

[86] Okoniewski, M.J., et al., High correspondence between Affymetrix exon and standard expression arrays. *Biotechniques*, 2007. 42(2): p. 181-5.

[87] Walton, S.P., M.N. Mindrinos, and R.W. Davis, Analysis of hybridization on the molecular barcode GeneChip microarray. *Biochem Biophys Res Commun*, 2006. 348(2): p. 689-96.

[88] Kress, W.J. and D.L. Erickson, A two-locus global DNA barcode for land plants: the coding rbcL gene complements the non-coding trnH-psbA spacer region. *PLoS ONE,* 2007. 2(6): p. e508.

[89] Pierce, S.E., et al., A unique and universal molecular barcode array. *Nat Methods,* 2006. 3(8): p. 601-3.

[90] Parameswaran, P., et al., A pyrosequencing-tailored nucleotide barcode design unveils opportunities for large-scale sample multiplexing. *Nucleic Acids Res*, 2007. 35(19): p. e130.

[91] Lodes, M.J., et al., Use of semiconductor-based oligonucleotide microarrays for influenza a virus subtype identification and sequencing. *J Clin Microbiol*, 2006. 44(4): p. 1209-18.

[92] Mavromatis, K., et al., Use of simulated data sets to evaluate the fidelity of metagenomic processing methods. *Nat Methods*, 2007. 4(6): p. 495-500.

[93] Karaman, M.W., et al., Comparisons of substitution, insertion and deletion probes for resequencing and mutational analysis using oligonucleotide microarrays. *Nucleic Acids Res*, 2005. 33(3): p. e33.

[94] Chee, M., et al., Accessing genetic information with high-density DNA arrays. *Science*, 1996. 274(5287): p. 610-4.

[95] Davignon, L., et al., Use of resequencing oligonucleotide microarrays for identification of Streptococcus pyogenes and associated antibiotic resistance determinants. *J Clin Microbiol*, 2005. 43(11): p. 5690-5.

[96] Sulaiman, I.M., et al., Evaluation of affymetrix severe acute respiratory syndrome resequencing GeneChips in characterization of the genomes of two strains of coronavirus infecting humans. *Appl Environ Microbiol*, 2006. 72(1): p. 207-11.

[97] Wong, C.W., et al., Tracking the evolution of the SARS coronavirus using high-throughput, high-density resequencing arrays. *Genome Res*, 2004. 14(3): p. 398-405.

[98] Sulaiman, I.M., et al., GeneChip resequencing of the smallpox virus genome can identify novel strains: a biodefense application. *J Clin Microbiol*, 2007. 45(2): p. 358-63.

[99] Kringen, P., et al., Evaluation of arrayed primer extension for TP53 mutation detection in breast and ovarian carcinomas. *Biotechniques*, 2005. 39(5): p. 755-61.

[100] Pandya, G.A., et al., A bioinformatic filter for improved base-call accuracy and polymorphism detection using the Affymetrix GeneChip whole-genome resequencing platform. *Nucleic Acids Res*, 2007. 35(21): p. e148.

[101] Okou, D.T., et al., Microarray-based genomic selection for high-throughput resequencing. *Nat Methods,* 2007. 4(11): p. 907-9.

[102] Spinola, M., et al., Genome-wide single nucleotide polymorphism analysis of lung cancer risk detects the KLF6 gene. *Cancer Lett*, 2007. 251(2): p. 311-6.

In: Oligonucleotide Array Sequence Analysis
Editors: M.K. Moretti and L.J. Rizzo, pp. 323-345

ISBN: 978-1-60456-542-3

Chapter 11

IMPROVING DETECTION OF DIFFERENTIALLY EXPRESSED GENES IN MICROARRAYS USING MORE EFFICIENT STATISTICS, THE LOCAL OPTIMAL FUDGE FACTOR AND A TWO-STAGE FDR CORRECTION

Shunpu Zhang
Department of Statistics, University of Nebraska Lincoln
Lincoln, NE 68583-0963

ABSTRACT

Nonparametric methods for detecting differentially expressed (DE) genes in microarrays have attracted much attention recently. One of the most widely-used methods is the Significance Analysis of Microarrays (SAM) method. Empirical studies have shown that SAM can provide results highly consistent with biological findings. However, it has been recently reported that SAM tends to over-estimate the true false discovery rate (FDR). The cause of over-estimation was attributed to the over-dispersed null scores obtained by applying the SAM statistic to the permuted observations of DE genes. More recently, questions about SAM's choice of fudge factor have also been raised. It has been demonstrated in the literature that the fudge factor used in SAM does not generally have good power properties. In this paper, we propose an improved method for detecting DE genes in microarrays. Our proposed method is based on a pair of recently-constructed test statistic and null statistic. Unlike the SAM statistic, we show that the new null statistic does not have the permutation-caused over-dispersion problem. We further argue that, instead of using a global fudge factor, it is advantageous to use a local fudge factor which depends on the number of significant genes. The local fudge factor that minimizes the FDR is called the local optimal fudge factor. Contrary to the common understanding that the over-dispersed null scores are solely caused by the permutations, we demonstrate that the over-dispersion is also an intrinsic problem caused by the use of the fudge factor. In other words, the use of fudge factor will necessarily lead to over-dispersed null scores, and hence results in over-estimation of FDR. A two-stage FDR correction method is proposed to resolve this issue. Through extensive simulations, we show that our proposed method consistently outperforms SAM in the sense that it tends to provide lower FDR given that they have identified the same number of significant genes. The improvement of our proposed method is achieved by the use of more efficient test and null statistics, the local optimal fudge factor and a two-stage FDR correction. A real data example also shows that our proposed method provides more reasonable results and better control of FDR than SAM and a symmetric cutoff method based on the SAM statistic.

INTRODUCTION

One of the most common applications of microarrays is to identify the DE genes in a microarray dataset. Numerous statistical methods have been proposed in the literature on how to detect the DE genes efficiently. Some of the well-known methods include the parametric methods such as the analysis of variance approach (Kerr et al. (2000)), the parametric empirical Bayes (EB) methods (Newton et al. (2001), Kendziorski et al. (2003) and Smyth (2004)) and the semi-parametric hierarchical Bayes method (Newton et al. (2004)), among others. Recently, nonparametric method has attracted much attention due to its robustness to model mis-specification. One of the most popular nonparametric methods for analyzing microarrays is SAM (Tusher et al. (2001)). Other well-known nonparametric methods include the nonparametric EB method (Efron et al. (2000)) and the mixture model method (MMM) (Pan et al. (2003)). More recently, some variations of SAM, which use the rank sums instead of the original data, have been proposed in Weil (2004) and Schwender et al. (2003).

Control of FDR has now become a standard used in comparing different methods for detection of DE genes. Given that two methods have identified the same number of significant genes, a method with lower FDR is considered to be superior to a method with higher FDR. This shows that, for a given method, one needs to try to minimize the number of genes falsely identified as significant among the significant genes in order to reduce the FDR. It has been demonstrated in Tusher et al. (2001) that the use of the fudge factor can efficiently reduce the FDR. However, the choice of fudge factor in SAM is generally not optimal due to the following two reasons: 1) it does not have direct control of FDR (Grant et al. (2005)); 2) it uses a global fudge factor regardless of the number of genes declared significant. The dependence of FDR on the specific number of genes declared significant implies that the optimal choice of fudge factor should be local.

In this paper, we propose an improved method for detecting DE genes in microarrays. The improvement is achieved by employing the more efficient test statistic and null statistic of Zhang (2006), the local optimal fudge factor and a new two-stage FDR correction method. The paper is organized as follows. In next section, we first give a brief introduction of the SAM statistic and the test and null statistics of Zhang (2006), followed by an introduction of the fudge factor. The necessity of the local optimal fudge factor and its induced over-dispersion are demonstrated in the same section. Then, we provide a local optimal fudge factor searching algorithm and a two-stage FDR correction method. Our proposed method is introduced in the Algorithm section. Extensive simulations are carried out in the Numerical results section to compare the performance of SAM and our proposed method with respect to different FDR correction methods. Finally we apply our proposed method to the well-known leukaemia data of Golub et al. (1999).

The Test Statistic, the Null Statistic and Permutation Caused Over-Dispersion

In this paper, we focus on the nonparametric methods which use the modified *t*-statistic or its variations as the test statistic. A well-known statistic of such type is the SAM statistic. Let X_{ij} be the expression level of gene *i* under experimental condition 1 in the *j*th replicate and Y_{ik} be the expression level of gene *i* under experimental condition 2 in the *k*th replicate, where $i = 1, \ldots, n$, $j = 1, \ldots, J$, and $k = 1, \ldots, K$. The SAM statistic is defined as follows:

$$d(i) = \frac{\overline{X}_i - \overline{Y}_i}{s(i) + s_0} \quad (1)$$

where $\overline{X}_i$ and $\overline{Y}_i$ are the sample means of the expression levels of gene *i* under experimental conditions 1 and 2. The gene-specific scatter $s(i)$ is defined as

$$s(i) = \sqrt{a\{\sum_{j=1}^{J}(X_{ij} - \overline{X}_i)^2 + \sum_{k=1}^{K}(Y_{ij} - \overline{Y}_i)^2\}}, \quad (2)$$

where $a = (1/J + 1/K)/(J + K - 2)$. The constant s_0 is called the fudge factor. The use of fudge factor is a simple, but very important concept for efficient reduction of FDR. Given a test statistic, the use of appropriate fudge factor can result in significantly smaller FDR than a method without the use of fudge factor. Since the true FDR is unknown, a common method for estimating the FDR is to apply the so-called null statistic to the permuted data to generate numerous sets of null scores, and then use the null scores to estimate the number of false positive (FP) and FDR. Such a method is called the permutation estimation method. By definition, a null statistic is the statistic, when being applied to the permuted data, provides the value of the test statistic of an equivalently expressed (EE) gene regardless of whether a gene is DE or EE. Hence, the null statistic is independent of experimental conditions. In SAM, the null statistic is chosen to be the same as the test statistic: the SAM statistic. This actually causes problems. For example, it is obvious that SAM uses its own test statistics as one set of the null scores since they can be viewed as the null scores obtained from using the original ordering of the data as the permutation (Zhang (2007)). Therefore, the null scores SAM uses are not independent of experimental conditions. Since the test statistics are more dispersed than the true null scores, the use of test statistics as null scores will cause over-estimation of FDR. A permutation similar to the original ordering of the data may also produce over-dispersed null scores. This shows that one has to find the correct null statistic of the SAM statistic in order to correct such permutation caused over-estimation problem. However, recent research shows that constructing the null statistic of the SAM statistic is impossible without modifying the SAM statistic itself (Pan et al. (2003), Zhao and Pan (2002), Zhang (2006, 2007)). A meaningful attempt on such construction was made in Pan (2003) in which the following test and null statistics were proposed:

$$Z^*(i) = \frac{\frac{\bar{X}_1 + \bar{X}_2}{2} - \frac{\bar{Y}_1 + \bar{Y}_2}{2}}{\sqrt{\frac{s_{11}/j_1 + s_{12}/j_2}{4} + \frac{s_{21}/k_2 + s_2/k_2}{4} + s_0}}, \tag{3}$$

$$z^*(i) = \frac{\frac{\bar{X}_1 - \bar{X}_2}{2} - \frac{\bar{Y}_1 - \bar{Y}_2}{2}}{\sqrt{\frac{s_{11}/j_1 + s_{12}/j_2}{4} + \frac{s_{21}/k_2 + s_2/k_2}{4} + s_0}}, \tag{4}$$

where $\begin{cases} j_1 = j_2 = J/2, & \text{if } J \text{ is even} \\ j_1 = j_2 - 1 = (J-1)/2, & \text{if } J \text{ is odd} \end{cases}$, and k_1, k_2 are similarly defined. The statistics $\bar{X}_1, \bar{X}_2$ are the sample means of the two sub-samples ($X_{ij}, j = 1, \ldots, j_1$ and $X_{ij}, j = j_1 + 1, \ldots, J$) of gene i under experimental condition 1, and $\bar{Y}_1$, $\bar{Y}_2$ are the sample means of the two sub-samples ($Y_{ij}, j = 1, \ldots, k_1$ and $Y_{ij}, j = k_1 + 1, \ldots, K$) of gene i under experimental condition 2, respectively. The statistics s_{11}^2, s_{12}^2, s_{21}^2 and s_{22}^2 are the sample variances of the four sub-samples of gene i under the two experimental conditions. It was shown that $z^*(i)$ is the null statistic of $Z^*(i)$ with respect to the normal data (Pan (2003)).

The test and null statistics (3) and (4) provide significant improvement over the previous constructions of test and null statistics (Zhao and Pan (2002), Pan et al. (2003)). However, as pointed out in Zhang (2006), the above construction ignored the fact that the observations under each experimental condition have the same probability distribution. Therefore, the replicates (or arrays) under the same experimental condition should be pooled together to estimate the individual variances of each gene under the two experimental conditions. With this observation, the following test and null statistics were proposed in Zhang (2006):

$$Z(i) = \frac{\frac{\bar{X}_1 + \bar{X}_2}{2} - \frac{\bar{Y}_1 + \bar{Y}_2}{2}}{\sqrt{\frac{1/j_1 + 1/j_2}{4} s_1^2 + \frac{1/k_2 + 1/k_2}{4} s_2^2 + s_0}}, \tag{5}$$

$$z(i) = \frac{\frac{\bar{X}_1 - \bar{X}_2}{2} - \frac{\bar{Y}_1 - \bar{Y}_2}{2}}{\sqrt{\frac{1/j_1 + 1/j_2}{4} s_1^2 + \frac{1/k_2 + 1/k_2}{4} s_2^2 + s_0}}, \tag{6}$$

where j_1, j_2, k_1, k_2 $\overline{X}_1, \overline{X}_2$ and $\overline{Y}_1$, $\overline{Y}_2$ are similarly defined as in (3) and (4). The statistics $s_1^{\,2}$ and $s_2^{\,2}$ are the pooled sample variances from the replicates of each gene under each conditions and are defined as

$$s_1^{\,2} = \frac{\sum_{j=1}^{j_1}(X_{ij} - \overline{X}_1)^2 + \sum_{j=j_1+1}^{J}(X_{ij} - \overline{X}_2)^2}{J - 2 + I(J = 2)}, \tag{7}$$

and

$$s_2^{\,2} = \frac{\sum_{k=1}^{k_1}(Y_{ij} - \overline{Y}_1)^2 + \sum_{k=k_1+1}^{K}(Y_{ij} - \overline{Y}_2)^2}{K - 2 + I(K = 2)}, \tag{8}$$

respectively. The indicator function I, which equals to 1 when $J\ (\text{or } K) = 2$ and 0 otherwise, is added to the denominator of $s_i^{\,2}$ to avoid the undefined situation when $J\ (\text{or } K) = 2$. Note that such modification made the minimum requirement on the numbers of arrays under one condition to be 2 if the number of arrays under the other condition is at least 3.

It should be pointed out that the test and null statistics (5) and (6) require the use of within-condition permutations. This is different from SAM which uses all or a subset of possible simple permutations. Since a permutation which does not result in a different split (or partition) of the data under each experimental condition produces the same null scores, one only needs to carry out permutations which provide different splits of the data under each experimental condition. Since there are $\binom{J}{j_1}$ and $\binom{K}{k_1}$ different ways of splitting the data under the two experimental conditions respectively, there are in total $\binom{J}{j_1} \bullet \binom{K}{k_1}$ sets of null scores available for the null statistic (6).

It can be easily seen that the null statistic (6) is independent of experimental conditions. Hence, the null scores calculated from the permuted data by using (6) do not have the permutation caused over-dispersion problem. For illustration, we carried out a simulation study under the following model for n (=5000) genes with J=4 and K=4:

$$X_{ij} = \mu_i + \varepsilon_{ij}, \text{ for } j=1, \ldots, J,$$

and (9)

$$Y_{ik} = \eta_i + \omega_{ik}, \text{ for } k=1, \ldots, K,$$

for i=1, ..., n.

In the simulation, we generated 2000 DE genes at the beginning and end of the gene sequence. The first 1000 DE genes were generated from $X_{ij} \sim N(\mu_i, 1)$ and $Y_{ik} \sim N(\eta_i, 1)$, where $\mu_i = 0$ and $\eta_i = 3$, respectively, and the last 1000 genes were generated similarly with $\mu_i = 0$ and $\eta_i = -3$. The 3000 EE genes were generated from the standard normal distribution regardless of experimental conditions. To avoid the confounding effect of fudge factor, the comparisons reported in Figures 1 and 2 were carried out with $s_0 = 0$.

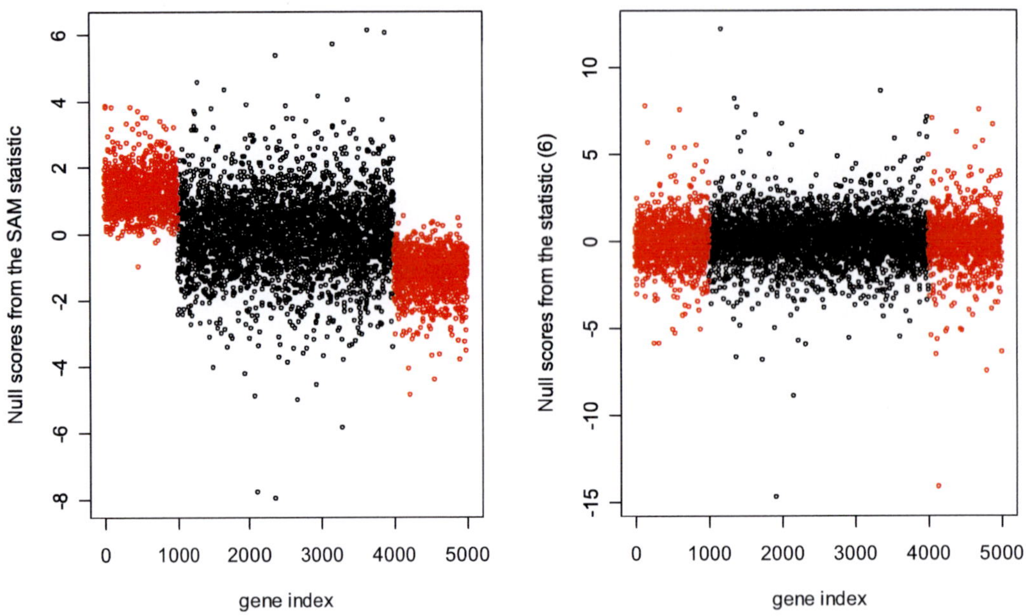

Figure 1. Plots of null scores from the SAM statistic and the null statistic (6).

Figure 1 shows the plot of one set of null scores calculated from the SAM statistic (left) and the plot of one set of null scores calculated from the null statistic (6). The red points are the permuted null scores from the DE genes. The black points are the permuted null scores from the EE genes. The permutation used for the SAM statistic was 1 1 1 2; 2 2 1 2, which simply swapped the group identity of the 4th array under experimental condition 1 with the 3rd array under experimental condition 2. The over-dispersion of the null scores of DE genes is obvious from the plot at the left of Figure 1 by looking at the deviation of the red points from the general pattern of the null scores of EE genes. It is obvious that such over-dispersion was the effect of the different mean expression levels of DE genes. In other words, the permutation did not clean out the effects of mean expression levels when generating the null scores using the SAM statistic. The plot at the right of Figure 1 shows a similar comparison using the null statistic (6) from one randomly selected permuted set of null scores. It can be seen that the effect of different mean expression levels has been removed in this case. We also examined the plots (not reported here) of the null scores from (6) with respect to different permutations. All such plots showed the same pattern as that observed in the right plot of Figure 1.

Despite the fact that the null statistic (6) is free of permutation caused over-dispersion, an obvious criticism of (5) and (6) is the reduced degree of freedom compared to that of the SAM statistic. For example, assume that the normality assumption holds, the degree of freedom of (5) and (6) when $J=K=4$ is only 4 while the degree of freedom for the SAM statistic is 6. The smaller degree of freedom of (5) and (6) may easily give an impression that a testing method based on (5) and (6) will have lower power than a testing method based on the SAM statistic. This would be true if the true null distribution were known to us. However, the fact is that the null distribution is unknown and has to be estimated from the null scores. Therefore, the performance of a testing procedure is determined by the quality of the estimated null distribution.

As discussed previously, the null scores used in SAM are more dispersed than the true null distribution if there are DE genes in the data. Given that a threshold (or cutoffs) has identified a certain number of genes as significant, such over-dispersed null scores will result in falsely higher estimate of the number of FP among the significant genes, and hence leads to over-estimation of FDR. This will cause researchers to regard some DE genes which contain important biological information about certain disease as EE genes and result in reduced power of the test. The loss of power caused by the over-dispersed null scores can be seen more clearly from the conventional significance level approach. Given a gene-specific level of significance, the over-dispersed null scores will cause falsely larger cutoffs than the true ones. Such cutoffs will make the declaration of a true DE gene as DE more difficult. As a result, the test will have lower power compared to a test without over-dispersed null distribution.

On the other hand, although the null distribution based on the null statistic (6) has lower degree of freedom than that of the SAM statistic, the null statistic (6) does not have the over-estimation problem caused by permutations. Hence, it does not have the reduced power problem due to permutations.

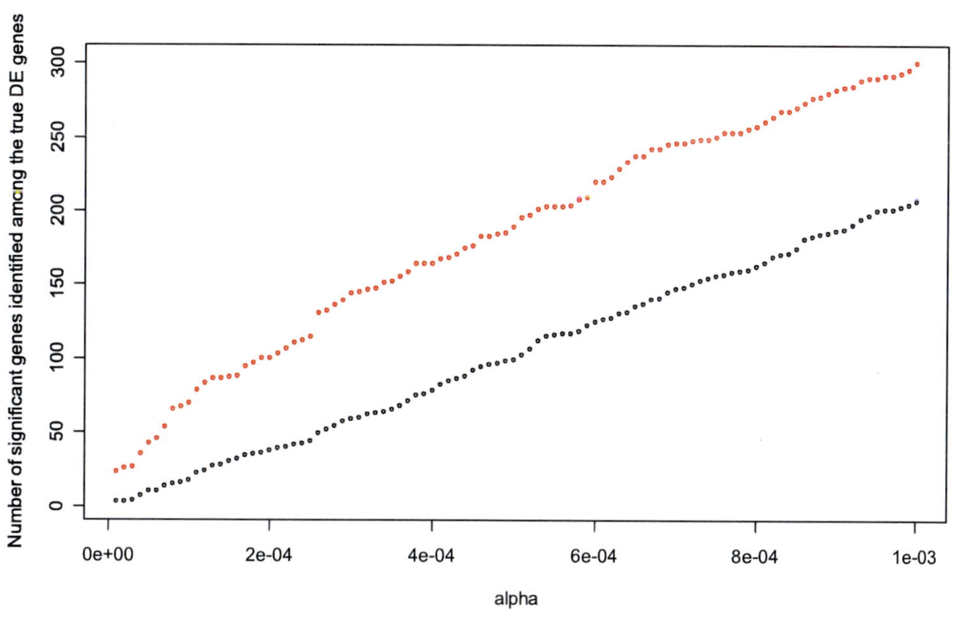

Figure 2. Number of DE genes identified as significant at different levels of significance.

The above discussion shows that, despite the fact that the SAM statistic has higher degree of freedom; its performance is compromised by its over-dispersed estimated null distribution. To better understand this, we plotted in Figure 2 the numbers of true DE genes declared as significant by the SAM statistic (black points) and the test statistic (5) (red points) at gene-specific level α=.00001, .0001, … , .1. The simulation was carried out under the same setup as that of Figure1 with $s_0 = 0$. For each α, the cutoffs were the quantiles obtained from the pooled sets of null scores calculated from the SAM statistic and the null statistic (6), respectively. At the same level, one would expect a method with higher power to identify more DE genes as significant genes than a method with lower power. It is obvious from Figure 2 that the testing method based on (6) has higher power.

THE FUDGE FACTOR AND ITS INDUCED OVER-DISPERSION

One crucial factor responsible for the success of the nonparametric methods such as SAM is the use of fudge factor. The fudge factor is a constant added to the denominator of the test statistic. The idea of using the fudge factor was pioneered in Tusher et al. (2001). A similar idea was also suggested in Efron et al. (2000). The importance of fudge factor in multiple comparisons has also been extensively studied in Grant et al. (2005). In SAM, the motivation of adding the fudge factor is to ensure the variance of the SAM statistic is independent of gene expression, and therefore to reduce the number of FP (or the FDR) caused by excessively small values of $s(i)$ among the genes declared significant. SAM achieved this by finding the fudge factor which minimizes the coefficient of variation of the test statistics. In the EB method of Efron et al. (2000), it was proposed to use an appropriate percentile of the sample standard deviations as the value of the fudge factor.

The choices of fudge factor in the aforementioned methods can be viewed as a global fudge factor since it does not concern the specific number of genes being declared significant at a given threshold. However, as mentioned in the previous paragraph, the purpose of using the fudge factor is to reduce the FDR. This reminds us to directly search for the fudge factor which minimizes the FDR for each given number of significant genes of interest. The simulation results reported in Figures 3-4 demonstrate the advantage of using the local fudge factor. The data used in the simulations were generated under Setup 1 (see next page). The plot at the left of Figure 3 displays the numbers of FP from SAM with respect to the global fudge factor and the local optimal fudge factor, respectively for different total numbers of significant genes (*TS*). The circles connected by red dashed line are the numbers of FP from the global fudge factor (from the output of the SAM R-package) and the circles connected by the solid black line are the numbers of FP from the local optimal fudge factor. The plot at the right of Figure 3 displays the values of the local optimal fudge factor corresponding to the values of *TS*. Similar results with the test and null statistics (5) and (6) are reported in Figure 4. The plot at the left shows the numbers of FP from using the global fudge factor (red points) and from using the local optimal fudge factor (black points) for different values of *TS*. The global fudge factor is the default value used in SAM. The plot at the right displays the values of the local optimal fudge factor corresponding to each value of *TS*. For each given value of *TS*, the cutoffs for (5) and (6) were obtained by finding the value (say, δ) such that there are *TS* genes with the absolute values of the test statistics of these genes exceeding δ.

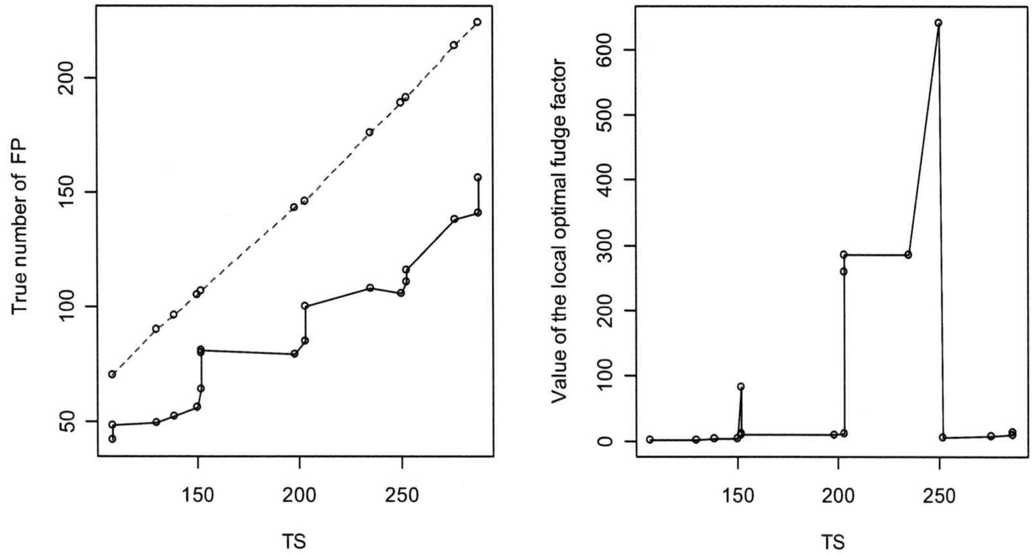

Figure 3. Numbers of FP for different values of TS for SAM and the corresponding values of the local optimal fudge factor.

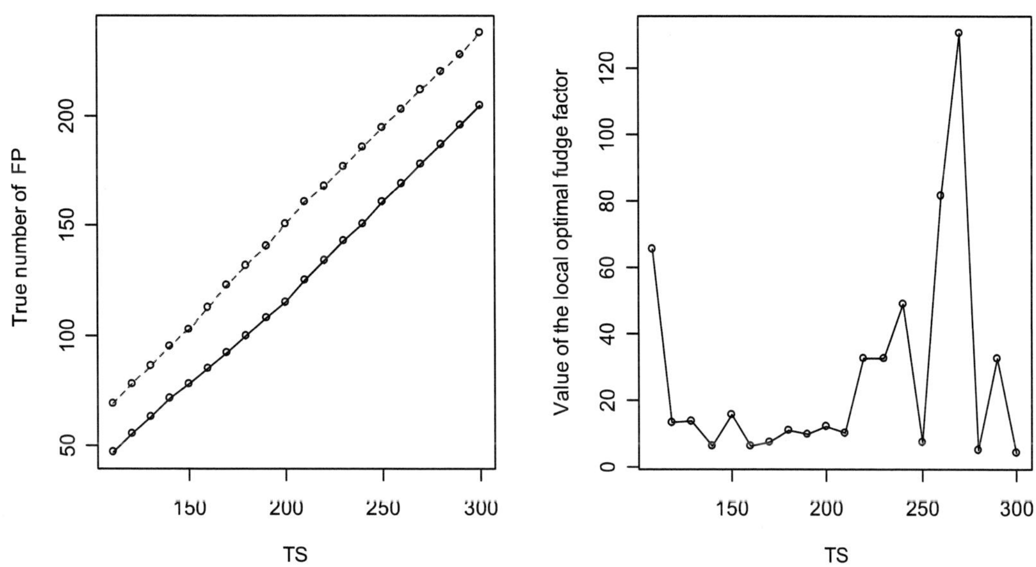

Figure 4. Numbers of FP for different values of TS for the test statistic (5) and the corresponding values of the local optimal fudge factor.

Figures 3 and 4 show that the use of local optimal fudge factor can lead to significant reduction of FP (hence, of FDR). Since the local optimal fudge factor is not directly available in practice, a practical way is to search for the fudge factor which minimizes the estimated FP and use that as the estimated local optimal fudge factor.

The permutation estimation method for estimating the number of FP in nonparametric analysis of microarrays is to generate multiple sets of the permuted null scores from all genes and then count the number of genes whose null scores exceed the cutoffs. The median number of FP (as in SAM) or the mean number of FP (this is what we will use for the null statistic (6)) is used as the estimated number of FP. It is obvious that the performance of a fudge factor searching algorithm depends on the performance of the method used to estimate the number of FP. Unfortunately, the following study shows that the introduction of fudge factor causes over-dispersion of the null scores calculated from the DE genes, even for the null statistic (6). Such over-dispersion will cause over-estimation of the number of FP and a fudge factor searching algorithm based on the over-dispersed null scores may provide biased estimate of the local optimal fudge factor.

To see the fudge factor induced over-estimation, it will be informative to compare the proportions of significant genes from the null scores of DE and EE genes, respectively. For this purpose, under each setup described below, we obtained 100 permuted sets of null scores from the SAM R-package. For each set of null scores, we counted the number of genes whose null scores exceed the cutoffs from both DE and EE genes. Then, the mean proportions of significant genes from all the permuted sets of null scores for both DE and EE genes are calculated. The cutoffs were obtained from the SAM R-package by fixing the total number of significant genes at *TS*=200. A similar simulation was carried out with the test and null statistics (5) and (6). Note that, in this case, there are only a total of 36 permutations available.

Setup 1: We generated the data for 5000 genes (200 are DE) from model (9) with *K*=*J*=4. The 200 DE genes were generated from $X_{ij} \sim N(\mu_i, \sigma_{i1}^2)$ and $Y_{ik} \sim N(\eta_i, \sigma_{i2}^2)$, where $\mu_i \sim N(0,5^2)$ and $\eta_i \sim N(0,5^2)$, respectively. For each DE gene, the variances σ_{i1} and σ_{i2} were generated individually from $gamma(5,1)$. The middle 4800 EE genes were generated from the normal distributions with $\mu_i = \eta_i \sim N(0,5^2)$ and $\sigma_{i1} = \sigma_{i2} \sim gamma(3,1)$.

Setup 2: The data for 5000 genes were generated from model (9) with *K*=*J*=4. Among them, 200 genes are DE, which were generated from $X_{ij} \sim N(\mu_i, 1)$ and $Y_{ik} \sim N(\eta_i, 2)$, with 100 each from ($\mu_i = 0$, $\eta_i = 3$) and ($\mu_i = 0$, $\eta_i = -3$), respectively. The 4800 EE genes were generated from the standard normal distribution regardless of the experimental conditions.

Setup 3: The same as Setup 2 except that the standard deviations of all the genes (both DE and EE) are chosen to be 1.

Figures 5-7 display the proportions of significant genes from the DE and EE genes under each setup. The plot at the left of each figure shows the proportions of significant genes from the null scores of DE genes (black dots connected by black solid line) and the proportions of significant genes from the null scores of EE genes (red dots connected by red dashed line) from SAM. The plot at the right shows the proportions of significant genes from the null scores of DE genes (black dots connected by black solid line) and the proportions of significant genes from the null scores of EE genes (red dots connected by red dashed line) from the test statistic (5). By checking these figures, we have discovered the following: 1) the fudge factor causes the higher proportion of significant genes among the null scores of DE genes than among those of EE genes; 2) a larger value of fudge factor helps reducing the proportion of significant genes among the EE genes; 3) No fudge factor induced over-dispersion was observed under Setup 3.

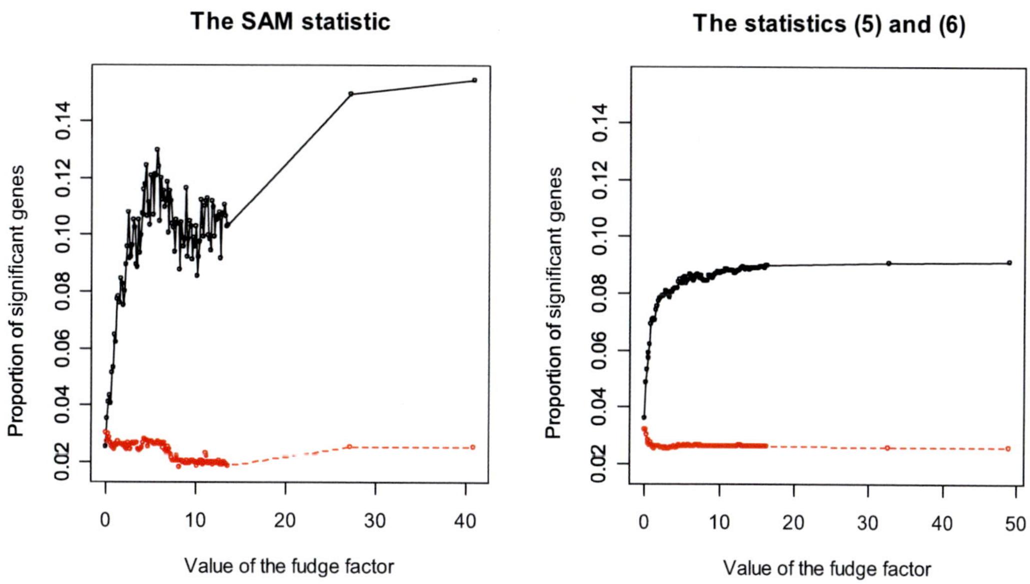

Figure 5. Proportions of significant genes from the null scores of DE and EE genes for different values of the fudge factor under Setup 1.

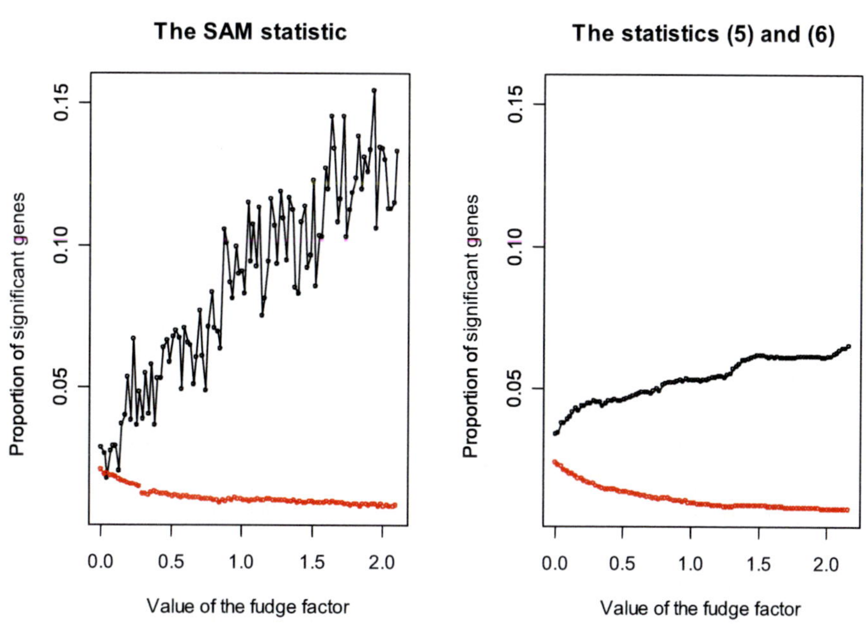

Figure 6. Proportions of significant genes from the null scores of DE and EE genes for different values of fudge factor under Setup 2.

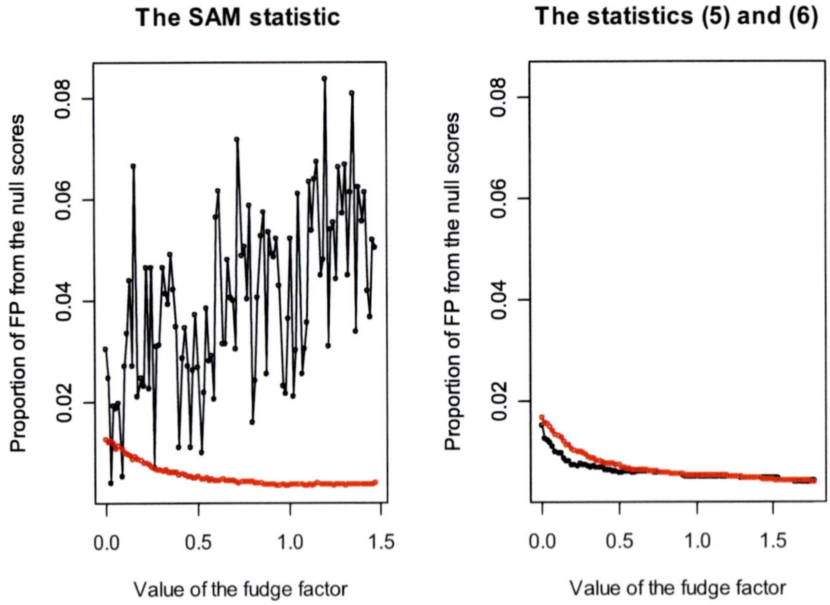

Figure 7. Proportions of significant genes from the null scores of DE and EE genes for different values of fudge factor under Setup 3.

It can be seen from these figures that the proportion of significant genes from the null scores of DE genes is similar to that from the null scores of EE genes when $s_0 = 0$, and the two proportions start to drift apart as s_0 increases with the proportion from the DE genes gets larger and the proportion from the EE genes gets smaller. The larger s_0 is, the larger difference between the two proportions become. This means that the over-dispersion problem among the null scores of DE genes becomes more severe as s_0 goes up.

It can also be seen from Figures 5-7 that the over-dispersion (or the deviation between the two proportions) is more dramatic for the SAM statistic than that of the test and null statistics (5) and (6). The larger discrepancy between the two proportions from SAM is the confounding effect of permutation caused over-dispersion and fudge factor induced over-dispersion.

What we observed in Figures 5-7 shows that it would be beneficial to use a large fudge factor if we were to know which genes are DE or EE and exclude them in the estimation of FDR. Since the true identity of a gene is unknown in practice, one needs to either reduce the effect of the DE genes or remove the genes we consider as DE in the estimation of FP. The former approach was adopted by Guo and Pan (2005) in which they proposed to use a weighted permutation approach to reduce the effect of DE genes. Other similar efforts in reducing the effect of DE genes can be found in Weil (2004) and Schwender (2003) in which they proposed to replace the original microarray data by the rank scores to avoid the over-estimation problem. The latter approach was adopted by Xie et al. (2005), in which they proposed to estimate the FDR only using the predicted EE genes.

It is clear from Figure 7 is that there is no fudge factor induced over-dispersion for the null statistic (6). This is due to the fact that all the genes (both DE and EE) under this setup were generated from the normal distributions with the same variance 1. In such a case, Figure 7 shows that, for the null statistic (6), the larger the fudge factor is, the smaller the number of FP becomes. As a matter of fact, the optimal value of fudge factor in this case is ∞. This means that the best separation of DE and EE genes is achieved by comparing the sample means if the gene expressions have the same variance. However, the same is not true for the SAM statistic. The plot at the left of Figure 7 shows that the over-dispersion problem of the SAM statistic still exists under Setup 3. In this case, the over-dispersion of the SAM statistic is solely the result of permutation.

THE TWO-STAGE FDR CORRECTION

The FDR correction method of Xie et al. (2005) is very promising due to its simplicity and effectiveness in correcting the over-estimation problem. It was proposed in Xie et al. (2005) to remove the predicted DE genes in the estimation of FDR. The intention is to get rid of the over-estimation caused by the over-dispersed null scores calculated from the DE genes. Note that the estimated FDR from SAM is obtained from multiplying the estimated proportion ($\hat{\pi}_0$) of DE genes to the ratio between the estimated number of FP ($\widehat{FP}_n$) assuming that all n genes are EE and the total number of significant genes TS. Using the same logic, Xie et al.'s correction can be stated as follows: 1) Obtain an estimate of the number of FP from n EE genes by $\widehat{FP}_n = \dfrac{n}{n-TS} FP^*$, where FP^* is the number of FP estimated from the genes after removing the TS significant genes; 2) using $1-\dfrac{TS}{n}$ as $\hat{\pi}_0$, then multiply it to the ratio $\widehat{FP}_n / TS$ to obtain an estimate of FDR. It is expected that FP^* is not as much affected by the over-dispersed null scores as SAM.

It is argued in Xie et al. (2005) that removing the predicted DE genes in the estimation of FP would provide improved estimate of FP than SAM. However, there are two obvious problems with this method. The first problem is under-estimation of FDR. The under-estimation can be caused by over-deletion when TS is larger than the true number of DE genes. It may also happen when the un-corrected estimated FDR does not have severe over-estimation problem (see the results from the first three cases under Setup 1 of Table 1). The second problem is over-estimation when TS is smaller than the true number of DE genes. The over and under-estimation can both be made worse by its use of $1-\dfrac{TS}{n}$ as an estimate of π_0, which ranges from 0 to 1 depending only on the value of TS without regard to the true value of π_0. In other words, $1-\dfrac{TS}{n}$ is in no sense an informative estimate of π_0 unless TS is close to the true number of DE genes. For illustration, let's use the leukaemia data Golub et al. (1999) (which will be discussed later) for example. There are 7129 genes in the dataset. The estimated proportion of DE genes obtained from the SAM-R package $\hat{\pi}_0$ is 68.92% while the estimated value of π_0 from Xie et al.'s correction changes as TS

changes, ranging from 100% (when *TS*=0) to 0 (when *TS*=7129). Take the case *TS* =240 for example, the output from the SAM-R package reports 0.0028 as the estimated FDR (Table 2). We applied Xie et al.'s correction to this case and obtained the estimated FDR as 0.0085, which is three times as large as the un-corrected estimated FDR. The computer output shows that the estimated number of FP assuming all genes are EE is 1 with $\hat{\pi}_0$=68.92% from SAM and 2.1214 with $\hat{\pi}_0$=96.63% from Xie et al.'s correction, respectively. This shows that the higher estimated FDR from Xie et al.'s correction is the combined result of increased estimate of FP from 1 to 2.1214 and the incorrect estimate of π_0 from Xie et al's correction. The increased estimated number of FP was caused by the use of the average in Xie et al.'s correction instead of the median as that used in SAM. It was pointed out in Zhang (2007) that the use of median is necessary to avoid permutation caused over-estimation when using the SAM statistic.

The above discussion shows that improvement on Xie et al.'s correction can be made through the following two aspects: 1) remove only the genes we are convinced to be DE among the significant genes; 2) abandon Xie et al's method of estimating π_0. This motivates us to propose the following two-stage FDR correction method:

Assume that a method has identified a total of *TS* genes as significantly expressed,

Step (a). Estimate the number of FP (say, $\widehat{FP}_0$) using all the null scores.

Step (b). Since $\widehat{FP}_0$ is an estimate of FP assuming that all genes are EE, a reasonable estimate of FP from the true EE genes in the microarray data is $\pi_0\widehat{FP}_0$, where π_0 is the true proportion of EE genes. Now remove the (TS- $\pi_0\widehat{FP}_0$) genes with the highest absolute values of test statistics from the gene list.

Step (c). Re-estimate the number of FP using the null scores from the remaining n- (TS- $\pi_0\widehat{FP}_0$) genes and denote it by $\widehat{FP}_1$.

Step (d). Since $\widehat{FP}_1$ is the estimated number of FP assume that all n- (TS- $\pi_0\widehat{FP}_0$) genes are EE, an estimated number of FP assuming all n genes are EE is

$$\widehat{FP}_2 = \frac{n}{n-(TS-\pi_0\widehat{FP}_0)}\widehat{FP}_1.$$

Step (e). The estimated FDR is $\widehat{FDR}_2 = \pi_0\widehat{FP}_2/TS$.

The proportion π_0 is unknown in practice. When the proportion of DE genes is small as it is usually the case in microarray experiment, π_0 in the two-stage FDR correction can be replaced by 1. Otherwise, it can be estimated by the natural spline estimator Storey and Tibshirani (2003).

Although the above proposed two-stage FDR correction aims to correct the two problems of Xie et al.'s correction, it is obvious that it may still over-estimate the FDR if *TS* is much smaller than the true number of DE genes in the experiment. Another scenario in which the two-stage FDR correction may still over-estimate the FDR is Setup 2 considered above.

Under Setup 2, all the DE genes were generated from a distribution with the same variance 2^2 while all the EE genes have the same variance 1. In such a case, the DE genes tend to produce large null scores which are not necessarily caused by small standard deviations. Since the fudge factor can not correct the large null scores of DE genes caused by relatively large standard deviation, $\widehat{FP}_0$ in Step (a) usually significantly over-estimate the true number of FP. This will cause a smaller value of TS- $\pi_0 \widehat{FP}_0$, and hence resulting in removing too few genes and leads to over-estimation of FP (and hence of FDR) in Step (e). In this case, Xie et al.'s correction may actually work better since it removes more genes (see the results under Setup 2 of Table 1).

THE ALGORITHM

In microarray data analysis, the FDR approach proposed in Benjamini and Hochberg (1995) has been widely accepted as a standard method for controlling the error rates. Assume that TS genes have been declared significant, the FDR and the estimated FDR are obtained as follows (Storey and Tibshirani (2003), Benjamini and Hochberg (1995)):

$$FDR = \frac{FP}{TS}, \text{ and } \widehat{FDR} = \frac{\widehat{FP}}{TS} \tag{10}$$

where FP is the true number of FP among TS significant genes and $\widehat{FP}$ is the estimated value of FP.

Combining the test and null statistics (5), (6), the local fudge factor and the two-stage FDR correction, we propose the following algorithm for detecting the significantly differentially expressed genes and estimating the associated FDR:

Step 0. Calculate the standard deviation for each gene by

$$s(i) = \sqrt{\frac{1/j_1 + 1/j_2}{4} s_1^{\ 2} + \frac{1/k_1 + 1/k_2}{4} s_2^{\ 2}} \tag{11}$$

where $s_1^{\ 2}$ and $s_2^{\ 2}$ are defined in (8) and (9).

For $s_0 = p \bullet \max_{1 \le i \le n}(s(i))$, where p=0, 1%, ..., 100%(=1), 2, ..., 50, repeat the following Steps 1-5:

1. Calculate the test scores $Z(1)$, ..., $Z(n)$ for all genes using (5).
2. Declare the top TS genes with the largest absolute test statistics significant.
3. Define the cutoffs $\delta_U = -\delta_L = v$, where v is the smallest absolute value of the test statistics of the genes declared significant in Step 2.
4. Estimate the number of FP by

$$\widehat{FP}_0 = \frac{\sum_{b=1}^{B} \#\{1 \le i \le n:\ z^b(i) > \delta_U \text{ or } z^b(i) < \delta_L\}}{B} \tag{12}$$

where $z^b(i)$, i=1, …, n, are the null scores calculated from (6) corresponding to permutation b and B is the total number of permutations used.

5. Denote the set of genes after removing the (TS-$\hat{\pi}_0\,\widehat{FP}_0$) genes with highest absolute values of test statistics by T. Re-estimate the number of FP using the null scores from the genes in T:

$$\widehat{FP}_1 = \frac{\sum_{b=1}^{B} \#\{i \in T:\ z^b(i) > \delta_U \text{ or } z^b(i) < \delta_L\}}{B}.$$

Hence, an estimate of FP by assuming all n genes are EE is

$$\widehat{FP}_2\,[p] = \frac{n}{n-(TS-\hat{\pi}_0\widehat{FP}_0)}\widehat{FP}_1,$$

where $\hat{\pi}_0$ is the estimate of the true proportion of EE genes π_0. As discussed in the previous sub-section, $\hat{\pi}_0$ can be replaced by 1 if we know the proportion of DE genes is very small.

6. With $\widehat{FP}_2\,[p]$, p=0, 1%, …, 100%(=1), 2, …, 50 available, find the value of p (denote it by p_0) which provides the smallest value of $\widehat{FP}_2\,[\bullet]$. The corresponding value of $s_0 = p_0 \bullet \max_{1\le i\le n}(s(i))$ is the estimated local optimal fudge factor.

7. Denote the smallest value of $\widehat{FP}_2\,[\bullet]$ by $\widehat{FP}_2$. The estimated FDR is $\widehat{FDR}_2 = \hat{\pi}_0\widehat{FP}_2 / TS$.

Note that the composition of the TS significant genes might be different for different values of s_0 in the search of the local optimal fudge factor. The above process can be viewed as a process of searching for a pool of TS significant genes which contain the smallest number of FP.

Numerical Results

Simulated Examples

In this section, we use simulations to compare the performance of SAM and our proposed method. The numerical comparisons were carried out under the previously defined three setups with different ratios of induced to repressed genes, different ratios of DE to EE genes and different values of *TS*. To account for the situations in which the distributions of induced and repressed genes are different, the DE genes under Setup 3 were generated from $X_{ij} \sim N(\mu_i, 1)$ and $Y_{ik} \sim N(\eta_i, 1)$ with $\mu_i = 0$, $\eta_i = 3$ for induced genes and with $\mu_i = 0$, $\eta_i = -1$ for repressed genes, respectively. Under each setup, we calculated the mean number of *TS*, mean true FDR, mean un-corrected estimated FDR, mean estimated FDR with Xie et al.'s correction and mean estimated FDR from our proposed method from 100 simulations.

Table 1. Simulation results

Setup	Genes Ind./EE/Rep.	Methods	Mean $\hat{\pi}_0$ / True π_0	Mean TP	True FDR	Estimated FDR		
						$\widehat{FDR}$	$\widehat{FDR_X}$	$\widehat{FDR_2}$
1	200/4800/0	SAM	0.9811	190.25	**0.6704**	**0.6751**	0.6466	0.6752
		Ours	0.9600		**0.5861**	0.6098	0.5459	**0.5862**
	150/4800/50	SAM	0.9810	192.27	**0.6678**	**0.6699**	0.6421	0.6687
		Ours	0.9600		**0.5864**	0.6078	0.5421	**0.5843**
	100/4800/100	SAM	0.9810	406.93	**0.7858**	**0.7918**	0.7208	0.7890
		Ours	0.9600		**0.7335**	0.7658	0.6225	**0.7377**
	1000/3000/1000	SAM	0.7963	499.26	**0.0909**	**0.1098**	0.1320	0.1173
		Ours	0.6000		**0.0304**	0.0825	0.0893	**0.0788**
2	200/4800/0	SAM	0.9704	198.99	**0.2012**	**0.2235**	0.2092	0.2197
		Ours	0.9600		**0.1648**	0.2409	0.1776	**0.1879**
	150/4800/50	SAM	0.9683	395.46	**0.5662**	**0.6158**	0.5311	0.5838
		Ours	0.9600		**0.5395**	0.6003	0.5150	**0.5531**
	100/4800/100	SAM	0.9655	198.12	**0.3334**	**0.3606**	0.3326	0.3477
		Ours	0.9600		**0.1608**	0.2312	0.1705	**0.1795**
	1000/3500/500	SAM	0.8106	499.70	**0.0465**	**0.0634**	0.0677	0.0482
		Ours	0.7000		**0.0156**	0.0180	0.0200	**0.0155**
3	200/4800/0	SAM	0.9661	495.87	**0.5982**	**0.6011**	0.5475	0.5861
		Ours	0.9600		**0.6044**	0.5861	0.5522	**0.5871**
	150/4800/50	SAM	0.9685	199.18	**0.2674**	**0.2868**	0.2694	0.2781
		Ours	0.8600		**0.2537**	0.2405	0.2393	**0.2409**
	100/4800/100	SAM	0.9712	197.17	**0.4911**	**0.5159**	0.4824	0.5009
		Ours	0.9600		**0.4287**	0.4121	0.4095	**0.4127**
	500/3500/1000	SAM	0.8166	499.75	**0.0468**	**0.0653**	0.0690	0.0574
		Ours	0.7000		**0.0164**	0.0182	0.0201	**0.0182**

Table 1 reports the results from the simulations. Column 2 of Table 1 shows the composition of genes. For example, 200/4800/0 under Setup 1 means that there are 200 induced genes, 4800 EE genes and 0 repressed genes. Column 4 displays the mean estimated proportions of DE genes and the true proportion of DE genes in the microarray data. Column 5 reports the mean of total number of significant genes from 100 simulations. Column 6 reports the mean true FDR values for SAM and our proposed method. The mean estimated FDR from SAM is reported in Column 8 (the number in bold under $\widehat{FDR}$). The mean estimated FDR from our proposed method is reported in Column 10 (the number in bold under $\widehat{FDR}_2$). For comparison, we also reported the mean estimated FDR with Xie et al's correction ($\widehat{FDR}_X$) and the mean estimated FDR with the two-stage correction ($\widehat{FDR}_2$), respectively for SAM. The results for SAM are directly obtained from the output of the SAM R-package. We also reported the mean un-corrected estimated FDR ($\widehat{FDR}$) and the mean estimated FDR with Xie et al's correction ($\widehat{FDR}_X$) for our proposed method. Note that our proposed method contains the two-stage FDR correction. The values of $\widehat{FDR}$ corresponding to our proposed method were obtained by removing the two-stage FDR correction in the estimation of FDR and the values of $\widehat{FDR}_X$ corresponding to our proposed method were obtained by applying Xie et al's correction to the values of $\widehat{FDR}$.

By checking the results under Setups 1-3, we have observed the following:

(1) Our proposed method (see Column 6 of Table 1) consistently has significantly smaller true FDR than SAM under all setups except Case 1 under Setup 3 in which SAM is marginally better. Note that all the DE genes under Case 1 of Setup 3 are induced genes. Such a scenario rarely happens in the real world. It was included in the simulation just to show that SAM might have some advantage over our proposed method in such an extreme case. Note that our proposed method uses symmetric cutoffs. Hence, it can be viewed as a symmetric cutoff method based on the test and null statistics (5) and (6). A symmetric cutoff method based on the SAM statistic was studied in Zhang (2007). It was reported Zhang (2007) that, when the DE genes are all induced or repressed, SAM consistently has significantly lower FDR than the symmetric cutoff method based on the SAM statistic. The results from Table 1 show that such advantage of SAM has been lost when compared to our proposed method. As a matter of fact, the other two extreme cases we considered (Case 1 under Setups 1 and 2) both show that our proposed method has significantly lower true FDR than SAM. This shows that our proposed method also has better performance than the symmetric cutoff method based on the SAM statistic. More importantly, our proposed method has the advantage of easy interpretation of results due to the use of symmetric cutoffs. As discussed in Zhang (2007), SAM's use of asymmetric cutoffs often brings difficulty in interpreting the results since there is no reasonable explanation why one should make the declaration of a gene as significant positive more difficult than as significant negative or vice versa.

(2) The un-corrected estimated FDR from SAM tends to over-estimate the true FDR. while the un-corrected estimated FDR from our proposed method may over-estimate the true FDR (under Setups 1-2) and may also under-estimate the true FDR slightly (under Setup 3).

(3) Xie et al.'s correction may cause severe under-estimation of FDR when the number of significant genes exceeds the true number of DE genes (see Case 3 under Setup 1; Case 2 under Setup 2 and Case 1 under Setup 3). Such under-estimation has been mostly corrected by our proposed two-stage FDR correction.

(4) If the DE genes form a significant portion of all the genes such as that in Case 4 under Setups 1-3, Xie et al.'s correction may severely over-estimate the true FDR when the value of TS is significantly smaller than the true number of DE genes. It can be seen that the over-

estimation problem has been alleviated by the use of our proposed two-stage FDR correction, especially in Case 4 under Setups 2-3.

Among the simulations we carried out, Setup 1 is the most representative of a real microarray data. The results in the first three cases under Setup 1 show that the un-corrected estimated FDR slightly over-estimate the true FDR. In a situation when the over-estimation is not severe, it can be seen that Xie et al.'s correction may cause under-estimation of FDR due to the removal of a large number of genes and our proposed method provides reasonably unbiased estimates of FDR. However, the results from Case 4 under Setup 1 (the bottom row under Setup 1) show that our proposed method is not able to correct the over-estimation problem efficiently in this case. Our study finds that the major cause of this is the over-estimation of the true proportion (π_0) of DE genes. The computer outputs show that the estimated π_0 from the natural cubic spline method consistently over-estimate the true proportion π_0 (see the values of mean $\hat{\pi}_0$ and π_0 in Column 4 of Table 1). The over-estimation becomes much more severe when the proportion of DE genes is high in a microarray dataset. For example, the estimated value of π_0 in Case 4 under Setup 1 is 79.63%, which is much higher than the true proportion 60%. Such over-estimation causes the two-stage FDR correction deleting too few genes and hence results in over-estimation of FDR.

The data generated under Setups 2 and 3 may shed some light in how different methods may perform under different scenarios. First, we see that the over-estimation of the un-corrected FDR estimates under Setup 2 is more severe for our proposed method than SAM. This seems contradictory to what we observed in Figure 6 in which we reported more severe over-dispersion problem of the null scores for SAM. Nevertheless, one has to remember that the comparison in Figure 6 was based on the same fudge factor for both methods while the results reported in Table 1 was based on the global fudge factor for SAM and the estimated local optimal fudge factor for our method. Under Setup 2, the estimated local optimal fudge factor is usually larger than s_0 from SAM, and hence we see a more severe over-estimation of the un-corrected FDR estimates for our method. Under Setup 3, the estimated local optimal fudge factor is similar to s_0 of SAM. As can be seen from Table 1, the over-estimation of the un-corrected FDR estimates is gone for our method and still exists for SAM.

Real Data Example

We applied SAM and our proposed method to the well-studied leukaemia data (Golub et al. (1999)). We used all the 27 of ALL and 11 of AML samples in our calculation. For each sample (or array), the original data were pre-processed by first subtracting its median and then this difference was divided by its quartile range (the difference between the first and third quartiles).

The value of s_0 from SAM is 0.01727888 and the estimated value of the local optimal fudge factor from our method is 0. This shows that the use of fudge factor is not beneficial for the leukaemia data. Since one does not know which genes are DE or EE for the leukaemia data, Zhang (2007) used the list of 15 informative genes provided in Golub et al. (1999) as a standard to compare the performances of different methods. Here is the list of genes: CD11c, CD33, MB-1, the leptin receptor, zyxin, Cyclion D3, Op18, MCM3, Rbap48, SNF2, TFIIEβ, c-Myb, E2A, HOXA9, TOP2B and MPO. Some of the genes in the list have been shown to provide useful information on the diagnosis of leukaemia while some others have been shown to be related to cancer pathogenesis.

Table 2. Results obtained for the leukaemia data

$\widehat{TP}$	**Method**	$\widehat{FDR}$	$\widehat{FDR}_X$	$\widehat{FDR}_2$	**# of genes from Golub et al.'s list**
612	Our proposed method	0.0202	0.0277	**0.0209**	15
	Symm. cutoff (SAM)	0.0230	0.0332	0.0221	15
	SAM	**0.0225**	0.0332	0.0221	15
398	Our proposed method	0.0051	0.0075	**0.0055**	15
	Symm. cutoff (SAM)	0.0086	0.0159	0.0091	15
	SAM	**0.0095**	0.0174	0.0091	14
312	Our proposed method	0.0044	0.0064	**0.0046**	15
	Symm. cutoff (SAM)	0.0044	0.0099	0.0046	15
	SAM	**0.0066**	0.0123	0.0069	14
240	Our proposed method	0.0028	0.0041	**0.0029**	15
	Symm. cutoff (SAM)	0.0028	0.0062	0.0029	15
	SAM	**0.0028**	0.0085	0.0029	13
199	Our proposed method	0.0034	0.0050	**0.0035**	13
	Symm. cutoff (SAM)	0.0034	0.0046	0.0035	14
	SAM	**0.0034**	0.0058	0.0035	11
124	Our proposed method	0	0	**0**	12
	Symm. cutoff (SAM)	0	0.0029	0	13
	SAM	**0**	0.0031	0	7
93	Our proposed method	0	0	**0**	11
	Symm. cutoff (SAM)	0	0.0017	0	12
	SAM	**0**	0.0024	0	7
54	Our proposed method	0	0	**0**	10
	Symm. cutoff (SAM)	0	0.0003	0	10
	SAM	**0**	0.0002	0	7
34	Our proposed method	0	0	**0**	10
	Symm. cutoff (SAM)	0	0	0	10
	SAM	**0**	0	0	5
29	Our proposed method	0	0	0	9
	Symm. cutoff (SAM)	0	0	0	8
	SAM	0	0	0	5

Table 2 reports the results we obtained by applying SAM, the symmetric cutoff method based on the SAM statistic Zhang (2007), and our proposed method to the leukaemia dataset.

Table 2 displays the un-corrected estimated FDR ($\widehat{FDR}$, column 3), the estimated FDR with Xie et al.'s correction ($\widehat{FDR}_X$, column 4) and the estimated FDR with the two-stage FDR correction ($\widehat{FDR}_2$, column 5) for the leukaemia data with respect to different values of *TS*. Column 6 reports the number of genes in Golub et al.'s list declared as significant by each method at different values of *TS*. The bold numbers under $\widehat{FDR}$ (column 3) are the estimated FDR from the SAM R-package. The bold numbers under $\widehat{FDR}_2$ (column 5) are the estimated FDR from our proposed method. The results for the method "Symm. cutoff (SAM)" were obtained from the symmetric cutoff method based on the SAM statistic.

It is obvious that SAM is the most inefficient on both the estimated FDR and the number of genes from the list included in the pool of significant genes each method identified. It can be seen from Table 2 that the symmetric cutoff method based on the SAM statistic has significantly smaller un-corrected estimated FDR than SAM for *TS*=312, 398. Such results are consistent with those reported in Zhang (2007). However, the un-corrected estimated FDR from the symmetric cutoff method based on the SAM statistic is slightly higher than that of SAM for *TS*=612. It was reported in Zhang (2007) that the un-corrected estimated FDR from the symmetric cutoff method based on the SAM statistic has a more serious over-estimation problem than SAM under certain situations. The higher estimated FDR observed in this case is very likely due to the effect of over-estimation.

It can also be seen from Column 4 of Table 2 that the estimated FDR values from Xie et al.'s correction is much higher than those from the un-corrected estimated FDR, especially for small *TS* values. Our discussion on the potential problems of Xie et al.(2005)'s correction (page 10-11) shows that this is caused by the large number of DE genes existed in the leukaemia dataset and the relatively small values of *TS* used in Table 2. Therefore, the estimated FDR values from Xie et al.'s correction are not reliable.

The estimated FDR with the two-stage correction reported in Column 5 of Table 2 provides a more consistent picture of the performance of each method. It is obvious from Table 2 that our proposed method performs the best, followed by the symmetric cutoff method based on the SAM statistic, and SAM is the worst among the three methods.

The last column of Table 2 shows that the number of significant genes each method takes to include all the 15 informative genes. For SAM, this number is 612 while it is only 240 for the symmetric cutoff method and our proposed method.

CONCLUSIONS

In this paper, we have proposed an improved method for detecting differentially expressed genes. It is demonstrated that the use of local optimal fudge factor can lead to significant reduction of FDR compared to the methods which use global fudge factors. We have also pointed out that, in addition to permutation caused over-estimation, the use of fudge factor may also cause over-dispersed null scores, too. To solve this problem, we have proposed a two-stage FDR correction method to improve the FDR estimation. Extensive simulations show that our proposed method provides improved control of FDR than SAM and the symmetric cutoff method based on the SAM statistic.

One concern for our proposed method is that it might be computationally expensive. Note that the most intensive computation in our method is the calculation of the permuted sets of null scores, which is necessary for any permutation method. Our experience shows that the computing time for our proposed method is still reasonable although it takes significantly longer than that required by SAM.

It is obvious from the simulation results that the performance of the two-stage FDR correction method depends on the quality of $\hat{\pi}_0$ as an estimate of π_0. An incorrect estimate of π_0 may lead to significant over or under-estimation of FDR when using the two-stage FDR correction. In addition to the natural cubic spline method for estimating π_0, a number of new methods for estimating π_0 have been proposed in the literature in the past few years (Newton et al. (2001, 2004), Smyth (2004), Lönnstedt. and Speed (2002), Langaas et al. (2005) and Cox and Wong (2004)). A comprehensive evaluation of these estimators may provide useful information on which to choose in application. Another issue directly associated with the current paper is to generalize our proposed method to the multiple experimental condition case. Some preliminary results on this issue have been obtained and will be reported in a subsequent paper.

REFERENCES

Benjamini, Y., & Hochberg, Y. (1995). Controlling the false discovery rate: a practical and powerful approach to multiple testing. *J. R. Statist. Soc. B,* **57**, 289-300.

Cox, D. R. and Wong, M. Y. (2004). A simple procedure for the selection of significant genes. *J. R. Statist. Soc. B*, **99**, 395-400.

Efron, B., Tibshirani, R., Goss, V., & Chu, G. (2000). Microarrays and their use in a comparative experiment. *Technical Report*, Department of Statistics. Stanford University. U.S.A.

Golub, T. R., Slonim, D.K., Tamayo, P., Huard, C., Gaasenbeek, M., Mesirov, J. P., Coller, H., Loh, M. L., Downing, J. R., & Caligiuri, M. A. (1999). Molecular classification of cancer: class discovery and class prediction by gene expression monitoring. *Science*, **285**, 531-537.

Grant, G., Liu, J., & Stoeckert, C. J. (2005). A practical false discovery rate approach to identifying patterns of differential expression in microarray data. *Bioinformatics*, **21**(11), 2684-2690.

Guo, X., & Pan, W. (2005). Using weighted permutation scores to detect differential gene expression with microarray data. *Journal of Bioinformatics and Computational Biology*, **3,** 989-1006.

Kendziorski, C. M., Newton, M., Lan, H., & Gould, M. N. (2003). On parametric empirical Bayes methods for comparing multiple groups using replicated gene expression profiles. *Stat Med.*, **22**, 3899-39144.

Kerr, M. K., Martin, M., & Churchill, G. A. (2000). Analysis of variance for gene expression microarray data. *J. Comput. Biol.*, **7,** 819-837.

Langaas, M., Lindqvist. B. H., & Ferkingstad, E. (2005). Estimating the proportion of true null hypotheses, with application to DNA microarray data. *J. R. Statist. Soc*. B, **67**, 555–572.

Lönnstedt, I. and Speed, T. (2002). Replicated microarray data. *Statist. Sin*., **12**, 31-46.

Newton, M. A., Kendziorski, C. M., Richmond, C.S., Battner, F. R., and Tsui, K.W. (2001). On differentially variability of expression ratios: improving statistical inference about gene expression changes from microarray data. *J. Comput. Biol.*, **8**, 37-52.

Newton, M., Noueiry, A., Ahlquist, P., & Sarkar, D. (2004). Detecting differential gene expression with a semiparametric hierarchical mixture method. *Biostatistics*, **5**(2), 155-176.

Pan, W. (2003). On the use of permutation in and the performance of a class of nonparametric methods to detect differential gene expression. *Bioinformatics*, **19**(11), 1333-1340.

Pan, W., Lin, J., & Le, C. (2003). A mixture model approach to detecting differentially expressed genes with microarray data. *Funct Integr Genomics*, **3**.117-124.

Schwender, H., Krause, A., & Ickstadt, K. (2003). Comparison of the empirical Bayes and the Significance Analysis of Microarrays. *Technical report*. Collaborative Research Center. University of Dortmund. Germany

Smyth, G. K. (2004). Linear models and empirical Bayes methods for assessing differential expression in microarray experiments. *Statistical Applications in Genetics and Molecular Biology*, **3**(1), Article 3.

Story, J. D., & Tibshirani, R. (2003). Statistical significance of genome-wide experiments. *Proc. Natl Acad. Sci. USA*, **(100)**, 9440-9445.

Tusher, V. G., Tibshirani, R., & Chu, G. (2001). Significant analysis of microarrays applied to the ionizing radiation response. *Proc. Natl Acad. Sci. USA*, **98**, 5116-5121.

Weil, M. A. van de (2004). Significance Analysis of Microarrays Using Rank Scores. *Kwantitatieve Methoden*, **(71)**, 25-37.

Zhang, S. (2006). An improved nonparametric approach for detecting differentially expressed genes with replicated microarray data. *Statistical Applications in Genetics and Molecular Biology*, **5**(1), Article 30.

Zhang, S (2007). A comprehensive evaluation of SAM, the SAM R-package and a simple modification to improve its performance. *BMC Bioinformatics*, **8**, 230.

Zhao, Y., & Pan, W. (2002). Modified nonparametric approaches to detecting differentially expressed genes in replicated microarray experiments. *Bioinformatics*, **19**(9),1046-1054.

Xie, Y., Pan, W., & Khodursky, A. (2005). A note on using permutation based false discovery rate estimate to compare different analysis methods for microarray data. *Bioinformatics*, **21**(23), 4280-4288.

This paper is reviewed by Professor R. J. Karunamuni of Department of Mathematical and Statistical Sciences, University of Alberta, and by Professor Fenghai Duan, Department of Biostatistics, College of Public Health, University of Nebraska Medical Center.

In: Oligonucleotide Array Sequence Analysis
Editors: M.K. Moretti and L.J. Rizzo, pp. 347-366
ISBN: 978-1-60456-542-3

Chapter 12

BASIC APPROACHES TO GENE/EXON LEVEL DATA ANALYSIS: 3' IVT AND EXON 1.0 ST MICROARRAY PLATFORMS

***Roberta Bosotti*[1], *Raffaele Calogero*[2] *and Vanessa Marchesi*[1]**
[1]Department of Biotechnology, Genomics Lab, Nerviano Medical Sciences, Nerviano (MI) Italy
[2]Department of Clinical and Biological Sciences, University of Torino, Torino Italy

ABSTRACT

Information on protein expression levels can be used to gain insights about putative physiological gene function, for identifying target proteins that show differential expression in tumor versus normal cells and for formulating hypotheses regarding possible side effects of target inhibition.

However, obtaining systematic information at protein level is not trivial and currently standardized high throughput protein methods are not available. Gene expression levels, analyzed with microarray (MA) technology producing genome-wide information at the mRNA level, can partially compensate for the lack of information at protein level. Indeed, microarray-based transcription profiling is a consolidated methodology commonly used in areas such as pharmacogenomics, diagnostics and drug target identification.

Several high-density oligonucleotide array platforms are currently commercially available, the best known and most-widely used being Affymetrix. Affymetrix offers the largest panel of gene expression microarrays, which are available for a number of different organisms. Besides classical expression arrays, Human Gene/Exon 1.0 ST is a new emerging technology, suited for identifying both differential gene and exon expression allowing to infer alternative splicing events. It has recently been applied to the discovery of potentially interesting new targets for diseases such as cancer.

In this chapter, we will review tools and methods for data analysis of both microarray transcription profiling and alternative splicing analysis.

INTRODUCTION

Microarrays, coupled with Quantitative Real Time PCR (QPCR), are emerging as the preferred method for analysis of gene expression on a genome-wide scale. Several platforms are currently commercially available, including the classical two channel technology [1,2], as offered by Agilent (http://www.home.agilent.com/agilent/home.jspx) (i.e. reference and sample labeled with two different fluorophors and hybridized on the same array), and high density oligonucleotide arrays (i.e. all samples labeled with the same fluorophor and each hybridized on a different array), available from a number of different companies, including Affymetrix (www.affymetrix.com), Illumina (http://www.illumina.com/), Amersham (http://www.amersham.com/) and Applied Biosystem (www.appliedbiosystems.com). Agilent, the first company to begin development of two channel technology, now also produces single channel microarrays. The commercially available single channel microarray platforms vary in many respects, such as printing technologies, length of oligonucleotides and possibilities for customization and automation.

Affymetrix is the best known and most-widely used among the commercial producers of microarrays. Affymerix has the largest panel of microarrays, which are available for a number of different organisms and the vast majority of publicly available data sets (www.ncbi.nlm.nih.gov/geo/; www.ebi.ac.uk/microarray-as/aer/?#ae-main[0]) have been produced on its platforms. Therefore, this review will focus primarily on the tools and methods available for Affymetrix arrays.

An Affymetrix GeneChip® microarray consists of thousands of different 25 nucleotide long oligonucleotides synthesized onto a small solid support. The oligonucleotides are produced one base at a time using photolithographic technology directly onto the glass slides, with each feature having a size of 5-11 μm. Arrays based on 11 μm features are also referred to as 3'based expression arrays (3'IVT arrays) since each transcript on these chips is queried by a probe set consisting of 22 probes that map to the most 3' end 600 bases of the transcript. These consist of 11 perfect match (PM) oligos and 11 mismatch (MM) oligos. The MM oligos contain a mismatch at position 13 and are used to measure non-specific binding. The Affymetrix algorithm (MAS5) uses probe pair intensities to generate a Detection p-value and assigns a Present, Marginal, or Absent call. Each probe pair in a probe set is considered as having a potential vote in determining whether the measured transcript is detected (Present) or not detected (Absent).

Affymetrix has also recently begun the production of arrays based on 5 μm features. The ability to manufacture these higher density arrays has facilitated whole-genome scale profiling at exon level using a single array (Exon 1.0 ST). In these arrays 4 probes in average are used to query each exon and approximately 40 probes are available for each gene. The technological improvements allowing the production of these arrays have also enabled the production of other arrays where 26 probes are used to query each transcript (Gene 1.0 ST arrays). Since these probes are spread across the full-length of the transcript, they should represent a more complete and accurate picture of gene expression than that obtained using 3' based expression arrays. A recently published study [3] comparing 3'IVT and the newer exon platforms has highlighted a high degree of similarity between the two, with little difference between the median relative sensitivity for both types of arrays.

Although much progress has been made in the field of microarray analysis, a number of issues must be addressed before data obtained from this methodology can routinely be used in the clinic. These include reproducibility, reliability and standardization of both microarray analysis and results.

Since gene expression levels are measured indirectly as dye signals in microarray experiments, data analysis is complicated by many factors that can alter levels of fluorescence and introduce possible errors. An additional major source of variability is the limited number of experimental replications, which is always lower than the number of variables. For these reasons experimental design has to be planned accurately to avoid bias, to optimize resources and to increase efficiency. Although automation and technological improvements have facilitated production of chips of extremely high quality, quality control steps are still a fundamental requirement to remove poor quality chips prior to the start of array preprocessing. Several open source and commercial software packages are available (Bioconductor (www.bioconductor.org), Partek Genomics Suite (http://www.partek.com/software)) providing different solutions for microarray data preprocessing. Graphical interface packages are also available to facilitate Bioconductor tools usage for researchers without experience in R programming [4].

QPCR often accompanies microarray analysis for validation of results, since it is an alternative technique that has good detection sensitivity, sequence specificity and large dynamic range [5-7]. Other interesting alternative technologies have also recently appeared on the market, including the Combimatrix arrays (www.combimatrix.com), which combine the high throughput of microarray technology with the design flexibility of QPCR.

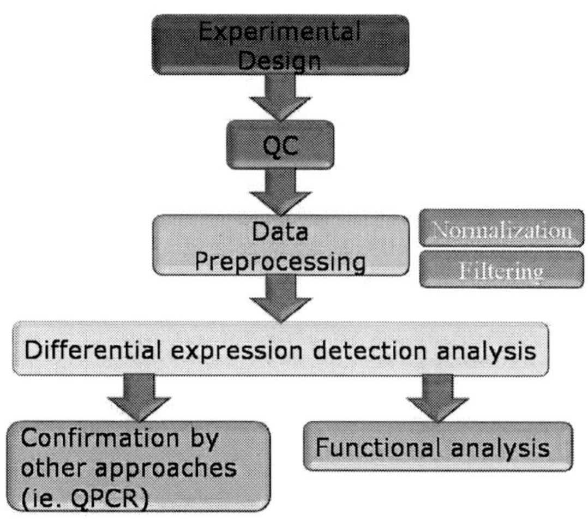

Figure 1. Microarray data analysis scheme. Schematic representation of microarray data analysis workflow.

In this chapter, we will review the main steps involved in Affymetrix microarray data analysis for both 3'IVT arrays and exon arrays, starting with the description of the experimental design, followed by quality checks and preprocessing and finally describing the techniques currently available to detect differentially expressed genes. We will also discuss how to make sense of the long list of genes obtained after statistical analysis, covering the current tools available for annotation purposes and pathway analysis (Figure1).

EXPERIMENTAL DESIGN

To obtain biologically significant results when using microarrays, good experimental design is vital to avoid bias. Therefore, the initial step in a microarray experiment is to develop an accurate plan and this should involve biologists, bioinformaticians and statisticians. Since the number of arrays is often a limiting step, a well-planned experiment should also aim to optimize resources and increase efficiency, resulting in reduced costs.

Biological and technical variations are the two main sources of variability in microarray experiments. Since the aim of these studies is to measure the biological variability, experimental design is focused on trying to control the technical variability. The data collection process is an area where a high degree of variability may be introduced. Therefore, the procedures used for this must be tightly controlled through standardization of methods used both for collection and storage of samples. RNA labeling and hybridization steps can also contribute to the technical variability. Preferably, for each experiment, all chips should be processed together by the same person using the same reagents. In fact, it has been shown that day-to-day variability is significant enough to mask real results and hinder data analysis [8]. However, processing large numbers of samples at the same time can be impractical. Therefore, a certain level of randomization has to be included in the experimental design. A complete randomization is not advisable due to the small number of samples compared to the conditions to be tested. Thus, a randomized block design associated with multiple readings may be a better choice.

The implications of sample pooling must also be addressed when designing experiments. RNA pooling is a common practice in situations where researchers are trying to reduce costs, i.e. in large experiments or in cases where there are insufficient amounts of RNA from the tissues being studied, common in clinical conditions where biopsies are being used. Consequently, it is important to assess whether pooling might cause inaccurate measurements from the introduction of bias [9]. It has been suggested that pooling may lower the efficiency in labeling and hybridization steps [10], affecting approximately 5% of the genes [9]. An alternative to pooling is RNA amplification. Two major types of RNA amplification procedures have been developed. One is a PCR-based method characterized by an exponential increase in copy number while the other uses T7-polymerase in-vitro transcription (IVT) to achieve a linear amplification of the target [11]. T7 based amplification techniques are reproducible but can introduce some bias. It has been reported that microarray analysis of amplified samples performs best at detecting differences in gene expression when these are large and when moderated statistics are used [12].

QUALITY CONTROLS (QC)

Before carrying on data analysis of results obtained from Affymetrix GeneChip® microarray experiments the quality of the results must be assessed. The first step is to eradicate poor quality chips or those with artifacts. This can be achieved by visual inspection of row data or using methods that highlight the presence of differences between slides. It should be noted that differentially expressed transcripts represent only a very small part of the transcripts queried on a genome-wide chip, therefore, they do not affect the overall characteristics of chip raw data distribution. Visual inspection of chips is an essential step to aid quality assessment of any microarray experiment. Often, small artifacts and blemishes are visible on chips but have little effect on probe set intensity estimation, since the probes are scattered over the array. However, it is better to remove slides that appear to be significantly different from others from the analysis. Experienced users can easily detect non-homogenous hybridization, e.g., dust or bubbles, masking feature intensities. A number of commercially available tools are also available to aid quality control. GCOS software, packaged within the Affymetrix platform, reports the intensity values for specific array areas and probes as control parameters and suggests thresholds to tag poor quality chips. Tools such as affyQCReport, affyPLM and simpleaffy, packaged in the open source software Bioconductor, are used to highlight microarray differences. AffyQCreport carries out a series of quality controls for raw intensities, e.g. boxplots of PM intensities and histograms of densities, giving a preliminary view of raw data homogeneity. It also calculates the percentage of "present calls", 3'/5' ratio of control genes and the signal background levels. In addition, it verifies the gridding on the chip and evaluates the correlation between chip pairs by checking control features localized around the border of the microarray. Simpleaffy is based on Affymetrix recommendations and exploits the different control features provided on each chip.

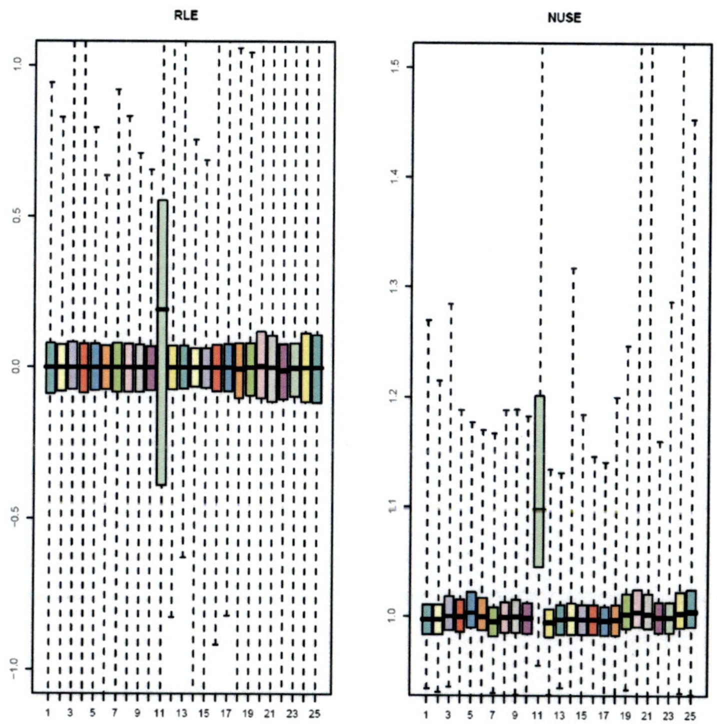

Figure 2. RLE and NUSE plots. Relative Log Expression (RLE) and Normalized Unscaled Standard Error (NUSE) plots of 25 arrays. A typical array of bad quality (n° 11) is shown with a larger spread and not centred at 0 in the RLE plot, and with a significantly elevated standard error and a wider distribution in the NUSE plot.

The AffyPLM package includes diagnostic procedures based on background adjusted and normalized probe-level data. Two different plots can be generated to identify the presence of strong array bias [13]: Relative Log Expression (RLE) and Normalized Unscaled Standard Error (NUSE) plots. The RLE plot displays the relative log expression distribution of each array calculated with respect to the average expression of all arrays as a box plot. Since differentially expressed transcripts do not affect the overall distribution of probeset intensities, the relative log expression distribution of each array should be centred on 0 and all arrays should share the same spread in the central part of the distribution. A low quality array will result in a larger spread and/or will not be centred at 0 (Figure 2A). NUSE plots are similar to RLE plots but in this case standard error is estimated for each probeset on each array. To account for the considerable variability between probesets, standard error estimates are standardized over the median standard error across arrays for each probeset. Thus, low quality arrays are those in which standard errors are significantly elevated or more spread out when compared to other arrays (Figure 2B).

To verify sample homogeneity, PCA (Principal component analysis) can be used [14]. PCA involves a mathematical procedure that transforms a number of (possibly) correlated variables into a (smaller) number of uncorrelated variables called principal components. The first principal component accounts for as much of the variability in the data as possible. Each following component accounts for as much of the remaining variability as possible. The components can be thought of as axes in n-dimensional space, where n is the number of components. Each axis represents a different trend in the data. The expected results from this type of analysis should show all the replicates grouped close together with the control samples far from the treated samples. An example of a PCA performed on triplicate samples is shown in Figure 3, where controls are compared with treated samples followed over time.

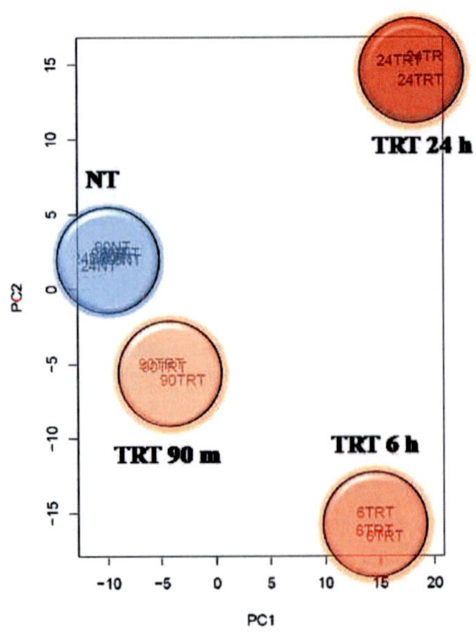

NT = untreated, TRT 90 m = treated for 90 minutes, TRT 6 hr = treated for 6 hours, TRT 24 m = treated for 24 hours

Figure 3. Example of a Principal component analysis (PCA). PCA performed on triplicate samples in a time course experiment (90 m, 6hr, 24hr) where the A2780 cell line was treated with a chemical compound or left untreated (NT).

Few tools are actually available for quality control on exon arrays, due to the implementation in these arrays of new labelling and hybridization protocols. Expression Console from Affymetrix (www.affymetrix.com) is a primary analysis tool that can be used to perform basic quality controls on exon arrays, such as box plots of probe data, correlation plots and visualization of internal microarray hybridisation/labelling quality controls. Due to the newness of the technology, manufacturers have not yet suggested quality thresholds for exon arrays. However, it is suggested that a certain level of homogeneity within the same experiment must be highlighted in all metrics.

Bioconductor tools, e.g., oneChannelGUI [4], facilitate sample homogeneity checks using PCA and hierarchical clustering [15].

Data Preprocessing

Preprocessing transforms the raw fluorescence signal detected by microarray staining into a normalized signal for experimental errors. The main steps involved in this process are background subtraction, experiment normalization and transcript intensity summarization. For Affymetrix 3'IVT arrays, these steps are usually combined in a single algorithm.

A number of different algorithms have been developed, moving away from MAS5 (or GCOS), which uses information from one microarray only, towards model-based fitting methods (RMA, GCRMA) that include information acquired from numerous arrays. Hybridizing samples on multiple arrays involves different hybridization reactions, each of which may be slightly variable. Therefore, in order to compare the data from each array, it is necessary to correct for the variability that this might introduce. It appears that the selection of a preprocessing algorithm should be based on the biological problem under investigation [16-19], e.g. MAS5 has been suggested to be the best method for use in reverse engineering studies of cellular networks [20], whereas RMA and GCRMA perform well in basic differential expression studies.

The MAS5 algorithm, developed by Affymetrix (Affymetrix 2002, Statistical Algorithm Description Document), is designed to adjust the PM probes for non-specific binding by subtracting the MM probe signals from the corresponding PM probe signals. Although this method has been criticized for producing high false positive rates due to exaggerated variance at low intensities, recently Pepper et al. have reevaluated the MAS5 algorithm as a selective and sensitive method when used with detection calls for identifying differentially expressed genes [21].

Other methods have recently been developed within the scientific community that perform background correction, normalization and summarization in a modular way. These include RMA [22], which does not take into account non-specific probe hybridization when calculating the background signals for probe sets, and GCRMA [23], an extension of RMA containing a background correction component that uses the probe sequence data. PLIER (Probe Logarithmic Error Intensity Estimate), an algorithm recently proposed by Affymetrix, takes into consideration the experimentally observed patterns in probe behavior and handle errors at low and high signal values. GCRMA provides a good level of accuracy and is also useful for limiting false positives resulting from poorly performing probe sets [24].

For multiple array experiments data normalization is required to subtract undesirable systematic variations arising from sample preparation, variability in hybridization, spatial effects or scanner settings following the calculation of gene expression. One way to do this is by global median normalization, as used by MAS5.0 and GCOS software, but this is not normally advised since the simple adjustment of median intensity value within each array does not take into account local intensity bias. A widely used alternative is the quantile normalization method [25], which aims to normalize probe intensity distribution for each array in a set. In this method, probe intensities in each array are ordered and the average value for each probe among the different arrays is calculated. Each value is then substituted with the corresponding average value and data is reordered. Recent results have suggested that this method of normalization can impact on the biological variability and, therefore, appears not to be an ideal method from this point of view [26].

Once the normalization process has been completed, samples can be compared to each other. However, identification of truly differentially expressed genes is complicated by the fact that the number of genes analyzed in a microarray experiment is disproportionate (thousand of genes) compared to the number of samples (from 2 to a few hundred). From a statistical point of view, this is an "atypical" situation and the risk is that some transcripts that are actually differentially expressed become hidden among those that are non-differentially expressed (false negative). On the other hand, evaluation of a high number of multiple statistical tests increases the number of false positives, i.e. transcripts non-differentially expressed that are present within the true positives. Thus, before applying any statistical approach, it is convenient to eliminate transcripts not expressed at detectable levels (1-3 RNA copies per cell) or those that do not change within the experimental conditions under analysis [27-28].

A variety of methods can be used for microarray data filtering. These include present-calls amount, average intensity signal, within-gene signal variance, estimated fold-change or combinations of different methods. The aim of all these filtering methods is to remove genes that are unlikely to carry information about the phenotypes under investigation, e.g. due to their low overall intensity or variability.

A threshold fraction of Present detection calls (derived by MAS5) [29] provides a simple method that effectively eliminates probe sets that are unlikely to be reliable while preserving those that are most significant. This results in an increased ratio of true positives to false positives. Intensity levels are also commonly used as a filter. Probes that display a non-expression in all or most chips should be removed to eliminate a background invariant signal that does not contribute any information to the result of the experiment. The IQR filter (genefilter) is another generally applicable filter available in the Bioconductor package. It eliminates genes showing insufficient expression variation across all samples since these provide little discriminatory power [30].

Application of different filters produces differences in stringency levels depending on the dataset being analyzed. When choosing a filter method, the number of samples included in the analysis should be considered since a more restrictive filtering is preferable for small experiments.

The previously mentioned IQR filter can also be used for gene level analysis of exon arrays. For these arrays, Affymetrix has developed a new approach to evaluate the significance of the signal associated to a specific gene-level probe set: DABG (detection above background) p-values. DABG is a p-value that indicates the probability that a specific signal is above background in a specific array. DABG could be used as a filter to remove those probe sets which do not pass a user defined threshold, e.g., a transcript is discarded if at least 50% of the samples do not show a DABG greater than 0.5.

Affymetrix XHYB and CROSSHYB annotations can be used to remove cross-hybridizing exon-level probe sets that may negatively affect the detection of alternative splicing events.

XHYB largely acts as an indicator of weak assignment between a transcript cluster and the assigned mRNA, suggesting a potential cross-hybridization. CROSSHYB is instead a measure of the promiscuity of the probes within a probe set among transcribed sequences. It is characterized by three levels:

1. unique. All probes in the probe set perfectly match only one sequence in the putatively transcribed array design content. The vast majority of probe sets are unique.
2. similar. All the probes in the probe set perfectly match more than one sequence in the putatively transcribed array design content.
3. mixed. The probes in the probe set either perfectly match or partially match more than one sequence in the putatively transcribed array design content.

OneChannelGUI package [4] contains a filter based on DABG.

DIFFERENTIAL EXPRESSION DETECTION ANALYSIS

The ultimate goal of a microarray experiment is to identify a list of genes with either up regulated or down regulated expression levels. Therefore, the main interest is to extract meaningful genes from a sub-population that presumably is representative of the whole population. However, due to the small number of biological replicates typically used in microarray experiments, the statistical power of differentially expressed gene detection is limited. Classical statistics are used for experiments dealing with few variables and several replicates and, therefore, analysis of data from microarrays needs to exploit methods that are more appropriate to its particular conditions.

In microarray experiments thousands of probes are evaluated simultaneously. When performing multiple statistical tests two types of errors can occur:

–Type I error (False positive)
–Type II error (False negative)

Reduction of type I errors increases the number of type II errors.

It is important to identify an approach that reduces false positives with the minimum loss of information (false negative).

Benjamini Hochberg is the most used method for the correction of type I errors in microarray analysis. However, it has some limitations due to the initial assumptions regarding independency of gene expression levels from each other and uniformity of raw distribution of p values in the non-significant range.

Statistical validation can use either parametric tests or non-parametric tests. Parametric tests are more sensitive and they are used if the populations being analyzed are normally distributed. Non-parametric statistics are used if no assumptions can be made regarding the distribution of sample data. Both methods are able to recognize some true signals and some false signals but not all. Each combination of methods will attain some but not all true signals. At the same time each combination of methods will produce some false signals. The trick is to find the best conditions to maximize true signals while minimizing false ones. However, the only way to define the best combination of methods is to know the differential expressed subset of transcripts, which is unknown and it is the aim of the differential expression analysis.

Among the most used methods are SAM (Significance Analysis of Microarray) [31] and LIMMA (Linear model analysis of microarrays). In addition, Conesa et al. have developed an R package called maSigPro [32-33] for the analysis of single and multi series time course microarray experiments.

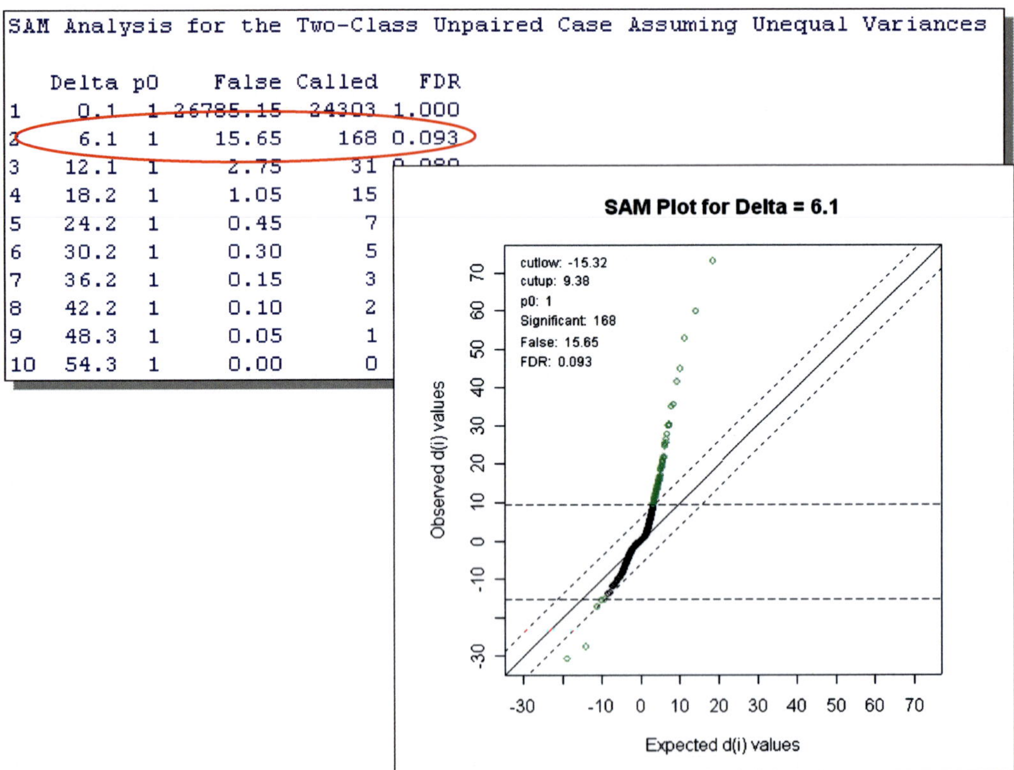
SAM Analysis for the Two-Class Unpaired Case Assuming Unequal Variances

	Delta	p0	False	Called	FDR
1	0.1	1	26785.15	24303	1.000
2	6.1	1	15.65	168	0.093
3	12.1	1	2.75	31	
4	18.2	1	1.05	15	
5	24.2	1	0.45	7	
6	30.2	1	0.30	5	
7	36.2	1	0.15	3	
8	42.2	1	0.10	2	
9	48.3	1	0.05	1	
10	54.3	1	0.00	0	

Figure 4. Example of SAM (Significance Analysis of Microarray) output. SAM associates the number of differentially expressed transcripts to a "False Discovery Rate" (FDR), which is calculated taking into account the average number of transcripts assigned as differentially expressed in the random permutations.

SAM uses a modified t-statistic called "relative difference" d(i) to evaluate the differential expression between two groups and compare its distribution to the mean $d_r(i)$, derived by random permutation of sample labels. SAM associates the number of differentially expressed transcripts to a "False Discovery Rate" (FDR), which is calculated taking into account the average number of transcripts assigned as differentially expressed in the random permutations (Figure 4). One drawback of SAM is that it is not particularly efficient at controlling FDR [34-37]. In particular, Larsonn et al. [38] have evaluated the effect of a variable FC setting, implemented in the Excel SAM, on the final gene list and have shown that it alters the list of differentially expressed genes. This effect can be so large that it substantially affects the results derived by biological interpretation, e.g. ontology overrepresentation analysis.

LIMMA is a linear modeling approach [13,39] implemented in the limma Bioconductor package [http://www.statsci.org/smyth/pubs/limma-biocbook-reprint.pdf]. This method decouples the mean–variance dependency by modeling the variance of the expression of a gene as a function of the mean expression of that gene. The first step is to fit a linear model, which fully models the systematic part of the data. In the second step, a matrix of contrasts is defined allowing comparisons to be made within the set of arrays being analyzed. The study published by Choe suggests that the empirical Bayes statistics [39] is probably the best suited method for identifying differential expression in small-scale experiments aiming to obtain a mechanicistical view of a biological treatment. However, alternative statistical approaches may be preferable if the aim of an experiment is to select features for classification purposes. The study by Jeffrey et al. [40] illustrates the influence of data set characteristics on the performance of the applied statistical methods. This report highlights that empirical Bayes statistics can be efficiently used for feature selection unless the datasets have high-pooled variance or low sample numbers.

Conesa et al. [32-33] have developed two statistical approaches for the analysis of time course experiments: maSigPro and ANOVA-SCA.

MaSigPro is part of Bioconductor and uses a two-step regression strategy to identify genes exhibiting significant temporal alterations in expression and significant differences between experimental groups. The first step used in this method is calculation of a regression fit for each gene and this is followed by calculation of a p-value associated to the F-statistic of the model corrected for multiple comparisons using the false discovery rate (FDR) procedure [41]. The second step involves the identification of the significant variables for each gene using a variable selection procedure [42].

ANOVA-SCA [32] aims to extract targeted signals from data whilst avoiding structural noise by combining ANOVA-modelling and a dimension reduction technique. Essentially, this method applies Principal Component Analysis (PCA) to the estimated parameters in each source of variation of an ANOVA model. It appears to be an effective way of extracting a signal of interest from noisy data through separation of data variability present in complex time course experiments. Two statistical methods, leverage and squared prediction error (SPE), are used to select genes with significant changes in expression. Leverage is a measure of the importance of a variable (i.e. transcript) in the PCA model and SPE is a measure of the fit of the model for that specific gene. Transcripts that vary according to the main trend and correspond to the key molecular functions affected by the treatment are associated with high leverage and low SPE. Transcripts with high SPE are model diverging data, corresponding to genes that respond to the treatment with a minority pattern. Transcripts with less specific responses and exhibiting low variance result in low leverage.

Following selection of differentially expressed genes, it is common practice to reduce the dimensionality of the list by only choosing genes that show higher variation between the control and the sample under examination. Intensity change between experimental groups (i.e. control versus treated) is called fold change. No rules exist to define the correct fold change (fc) threshold for differential expression but typically a $|fc|>2$ is arbitrarily chosen. Fc threshold estimation is dependent on the percentage of fc fluctuations due to experimental factors. However, from a biological point of view, many small variations, when combined within a specific biological scenario, can be functionally important, although they are difficult to extract from the experimental noise for further experimental validation.

QPCR is often referred to as the "gold standard" for gene expression measurements due to its detection sensitivity, sequence specificity, large dynamic range, as well as high precision and reproducible quantitation. For these reasons, it is good practice to use it for independent validation of microarray results.

A large-scale real-time validation experiment has been published [43] in which results from Applied Biosystem and Agilent Whole Human Genome Oligo Microarrays were confirmed in parallel by QPCR. This showed a reasonable coherence between the two types of data for both platforms.

Although many authors underline the importance of verifying microarray results using QPCR as a reliable independent technology for gene expression measurement, this approach is not always straightforward since it is expensive and time consuming, therefore allowing the reconfirmation of a very small fraction of the microarray data. Recently, the throughput of the technology has improved due to the development of Taqman QPCR 384 well cards. The availability of QPCR Gene Expression Assays (Applied Biosystems) eliminates the need for careful primers design, reducing time but not cost. We and others [44-45] have proposed as an alternative the use of two combined microarray platforms as a built-in cross validation of a huge fraction of the results of the experiments. Our data strongly suggests that the use of an approach based on two single channel microarray platforms combined within an analytical pipeline can achieve this objective. Indeed, the confirmation rate we obtained of 62/66 genes is particularly good taking into account that these genes were selected from the list of differentially expressed genes with the aim to cover the whole range of log2(Fold Change) and Average Intensity values observed in each platform [44].

It is worth noting that other interesting alternative technologies are also appearing on the market, including the Combimatrix arrays that combine the throughput of microarray technology with the design flexibility of QPCR. Although an extensive validation of the platform is needed, this kind of technology could be an attractive alternative in the near future.

Since Exon 1.0 ST® arrays are a novel technology, very little is known on the efficacy of statistical methods used with this technology to detect alternative splicing events.

Affymetrix has suggested that (exon_alt_transcript_analysis_whitepaper.pdf) ANOVA based methods applied to a Splicing Index equal to the ratio of exon signal to gene signal works well in two test data sets. The Splicing Index captures the basic metric for the analysis of alternative splicing. Specifically, it measures the difference in exon specific expression (with gene induction factored out) between two samples. The first step in this method is to normalize the exon level signals to the gene level signals. The splicing index is then defined as the ratio of normalized exon signal estimates from one sample or set of samples relative to another [46]. MiDas (Microarray Detection of Alternative Splicing) is the statistical method proposed by Affymetrix and implemented in their Expression Console software. Before applying MiDAS, Affymetrix suggests using the iter-PLIER summarization method to generate a robust estimate of gene-level signal by using data from all features in all exons in the gene. This signal has the advantage that it will be robust against exons that exhibit anomalous signals across samples, whether they are non-expressing probe sets, probe sets that are incorrectly assigned to a gene or true alternative splice exons. MiDAS is designed to analyze gene-level and exon level differential expression. Gene-level analysis estimates if there are effects other than exon effects present across samples and exons. Exon-level MIDAS assesses a single exon at a time and tests the hypothesis of no alternative splicing over all samples. Tests performed on simulations suggest that exon-level detection performs better than gene-level detection for identification of alternative splicing products.

FUNCTIONAL ANNOTATION AND PATHWAY ANALYSIS

Statistically significant differential expression analysis usually results in a list consisting of hundreds of probe sets. To make sense of these by looking at each transcript individually is impractical and also cause loss of information about processes involving several transcripts. Therefore, at this point begins the complicated job of trying to understand the biological meaning of this long list through correct annotation of the probes and analysis of the pathways in which the transcripts are involved.

A critical issue in the analysis of microarray data is the specific association of probe set identifiers to genome annotated transcripts. This can be achieved in different ways and it is complicated by the lack of annotation for all the probe sets, since most of the genes have been identified by sequence only and have not been functionally characterized. Of the ~25000 human genes known so far, only a fraction have been annotated. As a result, there is only a partial idea of what is really happening in cells.

The official Affymetrix annotation repository is available in NetAffx (www.affymetrix.com). This can also be used to assess the correctness of probe set mapping over genome data and to view the sequence region used to design the probe. Annotation libraries are also available from Bioconductor. These are built using a specific Bioconductor package, AnnBuilder, which uses the association between a probe set identifier and a GeneBank accession number (i.e. the primary target for probe design) [13].

Another available annotation source, which is particularly useful for comparisons between different platforms, is RESOURCERER [47]. This tool, from the TIGR centre, uses EST and gene sequences from their databases (www.tigr.org/tdb/tgi.shtml). Through analysis of publicly available EST and gene sequence data, RESOURCERER identifies transcripts and places them into a genomic context. In addition, this database also identifies, when possible, orthologs and paralogs.

Bioconductor and TIGR do not operate at probe level, neither do they consider the limited reliability of some probe sets due to probe cross-hybridization or incorrect probe/transcript annotation.

An additional tool from Ensembl annotation (www.ensembl.org) acts by directly mapping Affymetrix probes within the contents of the Ensembl sequence database. The weak point of this method is that it requires only 50% of the probes belonging to a specific probe set to match to an Ensembl gene to consider the association significant.

Once differentially expressed probe sets have been annotated, rather than solely concentrating on the list of genes, it is important to attempt to understand the biological functions they are involved in. Although the composition of the list can vary depending on the different statistics used for the analysis, it has been shown that relevant biological themes within the set of differentially expressed transcripts remain the same and are a stable representation of the biology underlying the experiment [48,49].

This concept has recently been illustrated in a study looking at several different gene signatures identified in breast cancer. Yu et al. [50] compared several published breast signatures and, while confirming a poor overlap at single gene level, they clearly show that the signatures had many pathways in common. This suggests that different prognostic gene signatures, despite the gene composition, represent common biological processes and, therefore, can be equally useful for predicting clinical outcome. This emphasizes that the predictive power of a signature comes from the overall pathways it touches and not from a single marker.

The Gene Ontology (GO) database is the database classically used for gene annotation [51]. It consists of three branches and it is used to annotate gene products based on their function (molecular function), the process they are involved in (biological process) and where they are localized (cellular component). Typically, genes are first ranked and selected based on statistically significant differential expression. GO categories are then examined to find those that are significantly overrepresented within the selected genes compared with the whole gene list. Various software tools have been developed for this analysis [49,52-55] that rely on thresholds for defining genes as differentially expressed genes. Other tools, such as GSEA (gene set enrichment analysis) [56] and global test [57], are instead based on continuous measures for gene expression. The criticism of the threshold-dependent methods is that the choice of the thresholds or the lengths of the lists for selecting differentially expressed genes can dramatically affect the identification of overrepresented functional categories [58-61].

	NCBI: H. sapiens genes (REF)	oligoD.48			
Biological Process	#	#	expected	+/-	ΔP value
Nucleoside, nucleotide and nucleic acid metabolism	3343	231	180.22	+	1.58E-03
Intracellular protein traffic	1008	83	54.34	+	4.07E-03
Cell cycle	1009	83	54.40	+	4.20E-03
Cell proliferation and differentiation	1028	82	56.42	+	1.17E-02
Intracellular signaling cascade	871	73	46.96	+	2.88E-02
Cell cycle control	418	41	22.53	+	3.84E-02
Oncogenesis	472	41	25.45	+	7.79E-02
Developmental processes	2152	146	116.02	+	8.57E-02
General vesicle transport	251	27	13.53	+	1.10E-01
Apoptosis	531	42	28.63	+	3.25E-01
Transport	1306	90	70.41	+	3.66E-01
Mesoderm development	551	46	29.70	+	4.36E-01
Cell structure and motility	1148	78	61.89	+	7.47E-01
Amino acid metabolism	230	20	12.40	+	8.65E-01
Carbohydrate transport	50	8	2.70	+	9.43E-01

	NCBI: H. sapiens genes (REF)	D.48			
Molecular Function	#	#	expected	+/-	ΔP value
Select regulatory molecule	1190	93	64.15	+	6.84E-03
Transporter	648	65	34.93	+	2.55E-02
Other miscellaneous function protein	427	41	23.02	+	6.47E-02
Phosphatase	263	26	14.18	+	8.46E-02
Hydrolase	736	58	39.68	+	9.53E-02
G-protein	288	29	15.53	+	2.15E-01
Membrane traffic protein	359	31	19.35	+	2.45E-01
Miscellaneous function	866	63	46.69	+	3.42E-01
Cytoskeletal protein	878	63	47.33	+	4.43E-01
Histone	86	12	4.64	+	4.81E-01
Nucleotide phosphatase	26	6	1.40	+	5.15E-01
Small GTPase	221	23	11.91	+	5.36E-01
Actin binding cytoskeletal protein	390	34	21.03	+	8.48E-01
Non-motor actin binding protein	165	18	8.90	+	9.21E-01
Nucleic acid binding	2850	176	153.65	+	9.42E-01

Figure 5. Example of Panther output. Panther (http://www.pantherdb.org) is a publicly available database developed by Applied Biosystem, whose annotation is based on data from Celera, the private institution involved in the initial sequencing of the human genome. Enriched themes in the category of Biological Process and Molecular Function are reported together with number of genes present in each category, number of genes modulated in the experiment, number of genes expected to be modulated by chance and significance values.

The use of Gene Ontology has a number of characteristic flaws due to a failure to address basic ontological principles [62,63]. These flaws are partially overcome by Panther (http://www.pantherdb.org), another publicly available database developed by Applied

Biosystem, which uses a more limited set of categories, thus strongly limiting the number of sub-branches (Figure 5). It has been designed for easy navigation and to partition proteins into biologically meaningful groups with a simplified ontology. This database has the additional advantage of using a more extensive annotation based on data from Celera, the private institution involved in the initial sequencing of the human genome.

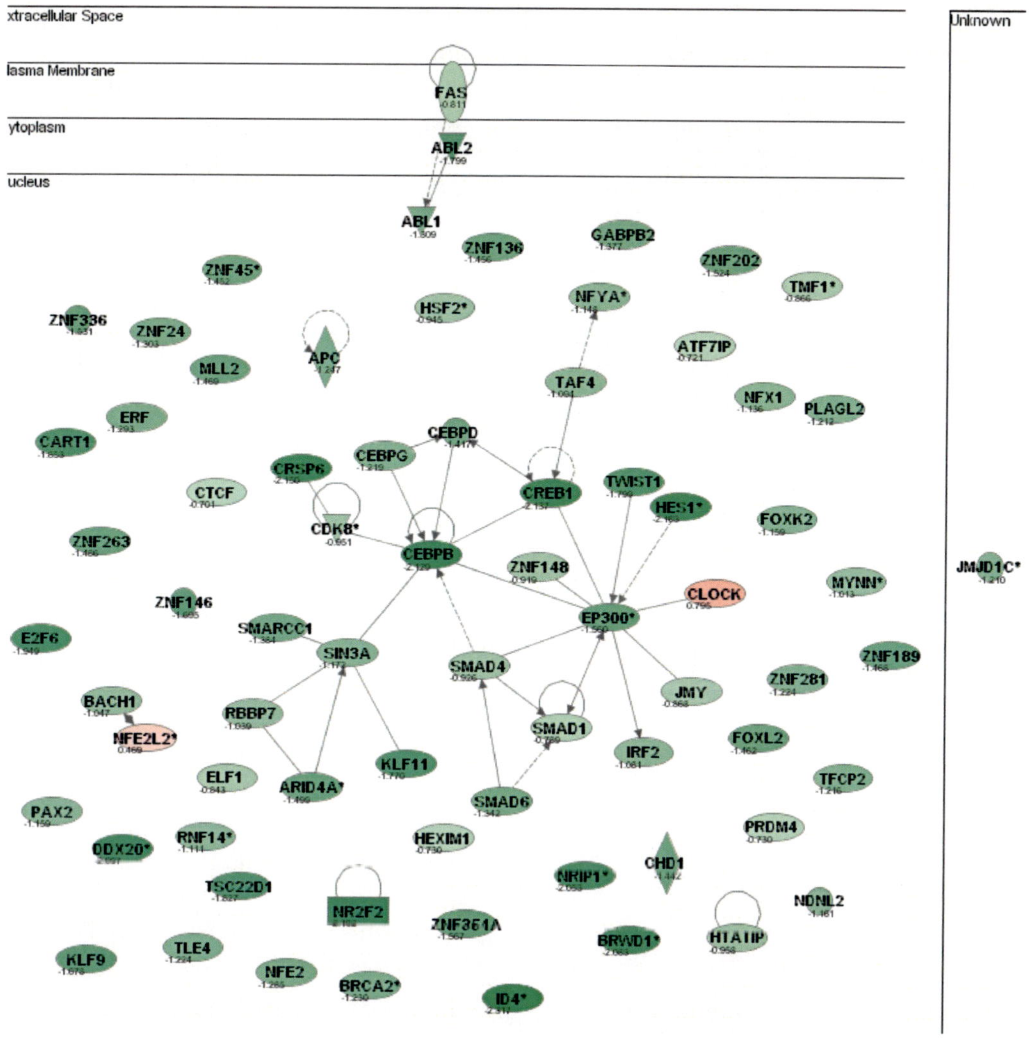

Figure 6. Example of Ingenuity network. Ingenuity (www.ingenuity.com) networks are based on manually curated and fully traceable data derived from primary literature sources. Gene expression data can be superimposed and are coloured according to modulation (green=down regulated, red=up regulated). Darker colours represent genes that have a stronger modulation.

Besides public sources of annotation, there are a number of commercial data mining databases consisting of highly structured knowledge ontologies. Some of these use text mining algorithms e.g., GeneGo' Metacore (www.genego.com) or Genomaticx Bibliosphere (http://www.genomatix.de) to extract biological knowledge in an automated manner whilst others are based primarily on manual curating, e.g., Ingenuity (www.ingenuity.com) or

Ariadne Genomics' PathwayStudio (http://www.ariadnegenomics.com/). The strength of databases such as Ingenuity is the availability of manually curated and fully traceable data derived from primary literature sources. It generates networks of protein/gene relations based on published literature findings (Figure 6).

Most of these tools provide a direct link to biology knowledge databases (e.g. GO, KEGG, Biocarta, PUBMED, etc.), allowing a better understanding of the pathway or pathways affected. Typically, these databases facilitate the identification of biological functions that are overepresented within the set of differentially expressed genes.

The identification of overepresented functions is typically based on Hypergeometric tests [64]. In probability theory and statistics, the hypergeometric distribution is a discrete probability distribution that describes the number of successes in a sequence of n draws from a finite population without replacement. The test based on the hypergeometric distribution is identical to the corresponding one-tailed version of Fisher's exact test. Reciprocally, the p-value of a two-sided Fisher's exact test can be calculated as the sum of two appropriate hypergeometric tests.

Another possible application could be the link of alternative splicing events, detected with the new exon oriented Affymetrix microarray platform, to functional pathways depicted by conventional differential expression analysis.

CONCLUSION

Microarray technology has become the most widely used technique for evaluating differential gene expression, allowing researchers to perform an extensive genome-wide analysis in a single experiment. Technically, microarray technology has attained a good level of accuracy and reproducibility. This has been aided by the automation of several steps involved, such as the washing and the hybridization procedures, resulting in reduced variability attributable to these areas.

In this chapter we have reviewed the main steps involved in data preprocessing and analysis, with an emphasis on the importance of experimental design and quality controls to limit the variability arising from technical issues rather than biological events. Despite the establishment of data analysis systems there is still no clear consensus on the best method to use. We have tried to cover the most commonly used techniques highlighting their strengths and limits. Most of them depend on command line tools primarily enclosed in Bioconductor software, which is becoming a standard tool used by the microarray community. However, some tentative moves are being made to simplify the use of statistical tools. One example is the oneChannelGUI [4] tool, which utilizes a graphical interface and uses the bioconductor libraries to perform a variety of microarray analyses.

The lack of a fully annotated genome remains an open issue. Sequencing of the human genome proceeded much faster than functional annotation of the newly identified genes, leaving a significant number of genes with unknown functions. With such limitations in mind, we have reviewed the main databases currently available for functional annotation and discussed the importance of not assessing only single genes but attempting to view them in the correct biological context using pathway-oriented analysis.

REFERENCES

[1] Shalon D, Smith SJ, Brown PO. A DNA microarray system for analyzing complex DNA samples using two-color fluorescent probe hybridization, *Genome Res* 1996;6:639-645.

[2] Schena M, Shalon D, Davis RW et al. Quantitative monitoring of gene expression patterns with a complementary DNA microarray, *Science* 1995;270:467-470.

[3] Abdueva D, Wing MR, Schaub B et al. Experimental Comparison and Evaluation of the Affymetrix Exon and U133Plus2 GeneChip Arrays, *PLoS ONE* 2007;2:e913.

[4] Sanges R, Cordero F, Calogero RA. oneChannelGUI: a graphical interface to Bioconductor tools, designed for life scientists who are not familiar with R language. *Bioinformatics*. 2007 Sep 17

[5] Wong ML, Medrano JF: Real-time PCR for mRNA quantitation. *Biotechniques* 2005, 39:75-85.

[6] Arya M, Shergill IS, Williamson M, Gommersall L, Arya N, Patel HR. Basic principles of real-time quantitative PCR. *Expert Rev Mol Diagn* 2005, **5**:209-219.

[7] Wilhelm J, Pingoud A. Real-time polymerase chain reaction. *Chembiochem* 2003, 4:1120-1128.

[8] Chen JJ, Delongchamp RR, Tsai CA, Hsueh HM, Sistare F, Thompson KL, Desai VG, Fuscoe JC. Analysis of variance components in gene expression data. *Bioinformatics*. 2004 Jun 12;20(9):1436-46.

[9] Mary-Huard T, Daudin JJ, Baccini M, Biggeri A, Bar-Hen A. Biases induced by pooling samples in microarray experiments. *Bioinformatics*. 2007 Jul 1;23(13):i313-8.

[10] Shih JH et al. Effects of pooling mrna in microarray class comparisons. *Bioinformatics* (2004), 20: 3318-3325

[11] Nygaard V, Hovig E. Options available for profiling small samples: a review of sample amplification technology when combined with microarray profiling. *Nucleic Acids Res*. 2006 Feb 9;34(3):996-1014.

[12] Diboun I, Wernisch L, Orengo CA, Koltzenburg M. Microarray analysis after RNA amplification can detect pronounced differences in gene expression using limma. *BMCGenomics* 2006, **7**:252

[13] R. Gentleman, V.J. Carey, W. Huber, R.A.Irizarry and S. Dudoit. *Bioinformatics and Computational Biology Solutions Using R and Bioconductor*. New York, NY: Springer; 2005

[14] Jolliffe, I.T. Principal Component Analysis, 2nd ed., Series: Springer Series in Statistics, 2002

[15] Brian S. Everitt, Sabine Landau, and Morven Leese. *Cluster Analysis*, 4th ed., Oxford University Press ed., 2001

[16] Millenaar FF, Okyere J, May ST, van Zanten M, Voesenek LA, Peeters AJ. How to decide? Different methods of calculating gene expression from short oligonucleotide array data will give different results. *BMC Bioinformatics* 2006, 7:137

[17] Barash Y, Dehan E, Krupsky M, Franklin W, Geraci M, Friedman N, Kaminski N. Comparative analysis algorithms for signal quantitation from oligonucleotide microarrays. *Bioinformatics* 2004, 20:839-846

[18] Choe SE, Boutros M, Michelson AM, Church GM, Halfon MS. Preferred analysis methods for Affymetrix GeneChips revealed by a wholly defined control datasets. *GenomeBiology* 2005, 6:R16

[19] Seo J, Bakay M, Chen YW, Hilmer S, Shneiderman B, Hoffman EP. Interactively optimizing signal-to-noise ratios in expression profiling: project-specific algorithm selection and detection p-value weighting in Affymetrix microarrays. *Bioinformatics* 2004, 20:2534-2544

[20] Lim WK, Wang K, Lefebvre C, Califano A. Comparative analysis of microarray normalization procedures: effects on reverse engineering gene networks, *Bioinformatics.* 2007 Jul 1;23(13):i282-8.

[21] Pepper SD, Saunders EK, Edwards LE, Wilson CL, Miller CJ. The utility of MAS5 expression summary and detection call algorithms. *BMC Bioinformatics.* 2007 Jul 30; 8:273.

[22] Irizarry RA, Bolstad BM, Collin F, Cope LM, Hobbs B, Speed TP. Summaries of Affymetrix GeneChip probe level data, *Nucleic Acids Res* 2003;31:e15.

[23] Wu Z, Irizarry RA. Preprocessing of oligonucleotide array data, *Nat Biotechnol* 2004; 22:656-658, author reply 658.

[24] Seo J, Hoffman EP. Probe set algorithms: is there a rational best bet?, *BMC Bioinformatics* 2006;7:395.

[25] Bolstad BM, Irizarry RA, Astrand M et al. A comparison of normalization methods for high density oligonucleotide array data based on variance and bias, *Bioinformatics* 2003;19:185-193.

[26] Boes T, Neuhauser M. Normalization for Affymetrix GeneChips, *Methods Inf Med* 2005; 44:414-417.

[27] Su,A.I., Cooke,M.P., Ching,K.A., Hakak,Y., Walker,J.R., Wiltshire,T., Orth,A.P., Vega,R.G., Sapinoso,L.M. et al. (2002) Large-scale analysis of the human and mouse transcriptomes. *PNAS*, 99, 4465–4470.

[28] Jongeneel,C.V., Iseli,C., Stevenson,B.J., Riggins,G.J., Lal,A., Mackay,A., Harris,R.A., O'Hare,M.J., Neville, A.M. et al. (2003). Comprehensive sampling of gene expression in human cell lines with massively parallel signature sequencing. *PNAS*, 100, 4702–4705.

[29] Liu WM, Mei R, Di X, Ryder TB, Hubbell E, Dee S, Webster TA, Harrington CA, Ho MH, Baid J, Smeekens SP. Analysis of high density expression microarrays with signed-rank call algorithms., *Bioinformatics* 2002;18:1593-1599.

[30] von Heydebreck A, Huber W, Gentleman R. Differential expression with the Bioconductor Project, Bioconductor Project Working Papers 2004.

[31] Tusher VG., Tibshirani R., and Chu G. Significance analysis of microarrays applied to the ionizing radiation response. *PNAS*, 2001 vol. 98(9)

[32] Nueda MJ, Conesa A, Westerhuis JA et al. Discovering gene expression patterns in time course microarray experiments by ANOVA-SCA, *Bioinformatics* 2007;23:1792-1800.

[33] Conesa A, Nueda MJ, Ferrer A et al. maSigPro: a method to identify significantly differential expression profiles in time-course microarray experiments, *Bioinformatics* 2006;22:1096-1102.

[34] Delmar P, Robin S, Daudin JJ. VarMixt. Efficient variance modelling for the differential analysis of replicated gene expression data, *Bioinformatics* 2005;21:502-508.

[35] Xie Y, Pan W, Khodursky AB. A note on using permutation-based false discovery rate estimates to compare different analysis methods for microarray data, *Bioinformatics.* 2005; 21:4280-4288.

[36] Guo X, Pan W. Using weighted permutation scores to detect differential gene expression with microarray data, *J Bioinform Comput Biol* 2005;3:989-1006.

[37] Zhang S. A comprehensive evaluation of SAM, the SAM R-package and a simple modification to improve its performance, *BMC Bioinformatics* 2007;8:230.

[38] Larsson O., Wahlestedt C. and Timmons JA. Considerations when using the significance analysis of microarrays (SAM) algorithm. *BMC Bioinformatics* 2005, 6:129

[39] Smyth GK. Linear models and empirical bayes methods for assessing differential expression in microarray experiments, *Stat Appl Genet Mol Biol* 2004;3:Article3.

[40] Jeffery IB, Higgins DG, Culhane AC. Comparison and evaluation of methods for generating differentially expressed gene lists from microarray data, *BMC Bioinformatics* 2006;7:359.

[41] Reiner A, Yekutieli D, Benjamini Y. Identifying differentially expressed genes using false discovery rate controlling procedures, *Bioinformatics* 2003;19:368-375.

[42] Draper N, Smith H. Applied Regression Analysis. New York, Wiley ed., 1998.

[43] Wang Y, Barbacioru C, Hyland F, Xiao W, Hunkapiller KL, Blake J, Chan F, Gonzalez C, Zhang L, Samaha RR. Large scale real-time PCR validation on gene expression measurements from two commercial long-oligonucleotide microarrays. *BMC Genomics.* 2006;**7**:59. doi: 10.1186/1471-2164-7-59.

[44] Bosotti R, Locatelli G, Healy S, Scacheri E, Sartori L, Mercurio C, Calogero R and Isacchi A. Cross platform microarray analysis for robust identification of differentially expressed genes. *BMC Bioinformatics*. 2007 Mar 8;8 Suppl 1:S5.

[45] Barnes M, Freudenberg J, Thompson S, Aronow B, Pavlidis P: Experimental comparison and cross-validation of the Affymetrix and Illumina gene expression analysis platforms. *Nucleic Acids Res* 2005, 33:5914-5923.

[46] Clark, T, Sugnet, C, Ares, M. Genome wide analysis of mRNA processing in yeast using splicing-specific microarrays, *Science* 2002 May 3;296(5569):907-10

[47] Tsai J, Sultana R, Lee Y, Pertea G, Karamycheva S, Antonescu V, Cho J, Parvizi B, Cheung F, Quackenbush J. RESOURCERER: a database for annotating and linking microarray resources within and across species. *Genome Biol.* 2001;2(11): SOFTWARE 0002.

[48] Jeffery IB, Higgins DG, Culhane AC. Comparison and evaluation of methods for generating differentially expressed gene lists from microarray data, *BMC Bioinformatics* 2006;7:359.

[49] Hosack DA, Dennis G Jr, Sherman BT, Lane HC, Lempicki RA. Identifying biological themes within lists of genes with EASE, *Genome Biol* 2003;4:R70.

[50] Yu JX, Sieuwerts AM, Zhang Y, Martens JW, Smid M, Klijn JG, Wang Y, Foekens JA. Pathway analysis of gene signatures predicting metastasis of node-negative primary breast cancer. *BMC Cancer* (2007) 7: 182

[51] Ashburner M, Ball CA, Blake JA, Botstein D, Butler H, Cherry JM, Davis AP, Dolinski K, Dwight SS, Eppig JT, Harris MA, Hill DP, Issel-Tarver L, Kasarskis A, Lewis S, Matese JC, Richardson JE, Ringwald M, Rubin GM, Sherlock G. et al. Gene ontology: tool for the unification of biology. The Gene Ontology Consortium, *Nat Genet* 2000;25:25-29.

[52] Al-Shahrour F, Díaz-Uriarte R, Dopazo J. FatiGO: a web tool for finding significant associations of Gene Ontology terms with groups of genes. *Bioinformatics.* 2004 Mar 1;20(4):578-80.

[53] Alexa A, Rahnenführer J, Lengauer T. Improved scoring of functional groups from gene expression data by decorrelating GO graph structure. *Bioinformatics*. 2006 Jul 1;22(13):1600-7

[54] Draghici S, Khatri P, Bhavsar P, Shah A, Krawetz SA, Tainsky MA. Onto-Tools, the toolkit of the modern biologist: Onto-Express, Onto-Compare, Onto-Design and Onto-Translate. *Nucleic Acids Res*. 2003 Jul 1;31(13):3775-81.

[55] Pehkonen P, Wong G, Törönen P. Theme discovery from gene lists for identification and viewing of multiple functional groups. *BMC Bioinformatics*. 2005 Jun 29;6:162.

[56] Subramanian A, Tamayo P, Mootha VK, Mukherjee S, Ebert BL, Gillette MA, Paulovich A, Pomeroy SL, Golub TR, Lander ES, Mesirov JP. Gene set enrichment analysis: a knowledge-based approach for interpreting genome-wide expression profiles, *PNAS*, 2005, 102: 15545-15550.

[57] Goeman JJ, van de Geer SA, de Kort F, van Houwelingen HC. A global test for groups of genes: testing association with a clinical outcome. *Bioinformatics*,2004, 20: 93-99.

[58] Ben-Shaul Y, Bergman H, Soreq H. Identifying subtle interrelated changes in functional gene categories using continuous measures of gene expression. *Bioinformatics.* 2005 Apr 1;21(7):1129-37

[59] Nilsson R, Peña JM, Björkegren J, Tegnér J. Detecting multivariate differentially expressed genes. *BMC Bioinformatics.* 2007 May 9; 8:150.

[60] Pan KH, Lih CJ, Cohen SN. Effects of threshold choice on biological conclusions reached during analysis of gene expression by DNA microarrays, *PNAS*, 2005, 102: 8961-8965.

[61] Da Yang, Yanhui Li, Hui Xiao, Qing Liu, Min Zhang, Jing Zhu,Wencai Ma, Chen Yao, Jing Wang, Dong Wang, Zheng Guo, Baofeng Yang. Gaining confidence in biological interpretation of the microarray data: the functional consistence of the significant GO categories. *Bioinformatics* 2007 Nov 15

[62] Köhler J, Munn K, Rüegg A, Skusa A, Smith B. Quality control for terms and definitions in ontologies and taxonomies, *BMC Bioinformatics* 2006;7:212.

[63] Park YR, Park CH, Kim JH. GOChase: correcting errors from Gene Ontology-based annotations for gene products, *Bioinformatics* 2005;21:829-831.

In: Oligonucleotide Array Sequence Analysis
Editors: M.K. Moretti and L.J. Rizzo, pp. 367-386
ISBN: 978-1-60456-542-3

Chapter 13

Simultaneous Qualitative and Quantitative Identification of Multiple Targets by Polymerase Chain Reaction on Microarray

Natalia V. Zakharova[1,*], Alexei L. Drobyshev[2,**], Dmitry A. Khodakov[2], Anna I. Myznikova[2] and Irina A. Gorban[3]

[1]State Research Institute of Genetics, 1st Dorozhny Pr. 1
117545 Moscow, Russian Federation
[2]Engelhardt Institute of Molecular Biology, Vavilov Str. 32
119991 Moscow, Russian Federation
[3]Moscow Institute of Physics and Technology, Institutsky Per. 9
141700 Dolgoprudny, Moscow Region, Russian Federation

Abstract

An approach for multiplex qualitative and quantitative microarray-based PCR analysis has been proposed. The characteristics of PCR executed on a gel-based oligonucleotide microarray with immobilized forward primers and a single common reverse primer in solution were investigated for several DNA targets. One-stage multiplex on-chip PCR was studied for simultaneous amplification of herpes simplex viruses types 1 and 2, cytomegalovirus DNA, and bacteriophage lambda DNA as an internal control. Additionally, the joint analysis of increased number of targets (with addition of *Chlamydia trachomatis*, *Mycoplasma hominis*, and *Ureaplasma urealyticum* DNA) was done in a two-stage version of assay: the first stage was in-tube PCR with target-specific primers, while the reverse ones contained 5'-adapter region; the second stage was on-chip amplification with immobilized target-specific forward primers and adapter as common reverse primer in solution. The possible application of one-stage reaction for human cDNA analysis was additionally demonstrated with utilization of a common poly-T-containing primer in solution. SYBR green I and Cy-5 labeled dUTP were used for real-time and end-point detection of specific PCR products. The efficiencies of both one-stage and two-stage reactions were shown to be strongly dependent on magnesium and primers concentrations. Quantitative PCR in both versions was studied with 10-fold serial

* E-mail address: NVZakharova@yandex.ru (to whom correspondences should be addressed)
** E-mail address: dro@newmail.ru (to whom correspondences should be addressed)

dilutions of phage lambda DNA. The method enabled detection of six DNA copies per reaction for both versions of assay. The quantitative interval for one-stage reaction covered eight orders of concentration. The revealed significant effect of gel pad size on microarray PCR effectiveness is discussed.

INTRODUCTION

As a standard tool of modern molecular biology, PCR is routinely used in many microarray-related protocols, mainly in the course of target preparation. It is aimed to increase the amount of analyzed target and decrease its excessive complexity. In the most simple scheme only one locus of target DNA is amplified and subsequently analyzed on microarray [1, 2]. For simultaneous analysis of several loci, the multiplex PCR with several pairs of primers could be used. However, in this case the degree of multiplexity is limited by potential primer dimmer formation. This effect is due to parasitic interactions in the pool of primers competing with primer-to-target interaction. It grows roughly like a square of number of primers and could be partially overcome by careful design of primer pairs [3]. Primer dimmer artifacts could be suppressed even further using various schemes with universal PCR.

Universal PCR is a multiplex polymerase chain reaction performed with a single or pair of universal primers to amplify a broad range of targets [4]. E. g., it could be designed as two-stage reaction, where the first stage occurs in the presence of chimerical primers consisting of locus-specific 3'-parts and universal sequences at their 5'-parts. Only universal parts are used as primers for the second stage amplification. One-stage reaction with chimerical primers could be performed with decreased annealing temperature at the first few cycles for amplification with specific 3'-parts of the primers, while at the subsequent cycles the temperature can be significantly increased enabling the annealing of only full-length chimerical primers [5]. These approaches selectively reduce the complexity of target by specific amplification of the loci of interest. The target complexity could be reduced in less specific manner without locus-specific sequences and without any risk of primer dimmer formation. In such an approach [6-8] genomic DNA is digested with restriction enzymes, common primer adapter is ligated and PCR is carried out with a single primer. As the restriction fragments longer than 1000 nucleotides are poorly amplified, the complexity of PCR product of digested DNA is lower than that of source DNA. Some modification of this procedure [9] can produce specific amplification of desired fragments. Also, whole-genomic DNA can be amplified without any reduction of complexity and the resultant product can be used e.g. for array-based comparative genomic hybridisation [10].

Universal PCR could be used for amplification of artificial sequences in whole genome genotyping experiments [11, 12]. These approaches imply the annealing and ligation of oligonucleotide probes at polymorphic sites of target DNA. The resultant probes contain sites complementary to universal primers as well as artificial sequences corresponding to particular site of polymorphism. Then the latter hybridize to complementary sequences of microarray upon the completion of PCR amplification of ligated probes with universal primers.

Solid phase PCR in microarray technology provides an elegant way for solution of many microarray-related problems but still has to be essentially elaborated for broad use in scientific community. In this approach at least some of immobilized oligonucleotides serve as primers in PCR carried out right over the microarray. Being immobilized on a solid phase, these primers cannot form dimmers regardless of their structure. For example single locus of target could be amplified with two liquid-phase primers while allele-specific solid-phase primers are elongated depending on target's sequence complementary to their 3′-end. The elongation of every solid-phase primer could be monitored either on-line [13] or upon the completion of reaction [14].

One of the more sophisticated versions of this approach implied the utilization of mineral oil to separate gel pads physically and to form thereby an array consisting of independent reaction volumes with immobilized primers [15]. This immobilization was reversible, i.e., 5′-immobilised solid phase primers containing 5′-triribonucleotide spacer were released to liquid phase by means of digestion of this spacer by ribonuclease A. Allele-specific primers were solid-phase-immobilized and internal in respect to ribonuclease-cleavable primers. The detection of allele-specific product was performed by means of post-PCR hybridization of fluorescently labeled oligonucleotides.

Another scheme with separation of reaction volumes implied an array of 3072 through-holes in 300μm thick stainless steel plate containing a set of gene-specific primer pairs for real time PCR-based expression profiling [16]. The higher parallelism (up to 300,000) was reached in the case of picoliter-scale plate manufactured by anisotropic etching of fiber optic face plate with resultant well density of 480 wells/mm^2 [17]. This microfabricated device enabled both liquid- and solid-phase PCR. In the latter case the solid-phase reverse primers are immobilized on Sepharose beads, while liquid phase contains both forward and reverse primers with 8-fold excess of forward ones. Later this technology was used (with substantial modifications) in novel method of massive parallel genome-wide DNA sequencing [18]. Nevertheless, the wide application of these arrays is still complicated due to some technical problems, particularly complication with their manufacturing.

Recently an original approach utilizing two-stage nested PCR and gel-based microarray for simultaneous quantitative analysis of three different viruses with specific pairs of primers was developed [19]. The approach was shown to be high-sensitive, specific, and its quantitative dynamic range covered up to six orders of concentrations. Nevertheless, the utilization of only specific primers essentially limits the number of targets for simultaneous analysis because of possible primer dimmer formation. In this respect, the present investigation is related to an improved approach, which enables quantitative analysis of increased number of nucleic targets by PCR on gel-based microarray with utilization of only one liquid-phase primer with common specificity. Direct one-stage reaction is studied for homologues viral DNAs and human cDNA. The two-stage indirect version of approach with introduction of universal nucleotide sequence via first-stage PCR with chimerical primers is developed for non-homologues targets. The detailed optimization of PCR conditions performed in the frame of the investigation has revealed several characteristics significant for high-effective amplification on gel-based microarray.

MATERIALS AND METHODS

Microarrays

Oligonucleotide microarrays were fabricated by spotting of gel polymerization solutions [20] containing different solid-phase primers onto glass slides (Corning 2947 Micro Slides, Corning, USA) using QArray spotter (Genetix, England). Glass slides treatment, gel polymerization and washing conditions were described earlier [19]. The fabricated microarrays consisted of spatially separated gel pads with varied 5`-immobilised forward primers specific to different DNA targets. In most of the experiments, gel pads with diameter of 600 μm were used (unless otherwise specified). Optimal primer concentrations are described in "Results".

Primers

Primers were designed using Oligo 6 (Molecular Biology Insights, USA) and DNA sequences available from public databases (Table 1). Primers were synthesized on an ABI-394 DNA/RNA synthesizer (Applied Biosystems, USA) by standard phosphoramidite method and purified by reverse phase high performance liquid chromatography on C18-Nucleosil (5 μm, 4.6 x 250 mm) columns (Sigma, USA). 5'-Amino-Modifier C6 (Glen Research, USA) was used for introduction of amino group into 5'-nucleotides of primers for on-chip immobilization (below referred as "solid-phase primers"). Primers without such amino groups are referred below as "liquid-phase primers".

The solid-phase forward primers were designed to contain highly specific oligonucleotide sequences free from any essential internal hairpin structures, the T_m of the primers were in the range of 55-60 °C (18-23 nucleotides). The varying nucleotides in allele-specific primers designed for HSV-1 and HSV-2 were located on 3`-ends.

Table 1. The sequences of primers used in the assay

Primer[a]	Sequence from 5′ to 3′	Position[b]
Solid-phase forward primers		
HSV-1_F_I	CGG ACT CCA TAT TTG TGC	3240
HSV-2_F_I	CGG ACT CCA TTT TCG TTT	2675
CMV_F_I	CGT GTT TGT CCG CTT TCG	2739
UU_F_I	ACA AAC CCA ACT ATT CCA TAT AC	1817
MH_F_I	CTA TTC TAG AAA GCC ACG ATG C	140
CT_F_I	GTT AAT ACC CGC TGG ATT TGA G	467
BL_F_I	AGG AGC TGG ACT TTA CTG ATG	1561
Actb_F_I	TAT TTT GAA TGA TGA GCC TTC GTG	1675
Hprt_F_I	TAA GAA GTT TTG TTC TGT CCT GG	1189
One-step solution-phase reverse primers		
HRP-R	CGC ACC AGA TCC ACG CCC TT	3456 (HSV-1) 2891 (HSV-2) 2951 (CMV)
BL-R	GTC AGG TGG CTC AAT CTC TTC	1670
T_R	GGC CAG TGA ATT GTA ATA CGA CTC ACT ATA GGG AGG CGG $(T)_{25}$	

Table 1. Continued

Primer[a]	Sequence from 5′ to 3′	Position[b]
Two-stage PCR primers		
HRP_OUT_F[c]	GCG GGT CAT CTA CGG GGA CAC GGA C	2712 (CMV) 3220 (HSV-1) 2655 (HSV-2)
HSV_u_R	*AGG TGG TGA ACG GGC TGT A*CG GTT GAT AAA CGC GCA GTT G	3483 (HSV-1) 2918 (HSV-2)
CMV_u_R	*AGG TGG TGA ACG GGC TGT A*GC CGA TGT AAC GTT TCT TGC	2901
UU_OUT_F	CAG AAG GTG CTG GTG GTG GAC A	1743
UU_u_R	*AGG TGG TGA ACG GGC TGT A*CT TCT GGA ACC TTA GGA TTT AAG TG	1910
MH_OUT_F	GAC TAC ATT ACA CCA GCT CGT TTA GAC	97
MH_u_R	*AGG TGG TGA ACG GGC TGT A*CA ACA ACG TTG ATT CCT CTG T	220
CT_OUT_F	GCG TGT GTG ATG AAG GCT CTA GGG TTG	407
CT_u_R	*AGG TGG TGA ACG GGC TGT A*AA CTA ACT TAC CTT TCC GCC TAC	608
BL_OUT_F	ATT GAG CGT GCA GCC AGT GA	1377
BL_u_R	AGG TGG TGA ACG GGC TGT ACG CTG T	1716
R_u	AGG TGG TGA ACG GGC TGT A	

[a] F, forward; R, reverse; I, immobilized; OUT, first stage PCR primers (outers); u, universal sequence; HSV-1, HSV-2, herpesviruses type 1 and 2, respectively; CMV, cytomegalovirus; UU, *U. urealyticum*; MH, *M. hominis*; CT, *C. trachomatis*; BL, bacteriophag lambda; Actb, human β-actin gene; Hprt, human hypoxanthine phosphoribosyltransferase gene.

[b] The figure indicates the 5′ terminus nucleotide position number according to Gen Bank accession no. M10792 (HSV-1, *pol* gene), AY038367 (HSV-2, *pol* gene), DQ180391 (CMV, *pol* gene), AF085728 (*U. urealyticum*, *ureA* gene), D13314 (*M. hominis*, *adi* gene), DQ019310 (*C. trachomatis*, 16S rRNA gene), J02459 (enterobacteria phage lambda, complete genome), Ensemble Transcript ENST00000331789 (human β-actin gene), and ENST00000298556 (human hypoxanthine phosphoribosyltransferase gene).

[c] The utilized common first stage forward *Herpesviridae* primer corresponds to CMV sequence while analogous HSV sequences contain some varied nucleotides close to 5′ terminus.

Two sets of unmodified liquid-phase primers were designed. A set for one-stage on-chip PCR of homologues targets (Table 1) consisted of a common reverse primer specific to HSV-1, HSV-2, and CMV and a separate reverse primer specific to lambda DNA (used as an internal control). Additionally, solid-phase primers for human β-actin and hypoxanthine phosphoribosyltransferase (Actb_F_I and Hprt_F_I, respectively, Table 1) were designed to be used in combination with common reverse liquid-phase primer T_R. Another set was designed for joint multiplex two-stage PCR of homologues HSV-1, HSV-2, CMV, and non-homologues *U. urealyticum*, *M. hominis*, *C. trachomatis*, and bacteriophage lambda DNAs (Table 1). Target-specific forward primers designed for in-tube first-stage reaction were outer relative to second-stage solid-phase forward ones. First-stage reverse primers were chimerical and contained 3′- target-specific and 5′- universal adapter sequences in order to apply the common reverse primer R_u for further second-stage on-chip PCR (Table 1).

Target Nucleic Acids

HSV-1, HSV-2, CMV, *U. urealyticum*, *M. hominis*, and *C. trachomatis* DNA preparations isolated from anonymous urea, cervical, and blood clinical samples were kindly provided by Scientific Center of Obstetrics, Gynecology and Perinatology, Russian Academy of Medical Sciences. Three samples of each target were utilized to validate the performance of the assay. Bacteriophage lambda DNA (Fermentas, Lithuania) with initial concentration of 6×10^{12} copy per 1 mL were used in optimization experiments and quantitative PCR studies. The 10-fold serial dilutions of lambda DNA was used to obtain concentrations from 6×10^{8} to 6×10^{0} copy per assay. Total RNA was isolated from K562 myelogenous leukemia cells with RNeasy Mini Kit (Qiagen, Germany) according to the manufacturer's protocol.

Reverse Transcription

The reaction mixture for reverse transcription (RT) of total RNA contained 1× StrataScript buffer (Stratagene, USA), 0.2 mM each dATP, dCTP, and dGTP, 0.6 mM dUTP, 2 μM RT-primer (T_R, Table 1), 10 units StrataScript® Reverse Transcriptase, 20 units RNasin® and 10 μL of total RNA preparation per 10 μL reaction. The reverse transcription was performed at 42 °C for 90 min with final incubation at 95 °C for 10 min. One μL of 4-fold diluted cDNA preparation was used for further on-chip PCR.

PCR on Microarray

One-step PCR was performed inside a microarray chamber formed by sealing the microchip slide with a cover glass slide by a Frame-Seal (Bio-Rad, USA). The reaction mixture consisted of 70 mM Tris-HCl (pH 8.6), 16.6 mM $(NH_4)_2SO_4$, 1.5 mM $MgCl_2$, 0.2 mM each dATP, dCTP, and dGTP, 0.6 mM dUTP (Sileks, Russia), 0.2 μM each solution-phase reverse primer (Table 1), 0.01% acetylated BSA (New England BioLabs, USA), 0.5 unit Platinum® *Taq* DNA Polymerase (Invitrogen, USA), 10 μL of template DNA, and 1× SYBR® green I (Molecular Probes, USA) or 2 μM Cy5-dUTP per 25 μl reaction. After initial incubation at 95 °C for 2 min, 60 cycles (by default, 40 s at 95 °C, 20 s at 55 °C and 45 s at 72 °C) were performed.

For two-stage semi-nested PCR the first stage of in-tube amplification with outer forward and chimerical reverse primers was performed with reaction mixture containing 70 mM Tris-HCl (pH 8.6), 16.6 mM $(NH_4)_2SO_4$, 2.5 mM $MgCl_2$, 0.2 mM each dATP, dCTP, and dGTP, 0.6 mM dUTP (Sileks, Russia), 0.02 μM each primer (Table 1), 0.5 units Platinum® *Taq* DNA Polymerase (Invitrogen, USA), and 10 μL of template DNA per 25 μl reaction. PCR was performed as follows: initial incubation for 2 min at 95 °C; 10 three-step cycles with denaturing for 20 s at 95 °C, annealing for 30 s at 55 °C, and elongation for 30 s at 72 °C; 25 three-step cycles with annealing at 68 °C and the same other conditions; and final incubation at 72 °C for 5 min. The second stage was performed on-chip in the conditions described above for one-step reaction. One μL of the first-stage PCR product was used as a template.

Data Acquisition and Processing

For real-time measurements of SYBR® green I fluorescence an original experimental setup consisted of fluorescent microscope equipped with thermotable, a CCD camera, a temperature controller [13] and specialized software ImageExpress® (Biochip-EIMB, Russia) was used. Images were acquired every cycle at the end of elongation step. The melting curves were obtained by measuring the fluorescence with 10 second interval upon the gradual increasing of temperature from 72 °C to 95 °C at the rate of 3 °C/min. The raw fluorescence data (F) were normalized according to formula $(F\text{-}F_{min})/(F_{max}\text{-}F_{min})$.

The end-point SYBR® green I and Cy5 fluorescence were measured with specialized software ImageWare® (Biochip-EIMB, Russia). SYBR® green I fluorescence was detected without removing of reaction mixture from microarray at 72 °C to avoid the contribution of non-specific hybridization. To measure Cy5 fluorescence the slides were unsealed, and unincorporated Cy5-dUTPs were removed by electrophoresis [21] in TAE-buffer at 15 V/cm for 5 min.

RESULTS

Design of the Assay

The scheme of PCR on a microarray with immobilized specific forward primers and a common liquid-phase reverse one is illustrated in Figure 1. Target-specific forward primers are immobilized in gel pads of microarray. This type of immobilization provides every element of microarray with primers in amount sufficient for PCR reaction [13, 19]. In this study we use a single universal primer matching all the targets under investigation (Figure 1A). This combination with spatially separated target-specific solid-phase primers enables PCR with (theoretically) unlimited multiplexity as immobilized solid-phase primers cannot form primer dimmers. In course of PCR, target DNA anneals to immobilized forward primers enabling their elongation by polymerase (Figure 1B). These elongated solid-phase anchored primers serve as templates for complementary strands synthesis by means of elongation of liquid-phase reverse primers (Figure 1C). The accumulated double stranded solid phase anchored product could be visualized by DNA stain (SYBR® green I) or by incorporation of fluorescently labeled dNTPs (not shown).

Such an approach seems to be a powerful tool for parallel analysis of homologous sequences, for example allele genes and gene mutations. The method could be also useful for analysis of targets possessing some common sequences, like poly-A tail in eukaryotic mRNA. The utilization of chimerical primers for introduction of a universal adapter oligonucleotide sequence by PCR (Figure 2) or combined restriction-ligation reactions gives another opportunity to apply on-chip PCR for analysis of complex mixtures of non-homologous nucleic targets. In this connection, the present study was focused on estimations of two on-chip PCR approaches: (i) direct one-stage PCR with one primer specific to several targets, and (ii) indirect two-stage reaction with introduction of universal sequence via in-tube PCR with chimerical primers (Figure 2).

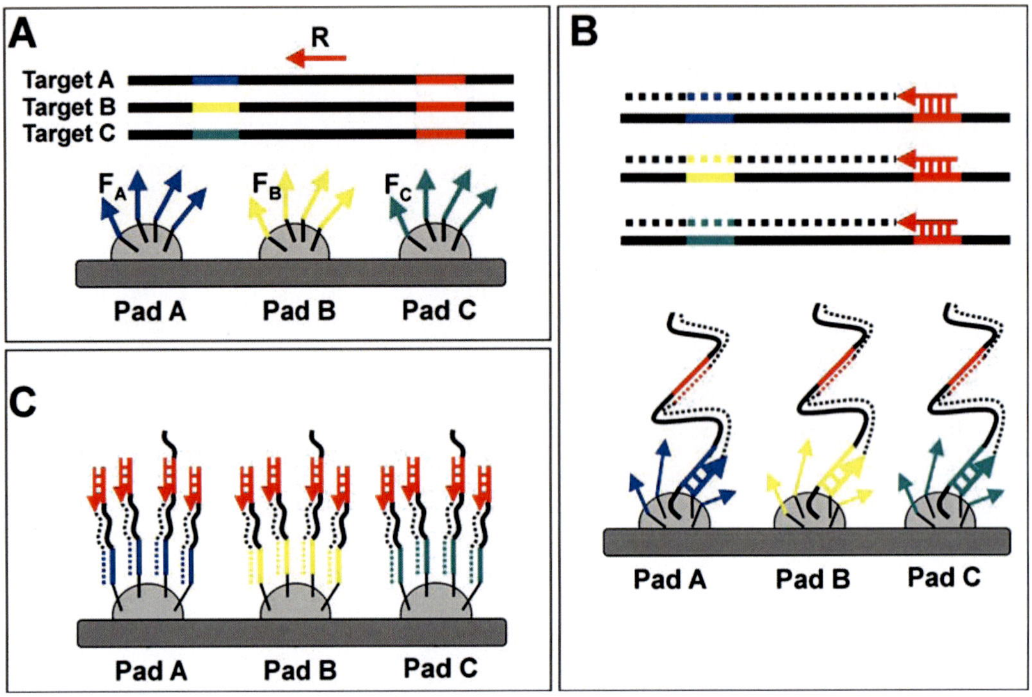

Figure 1. Scheme of PCR on gel pad oligonucleotide microarray. (A) A set of homologues DNA targets and array of gel-immobilised solid-phase forward primers for their identification. Every target contains the sequence specific to immobilized forward primer (both depicted with the same colour). Liquid phase contains only one reverse primer (red coloured) specific to all the targets in pool. (B) Target DNAs anneal to specific solid-phase forward primers enabling their elongation and formation of a complementary second strand. These copies serve as templates for subsequent amplification. (C) Elongated solid-phase anchored strands serve as templates for liquid-phase reverse primers complementary elongation.

The set of six pre-natal/new-born pathogen DNAs (HSV-1, HSV-2, CMV, *U. urealyticum*, *M. hominis*, and *C. trachomatis*) and bacteriophage lambda DNA (included for imitation of internal control) was considered as a model system since it contain both homologous and non-homologous targets. Furthermore, the detailed studies on some on-chip PCR characteristics required several comparative experiments with different nucleic targets (e.g. eukaryotic cDNA) and individually designed microarrays (see bellow).

Two different dyes were investigated for identification of specific on-chip PCR products. Non-specific dsDNA stain SYBR® green I was suitable for both real-time and end-point detection schemes as optimal measurements condition (72 °C) provided melting of non-specific short-chain structures and enabled the reliable detection of stable specific long-chain DNA products. The real-time monitoring of PCR by means of Cy5-dUTP was technically difficult because of strong fluorescence of unincorporated Cy5-dUTP and essential decrease of Cy5 fluorescence under PCR temperature conditions. Nevertheless, Cy5-dUTP was convenient for

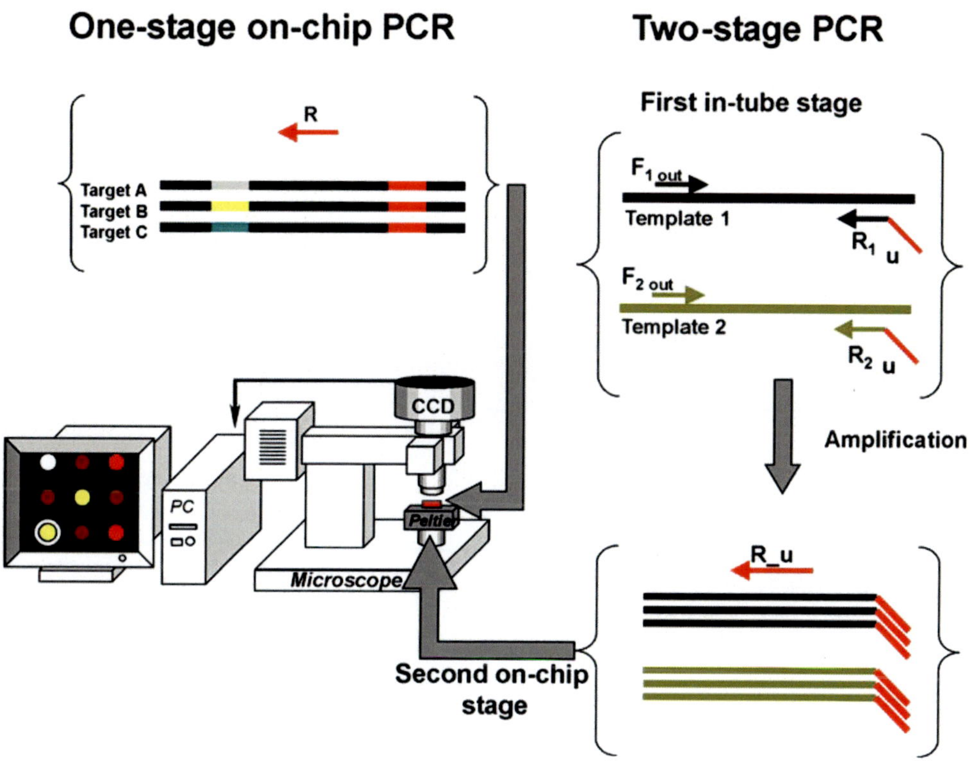

Figure 2. The scheme of one-stage and two-stage analysis via on-chip PCR. Homologues DNA targets can be analysed directly by one-stage on-chip PCR (left) using single liquid-phase primer (R) with common specificity. Non-homologues targets analysis requires two stages PCR (right). First in-tube stage of PCR is necessary for introduction of a universal sequence (u) into DNA products amplified from specific targets. The second on-chip stage of PCR can be performed with a common liquid-phase primer (R_u).

end-point detection of accumulated Cy5-labeled specific products covalently attached to microarray. The electrophoretical removal of unincorporated Cy5-dUTP from microarrays after the PCR enables improved discrimination between positive and negative gel pads. Figure 3 A and B represents the end-point results obtained with the same DNA targets in two parallel on-chip experiments performed with Cy5-dUTP and SYBR® green I, respectively.

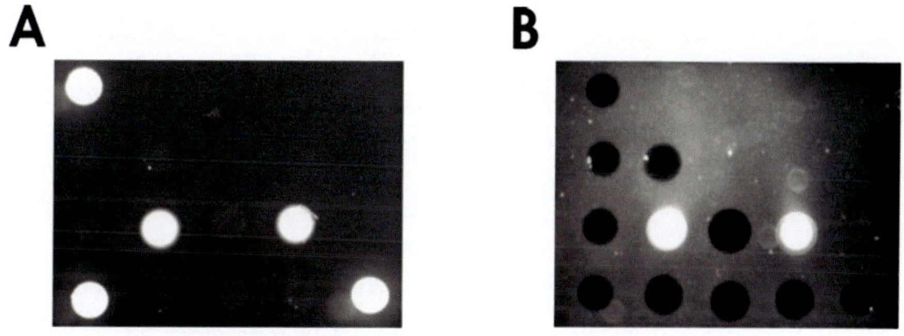

Figure 3. Continued on next page.

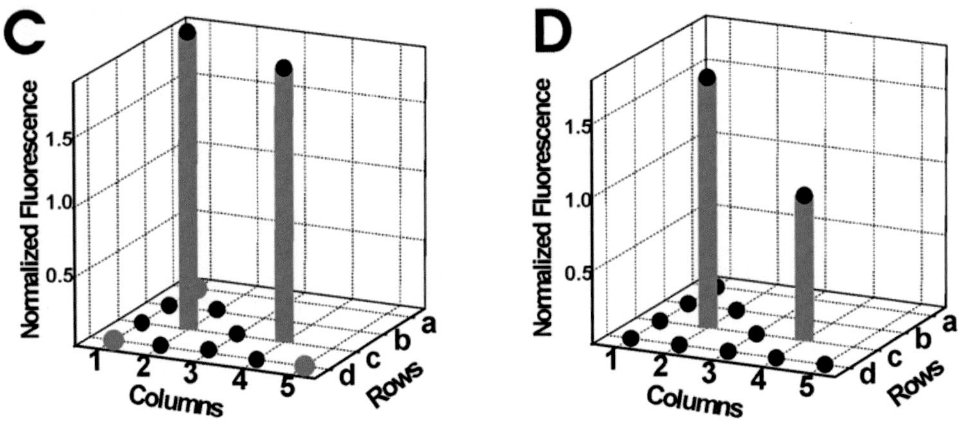

Figure 3. The visualisation of on-chip PCR results with fluorescent dyes. (A) Fluorescent image of microarray after PCR of HSV-1 and phage lambda DNAs performed in the presence of Cy-5-labelled dUTPs, taken after electrophoretical washing and subsequent drying. Bright spots in the corners contain Cy5-labelled immobilized marker oligonucleotide. (B) Fluorescent image of microarray after PCR with the same targets in presence of SYBR green I® taken at 72°C. (C) Quantification of data from panel A. Marker spots were not quantified. (D) Quantification of data from panel B.

Optimization of On-chip PCR

One of significant disadvantage of solid-phase PCR systems described so far is their low efficiency in comparison with traditional in-tube PCR [13, 22-24]. In order to improve this characteristic several factors were varied in this study. The use of SYBR® green I for real-time monitoring of on-chip PCR enables the direct estimations of effects of different parameters on PCR (Figure 4). No significant change was found when 70 mM Tris-HCl (pH 8.6) and 16.6 mM $(NH_4)_2SO_4$ PCR buffer was replaced by 10 mM Tris-HCl (pH 8.6) and 100 mM KCl one, and dNTPs concentrations were varied in the range from 0.1 to 0.4 mM for dATP, dCTP, and dGTP (0.3-1.2 mM dUTP, respectively) (data not shown). The concentration of liquid-phase primers strongly affected the threshold cycle number although did not significantly change the amplification coefficient (Figure 4A, all the curves have the same slope). However the use of solid-phase primers in concentrations less than 100 μM essentially influences the both characteristics (Figure 4B).

The simultaneous presence of more then one liquid-phase primer in reaction mixture significantly increases the chance of primer dimmers formation and leads to accumulation of non-specific double stranded PCR products competing with specific ones for SYBR® green I binding. This possibly caused the decrease of specific solid-phase PCR signals and accordant increase of liquid-phase fluorescence around the gel pad (Figure 4C, solid and opened circles, respectively) observed after 30-35th cycle for the reaction with two (HPR-R and BL-R; Table 1) liquid-phase primers and 10^7 copies of phage lambda DNA. The effect was not observed when only one liquid-phase primer (BL-R, Table 1) was used (Figure 4C, squares). The addition of PCR enhancer solution to reaction mixture with two primers eliminated this effect as well, but some increase of threshold cycle number was observed (Figure 4C, triangles).

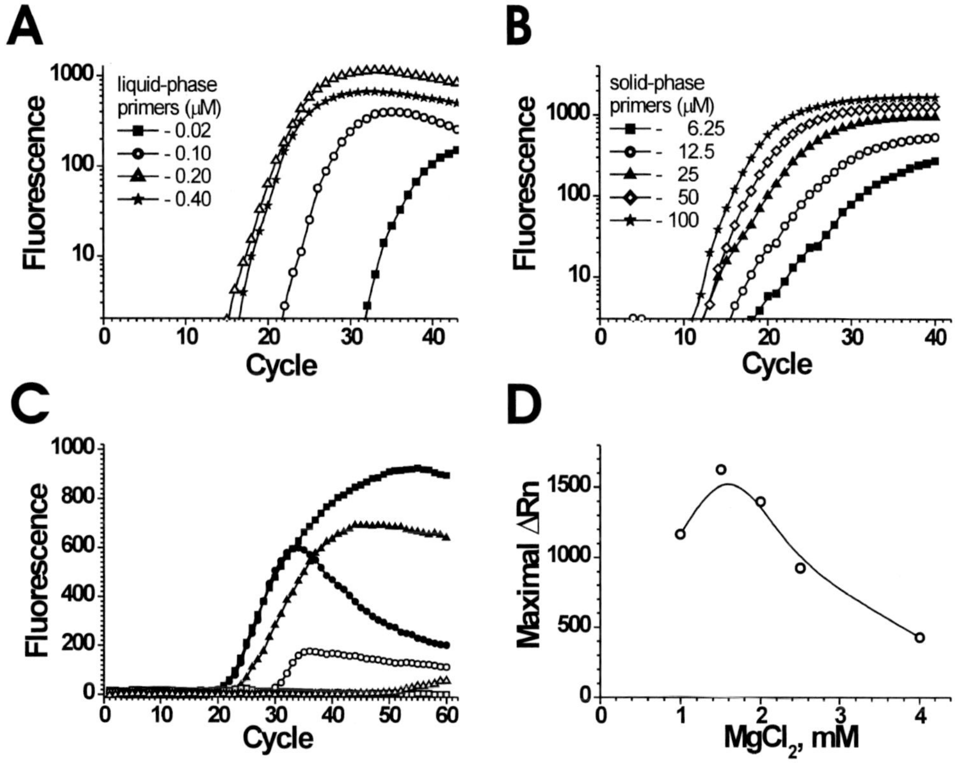

Figure 4. Effects of primers and magnesium concentrations on on-chip PCR characteristics. (A) The effect of liquid-phase primers concentrations on one-stage PCR. Liquid phase contains BL-R and HRP-R primers (Table 1) in equal concentrations indicated on the plot. (B) The effect of solid-phase primer concentration on second stage of two-stage PCR in the presence of 0.2 μM R_u primer (Table 1). (C) The effect of number of different primers in liquid phase on one-stage PCR. Closed symbols depict the fluorescence in specific pad; opened same-shaped symbols show corresponding background fluorescence measured on pads without any immobilized primers: squares – the PCR with one liquid-phase primer (BL-R); circles - the PCR in the presence of two liquid-phase primers (BL-R and HRP-R); triangles – PCR with the two liquid-phase primers in the presence of 3× PCR enhancer solution. (D) The effect of magnesium concentration on one-stage PCR. Solution phase contained one 0.2 μM BL-R primer. Lambda DNA was used as a target for all of the experiments: 6×10^7 copies per reaction for A, C, and D; 6×10^2 copies was the initial concentration in the case of two-stage reaction (panel B).

Magnesium concentration was varied from 1mM to 4mM and 1.5mM was found to be optimal in terms of maximal fluorescent signal (Figure 4D), while the threshold cycle was not significantly affected (not shown). Still the optimal concentration could be various for different sets of simultaneously amplified targets and must be adjusted individually. Interestingly that the optimal magnesium concentration obtained for on-chip reaction was not optimal for in-tube reaction performed in identical conditions (1× SYBR® green I) with the same DNA target (phage lambda DNA). The known inhibitory effect of SYBR® green I [25] requires the increase of magnesium concentration up to 4 mM in case of in-tube reaction only. This discrepancy for on-chip and in-tube reactions could imply some anti-inhibitory properties of gel pads. Nevertheless, SYBR® green I in concentrations higher than one fold inhibits on-chip reaction as well (data not shown).

Quantitative On-Chip PCR

Figure 5A demonstrates the typical PCR curves obtained with SYBR® green I in one-stage on-chip amplification of 10-fold dilution series (from 6 to 6×10^8 copies) of phage lambda DNA. The threshold cycle (C_T) values were determined from the interceptions of the curves with the threshold line (corresponded to 1/20 of maximal increase of fluorescence) and were used to obtain the calibration curve (Figure 5B) for further quantitative determinations of initial target amount. The calibration curve for the two-stage on-chip PCR was generated in the same way with first-stage DNA products as templates.

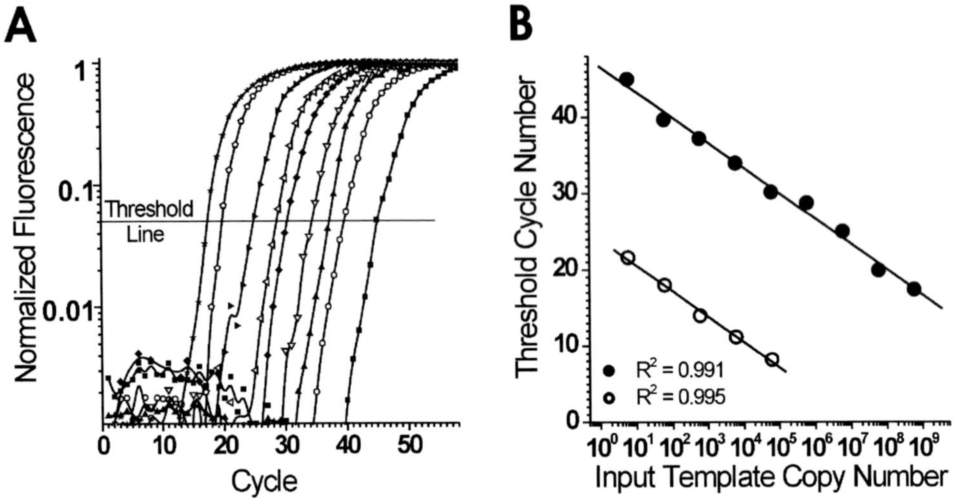

Figure 5. Quantitative on-chip PCR. (A) The amplification curves obtained for a series of 10-fold dilutions ($6 \div 6\times10^8$ copies per reaction, from left to right) of lambda DNA visualised with SYBR green I® fluorescence during one-stage on-chip PCR. (B) The calibration curves for DNA quantification. Closed circles – the calibration curve for one-stage on-chip reaction generated with C_T values calculated from PCR curves of Figure A. Opened circles – the calibration curve for second-stage on-chip PCR after 35 amplification cycles of first-stage in-tube PCR with $6 \div 6\times10^4$ initially input copies of lambda DNA.

The standard deviations of experimental data from linear fit (C_T versus logarithm of input copy number) were 0.85 and 0.37 cycles (with maximal deviations of 1.32 and 0.6 cycles) for one-stage and two-stage experiments, respectively. Both calibration curves have virtually the same slope (3.32 and 3.36 cycles per decade of input copy number, Figure 5B, solid and open circles, respectively) but the two stages scheme provides relatively narrow dynamic range (4 orders of magnitude) as accurate C_T measurements were technically difficult for concentrations of $\geq10^5$ copies per reaction after 35 amplification cycles performed during the first (in-tube) stage. The dynamic range could be broadened by decreasing the number of in-tube stage cycle amount [19]. Maximally it tends to 8 orders of magnitude, like in the case of direct one-stage on-chip PCR (i.e. 0 cycles of preliminary amplification, Figure. 5B). The certain inhibition of PCR was observed for 6×10^9 copies (data not shown), while the precise detection limit value was not determined in the frame of the investigation.

Multiplex On-Chip PCR

Figure 6 shows the end-point results of five individual on-chip experiments on detection of several simultaneously amplified DNA targets. Bacteriophage lambda DNA was added to all the samples analyzed to imitate the internal control. For higher multiplexity two of the shown experiments were performed with artificially prepared mixtures of different DNA targets (Figure 6 B and C). The results shown were obtained both with SYBR® green I (Figure 6 B and C) and Cy5 labelled dUTP (Figure 6 D-F), in one-stage (Figure 6 B and D) and second-stage (Figure 6 C, E, and F) on-chip PCR. Totally three samples of each target were utilized to validate the performance of the assay. The amplification of all analyzed samples was highly specific both in one-stage and two-stage reactions. This was especially important for related *Herpesviridae* targets, which were additionally verified by co-amplification in all possible combinations (Figure 6 B). No false-positive results were obtained with non-specific DNAs utilized in control experiments.

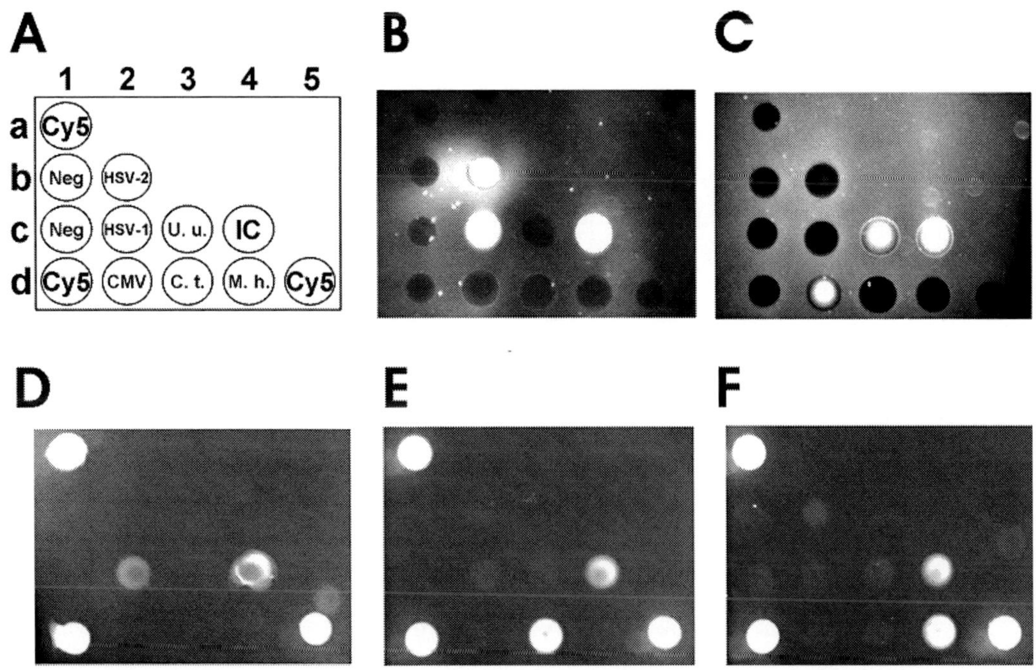

Figure 6. Individual results of multiplex on-chip PCR. (A) The template of an oligonucleotide microarray pads with solid-phase primers for identification of herpes simplex viruses types 1 and 2 ('HSV-1' and 'HSV-2', respectively), cytomegalovirus ('CMV'), *C. trachomatis* ('C. t.'), *M. hominis* ('M. h.'), *U. urealyticum* ('U. u.'), and bacteriophage lambda as an internal control ('IC'). The reference pads: 'Neg' – empty gel pads without immobilized primers; 'Cy5' – pads with 3′-immobilized Cy5-labelled non-specific oligonucleotides. (B, C) The detection of PCR results with SYBR green I®, one-stage (HSV-1 and HSV-2 positive) and second-stage PCR (CMV and *U. urealyticum* positive), respectively. (D-F) The detection of PCR results with Cy-5 labeled dUTP: D – one-stage reaction (HSV-2 positive), E, F – second-stage reactions (*C. trachomatis* and *M. hominis* positive, respectively).

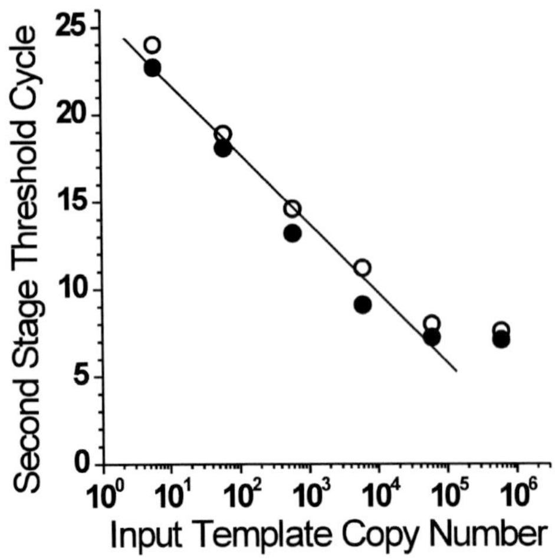

Figure 7. The effect of simultaneous amplification of lambda and HSV-2 DNAs on characteristics of second-stage calibration curve generated for lambda DNA quantification. The C_T values were calculated for the second-stage on-chip PCR after 35 cycles of first-stage in-tube amplification. Closed circles – data for individual PCR of lambda DNA (6 - 6×10^5 copies per reaction); opened circles – PCR of lambda DNA with simultaneous amplification of abundant (about 10^6 copies) HSV-2 DNA.

To estimate the possible competitions between simultaneous PCR of different DNA targets, the results of individual amplification of lambda DNA were compared with results of multiplex reactions. Figure 7 illustrates the reasonable concordance of PCR results obtained for series of lambda DNA concentrations (from 6 to 6×10^5 copies) in the absence or upon simultaneous amplification of abundant HSV-2 DNA (10^6 copies per reaction). These data as well as similar results obtained with several sets of three simultaneously amplified specific targets (data not shown) imply no essential competition between accumulations of individual PCR products upon joint multiplex analysis.

On-Chip PCR with Eukaryotic cDNA and Gel Pad Size Impact

Eukaryotic cDNA is another suitable object for application of the approach proposed, and its one-stage format, in particular. Poly-A tail provides a natural site for common reverse primer binding, semi-quantitative results of real time solid-phase PCR on microarray could provide biologically valuable information on abundance of particular mRNA species. However, in comparison with applications described above, this one has some additional challenges.

The first challenge is related to low melting temperatures of poly-A structures. This complicates the selection of gene-specific forward primers as they should have the same melting temperature as poly-T reverse primer. For this reason the reverse transcription in this investigation was carried out with the primer consisting of 3′-T_{25} part and 5′- T7 promoter part (primer T_R, Table 1). This primer was also used as a reverse liquid-phase primer in a subsequent one-stage on-chip PCR reaction.

Second point relates to high complexity of eukaryotic cDNA. It is a pool of thousands of individual cDNA species possessing common liquid-phase reverse primer annealing site.

Only tiny share of individual cDNA in this pool are specific to solid-phase forward primers. This circumstance challenges the very possibility of successful solid phase on-chip PCR.

Third challenge relates to high number of potentially interesting genes. In long term, the expression profiling could require a high density microarray with thousands of elements. In this case the decrease of gel pads physical size could be advantageous.

In order to address these challenges, microarray of gel spots with diameters of 600μm, 450μm, and 250μm containing immobilized forward primers specific to *Actb* and *Hprt* genes was manufactured and investigated in one stage on-chip PCR. Sixty cycles of PCR were carried out with T_R primer (Table 1) and cDNA from K562 cells as follows: 20s at 95°C, 40s at 54°C, 45s at 68°C. Microarray fluorescent images were taken at the end of elongation stage. PCR curves revealed a good sigmoid shape (not shown) with threshold cycles indicated at Table 2 (average from 3 experiments). Threshold cycles were found to be strongly dependent on the size of spot. For 250μm spots C_T was unacceptably high hampering accurate assessment of low abundance genes. For 600μm spots C_T was the lowest, and the discrimination between high and low abundance genes was decreased (26 and 28 cycles for *Actb* and *Hprt*, respectively). Post-PCR melting curves revealed sharp transition with melting temperature 87°C and 90°C for *Actb* and *Hprt*, respectively.

Table 2. Threshold cycle numbers for different spot diameters in human cDNA experiments

Spot diameter, μm	C_T for *Actb*	C_T for *Hprt*
600	26	28
450	30	40
250	39	50

Discussion

Microarray platforms have been widely investigated as a tool for performing of complicated multiplex analyses of nucleic acids [4]. Several microarray-based PCR approaches developed so far provide just qualitative methods which often suffer from low efficiency in comparison with traditional in-tube PCR [13-15, 22-24, 26-30]. Thereupon, the present investigation was focused on a study of several characteristics of PCR on a microarray of gel-based elements in order to develop the improved approach for multiplex DNA analysis. This chapter considers the application of microarray based solid phase PCR with a common liquid-phase primer to several relevant fields of modern life science, like detection of microorganisms, DNA quantification, eukaryotic expression profiling.

Previously the effective PCR on a microarray of gel elements was shown to occur only in the presence of both pairwise primers in liquid phase [13]. But for "two primers per target" scheme the probability of primer dimmer formation increases rapidly with the number of simultaneously analyzed targets. The recently developed two-stage nested PCR approach for simultaneous quantitative analysis of three blood-born viruses [19] also utilized only specific pairs of primers for each nucleic target: the first-stage reaction with outer primers was performed in tube, the second stage was carried out on-chip with immobilized inner forward primers and liquid-phase inner reverse ones.

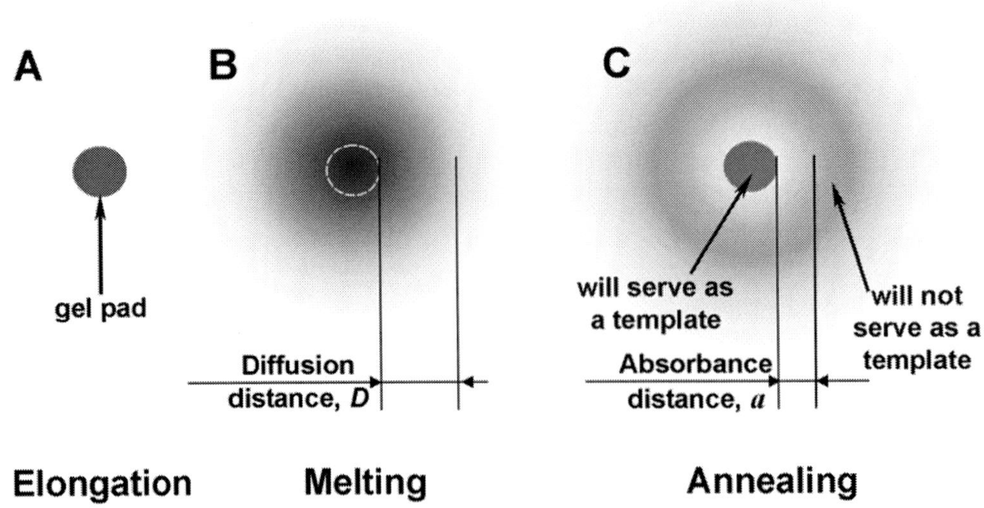

Figure 8. Top view of PCR product diffusion into and out of gel pad at different stages of PCR cycle. The concentration of PCR product is given in gradation of grey. (A) Elongation stage. Newly synthesized PCR product is located in the gel pad. (B) Melting stage. PCR product diffuses out of gel pad. (C) Annealing stage. Some portion of PCR product is absorbed by gel pad and can serve as a template for the next PCR cycle. The rest of PCR product cannot participate in the next cycle of reaction because it is located out of gel pad.

In contrast to previous studies [19], here we detect and quantify DNA of several microorganisms in a single stage format of PCR (Figure 3A, Figure 6B and D, and Figure 5B, solid circles, respectively). This format minimizes the risk of contamination and maximizes the dynamic range of assay. The key advantage of the investigation is utilization of only one universal liquid-phase primer, what (theoretically) enables the significant increase of amount of nucleic targets for simultaneous analysis. This goal was achieved in two ways: (i) directly by design of the primer possessing the common specificity to homologues DNAs, and (ii) indirectly by introduction of universal sequence into non-homologues targets via PCR with chimerical primers. The second way represents the version of semi-nested PCR format, which is still advantageous in comparison with approaches described for the gel element microarray earlier [13, 19] since it enables the analysis of increased number of targets. At the same time two-stage PCR provides more accurate results, as it inherent in conventional two-stage in-tube format. The effectiveness the on-chip PCR under conditions optimized in the frame of the investigation is shown to be very high and close to conventional in tube methods one.

One of the intriguing results of the study is the demonstration of the possibility of performing of microarray based PCR with eukaryotic cDNA. However, the effectiveness of the reaction (C_T value) was found to be strongly dependent on the gel pad size. To understand this phenomenon the difference between conventional liquid-phase (in tube) and microarray-based solid-phase PCR should be considered. Conventional liquid-phase PCR includes two pairwise primers uniformly distributed across the reaction volume. While in the case of microarray-based PCR the only one of the primers is distributed in liquid phase, and the opposite one is immobilized within the gel pad, which is just a tiny fraction of total reaction volume. This circumstance complicates the interaction of target DNA strands with complementary solid-phase primers. After the elongation step, target DNA copies diffuse out of gel pad (Figure 8A and B). To serve as templates again they have to diffuse back into the gel pad to anneal with forward primer during the next annealing stage (Figure 8C). Liquid-phase PCR is free from this limitation as every copy of the target could serve as a template every next cycle. In the case of gel-based microarray PCR this effect must be more expressive for small sized gel pads. To demonstrate it a following simple model of gel pad with hybridized PCR product after elongation stage can be considered (Figure 8A). If template DNA at the melting stage diffuse out of gel pad to some effective area (located at range *D* from the surface of gel pad, Figure 8B) and only molecules from the absorbance area (located at range *a* from the surface of gel pad, Figure 8C) are able to diffuse back upon the annealing completion, then the portion *p* of PCR product available as template at the next cycle could be estimated as the ratio of absorbance and diffusion areas:

$$p=(a+r)^2/(D+r)^2,$$

provided $(a+r)>>h$ and $(D+r)>>h$, where h=250μm is the thickness of solution layer over the micoarray, and r is the pad radius. Obviously, the higher r gives the higher p causing better yield of PCR cycle and lower C_T value, in accordance with data from Table 2. Also, the higher concentration of immobilized forward primer should result in better absorbance ability of a gel pad and lower C_T (in accordance with results of Figure 4B). The lower yield of PCR cycle could be the reason for higher discrimination between C_T value determined for *Actb* and *Hprt* genes in the case of small (450μm and 250μm) gel pads (Table 2). The observed discrimination for 600μm pads (ΔC_T=28-26=2) is close to the data obtained earlier with Taqman PCR for K562 cell line ([31]: ΔC_T=2.7 with C_T=18.9 and 21.6 for *Actb* and *Hprt*, respectively). Nevertheless, the smaller sized pads could be preferable if the higher ΔC_T discrimination is desired for some kind of reasons.

CONCLUSION

This chapter presents a new powerful approach applying the gel element based microarrays for high-effective multiplex PCR with utilization of multiple solid-phase immobilized primers and a common universal liquid-phase one. The results obtained demonstrate high sensitivity (<10 DNA copies per reaction) and specificity of the method. The approach enables the reliable qualitative and quantitative determinations of multiple nucleic acid targets analyzed simultaneously. The quantitative dynamic range covers up to eight orders of concentrations. The approach is shown to be suitable for direct application for parallel analysis of relevant genes possessing homologous sequences (as exemplified with DNA polymerase (*pol*) gene of three *Herpesviridae* viruses). Furthermore, non-homologous sequences can be analyzed by the microarray PCR indirectly if two-stage format with introduction of a universal sequence is applied (as shown for the set consisting of HSV-1, HSV-2, CMV, *U. urealyticum*, *M. hominis*, *C. trachomatis*, and bacteriophage lambda DNAs). The direct eukaryotic mRNA expression profiling has been considered as one of the advantageous fields for application of PCR on a microarray and requires further particular investigation.

ACKNOWLEDGMENTS

We are grateful to members of Scientific Center of Obstetrics, Gynecology and Perinatology, Russian Academy of Medical Sciences for providing DNA preparations isolated from human clinical samples, and Dr. Svetlana Strunina for preparation of total RNA from K562 cells. We also thank EIMB RAS colleagues for their technical support.

REFERENCES

[1] Yershov, G., Barsky, V., Belgovskiy, A., Kirillov, E., Kreindlin, E., Ivanov, I., Parinov, S., Guschin, D., Drobishev, A., Dubiley, S. & Mirzabekov, A. (1996) DNA analysis and diagnostics on oligonucleotide microchips, *Proc Natl Acad Sci U S A,* **93**, 4913-4918.

[2] Drobyshev, A., Mologina, N., Shik, V., Pobedimskaya, D., Yershov, G. & Mirzabekov, A. (1997) Sequence analysis by hybridization with oligonucleotide microchip: identification of beta-thalassemia mutations, *Gene,* **188**, 45-52.

[3] Vallone, P. M. & Butler, J. M. (2004) AutoDimer: a screening tool for primer-dimer and hairpin structures, *Biotechniques,* **37**, 226-231.

[4] Fan, J. B., Chee, M. S. & Gunderson, K. L. (2006) Highly parallel genomic assays, *Nat Rev Genet,* **7**, 632-644.

[5] Brownie, J., Shawcross, S., Theaker, J., Whitcombe, D., Ferrie, R., Newton, C. & Little, S. (1997) The elimination of primer-dimer accumulation in PCR, *Nucleic Acids Res,* **25**, 3235-3241.

[6] Lisitsyn, N. & Wigler, M. (1993) Cloning the differences between two complex genomes, *Science,* **259**, 946-951.

[7] Lucito, R., Nakimura, M., West, J. A., Han, Y., Chin, K., Jensen, K., McCombie, R., Gray, J. W. & Wigler, M. (1998) Genetic analysis using genomic representations, *Proc Natl Acad Sci U S A,* **95**, 4487-4492.

[8] Lucito, R., Healy, J., Alexander, J., Reiner, A., Esposito, D., Chi, M., Rodgers, L., Brady, A., Sebat, J., Troge, J., West, J. A., Rostan, S., Nguyen, K. C., Powers, S., Ye, K. Q., Olshen, A., Venkatraman, E., Norton, L. & Wigler, M. (2003) Representational oligonucleotide microarray analysis: a high-resolution method to detect genome copy number variation, *Genome Res,* **13**, 2291-2305.

[9] Shapero, M. H., Zhang, J., Loraine, A., Liu, W., Di, X., Liu, G. & Jones, K. W. (2004) MARA: a novel approach for highly multiplexed locus-specific SNP genotyping using high-density DNA oligonucleotide arrays, *Nucleic Acids Res,* **32**, e181.

[10] Tanabe, C., Aoyagi, K., Sakiyama, T., Kohno, T., Yanagitani, N., Akimoto, S., Sakamoto, M., Sakamoto, H., Yokota, J., Ohki, M., Terada, M., Yoshida, T. & Sasaki, H. (2003) Evaluation of a whole-genome amplification method based on adaptor-ligation PCR of randomly sheared genomic DNA, *Genes Chromosomes Cancer,* **38**, 168-176.

[11] Fan, J. B., Oliphant, A., Shen, R., Kermani, B. G., Garcia, F., Gunderson, K. L., Hansen, M., Steemers, F., Butler, S. L., Deloukas, P., Galver, L., Hunt, S., McBride, C., Bibikova, M., Rubano, T., Chen, J., Wickham, E., Doucet, D., Chang, W., Campbell, D., Zhang, B., Kruglyak, S., Bentley, D., Haas, J., Rigault, P., Zhou, L., Stuelpnagel, J. &

Chee, M. S. (2003) Highly parallel SNP genotyping, *Cold Spring Harb Symp Quant Biol,* **68**, 69-78.

[12] Hardenbol, P., Baner, J., Jain, M., Nilsson, M., Namsaraev, E. A., Karlin-Neumann, G. A., Fakhrai-Rad, H., Ronaghi, M., Willis, T. D., Landegren, U. & Davis, R. W. (2003) Multiplexed genotyping with sequence-tagged molecular inversion probes, *Nat Biotechnol,* **21**, 673-678.

[13] Strizhkov, B. N., Drobyshev, A. L., Mikhailovich, V. M. & Mirzabekov, A. D. (2000) PCR amplification on a microarray of gel-immobilized oligonucleotides: detection of bacterial toxin- and drug-resistant genes and their mutations, *Biotechniques,* **29**, 844-848, 850-842, 854 passim.

[14] Huber, M., Mundlein, A., Dornstauder, E., Schneeberger, C., Tempfer, C. B., Mueller, M. W. & Schmidt, W. M. (2002) Accessing single nucleotide polymorphisms in genomic DNA by direct multiplex polymerase chain reaction amplification on oligonucleotide microarrays, *Anal Biochem,* **303**, 25-33.

[15] Tillib, S. V., Strizhkov, B. N. & Mirzabekov, A. D. (2001) Integration of multiple PCR amplifications and DNA mutation analyses by using oligonucleotide microchip, *Anal Biochem,* **292**, 155-160.

[16] Morrison, T., Hurley, J., Garcia, J., Yoder, K., Katz, A., Roberts, D., Cho, J., Kanigan, T., Ilyin, S. E., Horowitz, D., Dixon, J. M. & Brenan, C. J. (2006) Nanoliter high throughput quantitative PCR, *Nucleic Acids Res,* **34**, e123.

[17] Leamon, J. H., Lee, W. L., Tartaro, K. R., Lanza, J. R., Sarkis, G. J., deWinter, A. D., Berka, J., Weiner, M., Rothberg, J. M. & Lohman, K. L. (2003) A massively parallel PicoTiterPlate based platform for discrete picoliter-scale polymerase chain reactions, *Electrophoresis,* **24**, 3769-3777.

[18] Margulies, M., Egholm, M., Altman, W. E., Attiya, S., Bader, J. S., Bemben, L. A., Berka, J., Braverman, M. S., Chen, Y. J., Chen, Z., Dewell, S. B., Du, L., Fierro, J. M., Gomes, X. V., Godwin, B. C., He, W., Helgesen, S., Ho, C. H., Irzyk, G. P., Jando, S. C., Alenquer, M. L., Jarvie, T. P., Jirage, K. B., Kim, J. B., Knight, J. R., Lanza, J. R., Leamon, J. H., Lefkowitz, S. M., Lei, M., Li, J., Lohman, K. L., Lu, H., Makhijani, V. B., McDade, K. E., McKenna, M. P., Myers, E. W., Nickerson, E., Nobile, J. R., Plant, R., Puc, B. P., Ronan, M. T., Roth, G. T., Sarkis, G. J., Simons, J. F., Simpson, J. W., Srinivasan, M., Tartaro, K. R., Tomasz, A., Vogt, K. A., Volkmer, G. A., Wang, S. H., Wang, Y., Weiner, M. P., Yu, P., Begley, R. F. & Rothberg, J. M. (2005) Genome sequencing in microfabricated high-density picolitre reactors, *Nature,* **437**, 376-380.

[19] Khodakov, D. A., Zakharova, N. V., Gryadunov, D. A., Filatov, F. P., Zasedatelev, A. S. & Mikhailovich, V. M. (2008) An oligonucleotide microarray for multiplex real-time PCR identification of HIV-1, HBV, and HCV, *Biotechniques,* **44**, 241-246, 248.

[20] Rubina, A. Y., Pan'kov, S. V., Dementieva, E. I., Pen'kov, D. N., Butygin, A. V., Vasiliskov, V. A., Chudinov, A. V., Mikheikin, A. L., Mikhailovich, V. M. & Mirzabekov, A. D. (2004) Hydrogel drop microchips with immobilized DNA: properties and methods for large-scale production, *Anal Biochem,* **325**, 92-106.

[21] Xiao, P. F., Cheng, L., Wan, Y., Sun, B. L., Chen, Z. Z., Zhang, S. Y., Zhang, C. Z., Zhou, G. H. & Lu, Z. H. (2006) An improved gel-based DNA microarray method for detecting single nucleotide mismatch, *Electrophoresis,* **27**, 3904-3915.

[22] Adessi, C., Matton, G., Ayala, G., Turcatti, G., Mermod, J. J., Mayer, P. & Kawashima, E. (2000) Solid phase DNA amplification: characterisation of primer attachment and amplification mechanisms, *Nucleic Acids Res,* **28**, E87.

[23] Shapero, M. H., Leuther, K. K., Nguyen, A., Scott, M. & Jones, K. W. (2001) SNP genotyping by multiplexed solid-phase amplification and fluorescent minisequencing, *Genome Res,* **11**, 1926-1934.

[24] Pemov, A., Modi, H., Chandler, D. P. & Bavykin, S. (2005) DNA analysis with multiplex microarray-enhanced PCR, *Nucleic Acids Res,* **33**, e11.

[25] Karsai, A., Muller, S., Platz, S. & Hauser, M. T. (2002) Evaluation of a homemade SYBR green I reaction mixture for real-time PCR quantification of gene expression, *Biotechniques,* **32**, 790-792, 794-796.

[26] Oroskar, A. A., Rasmussen, S. E., Rasmussen, H. N., Rasmussen, S. R., Sullivan, B. M. & Johansson, A. (1996) Detection of immobilized amplicons by ELISA-like techniques, *Clin Chem,* **42**, 1547-1555.

[27] Sjoroos, M., Ilonen, J. & Lovgren, T. (2001) Solid-phase PCR with hybridization and time-resolved fluorometry for detection of HLA-B27, *Clin Chem,* **47**, 498-504.

[28] Turner, M. S., Penning, S., Sharp, A., Hyland, V. J., Harris, R., Morris, C. P. & van Daal, A. (2001) Solid-phase amplification for detection of C282y and H63D hemochromatosis (HFE) gene mutations, *Clin Chem,* **47**, 1384-1389.

[29] Onodera, K., d'Offay, J. & Melcher, U. (2002) Nylon membrane-immobilized PCR for detection of bovine viruses, *Biotechniques,* **32**, 74-76, 78, 80.

[30] Liu, H., Wang, H., Shi, Z., Wang, H., Yang, C., Silke, S., Tan, W. & Lu, Z. (2006) TaqMan probe array for quantitative detection of DNA targets, *Nucleic Acids Res,* **34**, e4.

[31] Wang, Y. L., Lee, J. W., Cesarman, E., Jin, D. K. & Csernus, B. (2006) Molecular monitoring of chronic myelogenous leukemia: identification of the most suitable internal control gene for real-time quantification of BCR-ABL transcripts, *J Mol Diagn,* **8**, 231-239.

In: Oligonucleotide Array Sequence Analysis
Editors: M.K. Moretti and L.J. Rizzo, pp. 387-402
ISBN: 978-1-60456-542-3

Chapter 14

USE OF ARRAYS WITH NON-MODEL ORGANISMS: A UK PERSPECTIVE

***Michael J. Allen*[1], *Andrew R. Cossins*[2], *Neil Hall*[2], *Mark Blaxter*[3], *Terry Burke*[4] and *Dawn Field*[5]**

[1] Plymouth Marine Laboratory, Prospect Place, Plymouth, PL1 3DH, UK
[2] NERC Molecular Genetics Facility, School of Biological Sciences
University of Liverpool, Liverpool, L69 3BX, UK
[3] NERC Molecular Genetics Facility, Institute of Evolutionary Biology
University of Edinburgh, Edinburgh, EH9 3JT, UK
[4] NERC Molecular Genetics Facility, Dept of Animal & Plant Sciences
University of Sheffield, Sheffield, S10 2TN, UK
[5] NERC Environmental Bioinformatics Centre
NERC Centre for Ecology and Hydrology, Mansfield Road, Oxford, OX1 3SR, UK

ABSTRACT

Historically, the majority of microarray work has been restricted to well-defined model organisms. This was primarily due to the limited availability of genomic or transcriptomic sequence data and the then high cost involved in developing microarrays. However, recent technological developments have greatly enhanced the speed of generating the underpinning sequence data for non-model species, and have opened up more cost-effective approaches for microarray production to make them far more affordable for researchers at the lower end of the budget range. These developments have been seized upon by the environmental genomics community within the UK. The creation of a network of closely integrated facilities for sequencing, microarray printing and bioinformatics has opened the gateway for the study of environmentally relevant organisms. Here, we describe the infrastructure for microarray development within the UK, and the diverse applications for which they are currently being used.

INTRODUCTION

The advent of the genomic era has heralded a new dawn in the study of biological systems. The intensive study of model species such as *E. coli*, yeast, *Drosophila* and the mouse has provided a wealth of information on the molecular workings inside the cell. These advances were largely dependent on large-scale sequencing projects which determined the entire genome sequence of several well studied model strains [1-4]. The development of microarray technology for massively parallel detection of sequence targets defined from large-scale genome and EST sequencing projects, has allowed transcriptional profiles from cells, tissues and organisms at various life stages to be analysed and mapped on a global scale with increasing precision [5].

It cannot be denied that the intensive study of model organisms is of great benefit. However, the study of ecologically important, non-model species in their natural environment is arguably of greater relevance when analysing the response of species to real-world environmental conditions. Model organisms are often chosen for reasons of experimental power and convenience, rather than for ecological interest or evolutionary significance [6]. In an era of increasing awareness of the environmental changes occurring through global warming, habitat destruction, deforestation, globalisation and ocean acidification there has never been a more important time to study the impact on the Earth's ecosystem, and this means adapting advanced genomic tools for this purpose.

Historically, tools such as microarrays have been restricted to model species due to limitations in the genomic sequence available and the high cost involved in developing microarrays. However, technical advances have led to a significant decrease in the costs involved, opening the field to researchers working on non-model organisms. The uptake of microarray technology by environmental researchers, in particular the marine sciences, has been relatively slow partly due to a lack of both focused funding and technical expertise in the community. In the UK this has been addressed by the predominantly NERC-funded environmental genomics community which has created a dedicated network of facilities and resources to aid the use and development of microarrays.

FUNDING

The United Kingdom Natural Environment Research Council (NERC) has invested over £26 million in two directed science programmes; (i) Environmental Genomics (EG) and (ii) Post-Genomics & Proteomics (PGP). To complement these initiatives, NERC has also funded the creation of dedicated technical support centres covering DNA sequencing, genotyping, microarray design and fabrication (through NERC's Services and Facilities branch), and bioinformatics (through the thematic programmes themselves). Each centre offers a strong technical capability built on the existing expertise in established laboratories, and also consultation, training, and support not only to undertake the lab work but also in the subsequent data analysis and management. Access is open to all UK-based scientists, but priority is given to NERC grant holders based on an appraisal of science quality. Applicants for NERC-funded research are advised to take advantage of these facilities whenever possible.

In addition, to promote the use of these advanced technologies in environmental science, an annual small projects competition has allowed researchers to conduct pilot studies in support of full-scale future project applications. National and international collaboration between environmental scientists often working in vastly different areas is promoted through this system. The diverse range of subjects for which new microarrays have been developed include: marine phytoplankton viruses (see Case Study 1); adaptive osmoregulatory physiology of euryhaline teleosts; ecotoxicological responses of terrestrial annelids; genetic control of avian plumage traits; sexual differentiation and endocrine disruption in fish; and responses to cold and hypoxia in carp (see Case Study 2).

INFRASTRUCTURE

Every microarray experiment is dependent on a number of steps, all of which are essential to produce results that are scientifically competent and of publishable quality. These steps are listed in Table 1, and can be grouped into 3 main categories: the acquisition of genomic sequence data, the design and construction of the array and the experimental use of the array. Of course, this is only the beginning of a researcher's challenges: correct experimental design with the appropriate number and type of replicates, in combination with the large volume of data produced, demands the use of complicated statistical analysis to provide meaningful results.

Table 1. Integration of different UK research facilities in the microarray research

What?	Who?				
	Researcher	**MGF-Edinburgh**	**MGF-Liverpool (AGF)**	**MGF-Liverpool (LMF)**	**NEBC**
DNA Sequencing					
Sample acquisition	✓				
Sequencing	✓	✓	✓		
Annotation	✓	✓	✓		✓
Microarray Construction					
Oligo design	✓			✓	✓
Oligo synthesis				✓	
cDNA clones and EST sequencing	✓	✓		✓	
Microarray printing or submission of *in silico* oligoprobes to 3rd part fabrication	✓			✓	
Microarray Experiment					
Experimental design	✓				✓
Sample acquisition	✓				
Array hybridisation	✓			✓	
Image acquisition and quantification	✓			✓	
Data analysis	✓			✓	✓
Interpretation	✓			✓	✓
Publication	✓	✓	✓	✓	✓

An environmental scientist working alone on a microarray project will need expertise in a range of disciplines, such as molecular biology, bioinformatics and statistics. In addition, ready access to sequencing, array printing and hybridisation facilities is essential.

It is important that the researcher has an understanding of the processes involved at each stage, and also that the correct expertise is available at each stage in line with current opinion in a rapidly developing field. Thus the use of microarrays is a daunting prospect to any researcher who lacks significant experience or expertise in the technologies involved. In particular this includes environmental researchers whose field has been historically dominated by non-molecular scientists (i.e., large numbers of field biologists). It is perhaps for these reasons that environmental scientists studying non-model organisms have been relatively slow to embrace microarray technology.

Fortunately for the UK-based environmental science community, NERC foresaw this problem and put in place the infrastructure required to facilitate the technical development of microarray platforms and their application to environmental science. The new provision was based on the pattern established by an existing NERC genotyping centre based in Sheffield (MGF-Sheffield) and consists of three additional Molecular Genetics Facilities (MGF) nodes: two facilities are involved with sequencing (MGF-Edinburgh and MGF-Liverpool through its Advanced Genomics Facility, AGF-Liverpool), and a third facility with microarray construction (MGF-Liverpool through its Liverpool Microarray Facility, LMF-Liverpool). A fourth but separate facility, The NERC Environmental Bioinformatics Centre (NEBC) provides the necessary bioinformatic expertise required at all levels of the process. Combined these facilities provide a comprehensive suite of technical platforms and expertise, and they interact seamlessly with one another to satisfy the needs of the researcher.

Genomic Sequencing

There are three currently applicable methods for generating sequence data for environmentally relevant organisms: capillary Sanger sequencing, massively parallel pyrosequencing, and massively parallel sequencing-by-synthesis. Each method has its benefits and drawbacks, and any particular environmental genomics project will choose the route(s) most suited to its needs. Both MGF-Edinburgh and AGF-Liverpool are dedicated sequencing facilities and thus generally not directly involved in physical sample acquisition: it is assumed that a researcher working on a particular environmental organism has the necessary expertise for culture and/or sample extraction to provide suitable genomic material (quantity and quality) for sequence analysis.

Capillary Sanger sequencing, the workhorse of DNA sequencing for over 30 years, is offered through the MGF-Edinburgh. This generates 500–700 bases per read, and is suited to both survey sequencing (such as expressed sequence tag sequencing, or ESTs) and also whole-genome sequencing. The MGF-Edinburgh offers medium-throughput (~1000 samples per day) Sanger sequencing using ABI3730 instruments. This suits projects from ~100 to 50,000 sequencing reads from targets such as PCR products, cloned inserts in plasmids and other vectors provided by the user. It also carries out the complete sequencing of large DNA fragments (>30 kb; for example from fosmid and BAC clones) using shotgun strategies. Bioinformatic support is offered by MGF- Edinburgh, including preliminary sequence trace processing, assembly and annotation. For EST projects, a suite of in-house developed sequence annotation tools can also be applied to sequence datasets using programs such as PartiGene (databasing tool for EST datasets) [7], trace2dbEST (for processing EST sequencing traces) [7], prot4EST (for prediction of peptides from neglected genome sequence datasets) [8], and annot8r (annotation of peptide or nucleotide sequences with GO, KEGG and EC tags based on sequence similarity).

The Advanced Genomic Facility (AGF) based at the University of Liverpool offers access to the ultra-high throughput pyrosequencing Genome Sequencer FLX system (GS FLX, Roche), which at present can generate in excess of 100 Mb of genome sequence in a single run with a mean length of 250 bases. The FLX pyrosequencing platform incorporates single-molecule amplification and parallel sequencing of up to ~400,000 individual molecules for 250 bases, making large-scale scientific projects feasible and more affordable. By mid-2008 this instrument is expected to yield up to 1 Gb with read lengths approaching 500 bp. FLX pyrosequencing is suited to deep surveying of transcriptomes, complete sequencing of large-insert clones and smaller genomes, resequencing of larger genomes, and metagenomic analyses. AGF-Liverpool also has the latest computational tools and trained staff to deliver bioinformatic services for data handling and analysis, in particular for assembling sequence reads (Newbler assembly) and sequence alignment (FlowMapper).

Sequencing by synthesis is also massively parallel, but generates ~30 million reads of ~35 bases per run, generating 1 Gb of sequence. This technology is ideal for resequencing of complex genomes, metagenomic surveys of mesocosms, digital transcriptomics and *de novo* sequencing of smaller genomes. MGF-Edinburgh also offers a sequencing-by-synthesis service using an Illumina Genome Analyser instrument.

The production of raw sequence data is only the start of the process: the genomic or transcriptomic sequence must be correctly assembled and annotated in order to provide meaningful information that can be used at the next stage for microarray design. As shown above, both facilities have their own particular expertise in certain areas. However, in addition to the bioinformatic facilities offered by MGF-Edinburgh and AGF-Liverpool, additional bioinformatic support for genomic annotation is offered by the NERC Environmental Bioinformatics Centre (NEBC). The bioinformatic support provided by MGF-Sheffield is focused on population genetic data analyses, including genomic mapping.

Microarray Design and Fabrication

The key requirement for generating a DNA array is access to suitable genomic resources. For arrays printed in-house using amplified DNA, this comprises cloned cDNAs or large-insert genomic fragments isolated from libraries [9]. Alternatively, arrays can be printed from oligonucleotides synthesised in bulk with defined sequences as probes against known DNA targets of interest. Both techniques are expensive in time (for cloning and amplification) and costs (for oligonucleotide purchase). However, for situations where sequence data already exist it is now preferable to design probes *in silico* against all known targets for a particular species using one of several design algorithms, followed by on-chip synthesis by one of several manufacturers, notably Agilent. Advancing fabrication techniques now makes this a relatively inexpensive yet highly effective approach to array provision.

The design of oligonucleotide probes from a known target DNA sequence can be achieved using any one of several open source or commercial packages [10]. The MGF-Liverpool uses a commercial package (*ArrayDesigner*, Premier Biosoft) as an alternative to the widely used open source package OligoArray2 [11]. Design algorithms continue to advance, and one algorithm might suggest different probe designs to another. Commercial providers of oligonucleotides or arrays offer a design service (e.g., Agilent's eArray system), but the precise algorithms employed remain confidential to the company and so one cannot be totally confident of design quality or the criteria used in its execution. It is also worth noting that the outputs of any design algorithm do not necessarily provide for optimal function as a hybridisation probe, since they are based on thermodynamic criteria that are surrogates of the hybridisation process. This is why the MGF-Liverpool uses an optimisation approach, comparing the properties of several competing probe designs generated from a single algorithm against stringent performance criteria [12] in order to define the best performing design.

All of these options are available at the MGF-Liverpool which has access to both contact and non-contact (piezo) printing robots for array fabrication, has service centre status for both Affymetrix and Agilent platforms, and has access to the Illumina BeadStation, this being the third major commercial platform. As a central array facility, it has negotiated special rates for purchasing large numbers of synthesised oligos and all of the other necessary consumables, including catalogue arrays. It is thus well able to provide support to users with all aspects of microarray design, construction and commissioning. In the past few years MGF-Liverpool has constructed cDNA arrays for a wide range of species (e.g., common carp, flounder, rainbow trout, ground squirrel and eel) all of which were preceded by large-scale cDNA sequencing projects. It has also been involved in a series of microbial metagenomics projects printing commercially synthesised oligoarrays directed against rRNA targets defined by the user. MGF-Liverpool has also some experience in constructing oligoarrays for environmental model species directly from public EST data, taking advantage of the rapidly increasing capacity of new platforms. Thus, an oligoarray directed against 22,000 BLAST-identified genes in the rainbow trout was fabricated by Oxford Gene Technology to a MGF-Liverpool design, and this has been expanded to 44,000 probes [12]. This design was optimised in the sense that multiple predicted probes were quantitatively compared using several different tests and the best performing probe identified. The resulting arrays displayed a dynamic range and sensitivity that far exceeded that of conventional cDNA arrays [12] .

Analysing the Results of Microarray Experiments

Once a microarray has been designed and fabricated the actual experimental biology can begin. The first important issue is the statistical design of the array experiment. The shift from the early reference-based designs to more complex ANOVA-based loop designs [13], and the need to account for multiple sampling corrections in such large-scale statistical comparisons, makes this the realm of specialist statistical advice. Only rarely do biologists who develop skills in laboratory technique have the numerical and statistical skills for high-level data analysis and pattern searching. The staff of the MGF-Liverpool and NERC Environmental Bioinformatics Centre (NEBC) provide specialist advice to define the most appropriate and cost-effective experimental plan to success. The researcher also needs to understand the very fine definition offered by array experiments to ensure that the experiment is adequately controlled and that sufficient independent biological replicates are generated for detailed statistical analysis. Dedicated microarray bioinformaticians at both the MGF-Liverpool and at NEBC keep abreast of the latest developments and issues with microarray analysis and are best positioned to advise on correct experimental design. The MGF-Liverpool also supports all of the stages of array usage, including sample preparation, hybridisation protocols, imaging/data acquisition and interpretation of the resulting data. This includes the training of research students and staff from NERC-funded laboratories.

The fundamental responsibility for sample acquisition and for extraction of DNA or RNA lies with the client. Indeed, sample processing, hybridisation and data acquisition can be performed at the researchers own laboratory if facilities are available. Alternatively the entire programme can be performed as a service by MGF-Liverpool staff or the latter can provide the necessary training for clients to undertake the work. Emphasis is placed on the researcher remaining in control of the experiment, but with the full benefit of state-of-the-art protocols and equipment.

MGF-Liverpool offers several scanning platforms for image acquisition including Agilent, Perkin-Elmer and Axon. It also advises on the critical issue of interpreting the image file in order to quantify the distribution of fluorescence intensities across the population of probes. Once microarray images have been processed and quantified, it is strongly recommended that the researcher interacts further with the MGF-Liverpool or the NEBC if they require help progressing the statistical analysis. NEBC offers access to the commercial *GeneSpring* (Agilent) and the open-source *maxD* analysis platforms, and both the NEBC and the MGF-Liverpool routinely use algorithms from within the *Bioconductor* package written in the R statistical language. This combined provision has three main advantages. Firstly, the researcher is directed to well established statistical and data visualisation algorithms. Second, he/she can be trained in analytical technique through interactions with the facilities. Third, he/she can be exposed to objective approaches for drawing inferences both in terms of which genes are responding to experimental treatment, but also how this relates to the biological pathways and processes, and to the questions being addressed. This avoids biasing the analysis to favoured genes or to genes that fit with particular hypotheses. Bioinformaticians at any NERC facility can explain the data output, its strengths and its limitations. However, it is entirely up to the researcher to interpret and publish the data. This involves the use of gene ontology profiling tools such as GOMatrix, GoMiner, Reactome etc., as well as the newer gene set analysis techniques originated in the Broad Institute, MIT.

At the final stage of publication, it is necessary for all microarray data to be submitted to a public repository such as GEO or ArrayExpress [14, 15]. This is aided by the development of the Biolinux based *maxdload2* data submission program, developed by NEBC partners at the University of Manchester, which ensures it is MIAME-compliant [16]. Indeed, in response to the issues associated with environmental microarrays, the NEBC has developed its own guidelines for MIAME for environmental projects, designated MIAME/Env [17]. Researchers funded by NERC are also required to submit their data to the NERC-supported database EnvBase (http://nebc.nox.ac.uk/).

DATA ANALYSIS AND LONG-TERM MANAGEMENT

While each of the sequencing facilities have specialist informatics services focused on specific sequencing platforms, the NEBC is focused on delivering multi-omic approaches to analysis and data management. NEBC was formed in 2002 to specifically provide support for the NERC Environmental Genomics Programme, and subsequently for the sister Post-Genomics and Proteomics thematic programme. NEBC's primary remit is to archive all NERC-funded data generated under these programs subscribing to the NERC data policy of spending up to 15% of the programme budget on data management and archiving activities. This was one of the earliest data policies operated by a major UK funding agency and its wider implementation is mediated by a set of seven designated data centres that receive core funding from NERC for long-term data management.

To guide the data management of the programme, and the stewardship of 'omic data in NERC as a whole, NEBC has also produced an extension of the NERC data policy to cover 'omic data. This policy includes very specific mechanisms for submission of data. All data are catalogued in the EnvBase portal and must be submitted to an appropriate public repository where one exists. EnvBase contains discovery-level information about all listed projects, and for each data-holding includes a list of relevant hyperlinked accession numbers leading to databases such as ArrayExpress [14, 15]. EnvBase also provides a mechanism to house all data for which a recognized database does not exist.

NEBC was designed to build informatic capacity within the newly forming NERC environmental genomic community. NEBC is therefore staffed by bioinformaticians and computer scientists, in addition to data managers, who provide direct consultation to the wider community. NEBC was established in part to realize the vision of providing the entire Environmental Genomics community with a powerful and uniform computing environment based on Linux. This system, called Bio-Linux, is one of a growing suite of highly customized Linux distributions [18]. It contains over 60 software packages and is distributed free of charge to all labs supported by the program. The Bio-Linux system has been used to build teaching labs and computer clusters in several locations across the UK and is used by the core facilities and researchers to coordinate research.

The primary focus of NEBC efforts under these programs has been support for research that is generating microarray and EST data. Microarrays were a focus because of their wide-spread use among funded researchers, the expense and effort of tooling up to produce and interpret them for new users, and the added complexity that several NERC-funded groups aimed to build custom arrays for novel species and sets of genes. In order uphold its remit to provide stewardship over the collective data holdings of these programmes; the centre has become progressively more involved in standardization activities. While at first, this simply meant monitoring emerging standards and compliance activities, over time this led to active participation and leadership in standardization activities that expressly benefit the environmental 'omics community and that have made submission of data to public repositories possible.

The NEBC spearheaded a standardization project in response to the difficulties NERC researchers were facing in complying to the MIAME standard, a requirement for submission to ArrayExpress [19]. The MIAME/Env project emerged out of work aimed at simplifying this process for researchers across the board. This initiative extended the core MIAME specification to capture additional information about the environmental context of samples (for example geographical location, biological interactions, and environmental conditions) [17]. Work on MIAME/Env has since been extended into the metabolomics and genomics domains [20]. Because of participation in such projects, and its focus on multi-omic data sets, NEBC is also a founding member of the "Minimum Information about a Biological and Biomedical Investigation" (MIBBI) community [21].

CONCLUSION

The UK has developed a powerful network of specialist facilities to promote the use of high-density microarrays that help researchers overcome the myriad challenges of working with non-model but environmentally-focussed organisms. Depending on the resources available to them and their level of expertise, environmental researchers can selectively choose the facilities while having as much or as little control in the development, construction and processing of the microarrays as they are comfortable with. The nature of the organisms involved can vary dramatically, as can the type of experiments undertaken, but help is available at every stage of the process. This has had a strong beneficial effect on the UK-based environmental community by promoting the use of microarrays and driving environmental sciences into the 21st century. To illustrate this, the following are two case studies, typical of the type of work that has been undertaken, by researchers in the environmental genomics community.

NON-MODEL ORGANISM CASE STUDY 1: THE COCCOLITHOVIRUS MICROARRAY

The Science

Emiliania huxleyi is a marine coccolithophorid with worldwide distribution capable of forming vast blooms that can cover up to 10 000 km^2 [22, 23]. Its production of calcium carbonate coccoliths and its role in CO_2 cycling and dimethyl sulphide (DMS) production makes *E. huxleyi* an important species with respect to past, present and future marine primary productivity.

Viruses have been shown to be a major cause of bloom termination [24-26]. *Emiliania huxleyi* virus 86, EhV-86, is a giant virus that was originally isolated from a bloom in the English Channel off the south coast of England in 1999 [27]. EhV-86 is the type species of the genus Coccolithovirus within the family *Phycodnaviridae* and is currently its largest fully sequenced member [28]. The 407,339 bp genome is the second largest virus genome sequenced and is predicted to encode around 472 protein coding sequences (CDSs) [29]. The majority of EhV-86 CDSs exhibit no similarity with proteins in the public databases; a mere 21% of the CDSs contain protein–protein basic local alignment search tool (BLAST) results that matched an E value lower than 0.01 [29]. The large size (and availability) of the EhV-86 genome and its highly unknown nature easily justified the construction of a microarray to aid in the molecular characterisation of this unique virus family.

The Array

The EhV-86-based coccolithovirus microarray has been through 2 stages of development. Version 1 was dominated by probes for the 472 predicted CDSs of EhV-86 and contained only a handful of *E. huxleyi* probes, and consisted of around 1,600 features. As the work has progressed and developed, and the nature of the questions asked of the array changed, so the latest version 2 of the array (printed at MGF-Liverpool) has around 25,000 features and contains probes for multiple coccolithovirus strains and for ESTs from the host species *E. huxleyi* (developed with the aid of MGF-Edinburgh).

The Research

The majority of the EhV-86 genome is composed of genes of which the function is totally unknown [29]. Indeed, unusually among sequenced genomes, a number of predicted genes fail to find any matches whatsoever in the public databases [30]. This novelty of the genome makes the task of annotation difficult: without functional information it is difficult to assess whether what is predicted to be a gene is actually a gene. Thus, the first use to which the coccolithovirus array was put was simply to detect the presence of viral transcripts. The likelihood of an open reading frame being a functional coding sequence and gene is vastly increased by detecting the presence of a mRNA transcript. This approach was used effectively during the annotation of the EhV-86 genome, where message for 65% of the 472 predicted genes of EhV-86 was confirmed from a lytic phase total RNA extraction [29].

Following the lytic phase transcriptional work the microarray was used to outline the transcriptional profile generated by EhV-86 during the early stages of infection. Using a highly sensitive amplification method, the precise expression profile of the EhV-86 infection of *E. huxleyi* was determined over the first 4 h of infection [31]. This strategy allowed the designation of EhV-86 CDSs into three groups on the basis of when their transcripts are first detected (1, 2 or 4 h post infection). The results obtained suggested two distinct phases to the infection process: a primary phase dominated by a group of CDSs localized to a sub-region of the genome, which have no database homologues and are associated with a unique promoter element; and a secondary phase during which CDSs are transcribed from the remainder of the genome [30-32]. Since the majority of EhV-86 CDSs are unknown, the designation of CDSs into transcriptional categories is a significant step forward in determining their function. The distinctive transcriptional profile has answered and raised many interesting questions about the infection strategy and life style of the virus, in particular the role and function of the virally encoded RNA polymerase.

It is important to note that a microarray based on sequence data for a particular strain is not restricted for sole use with that strain. The Plymouth Virus Collection (PVC) contains 12 virus strains all capable of infecting *E. huxleyi* [33]. These strains were collected over the 4-year period between 1999 and 2003 from sampling sites in the English Channel and in a Norwegian Fjord [25, 27, 34]. The construction of the coccolithovirus microarray also allowed the genetic diversity of the viruses in the PVC to be assessed. A major disadvantage of using a specific strain/species microarray for this purpose is that the microarray can only tell you what is highly variable or not present in a particular genome. Genomic additions go undetected since there is no probe on the array to detect it. However, the use of the array for screening for genomic content can provide a wealth of information on genetic diversity within a family of closely related species. At least 70 genes found in EhV-86 are absent or highly variable in one or more of the genomes of the coccolithoviruses found in the PVC [33]. The distinctive pattern of genetic content displayed by each of the strains (essentially a genomic fingerprint) suggested a complicated series of gene loss/gain events. In particular, a high frequency of deleted or variable genes was found in a 100 kb region of unknown function suggesting this section of the genome may be under a high selection pressure [30, 33]. Thus the microarray method is a relatively cheap, efficient and quick way to glean an insight into the genetic content of a family of closely related species. Genes of interest that are conserved among all strains or are lacking and/or variable in some strains can be identified rapidly for further downstream analysis.

In summary, over a relatively short period of time, a predominantly uncharacterized virus has had its genome annotated, the kinetic profile of its transcripts determined and the diversity within its family assessed through the use of microarrays [29, 31, 33, 35]. The results have indicated a plethora of new and novel genes, a distinctive biphasic transcriptional pattern during the infection process and an unexpectedly large degree of genomic content variability among the coccolithoviruses. The construction of the microarray rapidly increased the understanding of the virus and its infection strategy and serves an excellent model for other researchers to follow.

Case Study 2: Gene Discovery and System-wide Overviews of Environmental Stress Responses in the Common Carp

The Science

Abiotic environmental stress, such as heat and cold, hypoxia, UV and pollution pose major problems for all forms of life, and organisms respond to stress by both protecting themselves from the damaging effects of extreme stress, but also by adjuting themselves to allow life processes continue unaffected by the change in environmental conditions. Animals from temperate climatic zones and which inhabit small freshwater ponds are routinely exposed to seasonal fluctuations of temperature and daily fluctuations of environmental oxygen concentrations, and survival depends upon the existence of powerful stress-coping mechanisms [36-38]. Understanding the underpinning mechanisms of stress responses and adaptation is important not only academically but practically in agriculture and in human society [39]. Cold presents special problems both in the form of reduced biochemical activity but also from the damaging effects of freezing [36]. Our understanding of response patterns to chronic cold in complex animals has been transformed by microarray analyses, from one composed of a few key or ‘candidate’ genes to one where the system-wide effects on large-scale networks of genes and proteins can be appreciated [40, 41]. Since 2000 NERC has funded a large-scale analysis of response patterns at the transcript level in the tissues of the common carp, *Cyprinus carpio*.

The Array

Given that there were no genomic resources for the carp, it was first necessary to generate a collection of cDNA clones, representing genes, all of which was accomplished by the MGF-Liverpool. Fish were exposed to the full range of environmental treatments under investigation, including hypoxia, and RNA isolated, normalised to rebalance representation the representation of genes, and then cloned directionally into a suitable vector. The inserts were sequenced from the 5' end, clustered clones into groups containing overlapping sequences and then their identity established by homology searching. The inserts of 13,500 clones were amplified by PCR and the amplicons were used to print a microarray on glass slides using a contact-printing robot. More recently, another 20,000 clones have been added to the collection which were 'cherry-picked' to reduce redundancy. The new, more representative collection consists of 26,000 clones, which have been printed using a non-contact, piezo-deposition robot. The evolution of the carp array has been detailed in Williams et al (2008) [42]

The Research

Our first experiment [9] followed changes in the expression of genes in 7 different tissues of the carp during long-term transition from warm (30°C) to cold conditions (10°C). The fish were cooled over a 3-day period and then held at 3 cooled temperatures (23, 17 and 10°C) for up to 21 days. On each sampling interval 6-8 animals were killed and the tissues dissected into a plastic ('sushi) tray for long-term freezing at -80°C. Samples of these tissues were used to generate target cDNA, which were hybridised to over 400 high-density cDNA arrays.

The data was subjected to refined statistical analysis to reveal a very complex picture of gene responses [9]. Over 3000 probes were found to be significantly different between warm and cooled specimens in at least one tissue. The first focus was on genes that were differentially expressed (DE) in all 7 tissues, which generated a list of 260 genes of which 180 were identified. This list was composed of genes that were largely involved in cell homeostasis, with the greatest and most consistent responses being for the Δ9-desaturase and CIRBP. The former gene is widely invoked in cold responses as being the key gene in transforming lipids from saturated to more unsaturated, thereby fluidising the cold-ordered lipids in cell membranes [43, 44]. The latter gene is an RNA-binding protein of unknown function, but its prominent responses indicate that it is important to all tissues during cold adaptation. A large number of genes involved in protein turnover were found, indicating strong up-regulation in all tissues.

The second focus was on the remaining ~1,700 identified genes that were regulated in one or more tissues. Clustering of this data using unsupervised techniques generated 23 groups of genes with characteristic tissue-specific responses. Each cluster was profiled by statistical analysis of the Gene Ontology annotations [45] and this pointed directly at the nature of gene responses in relation to tissue function. For example, cluster 5 was composed of genes that were up-regulated exclusively in the intestinal mucosa. This cluster included an unusually large number of transport-associated genes, which is entirely consistent with the role of the intestinal epithelium in nutrient absorption and ion/pH regulation. This gene profiling technique allows the broader gene expression properties across treatment space to be interpreted within the context of tissues function.

Comparing between clusters allowed the identification of major effects in the pathways of intermediary metabolism [9][46]. Thus, it was found that genes of glycolysis were generally up-regulated in heart and brain, but down-regulated in the other tissues, indicating a shift between tissues in the patterns of energy provision and storage. Consistent responses were identified in all of the major tissues, despite them being mediated by different genes or isoforms. Thus brain, liver, heart and liver all showed a very high representation of genes involved in energy pathways and carbohydrate metabolism in the responding gene clusters, which is consistent with a cold-induced modification of the substrates and sources of energy metabolism, and of the pathways involved.

A second major experiment involved the exposure of carp to chronic hypoxia [47]. Carp is unusually tolerant to hypoxia which allows it to prosper in natural ponds. This experiment involved over 700 microarrays, comprising responses to hypoxia in 7 tissues but also at two environmental temperatures, 17°C and 30°C. Again this revealed strong patterns within and between tissues which were surprisingly different between the two exposure temperatures. Curiously, a transcript corresponding to myoglobin was substantially up-regulated by hypoxia in carp liver. Myoglobin is an oxygen-binding protein which was famous for its role in protecting oxidative muscle cells including cardiac myocytes. Expression of this gene in non-muscle tissues is linked to the presence of myoglobin protein in the liver of hypoxia-treated fish [48] and that the amount of protein increases in carp exposed to chronic hypoxia. It was also found that carp possess a second isoform of this important gene, whose expression was limited to the brain, this being the only known vertebrate species possessing two myoglobin isoforms. This fact is likely to be linked to a whole genome duplication event some 15MYA in a clade of fish that includes the goldfish and crucian carp [49].

This discovery of non-muscle expression of this gene was an entirely serendipitous and unexpected finding, arising directly from an open microarray screen that was directed at the discovery of both genes and their environmental responses. It is notable because it flies in the face of conventional respiratory physiology of vertebrates animals. Similar myoglobin expression has been found in zebrafish and the mouse system is now being explored. This phenomenon is likely to have major implications for understanding hypoxia responses in vertebrate animals generally but particularly in humans where hypoxia injury caused by ischaemia and infarct are major clinical conditions in the developed world.

The combination of data from both cold and hypoxia treated carp constitutes a very large data set which can be used to explore the correlation of gene expression properties between pairs of genes. This allows the regulatory relationships between genes to be explored, as whether the correlated profiles between genes are conserved between species. More than 400 gene pairs have been identified whose close expression relationships are conserved between carp and human (Li, W and AR Cossins, to be published), and thus by extension across all vertebrate animals. Finally, the expression data collected on such a large scale and across several different tissues and stress modalities, constitutes another resource with which to differentiate the identities of closely-related members of gene families, which completes the more usual sequence-based approaches. This is a particular problem in recently duplicated genomes, such as the common carp [49] and the salmonid fish. Indeed, genes given a common identity by homology alignment proved to have distinguishable expression properties. On further investigation these expression forms have been found to constitute separate genes.

REFERENCES

[1] Blattner, F.R., et al., The complete genome sequence of Escherichia coli K-12. *Science*, 1997. 277(5331): p. 1453-74.

[2] Mewes, H.W., et al., Overview of the yeast genome. *Nature*, 1997. 387(6632 Suppl): p. 7-65.

[3] Adams, M.D., et al., The genome sequence of Drosophila melanogaster. *Science*, 2000. 287(5461): p. 2185-95.

[4] Waterston, R.H., et al., Initial sequencing and comparative analysis of the mouse genome. *Nature*, 2002. 420(6915): p. 520-62.

[5] Faith, J.J., et al., Large-scale mapping and validation of Escherichia coli transcriptional regulation from a compendium of expression profiles. *PLoS Biol*, 2007. 5(1): p. e8.

[6] Gracey, A.Y. and A.R. Cossins, Application of microarray technology in environmental and comparative physiology. *Annu Rev Physiol*, 2003. 65: p. 231-59.

[7] Parkinson, J., et al., PartiGene--constructing partial genomes. *Bioinformatics*, 2004. 20(9): p. 1398-404.

[8] Wasmuth, J.D. and M.L. Blaxter, prot4EST: translating expressed sequence tags from neglected genomes. *BMC Bioinformatics*, 2004. 5: p. 187.

[9] Gracey, A., et al., Coping with cold: An integrative, multitissue analysis of the transcriptome of a poikilothermic vertebrate. Proceedings of the National Academy of Sciences USA, 2004. 101: p. 16970-16975.

[10] Li, W., et al., Application of ESTs in microarray analysis in Expressed sequence tags: *Generation and Analysis*, J. Parkinson, Editor. 2008, Humana Press: Totowa, NJ.

[11] Rouillard, J., M. Zuker, and E. Gulari, OligoArray 2.0: design of oligonucleotide probes for DNA microarrays using a thermodynamic approach. *Nucleic Acids Res.*, 2003. 31: p. 3057-62.

[12] Olohan, L., et al., Detection of anoxia-responsive genes in cultured cells of the rainbow trout, Oncorhynchus mykiss, using an optimised, genome-wide oligoarray. *J. Fish Biology*, 2008. (In submission).

[13] Wit, E., A. Nobile, and R. Khanin, Near-optimal designs for dual-channel microarray studies. *Applied Statistics,* 2005. 54: p. 817-830.

[14] Parkinson, H., et al., ArrayExpress--a public repository for microarray gene expression data at the EBI. *Nucleic Acids Res*, 2005. 33(Database issue): p. D553-5.

[15] Parkinson, H., et al., ArrayExpress--a public database of microarray experiments and gene expression profiles. *Nucleic Acids Res,* 2007. 35(Database issue): p. D747-50.

[16] Hancock, D., et al., maxdLoad2 and maxdBrowse: standards-compliant tools for microarray experimental annotation, data management and dissemination. *BMC Bioinformatics*, 2005. 6(1): p. 264.

[17] Morrison, N., et al., Annotation of environmental OMICS data: application to the transcriptomics domain. *Omics*, 2006. 10(2): p. 172-8.

[18] Field, D., et al., Open software for biologists: from famine to feast. *Nat Biotechnol*, 2006. 24(7): p. 801-3.

[19] Brazma, A., et al., Minimum information about a microarray experiment (MIAME)-toward standards for microarray data. *Nat Genet*, 2001. 29(4): p. 365-71.

[20] Sansone, S.A., et al., The metabolomics standards initiative. *Nat Biotechnol,* 2007. 25(8): p. 846-8.

[21] Taylor, C.F., Standards for reporting bioscience data: a forward look. *Drug Discov Today*, 2007. 12(13-14): p. 527-33.

[22] Westbroek, P., et al., A Model System Approach to Biological Climate Forcing - the Example of Emiliania huxleyi. *Global and Planetary Change*, 1993. 8(1-2): p. 27-46.

[23] Westbroek, P., et al., Emiliania huxleyi as a key to biosphere-geoshere interactions., in *The Haptophyte Algae*., J.C. Green and B.S.C. Leadbeater, Editors. 1994, Clarendon Press: Oxford. p. 321-334.

[24] Jacquet, S., et al., Effects of inorganic and organic nutrient addition on a coastal microbial community (Isefjord, Denmark). *Marine Ecology-Progress Series*, 2002. 228: p. 3-14.

[25] Schroeder, D.C., et al., Coccolithovirus (Phycodnaviridae): Characterisation of a new large dsDNA algal virus that infects Emiliania huxleyi. *Archives of Virology*, 2002. 147(9): p. 1685-1698.

[26] Wilson, W.H., G. Tarran, and M.V. Zubkov, Virus dynamics in a coccolithophore-dominated bloom in the North Sea. Deep-Sea Research Part Ii-Topical Studies in *Oceanography*, 2002. 49(15): p. 2951-2963.

[27] Wilson, W.H., et al., Isolation of viruses responsible for the demise of an Emiliania huxleyi bloom in the English Channel. *Journal of the Marine Biological Association of the UK*., 2002. 82: p. 369-377.

[28] Wilson, W.H., et al., Family: Phycodnaviridae., in *Virus Taxonomy*, VIIIth ICTV Report, C.M. Fauquet, et al., Editors. 2005, Elsevier/Academic Press: London. p. 163-175.

[29] Wilson, W.H., et al., Complete Genome Sequence and Lytic Phase Transcription Profile of a Coccolithovirus. *Science*, 2005. 309(5737): p. 1090-2.

[30] Allen, M.J., et al., Evolutionary History of the Coccolithoviridae. *Mol Biol Evol*, 2006. 23(1): p. 86-92.

[31] Allen, M.J., et al., Locus-specific gene expression pattern suggests a unique propagation strategy for a giant algal virus. *J Virol*, 2006. 80(15): p. 7699-705.

[32] Allen, M.J., D.C. Schroeder, and W.H. Wilson, Preliminary characterisation of repeat families in the genome of EhV-86, a giant algal virus that infects the marine microalga Emiliania huxleyi. *Arch Virol*, 2006. 151: p. 525–535.

[33] Allen, M.J., et al., Use of microarrays to assess viral diversity: from genotype to phenotype. *Environ Microbiol*, 2007. 9(4): p. 971-982.

[34] Schroeder, D.C., et al., Virus Succession Observed during an Emiliania huxleyi Bloom. *Applied and Environmental Microbiology*, 2003. 69(5): p. 2484-2490.

[35] Allen, M.J. and W.H. Wilson, The coccolithovirus microarray: an array of uses. *Brief Funct Genomic Proteomic*, 2006. 5(4): p. 273-9.

[36] Cossins, A.R. and K. Bowler, *The Temperature Biology of Animals*. 1987, London: Chapman & Hall. 339.

[37] Wendelaar Bonga, S., The stress response in fish. *Physiological Reviews*, 1997. 77: p. 591-625.

[38] Bickler, P.E. and L.T. Buck, Hypoxia tolerance in reptiles, amphibians, and fishes: Life with variable oxygen availability. *Annual Review of Physiology*, 2007. 69: p. 145-170.

[39] Feder, M., Key issues in achieving an integrative perspective on stress. *J. Biosciences*, 2007. 32: p. 433-440.

[40] Gracey, A.Y. and A.R. Cossins, Application of microarray technology in environmental and comparative physiology. *Annual Reviews of Physiology*, 2003. 65: p. 231-259.

[41] Cossins, A.R. and D.L. Crawford, Opinion - Fish as models for environmental genomics. *Nature Reviews in Genetics*, 2005. 6(4): p. 324-333.

[42] Williams, D., et al., Genomic resources and microarrays for the common carp (Cyprinus carpio L). *J. Fish Biology* 2008: p. In press.

[43] Tiku, P.E., et al., Cold-induced expression of Δ9-desaturase in carp by transcriptional and posttranslational mechanisms. *Science*, 1996. 271: p. 815-818.

[44] Polley, S.D., et al., Differential expression of cold- and diet-specific genes encoding two carp liver delta 9-acyl-CoA desaturase isoforms. American Journal of Physiology; Regulatory, *Integrative and Comparative Physiology*, 2003. 284(1): p. R41-50.

[45] Ashburner, M., et al., Gene Ontology: tool for the unification of biology. *Nature Genetics*, 2000. 25: p. 25-29.

[46] Cossins, A., et al., Post-genomic approaches to understanding the mechanisms of environmentally induced phenotypic plasticity. *Journal of Experimental Biology*, 2006. 209: p. 2328-2336.

[47] Nikinmaa, M. and B.B. Rees, Oxygen-dependent gene expression in fishes. *Am J Physiol Regul Integr Comp Physiol*, 2005. 288(5): p. R1079-90.

[48] Fraser, J., et al., Hypoxia-inducible myoglobin expression in nonmuscle tissues. *Proceedings of the National Academy of Sciences* USA, 2006. 103: p. 2977-2981.

[49] David, L., et al., Recent duplication of the Common Carp (Cyprinus carpio L.) genome as revealed by analyses of microsatellite loci. *Molecular Biology and Evolution*, 2003. 20: p. 1425-1434.

In: Oligonucleotide Array Sequence Analysis
Editors: M.K. Moretti and L.J. Rizzo, pp. 403-422
ISBN: 978-1-60456-542-3

Chapter 15

POSITIONAL WEIGHT MATRIX AS A SEQUENCE MOTIF DETECTOR*

Youlian Pan and Sieu Phan
Institute for Information Technology, National Research Council Canada
1200 Montreal Road, Ottawa, Ontario, Canada K1A 0R6

ABSTRACT

In biological sequence research, the positional weight matrix (PWM) is often used for motif signal detection. A set of experimentally verified oligonucleotides known to be functional subsequences, which can be bound by a transcription factor (TF), as translational initiation sites or pre-mRNA splicing sites, are collected and aligned. The frequency of each nucleotide A, C, G, or T at each column of the alignment is calculated in the matrix. Once a PWM is constructed, it can be used to search from a nucleotide sequence for the subsequences that possibly perform the same function. The match between a subsequence and a PWM is usually described by a score function, which measures the closeness of the subsequence to the PWM as compared with the given background. However, selection of threshold scores that legitimately qualify a functional subsequence has been a great challenge. Many laboratories have attempted to tackle this problem, but there is no significant breakthrough so far. In this chapter, we discuss the characteristics of a PWM and factors that affect motif predictions and propose a new score function that is tied into information content and statistical expectation of a PWM. We also apply this score function in the PWMs from public databases and compare it favorably with the broadly-used Match method.

1. INTRODUCTION

The term motif is broadly used to describe a conserved pattern, which include a conserved domain (Pérez-Arellano et al., 2007), a functional structure (Schreiter and Drennan, 2007; Kiedzierska et al., 2007), a specific architecture in biological networks (Zhu et al., 2007) and a string pattern in protein or nucleotide sequences (Pan, 2006). A nucleotide sequence motif often refers to a short functional pattern in the DNA or RNA sequence and is

* Adapted from NRC 50391 of the National Research Council Canada

often used to characterize the interaction between a DNA and a protein, such as the binding sites of nucleases and transcription factors. Others at RNA levels are involved in translation, such as ribosome binding, and mRNA processing, such as splicing between exons and introns. Some motifs are involved in transcription termination, such as microRNAs binding. These motifs are usually conserved in the nucleotide sequence pattern.

Many proteins are able to bind to a DNA subsequence with alternative nucleotides at one or more positions in a motif. Among a set of functional subsequences that are known to have the same function, the most dominant ones are called *consensus* for that function. The term of *consensus* has later been relaxed to include degenerate, yet functional subsequences that match all example sites closely, but not necessarily exactly. A consensus describes statically the interaction between a protein and a DNA, but is unable to describe precisely the significance of one nucleotide over the others at certain position of a motif. In this regard, a positional weight matrix (PWM) is often more powerful to express a motif (Stormo, 2000). In the literature, a PWM is also referred as frequency matrix, position specific weight matrix, and positional probability matrix (see Pan, 2006 and references therein).

A PWM is a probabilistic description of a binding motif for a given protein. Due to its nature of pattern conservation, many similar subsequences could be discovered through computational pattern recognitions. A significant number of such subsequences have been experimentally validated in the biology laboratories across the world. Through the multiple alignments of a set of short nucleotide sequences that are known to perform the same function, probability of each nucleotide appears at certain position of the motif is derived simply by counting its occurrences and can be expressed either in the form of motif logo (Schneider and Stephens, 1990) or a probability matrix (Stormo, 2000).

Over the past two decades, many computational approaches are developed to discover conserved motifs with certain degree of success. Computational motif discovery process can be considered in two categories, unsupervised *de novo* motif discovery and supervised known motif prediction (Pan, 2006). In the unsupervised *de novo* motif discovery processes, numerous DNA subsequences of a given length w are locally aligned. The idea behind these methods is that each sequence in the input dataset contains one or more subsequences that could be instances of a conserved motif, but there is no prior knowledge with regard to their binding agent, motif pattern structure, or probability of each nucleotide at certain position. The conserved motifs are iteratively evolved through various optimization algorithms.

The optimization algorithms in *de novo* motif discovery can be subdivided into three distinct approaches: *deterministic optimization*, *stochastic optimization*, and *enumeration* (Pan, 2006, D'haeseleer, 2006). Information content is one of the objective functions applied to measure the quality of a local alignment (see Section 2). Optimization of the information content of a matrix for a fixed number of occurrences has been widely used in matrix-based motif discovery.

While performing local alignment, *deterministic optimization* methods simultaneously optimize a positional weight matrix describing a motif and probabilities of its associated sites. Expectation maximization (EM, Lawrence and Reilly, 1990) is one of such methods. The EM algorithms are named for their two iterative steps, expectation and maximization, which are repeated until a convergence criterion is satisfied. The expectation step evaluates the likelihood of each subsequence of length w to fit the PWM profile with respect to the background of the input sequence. A likelihood value Z_{ij} is computed for each position i of input sequence j. The maximization step puts the best subsequence (with maximum likelihood value) from each input sequence j and builds a new alignment PWM profile. A popular implementation of these algorithms is in MEME software (Bailey and Elkan, 1994). However, the result of the EM algorithm is sensitive to the initial subsequence selection from each input sequence and, therefore, it could be trapped by local maxima.

Stochastic optimization applies stochastic sampling instead of exhaustive survey of all subsequence of length w. Initially, one such subsequence is selected randomly from each of the k input sequences and a $4 \times w$ PWM profile is built on subsequences from k-1 input sequences. Similar to the EM algorithm, the likelihood value of each possible w-mer (Z_i) from the remaining sequence is evaluated to represent how well the w-mer fits to the model. A w-mer is sampled stochastically according to the relative likelihood values ($Z_i / \sum_i Z_i$, Lawrence et al. 1993). This relative likelihood value indicates that a w-mer that fits well (with high Z_i value) is more likely sampled. This procedure is repeated for each sequence until all input sequences have been sampled. The whole procedure is repeated until a convergence criterion is satisfied. Theoretically, the relative information is maximized only after an infinite number of cycles. Practically, this algorithm often converges very quickly (Zhang, 2002). Similar to the EM approach, stochastic sampling could be locked to a local maximum. Inserting another step into the algorithm solves this problem. After a specific number of cycles, one can automatically shift (called a phase shift) all aligned subsequences to the left or right by a certain number of nucleotides and a $4 \times w$ PWM profile is built after the shift. The iterative sampling is repeated. This approach has been implemented in the Gibbs Sampling family of tools that are very popular in motif discovery (Liu et al. 1995; Hughes et al., 2000; Liu et al.., 2001; Thijs et al., 2001; Thompson et al., 2003, 2007; Favorov et al., 2005).

Enumerative algorithms search exhaustively for all possible combinations of nucleotides and select statistically the top few overrepresented motifs from a set of input sequences. Subsequently, motif discovery is treated as a feature selection problem; a motif is treated as a feature of input sequences that discriminate the potential motifs from background sequences (Sinha, 2003). Instead of counting the frequency of overall occurrences of the motifs in a set of sequences, information about motif distribution in individual sequences is used to evaluate motif overrepresentation. This kind of methods has been considered too slow due to the fact that their time complexity grows exponentially with the length of the motif. Given a motif length w, these methods need to evaluate 4^w candidate patterns before an optimal solution is found. However, by indexing the input sequences with a suffix tree, the execution time becomes exponential with respect to the number of allowed substitutions, which is usually small, instead of motif length (Pavesi et al., 2004). These methods are guaranteed to find the globally most overrepresented motifs (Sinha and Tompa, 2002). Some of these methods compare favorably with others with regard to accuracy (Tompa et al., 2005).

Biological methods for motif discovery also become significant and more reliable. These include "photofootprinting" (Becker and Wang, 1984), chromatin immunoprecipitation (ChIP, Solomon and Varshavsky, 1985), and a combination of a modified version of the ChIP procedure (Orlando, 2000) with DNA microarray technology (chip) called ChIP-chip (Horak and Snyder, 2002), protein binding microarray (Bulyk et al., 2001; Mukherjee et al., 2004) and others reviewed in Pan (2006). These methods brought the motif discovery to a new stage and many computationally discovered motifs are validated.

Additionally, motif discovery are being advanced through phylogenetic footprinting. Mainly, two phylogenomic approaches have been considered. The first is to find motifs that are common from multiple orthologous sequences. Footprinter (Blanchette and Tompa, 2002) and OrthoMEME (Prakash et al., 2004) are two examples in this category. The second approach is to globally align the promoter sequences of the orthologous genes and then identify conserved windows. For example, EMnEM (Moses et al., 2004), CompareProspector (Liu et al., 2004), PhyME (Sinha et al., 2004), and PhyloGibbs (Siddharthan et al., 2005), Phylo-HMMs (Siepel and Haussler, 2005), phylogenetic Gibbs sampler (Newberg et al., 2007) and others reviewed in Pan (2006) falls into this category.

Through both computational and biological approaches, or through a combination of both approaches, many motif profiles in the form of PWM became known and validated. Many (trained) PWMs are available in public databases, such as TRANSFAC (Matys et al., 2006) and JASPAR (Sandelin et al., 2004). Such PWMs are extensively used for *supervised* prediction of putative motifs. Many PWM-based motif prediction tools have been developed over the past decades. These include Match (Kel et al., 2003), P-Match (Chekmenev, 2005), MatInspector (Cartharius et al., 2005) and GAPWM (Li et al., 2007).

The main challenge in *supervised* PWM-based motif prediction methods is the objective score function and determination of a threshold score. The score functions usually depend on PWM parameters such as its length and information content. Therefore, a threshold scores that legitimately qualify a functional motif is very hard to select without subjectivity. This chapter is to discuss the characteristics of a PWM, the existing score functions, the factors that affect motif predictions and finally propose a new score function that is tied into information content and statistical expectation of a PWM. We also apply this score function in the PWMs from public databases and compare it with the broadly used Match method.

In the following sections, we describe the existing objective functions, their strengths and weaknesses in Section 2, and ways to improve the score functions based on the existing ones in Section 3. We apply the proposed score function in two datasets in Section 4, discuss the result in Section 5, and finally conclude in Section 6.

2. EXISTING OBJECTIVE (SCORE) FUNCTIONS

Before describing a score function, it is necessary to review the information theory (Schneider et al., 1986) that is the backbone of most score functions. Information theory describes the uncertainty of a discrete random variable $X = \{x_1, x_2, \ldots, x_k\}$ as

$$H(X) = \sum_{b=1}^{k} p_b \log_2(1/p_b) = -\sum_{b=1}^{k} p_b \log_2(p_b) \text{ (bit)} \tag{1}$$

where p_b is the probability of x_b and $\sum_{b=1}^{k} p_b = 1$. One bit of information resolves the uncertainty of choice between 2 equally likely symbols. For a nucleotide sequence, there are 4 bases ($X = \{A, C, G, T\}$, $k = 4$). The maximum uncertainty corresponds to the case that all 4 bases have equal probability, $p_b = 0.25$; thus the maximum uncertainty is 2 bits. If only one base, say G, appears at certain position of a PWM, we have $p_A = 0$, $p_C = 0$, $p_G = 1$, $p_T = 0$. The uncertainty of that column is 0 bit (0 log 0 = 0 and 1 log 1 = 0). Therefore the uncertainty value for each column of a PWM is between 0 and 2.

The information content at column i of a PWM can be measured by a Kullback-Leibler (Kullback and Leibler, 1951) distance (or relative entropy), which is defined as

$$I_i = \sum_{b=1}^{4} p_{i,b} \log_2 \frac{p_{i,b}}{f_b} \tag{2}$$

or

$$I_i = -\sum_{b=1}^{4} p_{i,b} \log_2 f_b + \sum_{b=1}^{4} p_{i,b} \log_2 p_{i,b} \tag{3},$$

where $p_{i,b}$ is the probability of nucleotide b found at position i of the site, and f_b is the probability of nucleotide b on sequences other than the sites (i.e. background). With the assumption of equally probable background model in a DNA sequence, Equation (3) becomes

$$I_i = -\log_2 f_b(\sum_{b=1}^{4} p_{i,b}) + \sum_{b=1}^{4} p_{i,b} \log_2 p_{i,b} = -(-2)(1) + \sum_{b=1}^{4} p_{i,b} \log_2 p_{i,b}.$$

This special version of Equation (3) is the traditional Shannon distance (Shannon, 1948)

$$I_i = 2 + \sum_{b=1}^{4} p_{i,b} \log_2 p_{i,b} \tag{4}.$$

Information content of a PWM is the summation of information on each column:

$$I_{matrix} = \sum_{i=1}^{w} I_i \tag{5}.$$

Based on information content in each column of a PWM, the PWM can also be expressed graphically as sequence logo (Schneider and Stephens, 1990). An objective function of each PWM-based motif discovery method is to discriminate a subsequence between a background model and the PWM model. There are mainly six categories of the objective functions, *log-odd score (also called log-likelihood ratio). MAP score*, *relative information*, *Z score, specificity score,* and *positional bias scores*.

The *log-odd score* is to measure the relative likelihood of a subsequence s being an instance of the PWM against the given background model. It is the summation of the score on each base and can be expressed as:

$$V_s = \log_2(\prod_{i=1}^{w} \frac{p(b_i)}{f(b_i)}) = \sum_{i=1}^{w} \log_2(\frac{p(b_i)}{f(b_i)}) \tag{6}.$$

The log-odd score has been extensively used in various domains. It is the core of Viterbi algorithm that is extensively used in sequence alignment, hidden Markov model (Krogh et al.,

1994) and many motif finding tools. However, the log-odd score is dependent on length of the motifs and PWM models.

The *MAP* (*maximum a posteriori probability*) score combines the information content of the PWM and the rareness of PWM with regard to the background model (Liu et al., 2002):

$$MAP = \frac{\log_2(n)}{w}[\sum_{i=1}^{w}\sum_{b=1}^{k} p_{i,b}\log_2(p_{i,b}) - \frac{1}{n}\sum_{j=1}^{n}\log_2 f(s_j)] \quad (7),$$

where n is the number of sites used to construct the PWM and s_j is an individual site

The *relative information* measures the relative information of a site S based on PWM

$$Score(S) = \sum_{i=1}^{w} I_i p_{i,b_i} \quad (8),$$

where b_i is nucleotide i in the site. The *relative information* score does not consider background information and applied by MatInspector (Cartharius et al., 2005), Match (Kel et al., 2003) and P-Match (Chekmenev et al., 2005). These applications take a max score and then take a ratio between a site score and the max score:

$$R = \frac{Score(S)}{Score(\max)} \quad (9),$$

but Match (Kel et al., 2003) scales between the maximum score and minimum score and then takes the ratio

$$R(Match) = \frac{Score(S) - Score(\min)}{Score(\max) - Score(\min)} \quad (10).$$

The *Z-score* measures the significance of a prediction based on its over-representation and is expressed as

$$Z(s) = \frac{obs(s) - E(s)}{\sigma(s)} \quad (11),$$

where $obs(s)$ is the predicted number of PWM instances s, E(s) is the expected number of s in the background model and $\sigma(s)$ is the standard deviation. However, the statistical over-representation as measured by *Z-score* does not necessarily mean binding preference (Li and Tompa, 2006).

The *group specificity score* measures the goodness of a given motif that is localized to the set of input sequences as compared to the background. The rational is that the true motif is a subsequence that is clearly more frequent in the selected sequence set than in all sequences. This score is used in AlignACE (Hughes et al., 2000).

The *sequence specificity score* measures the likelihood of the set of input sequence potentially contains a functional motif. A prediction having multiple motifs in one sequence is less significant than a prediction having a balanced number of motifs in each sequence (Li and Tompa, 2006). This score is applied in ANN-Spec (Workman and Stormo, 2000) and Weeder (Pavesi et al., 2004).

Due to the uneven distribution of certain motifs (Pan et al., 2004, Smith et al., 2007), *positional bias* (Hughes et al., 2000) and *local positional bias* (Friberg et al., 2005) scores are being used to measure the concentration of the motifs within certain local regions of the input sequence set.

Most of the above described objective functions are exclusively used for finding a PWM representing a motif, such as *MAP score* and *group specificity*. Great effort has been placed in discovering PWMs through both biological and computational approaches and hundreds of known PWMs became available. The question is how to make use of these resources in predicting new instances of known PWMs. There is always an issue of a threshold selection for each PWM in predicting the new instance of the PWM. In Match and P-Match, three threshold values (minimum false positive value (minFP), minimum false negative value (minFN), and minimum of FP+FN) are generated for each PWM to satisfy users with different perspectives based on the sequences used to construct the PWM. This chapter investigates this problem by incorporating various sources of available information.

3. Proposal of an Integrated Approach That Has Three Components

3.1. The Log Odd Score

The log odd score as described by Equation (6) is motif length dependent. For a consensus motif c of the PWM, the log-odd score

$$V_c = \sum_{i=1}^{w} \log_2\left(\frac{p(b_{c,i})}{f(b_{c,i})}\right) \quad (12).$$

V_c is the best possible log-odd score of the PWM. For a subsequence s, $V_s = V_c$ means s is a best fit with the PWM and most likely a TFBS that the PWM specifies, while a $V_s = 0$ means s is identical to the background and very unlikely to be a TFBS that the PWM specifies. Thus, we defined the goodness (G, Pan and Phan, 2008) of fit between s and the PWM as:

$$G = \begin{cases} V_s / V_c, & \text{while } V_s \geq 0 \\ 0, & \text{while } V_s < 0 \end{cases} \quad (13).$$

The value of G is between 0.00 and 1.00 since $V_s \leq V_c$ and is independent of pattern length. From the G value, one could easily tell the confidence level of the putative TFBS that is found by a PWM regardless of the length of the subsequence.

3.2. Statistical Expectation

For a given PWM of length w, the theoretical number of motif variants is 4^w based on background model. Each motif variant has a probability of $4^{-w} \times (L - w + 1)$ to appear in a sequence S of length L. We can enumerate each motif variants in the PWM space and calculate its log odd-score and G value. At a given threshold, assume n motif variants satisfy the threshold, thus the probability of the PWM appearing on S is

$$p = 4^{-w} \times n(L - w + 1) \quad (14).$$

3.3. Information Content of the PWM

MAP score measures relative information content as compared with background and has been widely used and maintains a good balance between PWM specificity and diversity (Liu et al., 2002). However, the MAP score, as we can see from Equation (7), takes average information of all columns in the PWM. This simplistic averaging effect implicitly penalizes longer PWMs that usually have higher proportion of degraded columns that have lower information content. Overall information content on the other hand, as described by Equation (5), captures all information in the PWM.

3.4. Proof of Concept

For proof of concept, we retrieved 107 PWMs from JASPAR core database (Sandelin et al., 2004) on September 19, 2007. Since these PWMs are trained from different organisms ranging from yeast, insects to human and other mammals, we set the background probability as default (f_b = 0.25, b= {A, C, G, T}) and calculated best log-odd score, statistical expectation at different thresholds of G, *MAP* score, and information content of the PWM.

For a given G threshold g_t, we are interested to know the statistical expectation of finding a motif variant having a G value higher than g_t. This probability is called p_value associated with the given g_t. Theoretically, we calculate the probability density function, $e(g)$, over the entire g span (from 0 to 1). It is established by counting the number of motif variants found in a given interval g. The probability distribution function, $E(g)$, is then determined by

$$E(g) = \int_0^g e(\tau)d\tau \quad (15).$$

And the p_value is determined by

$$p\_value(g) = 1 - E(g) \quad (16).$$

Figure 1 shows the *p*-value function of the human PWM PAX2_01 (from TRANSFAC). With knowledge of *p*_value, we can now select a proper threshold with a desired level of confidence. This in turn suggests that for every PWM, we establish an accompanying *p*_value table. The *G* threshold can be selected from the table according to the desired confidence level.

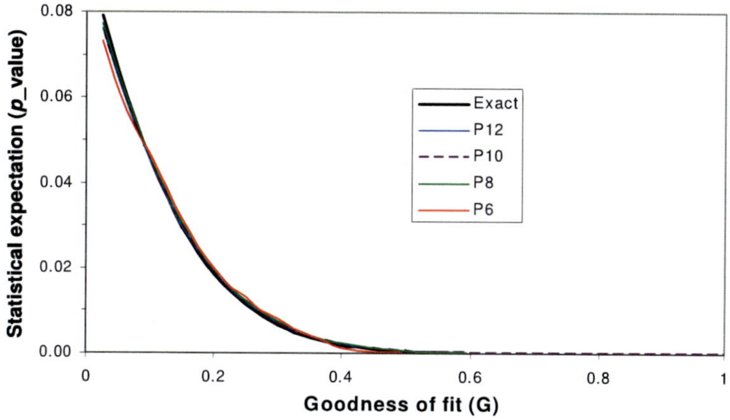

Figure 1. *p*_value function of human PWM PAX2_01. Exact: all instances in the PWM space are evaluated, Px: 4^x instances from the PWM space are sampled.

The establishment of the *p*-value function for a lengthy PWM is computationally expensive. For example, a PWM of 24 columns would take more than 100 days to complete the calculation under the current state of technology. To alleviate this problem, we devised a random sampling scheme to perform the calculation. Intuitively,

$$e(g) = \lim_{n \to N} e_n(g) \qquad (17),$$

where n is the number in the sampling subspace from the entire PWM space of $N = 4^w$ (w is the length of PWM) and $e_n(g)$ is the density function obtained by evaluating the G value of Equation (13) over the selected n variants. We verified that our random sampler and ensured distinct motif variants with no single repeat within the PWM space before it was used for random sampling. The *p*_value functions based on random sampling were verified to be technically identical as if the entire PWM space is sampled as long as sampling size was over 4^{10} (Figure 1). It is feasible to perform exact calculation for a PWM of $w <= 18$. For a PWM of $w > 18$, we applied the random sampling scheme to perform the computation.

The result indicates that the statistical expectation is highly correlated ($p < 0.0001$) with PWM length (w), information content, and the best log-odd score (Table 1). It appears that *MAP* score is not able to reveal the statistical expectation of a PWM since *MAP* score implicitly penalized longer PWM as described above. In this regard, we conducted the correlation analysis for the 19 PWMs that have equal length of 10. The correlation (*MAP* for L=10) appears to increase as result of this selection, but still not comparable with others such as best log-odd score, information content and PWM length. Additionally, in order to eliminate the effect of PWM length, we also calculated the cumulative information of 5 columns in the PWM that have the highest information content and named it as *core information* and analyzed its correlation with *p*_value regardless of the length of PWM. This correlation appears to be significant at lower G threshold and is not helpful in selection of a threshold value. Thus, the core information is not considered further. At a given length of PWM, the *p*_values significantly correlated with threshold G values (Figure 2).

Table 1. Correlation between the statistical expectations values and various properties of the 107 PWMs from JASPAR core database at a given threshold of G value

G>	0	0.1	0.2	0.3	0.4	0.5	0.6	0.7	0.8	0.9	0.95
Length	0.57	0.56	0.65	0.71	0.77	0.83	0.88	0.91	0.95	0.97	0.98
Information	0.93	0.95	0.97	0.98	0.99	0.98	0.96	0.93	0.88	0.83	0.80
Core Info	0.73	0.72	0.70	0.66	0.61	0.53	*0.46*	*0.39*	*0.30*	*0.22*	0.19
Best Log-odd	0.78	0.81	0.84	0.89	0.93	0.97	0.99	1.00	0.99	0.98	0.96
MAP score	0.10	0.09	0.08	0.05	0.00	0.12	0.12	0.19	*0.28*	*0.39*	*0.44*
MAP(L=10)	0.23	0.29	0.36	0.43	*0.52*	*0.58*	*0.64*	*0.68*	*0.61*	*0.45*	0.06

Note: The underlined correlation coefficients are at $p < 0.0001$. The italicized correlation coefficients are at $p < 0.05$. The row of "MAP (L=10)" has n = 19. All other rows have n = 107.

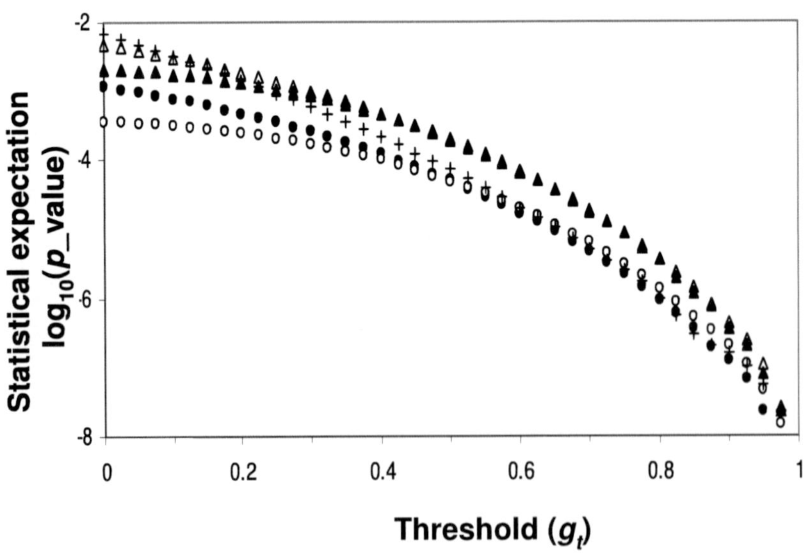

Figure 2. Relationship between probability and threshold G value of the 5 PWMs (MA0010, MA0017, MA0046, MA0069, MA0082). $w = 14$.

3.5. Integration

It is obvious that a selection of threshold should be based on a selection of G value in combination of either of PWM length (w), information content, or the best log-odd score. Since length factor implicitly incorporated in both information content and the log-odd score, we select the better one between information content and the best log-odd score. Figure 3 indicates that a combination of G values with the best log-odd scores performs better in representing the statistical expectation than the combination of G values with information contents. Thus, a combination of *G* values with the best log-odd scores is used as a guide to selecting a *G* threshold. The combination simply revert *G* values back to log-odd score.

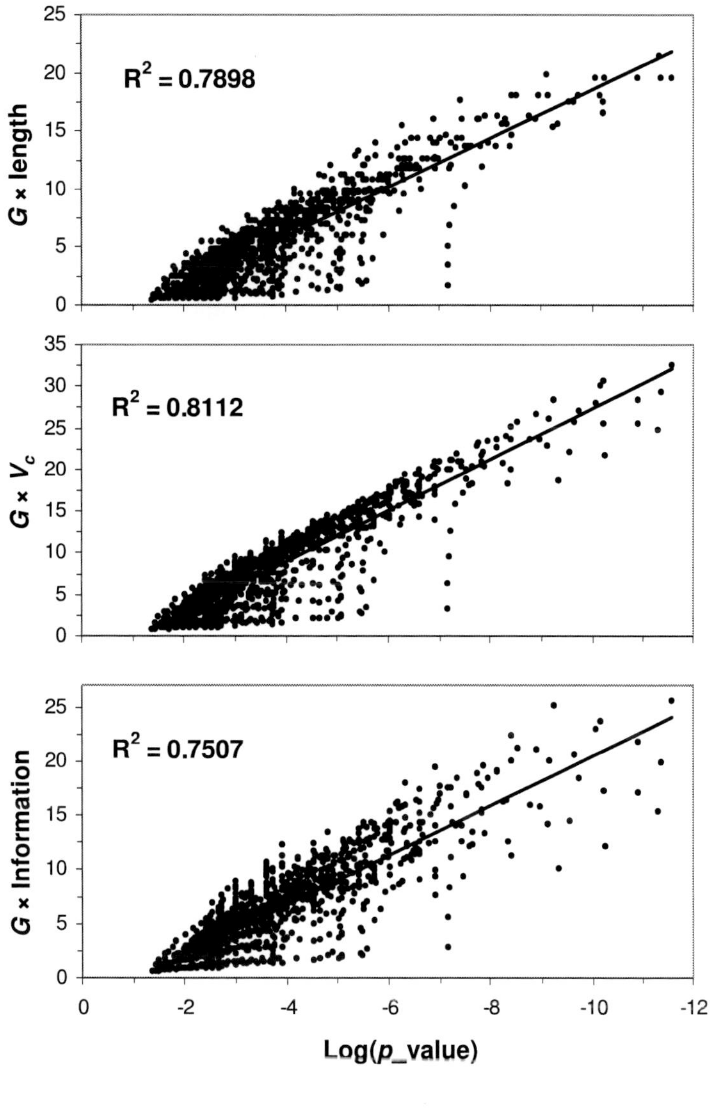

Figure 3. Correlation between statistical expectation and the combination of *G* value with various properties of the PWM.

4. Applications

We applied the above method to generate the statistical expectation function (over G) for each of the 585 vertebrate PWMs and 56 yeast PWMs from TRANSFAC database (Matys et al., 2006). The background probabilities are generated from enumeration of the nucleotides in all known genes' promoters of each species. We also searched for the putative binding sites of yeast transcription factors GCN4 and ROX1 using the TRANSFAC yeast PWMs, GCN4_C and ROX1_Q6, respectively. Finally, we searched for putative binding sites of the human tumor suppressor HIC-1 using the two vertebrate PWMs, HIC1_02 and HIC1_03, from TRANSFAC. In order to keep simplicity in calculation, no pseudo-count is applied in these applications. Therefore, all subsequences that contain a nucleotide with a 0.00 probability in the PWM are not considered in this study.

4.1. Vertebrate and Yeast PWMs

We used both species specific background and default background for comparison. The results indicate there is no perceivable difference between the two with regard to correlation between a combination of *G* value with PWM length, best log-odd score, or information content on one hand and the statistical expectation on the other (Table 2). Both vertebrate and yeast data confirms that a combination of *G* value with best log-odd score is the best in representing the statistical expectation.

Table 2. Correlation between the statistical expectations with various properties of PWMs from three different databases

PWM data Scource	JASPAR	TRANSFAC Vertebrate		TRANSFAC Fungi	
Background*	**default**	**human**	**default**	**yeast**	**default**
G x best log-odd	0.9007	0.8726	0.8752	0.8617	0.8637
G x Length	0.8887	0.8352	0.8359	0.8611	0.8637
G x Information content	0.8664	0.8403	0.8416	0.8120	0.8096

* Background: default = {0.25, 0.25, 0.25, 0.25};
human = {0.2259, 0.2716, 0.2767, 0.2258};
yeast = {0.3096, 0.1932, 0.1893, 0.3079}.

4.2. Yeast Genes

Among the available data, the yeast genome (*Saccharomyces cerevisiae*) is best studied. From the TRANSFAC database (Version 10.4), we retrieve two PWMs (GCN4_C and ROX1_Q6), which model the binding sites for transcription factors GCN4 and ROX1, respectively. The promoter sequences of all 5769 genes from SGDgene table were retrieved from the UCSC Genome Browser (http://genome.ucsc.edu/). The promoter sequences contain 600 bp upstream of the transcription start site (TSS). In order to validate the result, we retrieved the known (documented) associations between the transcription factors and their respective target genes from YEASTRACT database (Teixeira et al., 2006).

We searched the promoter sequences by using the two PWMs and setting different thresholds of *G* values ranging from 0.40 to 1.00. At each threshold, we considered a putative association between a TF and a target gene if a putative binding site of the TF is found in the promoter sequence of this gene. The result of such putative association is validated by known associations obtained from the YEASTRACT database.

For the purpose of comparison, we apply the terminology of *Sensitivity* (*Sn*) and *Positive Predictive Value* (*PPV*) as defined in (Altman and Bland, 1994a, 1994b). *Sensitivity* is the proportion of all known associations (*TP+FN*) that are accurately predicted (*TP*); and the *Positive Predictive Value* is the proportion of predicted associations (*TP+FP*) that are true (*TP*). Traditionally, *Specificity* (proportion of negatives that are predicted false) is used in the evaluation of a method. In genomic sequences, true negative (*TN*) is predominately higher than any of *FP, TP* or *FN*. *Specificity* as defined in Altman and Bland (1994a) would not be able to reveal signals effectively as its value would be very close to 1.00 in almost all instances (Burset and Guigo, 1996). Therefore, we adopted the *Positive Predictive Value* instead of *Specificity* in this study so that the signals can be comparable.

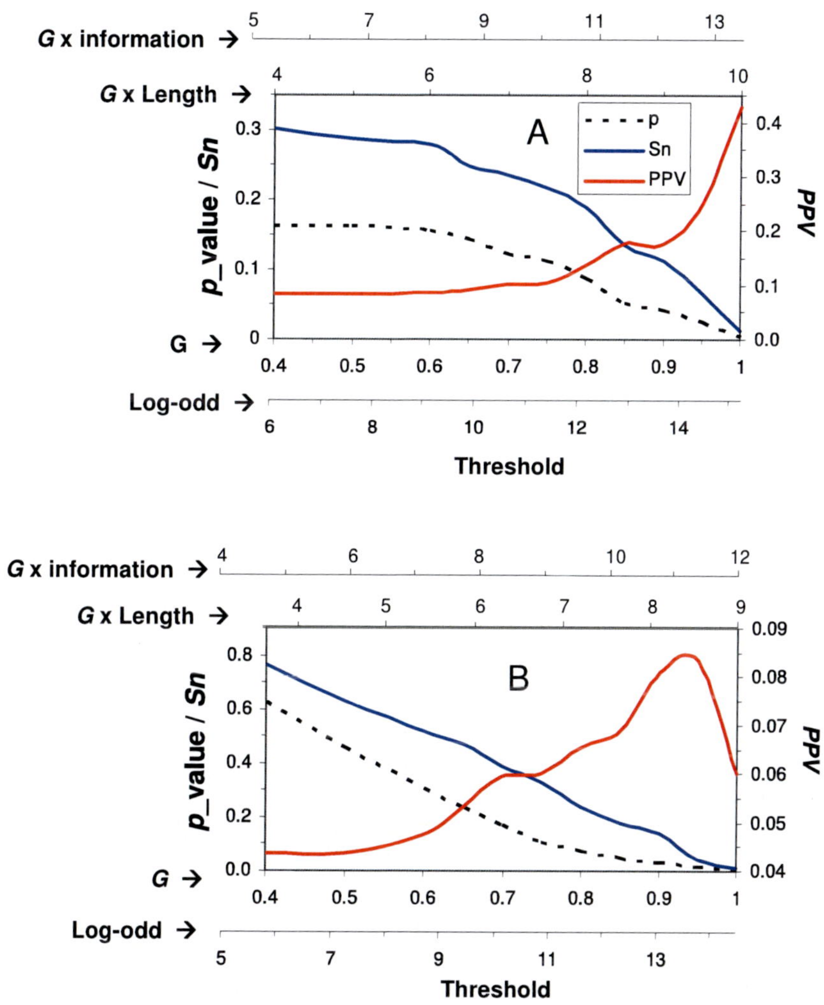

Figure 4. Performance evaluation on yeast data application case. The plot legends in panel A (GCN4_C) also apply to panel B (ROX1_Q6). *Sn*: sensitivity, *PPV*: positive predictive value, *p*: probability.

By decreasing the threshold, more known associations between TFs and their targets are found by the corresponding PWM as reflected by the *Sensitivity* values (Fig. 4). However, the number of false positives increases as reflected by the *Positive Predictive Values*. Based on the *p*_value, users will be able to find a corresponding threshold. For example, the log-odd threshold ($p \leq 0.05$) should be 13 for GCN4_C and 12.25 for ROX1_Q6.

4.3. Human Cancer Genes

We retrieved 406 cancer gene entries from the CancerGenes Resequencing Resource (Higgins et al., 2006). These 406 entries represent 385 distinct genes. We retrieved promoter sequences of these cancer genes from the UCSC Genome Browser (http://genome.ucsc.edu/). The promoter sequences cover the range of 1000 bp upstream and 200 bp downstream of TSS with a total length of 1200 bp. We included the 200bp downstream sequences in this study to account for potential alternative TSS annotations that way appear in some genes. We retrieved two PWMs (HIC1-02, HIC1_03) for a tumor suppressor gene HIC-1 from the TRANSFAC database and searched the promoter sequences for putative TFBSs that fit the two PWMs over the *G* threshold between 0.40 and 0.90. There is no similar database to YEASTRACT for known associations between the transcription suppressor and its target genes in the human genome.

We searched promoters of human cancer genes and predicted about 20% of these cancer gene have potential association with the transcription suppressor HIC-1 (Fig. 5, $p \leq 0.05$). Since our promoter sequences are from a normal human individual (not a cancer patient), it is not unreasonable that 20% of the cancer genes are associated with the transcription suppressor.

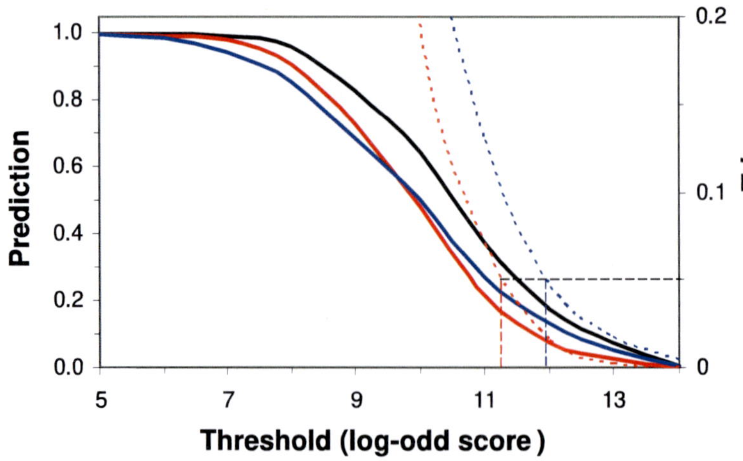

Figure 5. Prediction of association of transcription factor HIC1 with target human cancer genes. Red: predicted by HIC1_02, Blue: predicted by HIC1_03, Black: union prediction of HIC1_02 and HIC1_03. Solid curves: predicted percentage, Doted curves: variation of p_values. Broken lines: marks of significant levels ($p<0.05$).

4.4. Comparison with Match

We identified 16 yeast genes (Table 3) that are known to be associated with both transcription factors GCN4 and ROX1 (YEASTRACT database, Teixeira, et al., 2006) and used them to validate our method and compared it with Match (Kel et al., 2003). While using the default similarity threshold, Match could only find one of the 32 known associations. All associations are found by Match when decrease the threshold because it considers pseudo-count, which would guarantee to find all instance at a low threshold and to have very high false positive predictions. Without knowing their statistical expectation, we did not include those misleading predictions in Table 3. Our method found 16 of the 32 associations, of which 3 are significant ($p < 0.05$).

Table 3. Validation and comparison with Match (Kel. et al., 2003). The values in the current method columns indicate that the associations are found at threshold of G / p_value.

Gene	Match		Current method	
	ROX1_Q6	GCN4_C	ROX1_Q6	GCN_C
ADH1			0.71 / 0.16	
ADH5			0.69 / 0.18	
BOP2		+	0.71 / 0.16	1.00 / 0.005
CWP1			0.81 / 0.07	
CWP2				
GAT2			0.64 / 0.24	
GID8			0.72 / 0.15	
HSP12			0.47 / 0.50	
HSP26			0.88 / 0.04	
HXT5				0.63 / 0.15
IDH1			0.77 / 0.09	
LYS1			0.55 / 0.39	
LYS9			0.47 / 0.50	
MUC1			0.47 / 0.50	
RAD16			0.53 / 0.41	
RPI1			0.91 / 0.03	

5. Discussions

Positional weight matrix has been extensively used in discovery of sequence motifs, such as transcription factor binding sites. The main challenge in using PWM is to find a threshold for various objective functions. There is no substantial breakthrough so far in solving this problem. Match (Kel et al., 2003) appears more appealing than the others and widely used. Match takes the maximum and minimum score and scales them between 0.00 and 1.00 and it also takes into consideration five consecutive nucleotides whose maximum score is the best in any region of the PWM space. P-Match (Chekmenev et al., 2005) combines pattern matching and weight matrix approaches and claimed to be more accurate. We tried to search for the GCN4_C and ROX1_Q1 motifs using P-Match, but unable to find any by the threshold setting defined by those authors. It is expected to find all associations at a lower threshold same as those predicted by Match. However, those potential associations would be buried in a large amount of false positive predictions. Without prior knowledge of these associations, it would be nearly impossible to distinguish them from false positives. For the methods such as Match and P-Match, without an indication of statistical expectation of the predicted motifs, users would be skeptical on the result.

Nucleotide frequency varies across different genomes, for example, the frequencies of A, C, G, T in human promoter sequences are 0.23, 0.27, 0.27 and 0.23, respectively, which are not too much different from the default frequency (0.25 for each nucleotide). But in the yeast promoter sequences, they are 0.31, 0.19, 0.19 and 0.31, respectively. For this reason, we use genome specific nucleotide frequency. Additionally, the nucleotide frequencies vary over various regions of genomics sequences (Pan et al., 2004, Smith et al., 2007). For more precise prediction, regional nucleotide frequencies should be applied.

Occasionally, the log-odd score of a motif could be dominated by one or a few positions because of their extremely high or low frequency values for certain nucleotide(s). Probably, one could argue that a log-odd value of 0.00 for a subsequence might not represent its identity with the background model because the influence of one or more high value(s) of the high frequency nucleotide(s) at certain position(s) is neutralized by the influence of one extremely low value derived from a low frequency nucleotide at another position. Nevertheless, because the overall log-odd score is close to 0.00, no matter whether it is caused by the neutralization of frequencies across various positions or by overall values close to 0.00, this subsequence is not likely a true TFBS. Therefore, a 0.00 log-odd score indicates the subsequence is most unlikely a TFBS.

It is important to consider the statistical expectation of a predicted motif instance. With the density function that we generated for each PWM, it is convenient to find the statistical expectation of each predicted motif instance base on its G value. This G value can easily be reverted back to log-odd score or relative information as can be seen in Fig. 4. Since log-odd value is closely related with statistical expectation as indicated in this study, people would question the necessity of going through the step of G value. The necessity of G value step is demonstrated through the generation of p_value distribution and through its value in measuring the distance between the generated log-odd value and the PWM as compared to the background.

A transcription factor usually binds on a DNA sequence through several positions. Numerous previous studies indicated that the positions inside a motif are somewhat interdependent. For this reason, PWMs are converted to high order hidden Markov models (Krogh et al., 1994). The log-odd score of a motif instance can be calculated based on the state transition probability (e.g., Viterbi score) of the high order hidden Markov model. The same calculation proposed in Equation (13) can be applied to scale the log-odd scores derived from hidden Markov models and statistical expectations can be calculated accordingly.

Many biological problems can not be easily revealed by simply measuring statistical expectation. For example, in our application of yeast genome, too stringent threshold would exclude many potential candidates, such as the threshold set in Match for the two yeast PWMs (Table 3) and our work in ROX1_Q6 (Figure 4B). With varying threshold incorporating both the G values and the distribution function of statistical expectation, we are able to find more motifs using the method proposed in this paper.

Generally, a less stringent threshold would incur higher false positive prediction (Figure 4). This can be complemented by incorporating other information such as microarray gene expression data (Hertzberg et al., 2007) or through comparative genomic approaches (e.g., Newberg et al., 2007). Incorporating gene expression data is certainly a boost in motif finding. However, such data are not always available. With comparative genomics approach, it is arguable even though some successes were shown. Our recent study indicates that promoters of most human genes are significantly different from their orthologues in mice or rats. Similarly, the divergence in the binding sites of the pseudohyphal regulators Ste12 and Tec1 among three closely related yeasts, *Saccharomyces cerevisiae*, *S. mikatae*, and *S. bayanus*, is significant (Borneman et al., 2007). In that case, substantial number of functional motifs in one organism may not appear in the promoter of orthologous genes in another organism, even if both are closely related. Such motifs could be excluded during the prediction process.

Based on the application cases, we suggest taking consideration of both objective function scores and statistical expectation in selecting a threshold. In choosing a threshold, log-odd score appears to be a good objective function as compared with other functions. It is highly correlated with statistical expectation.

In the application of cancer genes, we predicted 20% of the 385 cancer genes are subjected to transcription suppression of HIC-1. Since the promoter sequences are from normal human genome, one would expect all cancer genes are suppressed one way or another. We need to realize that many other cancer gene suppressors are not included in this study. For example, the popular cancer suppressor gene P53 is one among many others.

6. Conclusions

In prediction of a sequence motif using positional weight matrix, it is important to find a statistically meaningful threshold. In this chapter, we described an approach to find a threshold that incorporates the goodness-of-fit of a subsequence to a PWM, statistical expectation, and information content. We demonstrated the strength of this approach through various application cases and compared the method favorably with the broadly used Match method using 16 yeast genes of known association with two transcription factors. We also predicted that 20% of the known cancer genes have association with the tumor suppressor gene HIC-1.

References

Altman DG, Bland JM, 1994a. Statistics Notes: Diagnostic Tests 1 - sensitivity and specificity. *British Medical Journal* **308**: 1552.

Altman DG, Bland JM, 1994b. Statistics Notes: Diagnostic Tests 2 - Predictive Values. *British Medical Journal* **309**: 102.

Bailey EL, Elkan C, 1994. Fitting a mixture model by expectation maximization to discovery motifs in biopolymers. In: *Proceeding of the Second International Conference on intelligent System for Molecular Biology*. AAAI Press, pp. 28-38.

Becker MM, Wang JC, 1984. Use of light for footprinting DNA *in vivo*. *Nature* **309**: 682-687.

Blanchette M, Tompa M, 2002. Discovery of regulatory elements by comparative method for phylogenetic footprinting. *Genome Res* **12**: 739-748.

Borneman AR, Gianoulis TA, Zhang ZD, et al., 2007. Divergence of transcription factor binding sites across related yeast species. *Science* **317**: 815-819.

Bulyk ML, Huang X, Choo Y, Church GM. Exploring the DNA-binding specificities of zinc fingers with DNA microarrays. *PNAS* **98**: 7158-7163.

Burset M, Guigo R, 1996. Evaluation of gene structure prediction programs. *Genomics* **34**: 353–367.

Cartharius K, Frech K, Grote K, et al, 2005. MatInspector and beyond: promoter analysis based on transcription factor binding sites. *Bioinformatics* **21**: 2933-2942.

Chekmenev DS, Haid C, Kel AE, 2005. P-Match: transcription factor binding site search by combining patterns and weight matrices. *Nucleic Acids Res* **33**: W432-437.

D'haeseleer P, 2006. How does DNA sequence motif discovery work? *Nature Biotechnology* **24**: 959-961.

Favorov AV, Gelfand MS, Gerasimova AV, Ravcheev DA, Mironov AA, Makeev VJ, 2005. A Gibbs sampler for identification of symmetrically structured, spaced DNA motifs with improved estimation of the signal length. *Bioinformatics* **21**: 2240-2245.

Friberg M, von Rohr P, Gonnet G, 2005. Scoring functions for transcription factor binding site prediction. *BMC Bioinformatics* **6**: 84.

Hertzberg L, Izraeli S, Domany E, 2007. STOP: searching for transcription factor motifs using gene expression. *Bioinformatics* **23**: 1737-1743.

Higgins ME, Claremont M, Major JE, Sander C, Lash AE, 2006. CancerGenes: a gene selection resource for cancer genome projects. *Nucleic Acids Res* **35**: D721-726.

Horak CE, Snyder M, 2002. ChIP-chip: A genomic approach for identifying transcription factor binding sites. *Methods In Enzymology* **350**: 469-483.

Hughes JD, Estep PW, Tavazoie S, Church GM, 2000. Computational identification of cis-regulatory elements associated with groups of functionally related genes in *Saccharomyces cerevisiae*. *J Mol Biol* **296**: 1205-1214.

Kel AE, Gossling E, Reuter I, et al, 2003. MATCH: A tool for searching transcription factor binding sites in DNA sequences. *Nucleic Acids Res* **31**: 3576-3579.

Kiedzierska A, Smietana K, Czepczynska H, Otlewski J, 2007. Structural similarities and functional diversity of eukaryotic discoidin-like domains. *Biochim Biophys Acta* **1774**:1069-1078.

Krogh A, Brown M, Mian IS, Sjolander K, Haussler D, 1994. Hidden Markov models in computational biology: application to protein modeling. *J Mol Biol* **235**: 1501-1531.

Kullback S, Leibler RA, 1951. On information and sufficiency. *Ann Math Stat* **22**: 79-86.

Lawrence CE, Reilly AA, 1990. An expectation maximization (EM) algorithm for the identification and characterization of common sites in unaligned biopolymer sequences. *Proteins Struct funct Genet* **7**: 41-51

Lawrence CE, Altschul SF, Boguski MS, Liu JS, Neuwald AF, Wootton JC, 1993. Detecting subtle sequence signals: a Gibbs sampling strategy for multiple alignments. *Science* **262**: 208-214.

Li L, Liang Y, Bass RL, 2007. GAPWM: a genetic algorithm method for optimizing a position weight matrix. *Bioinformatics* **23**: 1188-1194.

Li N and Tompa M, 2006. Analysis of computational approaches for motif discovery. *Algorithms for Molecular Biology* **1**:8.

Liu JS, Lawrence CE, Neuwald A, 1995. Bayesian models for multiple local sequence alignment and its Gibbs sampling strategies. *J Amer Statist Assoc* **90**: 1156-1170.

Liu X, Brutlag DL, Liu JS, 2001. BioProspector: discovering conserved DNA motifs in upstream regulatory regions of co-expressed genes. *Pac Symp Biocomput* 2001: 127-138.

Liu X, Brutlag DL, Liu JS, 2002. An algorithm for finding protein-DNA binding sites with applications to chromatin-immunoprecipitation microarray experiments. *Nature Biotech* **20**: 835-839.

Liu Y, Liu XS, Wei L, Altman RB, Batzoglou S, 2004. Eukaryotic regulatory element conservation analysis and identification using comparative genomics. *Genome Res* 14: 451-458.

Matys V, Kel-Margoulis OV, Fricke E, et al, 2006. TRANSFAC and its module TRANSCompel: transcriptional gene regulation in eukaryotes. *Nucleic Acids Res* **34**: D108-110.

Moses AM, Chiang DY, Eisen MB, 2004. Phylogenetic motif detection by expectation-maximization on evolutionary mixtures. *Pac Symp Biocomput* 2004: 324-35.

Mukherjee S, Berger MF, Jona G, et al, 2004. Rapid analysis of the DNA-binding specificities of transcription factors with DNA microarrays. *Nat Genet* **36**: 1331-1339.

Newberg LA, Thompson WA, Conlan S, Smith TM, McCue LA, Lawrence CE 2007. A phylogenetic Gibbs sampler that yields centroid solutions for cis-regulatory site prediction. *Bioinformatics* **23**:1718-1727

Orlando V, 2000. Mapping chromosomal proteins *in vivo* by formaldehyde-crosslinked-chromatin immunoprecipitation. *Trends Biochem Sci* **25**: 99-104.

Pan Y, 2006. Advances in the Discovery of cis-Regulatory Elements. *Curr Bioinfo* 1:321-336

Pan Y, Phan S, 2008. Guide to Threshold Selection for Motif Prediction Using Positional Weight Matrix. *Proceedings of the IAENG International Conference in Bioinformatics* (ISBN: 978-988 98671-8-8), March 19-21, 2008, Hong Kong; pp 151-156.

Pan Y, Smith B, Fang H, Famili FA, Sikorska M, Walker PR, 2004. Selection of putative cis-regulatory motifs through regional and global conservation. In: *Proceedings of the 2004 IEEE Computational System Bioinformatics Conference (CSB2004),* 16-19 August 2004, Stanford, CA, USA 2004; pp 684-685.

Pavesi G, Mereghetti P, Mauri G, Pesole G, 2004. Weeder web: discovery of transcription factor binding sites in a set of sequences from co-regulated genes. *Nucleic Acids Res* **32**: W199-203.

Prakash A, Blanchette M, Sinha S, Tompa M, 2004. Motif discovery in heterogeneous sequence data. *Pac Symp Biocomput* 2004; 248-259.

Sandelin A, Alkema W, Engstrom P, Wasserman WW, Lenhard B, 2004. JASPAR: an open-access database for eukaryotic transcription factor binding profiles. *Nucleic Acids Res* **32**: D91-94.

Schneider TD and Stephens RM, 1990. Sequence logos: a new way to display consensus sequences. *Nucleic Acids Res*. **18**: 6097-6100.

Schneider TD, Stormo GD, and Gold L, 1986. Information content of binding sites on nucleiotide sequences. *J Mol Biol* **198**: 415-431.

Schreiter ER, Drennan CL, 2007. Ribbon-helix-helix transcription factors: variations on a theme. *Nat Rev Microbiol*. **5**:710-720.

Shannon CE, 1948. A mathematical theory of communication. *Bell System Tech J* **27**: 379-423, 623-656.

Siddharthan R, Siggia ED, van Nimwegen E, 2005. PhyloGibbs: a Gibbs sampling motif finder that incorporates phylogeny. *PLoS Comput Biol* 2005; 1: 534-56.

Siepel A. Haussler D, 2005. Phylogenetic hidden Markov models. In: Nielsen R Ed, *Statistical Methods in Molecular Evolution*. Springer, New York 2005; pp 325–351.

Sinha S. Discriminative motifs, 2003. *J Comput Biol* **10**: 599-615.

Sinha S, Blanchette M, Tompa M, 2004. PhyME: a probabilistic algorithm for finding motifs in sets of orthologous sequences. *BMC Bioinformatics* **5**: 170.

Sinha S, Tompa M, 2002. Discovery of novel transcription factor binding sites by statistical overrepresentation. *Nucleic Acids Res* **30**: 5549-5560.

Smith B, Fang H, Pan Y, Walker PR, Famili AF, Sikorska M, 2007. Evolution of motif variants and positional bias of the cyclic-AMP response element. *BMC Evol Biol*. **7**: S15.

Solomon MJ, Varshavsky A, 1985. Formaldehyde-mediated DNA-protein crosslinking: a probe for *in vivo* chromatin structures. *PNAS* **82**: 6470-6474.

Stormo GD, 2000. DNA binding sites: representation and discovery. *Bioinformatics* **16**: 16-23.

Teixeira MC, Monteiro P, Jain P, *et al*. 2006. The YEASTRACT database: a tool for the analysis of transcription regulatory associations in *Saccharomyces cerevisiae*. *Nucleic Acids Res*. **34**: D446-451.

Thijs G, Lescot M, Marchal K, et al, 2001. A higher order background model improves the detection of regulatory elements by Gibbs Sampling. *Bioinformatics* **17**: 1113-1122.

Thompson W, Rouchka EC, Lawrence CE, 2003. Gibbs recursive sampler: finding transcription factor binding sites. *Nucleic Acids Res* **31**: 3580-3585.

Thompson WA, Newberg LA, Conlan S, McCue LA, and Lawrence CE 2007. The Gibbs Centroid Sampler. *Nucleic Acids Res* **35**: W232-237.

Tompa M, Li N, Bailey TL, et al, 2005. Assessing computational tools for the discovery of transcription factor binding sites. *Nat Biotech* **23**: 137-144.

Workman CT and Stormo GD, 2000. ANN-Spec: a method for discovering transcription factor binding sites with improved specificity. *Pac Symp Biocomput* 2000: 467-478.

Zhang MQ, 2002. Computational methods for promoter recognition. In: Jiang T, Xu Y, Zhang MQ Eds, *Current Topics in Computational Molecular Biology*. MIT Press, Cambridge, Massachusetts. pp 249-268.

Zhu X, Gerstein M, Snyder M, 2007. Getting connected: analysis and principles of biological networks. *Genes & Dev* **21**: 1010-1024.

In: Oligonucleotide Array Sequence Analysis
Editors: M.K. Moretti and L.J. Rizzo, pp. 423-442
ISBN: 978-1-60456-542-3

Chapter 16

QUALITY CONTROL METHODS FOR OLIGONUCLEOTIDE MICROARRAYS

Manuel Nieto-Díaz [1,2]**, *Wolfgang Pita-Thomas*** [1,2]**,**
Rodrigo Martínez-Maza [1] ***and Manuel Nieto-Sampedro*** [1,2]

[1]Unidad de Neurología Experimental, Hospital Nacional de Parapléjicos (SESCAM)
Ctra. La Peraleda s/n 45071 Toledo, Spain.
[2]Instituto Cajal de Neurobiología (CSIC) Avda. Dr. Arce 37, 28002 Madrid, Spain.

ABSTRACT

The use of oligonucleotide microarrays to obtain gene expression measurements is a complex process with many potential sources of variability. Leaving apart those variations due to the processes or treatments under study, variability may arise from the sample preparation and hybridization or even from the analysis system itself. Subsequent changes in the hybridization values may obscure those changes due to the biological processes under study and thus lead the researchers to attain erroneous conclusions about gene expression phenomena. It becomes evident that any gene expression analysis should be preceded by a quality control of the hybridization data ensuring that following analyses are developed with appropriate and comparable data. To deal with this necessity, a number of QC methods have been developed, both to be applied before or after hybridizing the RNA to the microarray. These methods have been shown to differ in the obtained results and in the source of variability explored, making it difficult to choose the most appropriate QC measurement. The present contribution will provide the reader with a detailed and up-to-date review of the most important quality control methods available for Affymetrix oligonucleotide microarrays, the most widely used microarray platform. The review begins with a description of all recognized sources of variability in microarray analysis followed by a brief review of the RNA quality analyses. Then follows a detailed description and discussion of the quality controls of the hybridization data, including information on the available bioinformatic tools.

1. INTRODUCTION

Microarray technology provides an unmatched tool to measure the expression of thousands of genes at the same time. As a consequence, the number of microarray studies in the scientific bibliography has increased exponentially since the implantation of this technology more than a decade ago. However, microarray results are far from being accepted by the scientific community. Different studies have identified problems of validity, reproducibility, or comparability between platforms (Chen et al., 2007; Garrett-Mayer et al., 2007), and some authors have proposed that the quality of gene expression data requires significant improvements (Ji & Davis, 2006). The importance of these problems have led important agencies in the US like the FDA to fund the investigation of groups like the MicroArray Quality Control (MAQC) project or the External RNA Controls Consortium (ERCC).

Among the different platforms currently available, Affymetrix GeneChips® is the most broadly used and cited platform (more than 10% of the published microarray articles use this technology). Genechips are oligonucleotide microarrays that derive gene expression measures –like the presence and abundance of a transcript– from the hybridization values of sets of 11 to 20 pairs of probes (probesets) for each transcript (for a detailed discussion on the functioning of GeneChips see Affymetrix, 2004). Thus, the GeneChips' raw data consist of the hybridization values of a very large amount of probes (up to millions), which, as stated by Dumur and co-workers (Dumur et al., 2004), make the errors associated with the experimental procedure to become specially significant As a consequence, there have been a growing interest in methods of analysis of data quality (Dumur et al., 2004; Wilson & Miller, 2005; Lee et al., 2006).

The present contribution deals with quality controls (QC) that should be performed before carrying out any expression analysis using Affymetrix GeneChips®. We will specifically treat the methods that have been developed for posthybridization data (i.e., data obtained from hybridizing the samples to the Affymetrix microarrays) although we will also talk about the methods using prehybridization information. Before discussing QC methods we will introduce the process of analysis of Affymetrix GeneChips and their associated sources of variability.

2. From Individuals to Profiles: Sources of Variability in Affymetrix Data

The use of GeneChip technology to generate gene expression data is a complex process (Affymetrix, 2004; see Figure 1) with many steps, each one representing a potential source of unwanted variation. In general, any microarray study starts with the researcher defining her/his objectives and designing the experiments to fulfil them. According to this design, the researcher chooses the kind of samples to be analysed as well as the number of replicates statistically needed. Once the samples have been collected, the RNA is extracted, purified, converted into cDNA, and converted back into biotinilated cRNA. Then, it is fragmented and mixed with spikes (oligonucleotides included to evaluate the performance of the different hybridization and scanning processes) yielding the hybridization cocktail. All these procedures from the RNA extraction to the preparation of the hybridization cocktail are together known as sample preparation (area 2 in Figure 1) and include incidental processes like sample or RNA storage. Once prepared, the hybridization mixture is hybridized to the GeneChip in an incubation oven, and later passed through several staining and washing steps in a fluidics station. Finally the GeneChip is scanned yielding an image file .DAT from which the Affymetrix software derives a file containing the hybridization intensity values for all probes (.CEL files area 3; Figure 1). The processing of the array is usually carried out by Affymetrix® instruments following standard protocols.

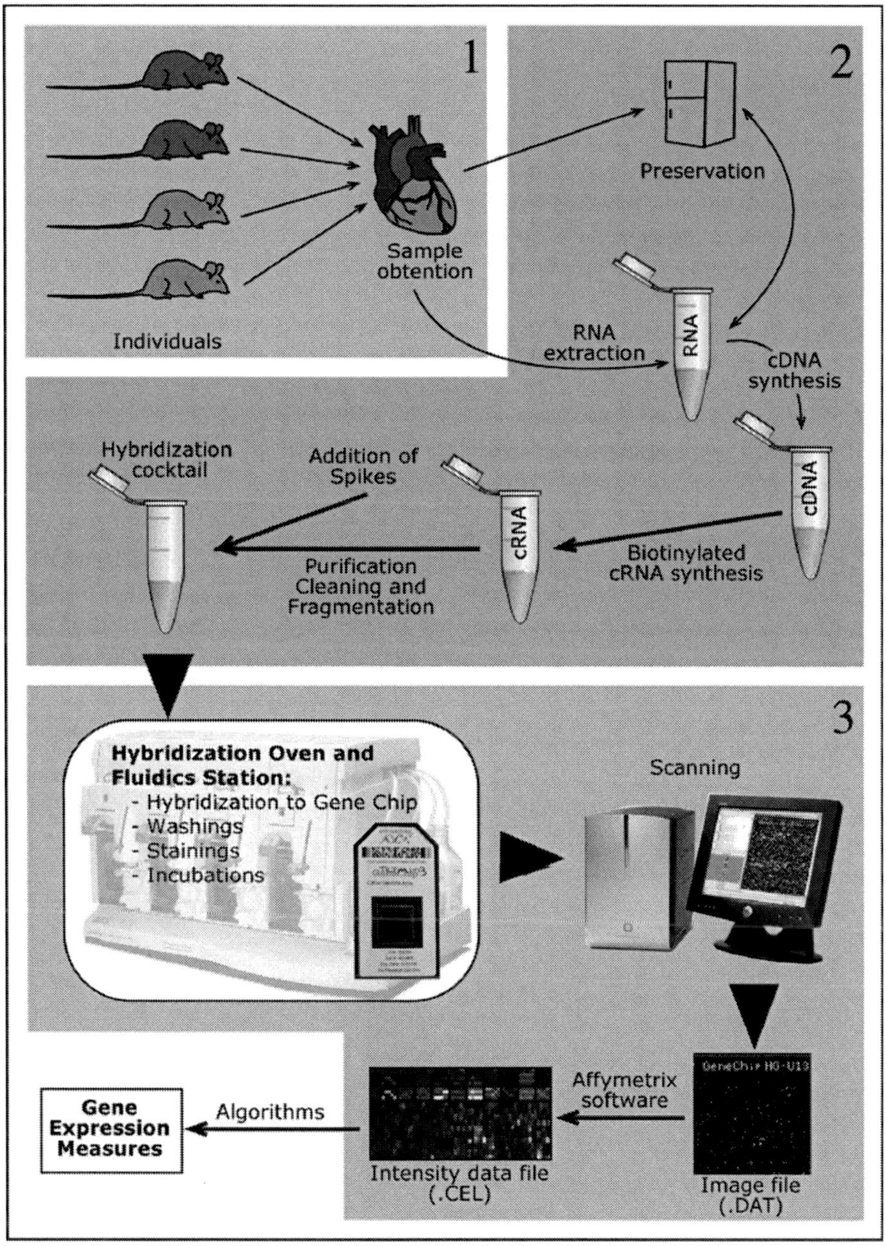

Figure 1. Processes involved in microarray analyses, with all steps detailed from the individuals to the gene expression measures. The area numbered 1 in the upper left corner includes the processes reponsible for the biological variation (mainly individual and tissue or cell heterogeneity). Surrounding this area, is area 2 that comprises all processes of sample preparation, while area 3 (below) corresponds to the system of analysis and includes all processes under direct control of the Affymetrix instruments.

Normally, all sources of variation of Affymetrix data are divided into two different groups. First of all, differences among individuals and samples make up what is known as biological variation. This variation reflects the differences among experimental units under study, but also arises from heterogeneities at different levels, from individuals to tissues or cells (Pritchard et al., 2001; Enard et al., 2002; Zakharkin et al., 2005). The rest of variation is known as technical variation, and includes the differences caused by the sample preparation (including RNA isolation and labelling) and those due to the system of analysis (instruments and arrays, figure 1) (see Affymetrix, 2004; Zakharkin et al., 2005; Garret-Mayer et al., 2007). Most authors, including the Affymetrix Company, consider biological heterogeneity as the major source of variation (Bakay et al., 2002; Novak et al., 2002; Han et al., 2004; Zakharkin et al., 2005), particularly aspects like the tissue heterogeneity (Bakay et al., 2002), differences within and among populations (Bakay et al., 2002; Oleksiak et al., 2002), or day-to day variations (Dumur et al., 2004). Biological variation may in turn depend on the algorithms being employed to calculate the gene expression measures (Zakharkin et al., 2005).

On the contrary, the majority of authors consider technical variation or experimental error a less-significant source of unwanted variability in expression profiling experiments (Bakay et al., 2002), or even negligible (Affymetrix, 2004; Klebanov & Yakovlev, 2007). However, some authors (Bolstad et al. 2004; Ji & Davis, 2006; Copois et al., 2007) defend that technical variation may affect the microarray results, recognizing up to 6 different non-biological potential sources of variability, including RNA preparation (from the RNA isolation to the production of the biotinilated RNA), hybridization, chip scanning, washing, image processing and chip defects. In particular, Copois et al. (2007) established that microarray quality data strongly depends on the quality of the RNA extracted, which in turn depends on the samples quality (see also Dumur et al., 2004; Weis et al., 2007). The stability of a given mRNA is regulated by environmental stimuli like the levels of nutrients, cytokines or hormones as well as by environmental stressors like tissue injury or anoxia. Differences in RNA degradation may be responsible for up to 75% of the changes of gene expression between two samples (Auer et al., 2003). Moreover, the degradation is not homogeneous among the different RNAs, being more intense in the 28S unit of the ribosomal RNA than in the 18S unit (Auer et al., 2003). Although sample or RNA preservation strategies may affect RNA degradation, an study by Mutter et al. (2004) have shown that this effect is minor or negligible using normal protocols (freezing or storing the samples in RNAlater). Other factors include imperfections of the fluidic station or the GeneChips, which may lead to quality problems spatially localized in the chip (Reimer & Weinstein, 2005) or even uniformly distributed over the entire array or affecting numerous arrays (like day effects identifiable by cluster analysis, Bolstad et al., 2004). Labelling seems to be a minor source of variation as compared to biological variation or other technical variations (Zakharkin et al., 2005). Differences between laboratories may also contribute to variation (Irizarry et al., 2005; Chen et al., 2007) although this effect can be considered minor when processing is done under optimal conditions (Irizarry et al., 2005; Wang et al., 2005). Technical variation may arise from very unexpected origins, as exemplified by the article of Zhang and colleagues (Zhang et al., 2006). In this study, the authors demonstrate that the same probes in different GeneChips (U133A and U133 plus2) yields different results indicating that the hybridization of a given probe depends on its context. A further source of variation can be added after the production of the .CEL files due to differences caused by the different available algorithms. Although these differences are clearly linked to the quality of the resulting microarray data, it is far beyond the scope of the present contribution.

3. Prehybridization Quality Controls: Measuring RNA Degradation

Affymetrix (2004) manuals state that RNA quality should be measured at two times, before starting to process it (i.e., extracted RNA) and before preparing the hybridization cocktail, that is, at the stage of biotinilated cRNA. All processes between these two times constitute what we have called RNA processing (see Figure 1) and the different assays of quality that can be performed during this stage are known as prehybridization quality controls. They differ from the posthybridization controls in the kind of data used. Posthybridization analyses mainly uses GeneChip data while prehybridization ones usually analyses the RNA quality directly (for example through electrophoresis methods). Most authors measure the quality (or degradation) of the RNA using the ratio of the rRNA 28S and 18S subunits. However other measures supplement this ratio and may inform the researcher about other problems, like possible protein/DNA contaminations (Affymetrix, 2004).

3.1. Analyses of the 28s and 18S rRNA Subunit Signals

Ribosomal RNA 18S and 28S subunits can be analysed in many different ways. The most simple is proposed by Affymetrix (2004). According to their protocols, the presence of discrete bands for each subunit in an agarose gel is enough indication of RNA integrity. On the contrary, the presence of smears such that the 18S and 28S bands can not be differentiated indicates RNA degradation.

A more elaborated way of measuring RNA quality is quantifying the 28S/18S ratio in agarose gels stained with ethidium bromide (a detailed protocol can be obtained from Sambrook and Russell, 2001) or by means of capillary electrophoresis (see below). The so-obtained ratio has been considered the gold standard to measure RNA degradation, being 1.8 the accepted lower threshold for an integer RNA (some authors like Copois et al. (2007) propose 1.67 as a more useful limit, although acknowledging the difficulty of establishing a precise threshold). However, this ratio may be a misleading indicator (Dumur et al., 2004; Miller et al., 2004; see also Copois et al., 2007), and in fact, Dumur and colleagues (op. cit.) have shown that some high quality samples may yield very low 28S/18S ratios (below 1) and low quality RNA yield high ratios. As a consequence, these authors consider the 28S/18S ratio obtained this way as the least indicative criteria of RNA quality, proposing that other methods should be used instead. They propose to compare the 18S and 28S area (that is, their quantity) with the total amount of RNA and to use the precentage of rRNA as an integrity index (only samples with more than 30% of rRNA can be considered integer).

3.2. RNA Electropherograms

The Bioanalizer 2100 (Agilent) has become the most widely used system to analyze the RNA integrity. It performs capillary electrophoresis of very small amounts of RNA (5 ng) and yield gel images and electropherograms (chromatograms of RNA size distribution; see Figure 2). Different software has been developed to use the data contained in the electropherograms to analyze the RNA integrity. In 2003, Auer and co-workers (Auer et al.,

2003) developed the **Degradometer**, a software implementing a mathematical model that measure the "degradation peak signals". The program yields an index, the **DegFact** and according to it classifies the RNA in 4 different categories of increasing RNA integrity.

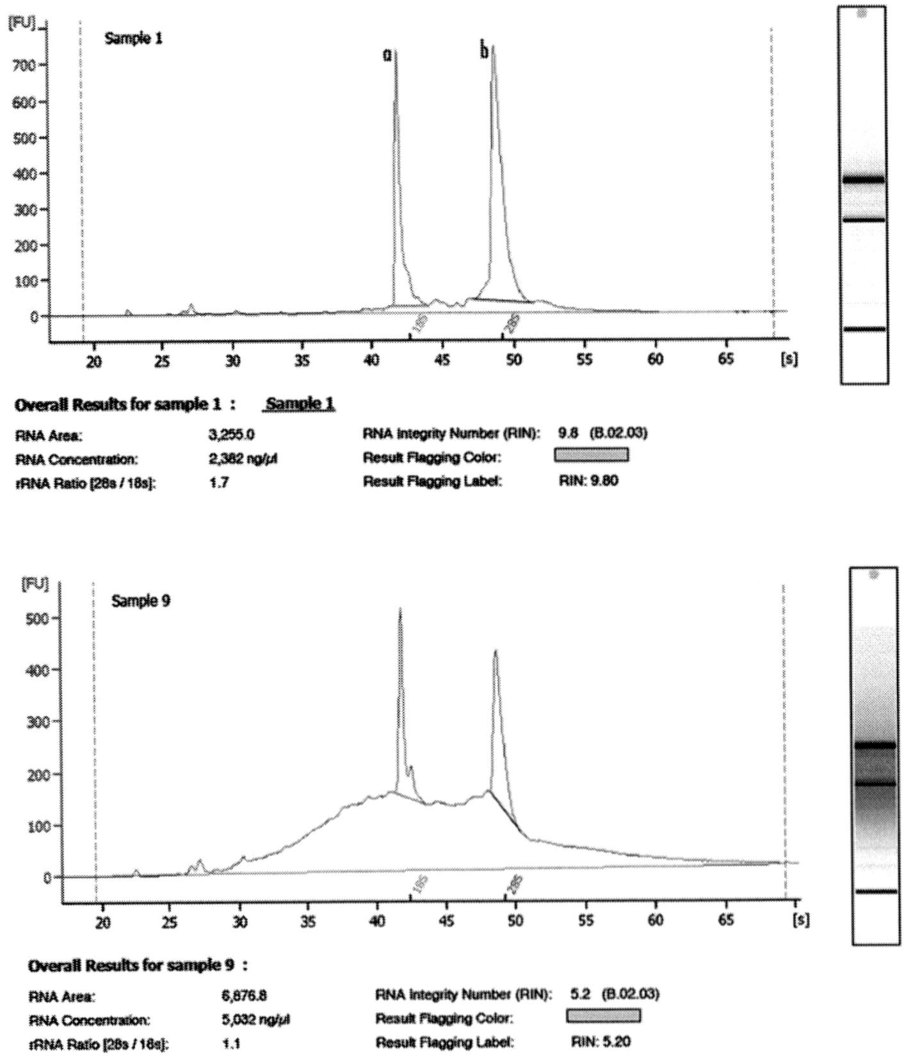

Figure 2. Electropherogram of RNA samples provided by the Bioanalizer 2100. The graphs show the size distribution of the RNA with peeks at 18S (a), 28S (b). On the right there is a gel-like representation of the data and some parameters detailed below each graph. The upper graph (sample 1) corresponds to a high purity RNA sample, presenting high 18S/28S ratio, high RIN number, and two clear ribosomal peeks in a low, flat baseline of RNA. On the contrary, sample 7 (lower) presents clear indications of RNA degradation, with lower 18S/28S ratio and RIN value and a higher, non-flat RNA baseline. Note also that the gel representation of sample 7 present a clear smear, not present in sample 1. Y axys in the electropherogram represents fluorescence intensity (estimation of quantity) and x axys represents increasing size (in arbitrary units). The lines below the 18S and 28S peeks of the electropherograms delimite the area used to estimate the abundance of both rRNA subunits.

According to its authors, the degradometer is able to detect RNA degradation changes due to apoptotic processes even before it could be detected using specific apoptotic markers like the Anexin-V (Auer et al., 2003). The Bioanalizer 2100 software estimates its own index, the **RNA integrity number** (**RIN**; Schroeder et al., 2006), using a neural network based algorithm. The software integrates data from up to 100 features of the electropherograms, including the height of the 18S rRNA subunit, the amount of total RNA. The so-obtained RIN values are then used to classify the RNA sample in one of 10 categories from 1 (completely degraded) to 10 (intact). Recently, Copois and co-workers have proposed an alternative qualitative scale, the **RNA quality scale** (**RQS**), based on the visual inspection of the Bioanalizer 2100 electropherograms (Figure 2). The scale contains 5 degrees of increasing RNA integrity according to the morphology of the 18S and 28S rRNA subunit signals and also the morphology of the region between the markers and the 18S peek (see Figure 2). In the same article, Copois et al. (2007) compared their RQS with the RIN index, the DegFact values and the 18S/28S ratio reaching the conclusion that the 18S/28S ratio was the least reliable method and that all other were more or less similar although the RQS was slightly better than the rest (although they aknowledge that the RQS may be more subjetive). They also proposed that RNA samples should have values of DegFact over 8.7, RIN over7.8 and RQS over 3 to be considered reliable for a micoarray analysis.

3.3. A260/A280 Absorbance

Besides the problems due to RNA degradation, other aspects of sample preparation like sample purity may affect microarray analysis. In this sense, Affymetrix (2004) recomends the use of espectrofotometry to assess the purity of the sample (total RNA) as well as of the biotinilated cRNA. The ratio of absorptions at 260nm and 280nm will inform about protein contamination (ratios below 1.8) and about possible degradaton or truncation of the RNA or even the presence of an excess of free nucleotides (ratios over 2.1). However, while the A260/A280 ratio is very informative for nucleic acid contamination in protein solution, it is no the same the other way around and only large amounts of protein contamination will significantly affect the 260:280 ratio.

4. Using Posthybridization Data for Quality Control

After carrying out the hybridization, and scanning and processing the image, each microarray still provides a wealth of information that can be used to evaluate the quality of the so obtained data. Different authors have explored this approach and many methods, parameters and software have been developed to use these data in quality control. Methods and used data vary fairly, from the study of the scanned images to the analysis of the spike intensity or the expression measures. In the following pages we will detail some of the most commonly used methods, grouped according to their theoretical basis. For example, we will describe together all methods that use the different preservation of the 3' and 5' ends of the RNA sequence to explore RNA degradation. For all of these groups, we will briefly explain their underlying logic, the source of variability that explore, the different approaches developed, and the software available to perform the analyses (information on all cited software is detailed in table 3).

Table 1. Sources of variation, their effects and relative importance according to the references commented in the text.

Source	Effect
Biological variation	The major component of variation
Individual/population heterogeneity	The most important source, includes the variaton under study
Sample (tissue, cell) heterogeneity	Relatively lesser than individual/population but still very important respect technical variation
Technical variation I: sample preparation	RNA degradation may bias the results. In some cases differences in RNA degradation may account for even 75% of the gene expression variations
RNA extraction	From negligible to very important, it may be confused with cell/tissue heterogeneity.
RNA preservation/sample preservation	Minor importance if conditions for RNA preservation are met.
RNA processing	From negligible to very important. Minor importance if enough care is taked in each step.
Technical variation II: system of analysis	Negligible according to Affymetrix but … … … Effects controlled with spikes and internal controls
Hybridization	Unknown, probably minor but even diiferent context make the same probes to yield different results
Fluidics station	May ocassionally cause local variations due to problems in washing although controlled by Affymetrix controls
Labelling	Minor effect according to published studies
Scanning	No studies found
Differences between labs	Minor importance if high quality standard are met, otherwise may be important
Other sources	
Algorithms	Variable, may change the effect of the other sources of variation

Table 2. Prehybridization quality controls. (*) Values of reference according to Copois *et* al. (2007) References of the methods are detailed in the text.

Analysis	Measures	Method	Observations
RNA Electrophoresis	RNA degradation	Agarose Gels	
mRNA 18S/28S ratio estimation	RNA degradation	Agarose Gel with Ethidium bromide staining	Intact RNA with ratio values above 1.67 or 1.8. Not very reliable according to most authors
		Capillary electrophoresis (Bioanalyzer 2100 software) (www.chem.agilent.com/Scripts/PDS.asp?lPage=35490)	
RNA distribution in electropherogram	RNA degradation	DegFact index calculated using the Degradometer (freely available for PC at www.dnaarrays.org/rna_quality.php)	Intact RNA above DegFact 8.7 (Copois *et al.*, 2007)

Table 2. Continued

Analysis	Measures	Method	Observations
Features of electropherogram	RNA degradation	RIN index calculated by the Bioanalyzer 2100 software (www.chem.agilent.com/Scripts/PDS.asp?lPage=35490)	Intact RNA above RIN 7.8 (Copois et al., 2007)
RNA distribution in electropherogram	RNA degradation	RQS (RNA quality scale), established by visual insection, 5 categories	Intact RNA above RQS category 3 (Copois et al., 2007)
A260/A280 ratio	Sample contamination with proteins or other	Spectrophotometry of samples	

Table 3. Software for posthybridization quality controls. All programs can be freely downloaded from the detailed URL. R designates a mathematical programming language and also the environment to run many informatic tools for the analysis of Affymetrix microarrays.

Tool	URL	Comments
GCOS	http://www.affymetrix.com/support/technical/product_updates/gcos_download.affx	
Affy	http://www.bioconductor.org	Needs R package installed (http://www.R-Project.org)
Simpleaffy	http://www.bioconductor.org	Needs affy and R installed
dCHIP	http://biosun1.harvard.edu/complab/dchip/	
Pixel Analyzer	http://www.people.vcu.edu/~kjarcher/Research/Data.htm	Needs R among others installed
AffyPLM	http://www.bioconductor.org	Needs R program installed

4.1. Microarray Image Inspection

Affymetrix (2004) recommends visually inspecting the microarray image in search for artefacts caused by the system of analysis. Though Affymetrix has standardized most of the procedures, still remains the possibility of bubbles, uneven hybridizations, poor wash, or imperfections of the chips that can cause areas of high or low intensity, high regional or total background or even scratches on the array surface (Reimers & Weinstein, 2005). Some of these problems may mean the exclusion of the array from an experiment or the necessity to filter the affected probes (Affymetrix, 2004).

Despite Affymetrix recommendations, visual inspection of the array only allows to recognize grosser faults and only to a trained eye. A more useful and objective methodology was developed by Reimers & Weinstein (2005). It consists on the comparison of each array image to a standard microarray image obtained after averaging all microarray images from the same tissue (or sample type) in the same experiment. The so obtained images and data can be used to quantify the regional bias as well as to visualize it in a much more clear way.

Bolstad and co-workers also developed an approach to this topic using the residuals of fitting the RMA model to the probe data (Bolstad et al., 2004). The fitted model will absorb most of the intensity variations, and as a consequence residuals will reflect image artefacts more clearly than intensity values. Different functions will allow the user to visualize the residuals' magnitude and sign, any combination of both or the probe weight The logic underlying this methodology is further detailed below, under the heading of analysis of residuals.

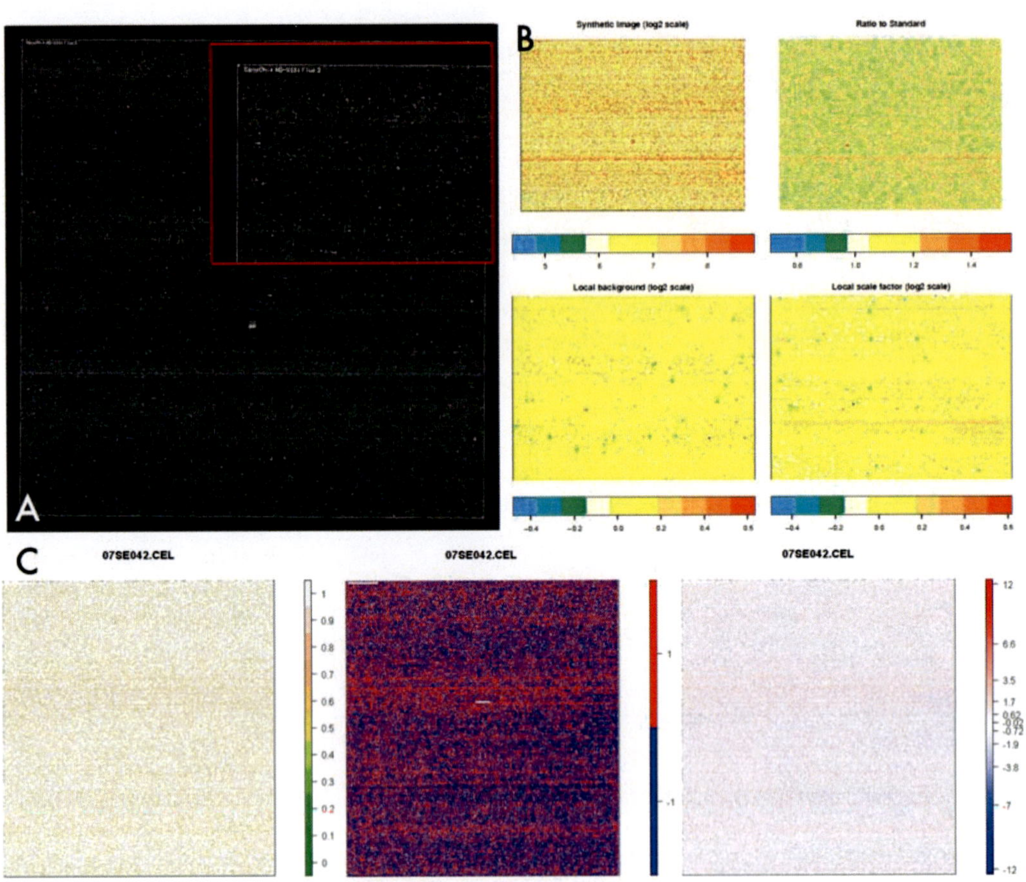

Figure 3. **A**. Scanned image of the GeneChip as provided by the Affymetrix GCOS software.
B. Representations of Affymetrix chips used to explore regional biases. These graphs are the result of applying affytools resources (http://www.discover.nci.nhi.gov/affytools) to our own data. The upper left image corresponds to an synthetic chip arranged as an average from the original chip images. The upper right image corresponds to the difference between the particular chip under scrutiny and the average image on the left. In the lower row are images for background and scale factor for the same chip.
C. Graphical representation of the array after the residual analysis (Bolstad et al., 2004) using affyPLM. Left: probes weights; center: residual signs; right: residuals magnitude and sign.

Implementations: GeneChip images can be visualized using the Affymetrix GeneChip Operating System software (GCOS) or the *image* function of the program Affy (information on these and following programs is provided in Table 3). Raw images are usually log transformed (at least by the image function of Affy) to help visualizing low intensity probes. Reimers and Weinstein methodology, including the assemblage of an standard chip image from our data, its comparison to individual chips (see figure 3) and the quantification of the regional biases, is implemented as a web resource at http://www.discovernci.nih.gov/affytools. Graphical inspection of the residuals can be carried out using the AffyPLM package of Bioconductor. (see Figure 3)

4.2. Noise and Background values

The noise or RawQ value is the pixel to pixel variation within each probe cell of the array (Jones et al., 2006). It is associated to electrical noise from the scanner or to sample quality (Jones et al., 2006) and it seems to be very sensitive to day-to-day variations in the reactives (Dumur et al., 2004). On the other hand, the background is a measure of the fluorescence signal due to non-specific binding and autofluorescence (Jones et al., 2006). According to these authors, high backgrounds may be caused by impurities such as cell debris or salts. A high background obscures the signals from low expressing mRNA and consequently reduces the sensitivity of the analysis.

There are no precise thresholds for these values. However, background values usually vary between 20 and 100 when using GeneArray Scanners calibrated with the new PMT settings and between 200 and 1000 for the old PMT settings (Affymetrix, 2004). Anyhow, values should be similar among the different chips of an experiment. A similar homogeneity should be expected in the rawQ values when samples have been scanned using the same scanner.

Implementation: Background and RawQ values can be obtained from the .CHP and .RPT files provided with the GeneChip results (.RPT files can be opened using a text editor). They may also be obtained from the "analysis info" in the GCOS program. Average background can be calculated using the QC function of simpleaffy program (Bioconductor) and displayed graphically using the plot (QC object) function of this program (see Figure 4). In all cases the values are the same and it is only a question of commodity to choose one option or another.

4.3. Spikes

Spikes are transcripts added to the hybridization cocktail at known concentrations in order to evaluate the system performance. Affymetrix (2004) has proposed that they may be used to measure data quality (in particular RNA quality) comparing the spike values of the group fof arrays scaled to a same value. Samples with less quality will present higher spike values than those samples with higher quality. However, according to Bolstad *et al.* (2004), their use is very problematical because the resulting measures tend to be very noisy and because these measures will only be able to detect very high differences among samples.

Following Affymetrix (2004), several of the sequences spiked during the standard protocols may be used for quality control including:

1. Oligo B2: this spike is included as a positive control and as a way to align the array image after scanning it. When inspecting the microarray image it should be seen forming the border of the array, as a chess pattern in the corners an forming the array name. Affymetrix

(2004) warns the researcher that it the intensity of B2 oligo may change among regions o the array although they do not explain the reason why.

2. bioB, bioC, bioD and Cre Spikes: these are hybridization controls. The first three sequences belong to *Escherichia coli* genes while Cre is gene from P1 phago (Affymetrix, 2004). Affymetrix protocols establish that they are added to the hybridization mixture in increasing concentrations (1.5, 5, 25, and 100 pM respectively). According to the manufactures protocols, bioB is at the detection capacity limit of the arrays and thus it should be present (P) in a 50% of the analyses. The rest of the spikes should be always present (P) with increasing signal values.

Implementation: microarray images can be visualized using the GCOS software to analyze the B2 oligo performance (see Figure 3). BioB, bioC, BioD and Cre data consist in their expression measures as provided by any of the common summarizing methods (MAS5, RMA, dChip, ...).

4.4. Percentage of Present Calls and Scaling Factor

The percentage of present calls is the ratio of the number of genes detected as being expressed and the total number of genes being evaluated. Similar samples or technical replicates should present similar percentages of present calls and thus it may be used to identify outliers within an experiment. Very low values might indicate problems (Heber & Sick, 2006) including bad RNA quality, although the parameter should be carefully evaluated. Dumur and colleagues (Dumur et al., 2004) point out that the percentage of present calls can be very useful in quality control when samples are very similar, that is, from the same tissue or cell type. In these cases this value can detect changes in the sample processing like the inclusion or exclusion of a RNA cleaning step (Dumur et al., op. cit.). According to Heber & Sick (2006), normal values of this parameter range from 20% to 50% although smaller figures can be obtained in cross-species analyses.

On the other hand, the scaling factor is the result of the global scaling process. To compare the different arrays in one experiment the total intensity of each array is computed and adjusted to a prefixed value (the Target Value). The scaling factor is the value that should be multiplied to the total intensity to obtain the Target Value. Following the same logic as with the previous parameter, they should present similar values in all arrays of an experiment. Heber & Sick (2006) propose that this parameter should be smaller than three within one experiment although no further details are provided in this respect.

Implementation: both parameters are detailed in the .RPT file from the GCOS results. They may be also estimated using the algorithms from Li and Wong (2001) implemented in the software dChip. The package simpleaffy of the Bioconductor environment allows computing these parameters using different algorithms (MAS5, RMA, ...). This software yields a graphical representation of both values (figure 4) within a complex figure that includes other typical QC parameters (Wilson & Miller, 2005).

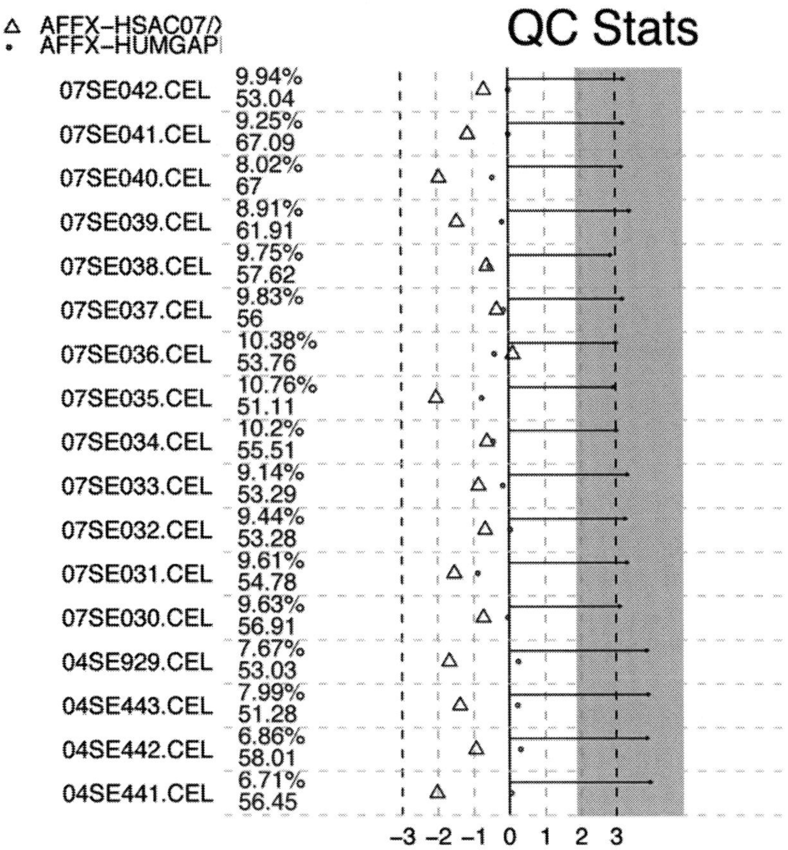

Figure 4. Graphical output from the QC function of the simpleaffy program. As explained by Wilson & Miller (2005), the graph represents several QC metrics for the desired arrays (each line represent an array), detailing: top value % present call; bottom value, background; circle, GAPDH ratio; triangle, beta actin ratio; line, scale factor. The shaded area represents a 3-fold range for the scale factor (centered around the mean). The arrays used are from our own experiments.

4.5. 3'/5' Ratio of House-Keeping Genes

GeneChips include probesets for the 3' and 5' ends of some housekeeping genes, typically GAPDH and beta actin genes. According to Affymetrix (2004), RNA degradation and inefficient transcription to cDNA or to biotinilated cRNA cause a decrease in the amount of the 5' end signals. This is caused because inverse transcriptions starts in the 3' end and proceeds towards the 5' end and thus any degradation or fail in the transcription will cause the process to finish before getting to the 5'end, decreasing the signal of these transcripts (Archer et al., 2006). Despite the common use of these ratios in QC, Bolstad and co-workers (Bolstad et al., 2004) pointed that their interpretation was complicated, as happens with the spike data.

3'/5' ratios can be calculated from two opposite ways. The first one uses the signal values of the 5' and 3' probesets of the chosen housekeeping genes to derive the ratio. Affymetrix (2004) states that the 3' values should not be 3 times greater than the 5' values. Discrepant values between beta actin and GAPDH are considered an indication of transcript specific problems or even image artefacts, more than the result of RNA degradation. Archer and co-workers (2006) have correctly indicated that this way of estimating 3'/5' ratio depends on the algorithm used to derive the expression measure (MAS5, RMA, dCHIP, ...). However, despite the differences in the ratio values obtained using the different algorithms, all agree in their classification of the samples among degraded and non-degraded RNA. Differences seem to be more related with sensitivity than with the discriminatory power of the ratio. Archer et al. (2006) also point that usually there is only one probeset for each end of the house keeping gene making impossible any estimation of the measurement uncertainty, making necessary the definition of a threshold between degraded and non-degraded RNA. To solve these problems, Archer et al. (2006) propose an alternative way of estimating the 3'/5' ratio, using directly the intensity of the probes from the scanned image. In this way, they avoid using gene expression measures and all possible noise arising during their calculation. PM intensity values from the 3' and 5' ends of beta actin, ISGF and GAPDH genes are analysed using mixed-effects models to obtain a mean value and a variance for the intensity values of these probes. The model also incorporates the effect of the sequence on the intensity of the probe using either the model proposed by Wu et al. (2004) or the content of GC nucleotides.

Implementation: beta actin and GAPDH 3'/5' ratios are provided by the GCOS software in the .RPT files. These values are calculated using the MAS5 algorithm. Estimation of these ratios using other algorithms can be accomplished with the QC function of the SimpleAffy program of Bioconductor. This function yields a graphical representation previously commented that includes the GAPDH and beta actin ratios (Figure 4). Archer et al. (2006) method is implemented in the software Pixel Analyzer (freely downloadable from the Internet), which yields the 3'/5' ratios and a 95% confidence interval for each array. The program allows the user to choose between the Wu et al. (2004) or GC estimation of the sequence effect (according to the authors, the GC estimation outperforms that of Wu and co-workers).

4.6. RNA Degradation Plots

This is a method proposed by Gauthier et al. (2004), based in the same principia as the 3'/5' ratio. Since in every probeset, the number of the individual probes (-at1, -at2, ...) indicates their relative position in the sequence (starting from the 5' end), the RNA degradation may be estimated comparing the values of the different positions. Relative intensity values at each position are plotted (see Figure 5) and used to estimate the slope of the curve. If the RNA is degraded, then the transcription will fall before reaching the 5' end leaving a significantly positive slope in the curve (Gauthier et al., 2005; Archer et al., 2006). Archer and co-workers have criticized this method, pointing that are difficult to interpret and not really effective. However, they only discuss the RNA degradation plots without any mention to the accompanying statistics (slope and significance level).

Implementation: RNA degradation plots as well as statisitics are implemented in the AffyRNAdeg, summaryAffyRNAdeg and plotAffyRNAdeg function of the Affy package (Bioconductor). Reimers and Weinstein web resource (http://www.discovernci.nih.gov/affytools) for analysis of regional biases also provides the user with a RNA degradation plot although it does not include the values for the slope and p level.

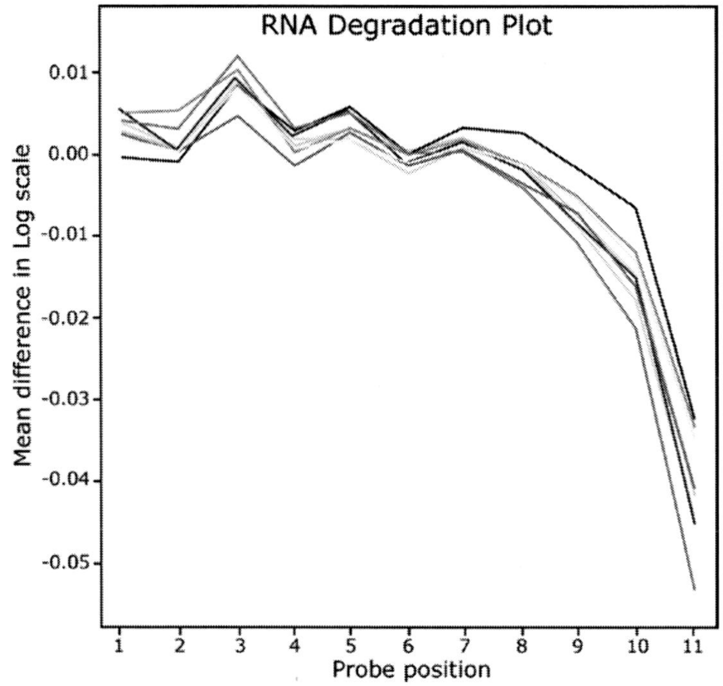

Figure 5. RNA degradation plot. Each colour line corresponds to an individual sample.

4.7. Analysis of Residuals

Bolstad and colleagues proposed that quality could be assessed analysing the residuals from the models fitted to estimate gene expression (Bolstad et al., 2004). According to these authors, the value of the residuals will increase when the data quality decreases and this relationship can be used to evaluate at all levels of analysis, from individual probes to complete chips. Following this logic, the authors developed a complete set of tools to analyze the residuals from fitting the RMA model in order to evaluate the quality of complete chips. This approach is particularly useful to explore the quality of individual arrays within an experiment with multiples arrays.

A first analysis is called NUSE for normalized Unscaled Standard Errors and associates an standard error to each probeset summary and normalized them (so that the median standard error for the probesets is 1). NUSEs are very sensitive to differences in the variability for the genes so that an array where there are elevated SE relative to the other arrays in the experiment is typically of low quality. Bolstad et al. (2004) compared the NUSE median to other QC parameters observing a high correlation between them but a greater sensitivity and resolution capacity in the NUSE.

Residuals can also be used to analyse spatial artefacts in the array. They are visualized in a way similar to cell fluorescence intensity. We have already commented the use and advantages of this chip visualization approach.

The authors also propose a quality assessment based on relative log expression (RLE) values. These values are computed by comparing for each probeset the expression value on each array against the median value of that probeset across all arrays in the experiment. The RLE is a measure of the expression measures variability independent from their expression level. Summarizing the information from the different chips of an experiment and the different probesets in the chips, it is possible to characterize each chip by a RLE vector that can be compared among all arrays in a experiment using boxplots. Although biological variations may affect these graphs, RLE values are very sensitive to technical sources of variability that are large compared to biological variation (Bolstad et al., 2004).

Implementation: all analyses of residuals, including their spatial analysis, are implemented as different functions of the affyPLM package of Bioconductor. NUSE function allows comparing the NUSE values of the different chip using a boxplot (Figure 6A) and also to displays their individual statistics. The RLE function does the same with the RLE values (Figure 6B).

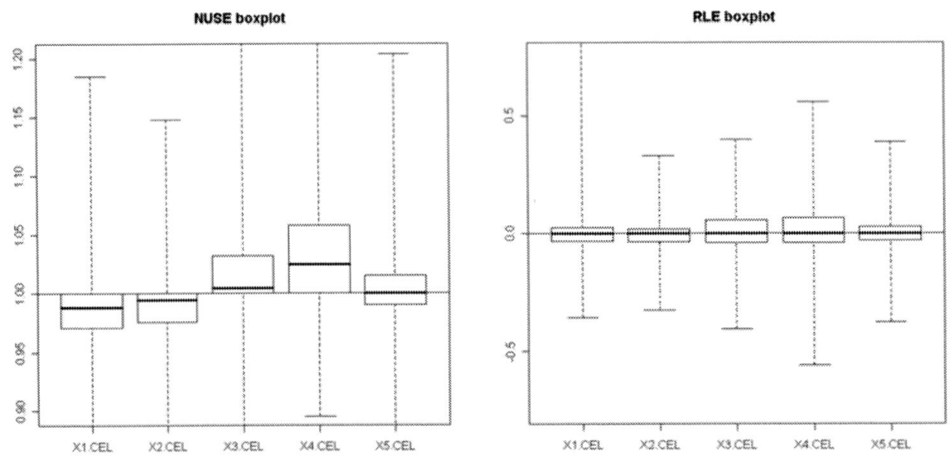

Figure 6. Graphical representation of different analyses of residuals. A (left). Boxplot of the NUSE values of different arrays. B (right). Boxplot of their RLE values.

4.8. Data Exploration

One of first steps in diagnosing quality problems can be plotting probe level data (Gauthier et al., 2005). These plots, including boxplots, histograms and MvAplots, are routinely used to monitor GeneChip data quality (Gentleman et al., 2005; Heber & Sick, 2006). Boxplots and histograms summarize the distribution of intensity values in the arrays, allowing their comparison easily and the identification of possible outliers (due to position or variability). MvA plots are comparisons of the intensity values from two arrays or from one array and a median chip obtained from all arrays in an experiment.

Implementation: affy package of Bioconductor incluyes functions plot, hist, and Maplot. This function displays the graphs as well as several statistics for each array (see Figure 7).

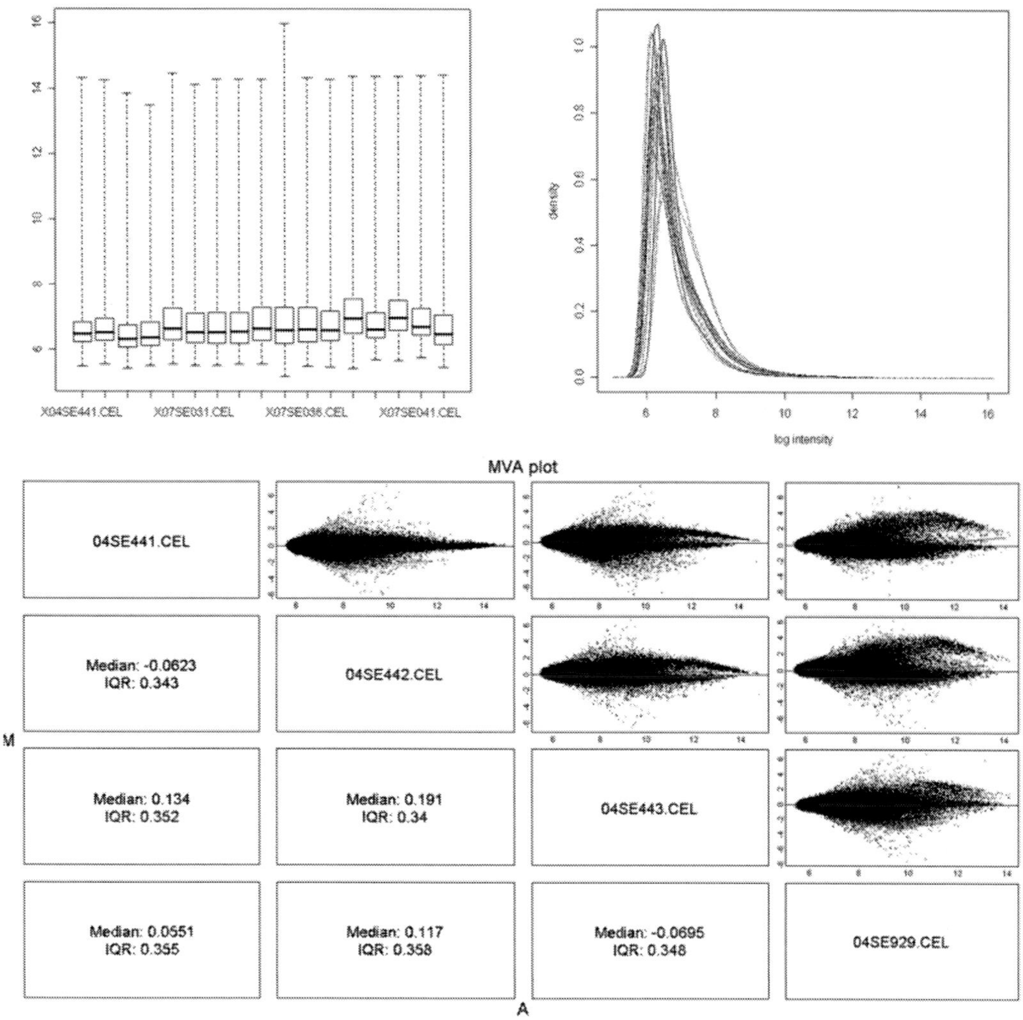

Figure 7. Exploratory graphs used to compare bewteen microarrays in a experiment. Upper left: boxplot of the intensity values from the probes of each microarray. Upper right: hystogram of the same values. Down: pairwise MvA plots, comparing the gene expression between pairs of arrays.

4.9. Other Methods

Many other authors have developed their own methods or software. Probably the most outstanding contributions comes from Li and Wong (2001) and their dChip program. They present methods for identifying and handling cross-hybridizing probes and contaminating array regions along with their model-based gene expression algorithm. These QC methods are implemented in their dChip program and fully described in Li and Wong (2001) publication. Other authors, like Lee et al. (2006), have developed methods and software for data exploration and graphical display with different microarray platforms (not specifically for Affymetrix arrays). These methods can be considered in most respects as variations of the ones already commented.

4.10. A Comment on the Software

We have shown several programs to analyse the quality of the arrrays but this is just a minor fraction of all available. Bioinformatics has found in microarray analysis a field that will spread and consequently there is plenty of it and it is highly interesting. Luckily enough the more interesting developments are freely accesible on the Internet and in fact most of this software is implemented in Bioconductor, a framework designed to develop free software for the analysis and comprehension of genomic data. Most programs within this framework are developed in a mathematical language, R, also freely available from http://www.r-project.org/. Most of these software developments are available for Windows, Linux or MacOS operating systems. Being free, the bioconductor and R software may be less friendly than commercal alternatives but beyond a hard first sight it is really easy to use this software and there is a lot of useful documentation.

5. Conclusion

Throughout this text, we have shown the reader many approaches and applications to analyse the quality of Affymetrix microarray data. Each method provides its own information that can be better than another in some respects and worst in others. Although some authors have compared some of the methods, the conclussions do not, at least in our opinion, clearly point to a particular method as the best one. Quality assessment should start with the analysis of the RNA quality before hybridization and proceed to image inspection and all the other methods detailed. None will provide the whole picture but the combination of all will provide researchers with very useful information for their later gene expression analyses. Although it may look like a big amount of work, the implementation of most analyses in bioconductor's packages make the analyses very easy and fast to carry out.

Let us end with a comment concerning the use of QC values. There is a natural tendency to look for thresholds that separate good and bad samples or data. However, QC values strongly depend on the sample, and quality values that can be acceptable for some samples (for example, those difficult to obtain) become unacceptable for other samples (for example, cell line RNAs). So, in most cases it is not possible to define absolute thresholds and the objective becomes to guarantee that all samples have similar enough QC parameters to be comparable (Wilson and Miller, 2005) because differences between arrays may cause false gene expresion changes if the normalization algorithms are unable to cope with that variability.

Acknowledgments

This work was supported by the Consejeria de Salud of the Junta de Comunidades de Castilla La Mancha (project 06025). WPT is a fellow of Project PAI07-0053-8620 from the Consejeria de Educación y Ciencia of the Junta de Comunidades de Castilla La Mancha.

References

Affymetrix. GeneChip® Expression Analysis: Data Analysis Fundamentals. 2004. Available from: http://www.affymetrix.com/support/downloads/manuals/

Auer, H., Lyianarachchi, S., Newsom, D., Klisovic, M.J., Marcucci, U. & Kornacker, K. (2003). Chipping away at the chip bias: RNA degradation in microarry analysis. *Nature Genetics,* **35**(4), 292-293.

Bakay, M., Chen, Y.M., Borup, R., Zhao, P:, Nagaraju, K. & Hoffman, E.P. (2002). Sources of variability and effect of experimental approach on expression profiling data interpretation. *BMC Bioinformatics,* **3**(1), 4

Bolstad, B.M., Collin, F., simpson, K.M., Irizarry, R.A. & Speed, T.P. (2004). Experimental design and low-level analysis of microarray data. *Int. Review Neurobiol.,* **60**, 25-58.

Chen, J., Hsueh, H.M., Deongchamp, R., Lin, C.J. & Tsai, C.A. (2007). Reproducibility of microarray data: a further analysis of microarray quality control (MAQC) data. *BMC Bioinformatics,* **8**, 412.

Copois, V., Bibeau, F., Bascoul-Mollevi, C., Salvetat, N., Chalbos, P., Bareil, C., Candeil, L., Fraslon, C., Conseiller, E., Granci, V., Maziere, P., Kramar, A., Ychou, M., Pau, B., Martineau, P., Molina, F. & Del Rio, M. (2007). Impact of RNA degradation on gene expression profiles: Assessment of different methods to reliably determine RNA quality. *J. Biotechnol.,* **127**, 549-559.

Dumur, C.I., Nasim. S., Best, A.M., Archer, K.J. Ladd, A.C., Mas, V.R., Wilkinson, D.S., Garret, C.T. & Ferreira-gonzalez, A. (2004). Evaluation of Quality-Control criteria for microarray gene expression analysis. *Clinical Chemistry* **50**, 11.

Enard W, Khaitovich P, Klose J, Zöllner S, Heissig F, Giavalisco P, Nieselt-Struwe K, Muchmore E, Varki A, Ravid R, Doxiadis GM, Bontrop RE, & Pääbo S. (2002). Intra- and interspecific variation in primate gene expression patterns. *Science*, **296**(5566), 340-343.

Garret-Mayer, E., Parmigiani, G., Zhong, X., Cope, L. & Gabrielson, E. (2007). Cross-study validations and combined analysis of gen expression microarray data. *Biostatistics*. Epub.

Gautier, L., Irizarry, R., Cope, L & Bolstad, B. Description of affy. 2005. Available from: http://www.bioconductor.org/

Gentleman, R., Huber, W., Carey, V.J., *et al.* (2005). *Bioinformatcis and computational biology solutions using R and Bioconductor.* New York, NY.: Springer.

Han, E.S., Wu, Y., McCarten, R., Nelson, J.F., Richardson, A. & Hilsenbeck, S.G. (2004). Reproducibility, sources of variation, pooling, and sample size: important considerations for the design of high-density oligonucleotide array experiments. *J. Gerontol. A. Biol. Sci. Med. Sci.,* **59**(4), 306-315.

Heber, S. & Sick, B. (2006). Quality assessment of Affymetrix GeneChip data. *OMICS,* **10**, 358-368.

Irizarry, R.A., Warren, D., Spencer, F., Kim, I.F. *et al.* (2005). Multiple-laboratory comparison of microarray platforms. *Nature Methods* **2**(5), 1-6

Jones, L:, Goldstein, D.R., Hughes, G., Strand, D.A., Collin, F., Dunnet, S.B., Kooperberg, C., Aragald, A., Olson, J.M., Augood, S.J., Faull, R.L.M., Luthi-Carter, R., Moskvina, V. & Hodges, A.K. (2006). Assessment of the relationship between pre-chip and post-chip quality measures for Affymetrix GeneChip expression data. *BMC Bioinformatics,* **7**, 211.

Klebanov, L. & Yakovlev, A. (2007). How high is the evel of technical noise in microarray data? *Biology Direct,* **2**, 9.

Lee, E.-K., Yi, S.-G. & Park, T. (2006). arrayQCplot: software for checking the quality of microarray data. *Bioinformatics,* **22**(18), 2305-2307.

Li, C. & Wong, W.H. (2001). Model-based analysis of oligonucleotide arrays: Expression index computation and outlier detection. *Proc. Natl. Acad. Sci. USA,* **98**, 31–36.

Miller, C.L., Diglisic, S., Leister, F., Webster, M. & Yolken, R.H. (2004). Evaluating RNA status for RT-PCR in extracts of postmorten human brain tissue. *Biotechniques,* **36**, 628-633.

Mutter, G.L., Zahrieh, D., Liu, C., Neuberg, D., Finkelstein, D., Baker, H.E. & Warrington, J.A. (2004). Comparison of frozen and RNALater solid tissue storage methods for use in RNA expression microarrays. *BMC Genomics,* **5**, 88.

Novak, J.P., Sladek, R. & Hudson, T.J. (2002). Characterization of Variability in Large-Scale Gene Expression Data: Implications for Study Design. *Genomics,* **79**, 104-113.

Oleksiak, MF., Churchill, G.A. & Crawford, D.L. (2002), Varaition in gene expression within and among natural populations. *Nat. Genet.,* **32**, 261-266.

Pritchard, C.C., Hsu, L., Delrow, J. & Nelson, P.S. (2001). Project normal: defining normal variance in mouse gene expression. *Proc Natl Acad Sci U S A.,* **98**(23), 13266-13271.

Reimers, M. & Weinstein, J.N. (2005). Quality assessment of microarrays: visualization of spatial artefacts and quantitation of regional biases. *BMC Bioinformatics,* **6**, 166.

Sambrook, J. & Russell, D.W. (2001). *Molecular Cloning: a Laboratory Manual.* (Volume 2). Cold Spring Harbor Laboratory Press.

Schroeder, A., Mueller, O., Stocker, S., Salowsky, R., Leiber, M., Gassmann, M., Lightfoot, S, Menzel, W., Granzow, M. & Ragg, T. (2006). The RIN: an RNA integrity nmber for assigning integrity values to RNA measurements. *BMC Mol. Biol.,* **7**, 3.

Schuchhardt, J., Beule, D., Malik, A., Wolski, E., Eickhoff, H., Lehrach, H. & Herzel, H. (2000). Normalization strategies for cDNA microarrays. *Nucleic Acid Research*, **28**, E47.

Wang, H., He, X., Band, M., Wilson, C. & Liu, L. (2005). A study of inter-lab and inter-platform agreement of DNA microarray data. *BMC Genomics*, **6**, 71.

Weis, S., Llenos, I.C., Dulay, J.R., Elashoff, M., Martínez-Murillo, F. & Miller, C.L. (2007). Quality control for microarray analysis of human brain samples: The impact of postmortem factors, RNA characteristic, and histopathology. *J. Neurosci. Methods,* **165**, 198-209.

Wilson, C.L. & Miller,C.J. (2005). Simpleaffy: a Bioconductor package for Affymetrix Quality Control and data analysis. *Bioinformatics,* **21**(16), 3683-3685.

Zakharkin, S.O., Kim, K., Mehta, T., Chen, L., Barnes, S., Scheiter, K.E., Parrish, R.S., Allison, D.B. & Page, G.P. (2005). Sources of variation in Affymetrix microarray experiments. *BMC Bioinformatics,* **6**, 214.

Zhang, L., Yoder, S.J. & Enkemann, S.A. (2006). Identical probes on different high-density oligonucleotide microarrays can produce different measurements of gene expression. *BMC Genomics*, **7**, 153

In: Oligonucleotide Array Sequence Analysis
Editors: M.K. Moretti and L.J. Rizzo, pp. 443-456
ISBN: 978-1-60456-542-3

Chapter 17

Applications of Qiagen QIAplex Technology in Clinical Multiplex Assays

John Brunstein [*]

Centre for Translational and Applied Genomics, PHSA Labs, Vancouver, BC;
Children's and Women's Health Centre of BC, PHSA Labs, Vancouver, BC;
and
Department of Pathology and Laboratory Medicine, University of British Columbia, Vancouver, BC

Abstract

Highly multiplex PCR or RT-PCR protocols have remained an elusive goal of molecular diagnostics almost since the introduction of nucleic acid amplification technologies. Implementation of these assays has generally faced technical challenges on three fronts: loss of sensitivity due to competition for limited reagents by multiple targets in a single reaction; loss of specificity due to an exponentially increasing number of possible primer interaction artifacts; and difficulty in distinguishing among multiple different end products. A recently developed methodology by Genaco (now a Qiagen company) and marketed under the 'QIAplex' name is a combination of technologies which look to overcome these difficulties. Assay parameters which limit issues of sensitivity and specificity shortcomings are coupled to liquid-phase hybridization oligonucleotide bead array detection on the Luminex platform. The approach lends itself well to mid-to-high multiplicity (10 - 50 targets or more) PCR and RT-PCR assays. Along with new capabilities, this form of assay brings its own specific challenges, particularly in selecting and implementing appropriate control systems. Early studies on the clinical application of this method have been highly successful. This review describes the underlying principles and technologies of the method, aspects of its clinical implementation in one context, ways to address the issue of appropriate controls, and future potential of the method. Two similar commercialized systems sharing the bead array hybridization approach are also briefly described.

[*] E-mail address: jbrunstein@cw.bc.ca. Phone: 1 604 875-2000 ext. 5283. Fax: 1 604 875-3777.

INTRODUCTION

The introduction of single-target ('Simplex') PCR and RT-PCR was shortly followed by attempts to simultaneously amplify and distinguish multiple distinct targets within a single reaction ('Multiplex' PCR or RT-PCR)[1, 2, 3]. In theory, multiplex reactions provide a method of reagents cost, labour, and time savings by allowing a single reaction to ascertain the presence or absence of multiple targets. An additional appealing feature in many cases is reduced requirement for starting assay specimen in contrast with an equivalent panel of simplex assays for the same targets. Altogether, these advantages make multiplex nucleic acid amplification based assays a logical goal in many clinical as well as some research settings. In practice however implementation of multiplex methods remains much less common than simplex methods, primarily due to technical hurdles to the effective real-life implementation of multiplex assays. These hurdles can generally be broken down into three categories: sensitivity, specificity, and discrimination of multiple possible assay products from one another.

Sensitivity

Simplex PCR or RT-PCR is characterized by extreme sensitivity levels; routine highly robust real-time assays frequently have detection limits of under 100 template copies per reaction, while more traditional endpoint analyzed methods commonly have detection limits on the order of 10 template copies and in well optimized examples are in the end really only limited by Poisson statistics on template distribution at high dilutions; certainly a single template copy can yield a positive assay result in some cases. In conventional multiplex assays, the inclusion of primer sets for each of the individual targets can lead to loss of sensitivity through two mechanisms. The first occurs through intra-assay competition for resources (for example polymerase, nucleotides) between each of the individual reactions [4]. Commonly referred to as 'squelching', this problem becomes most important when more than one assay target is present in the starting template. In this case, each individually positive reaction may reach a plateau phase in amplification much sooner and at lower product amplicon levels than it would on its own. Even in cases where only a single one of the target sequences is present in the starting template, this may remain a problem as normally infrequent side reactions such as primer dimers can occur much more readily with the large number of primer sets present than in a simplex reaction [5, 6]. The second source of lower sensitivity arises from constraints on assay conditions imposed by the need to compromise between what would be truly optimal conditions for each individual assay. Even with a great deal of effort on primer designs to ensure similar annealing temperatures, amplicon lengths, and target GC content, small differences in amplification kinetics between individual reactions are rapidly accentuated by the exponential growth rates inherent to the method and a marginally better reaction will rapidly out-compete its competitor and through simple mass action cause reagent starvation or squelching of the weaker reaction(s).

Specificity

In a simplex PCR reaction, it is generally not overly difficult to analyze proposed primers (and probes, if used) for spurious cross-interactions and alter primer / probe design as needed to minimize or eliminate interactions likely to lead to artifactual products such as primer dimers. As the number of primers / probes in a multiplex assay increase however, this challenge grows exponentially; a five target assay without probes has 100 possible primer interaction pairs to consider; the same five target assay but with probe or beacon real-time detection methods has 225 potential oligonucleotide interactions. For a ten target assay, the respective values would be 400 and 900; and for a 30 target assay the values are 3600 and 8100. Clearly, this puts enormous constraints on assay design; when combined with other constraints on primer and probe selection (Tm, secondary structure, amplicon length) the true scale of the problem becomes apparent. The related issue of ensuring each of the primers and probes is specific for its target and will not cross-anneal to another sequence also increases with increasing multiplicity, thereby compounding the problem. These design considerations and final assay specificity issues despite the best of designs are an enormous barrier to the successful implementation of high multiplicity (RT)-PCR assays for routine use.

Product Discrimination

The third class of technical hurdles to multiplex nucleic acid amplification assays comes in discriminating the potential end amplification products from one another as well as from any spurious products. Generally, detection of (RT)-PCR products falls into two methodological categories: size-resolved endpoint product analysis techniques, and fluorescent based real time techniques. Endpoint product amplicon size analysis is the older of the two methods and in its simplest form is classically performed by agarose gel electrophoresis. Product amplicon sizes can be compared against a size standard lane, and a positive score for a target amplification requires both the presence of an amplicon and the correct size of amplicon, giving additional assurance of assay specificity. While gel electrophoresis is generally slow, not well suited to automation or large throughput, and carries inherent risks of amplicon contamination of lab space, it remains a simple and effective technique for simplex (RT)-PCR analysis in the research and in some cases the clinical laboratory. The method does not lend itself well to highly multiplex reactions however; the need to have clearly distinguishable end products for each of the targets requires at minimum approximately 10% length difference between each product, and realistically, much larger differences (closer to 50%) are desirable for clear identification of each potential product band. As a limited range of amplicon lengths (perhaps 100 - 1200 bp) are suitable for diagnostic applications the maximal number of targets possible in this multiplex setting is on the order of 8 – 12. Differential fluorescent labeling of primer sets yielding similar product sizes can be used to increase this number somewhat, but adds expense and complexity. Two problems are inherent in amplicon size resolution as well; the first being kinetic differences in amplification of shorter versus longer products leading to increased inter-assay competition or squelching as described above, and the second being that the large number of expected possible 'real' product sizes greatly limits the method's ability to detect and reject spurious products based on irregular sizes. While a possible technical solution to some of these would be to use a higher-resolution product sizing method such as employed in DNA sequencing, the cost and complexity of doing so has generally made this an unviable solution for routine use except in some forensics applications such as STR sample typing (e.g. the Promega PowerPlex 16 and similar systems).

Fluorescence based real-time techniques offer a number of advantages over endpoint analysis. In particular, these methods are readily amenable to automation and mid to high throughput analysis while in most protocols greatly reducing or eliminating the risk of laboratory template contamination. An additional benefit of simplex real-time methods is the potential to extract quantitative data on starting template values. (While methods exist for extracting quantitative measures from endpoint gel-resolved (RT-)PCR approaches as well, both in simplex and multiplex formats [2], these approaches are generally accepted as being much less accurate than real-time approaches [7]). In a multiplex setting this ability to quantitate can be compromised to an unknown degree by inter-assay competition (squelching), but in most cases the quantitation values obtained remain at least generally indicative of relative template loads and may be of interpretive value. While a number of variations on the detection method exist such as DNA binding dyes [8], 5'-fluorgenic nuclease ("TaqMan") assays [9], molecular beacons [10], or scorpions [11], all rely on detection of an alteration in fluorescence properties of the reaction in response to product formation. While a large number of potential fluorophores are available for use, their application in a multiplex setting is limited firstly by the requirement they have distinct spectral properties to avoid possible unwanted fluorescence resonant energy transfer (FRET) between reporters and secondly by the need to be able to differentiate between the reporters for each individual target. It is this second restriction which is the most limiting, as most commercially available instruments on the market today cannot effectively distinguish between more than five fluorophores (six for the Corbett RotorGene) and are thus limited to at most a 5 (6)-plex assay. In many cases one of the fluorescence channels is dedicated to an internal normalizer signal which reduces the instrument capacity by one of these channels.

A review of the literature supports the effective limits to both endpoint size-analyzed and real-time assay multiplicity. For the first of these, while there exist reports of rare examples with as many as 13 targets [12] very few published reports exist of assays with more than 9 targets and the majority described for microbial detection in a recent comprehensive review are for 5 targets or fewer [3]; for real-time assays, the author was unable to find any published references to more than 5-plex assays (although Corbett's website indicates a report of a 6-plex assay developed in the Vermont Forensic lab is nearing publication [13]).

In the clinical diagnostic setting, the appeal of a multiplex test is especially strong in the setting of a single clinical presentation with multiple possible etiologies requiring different treatments. A particularly cogent example is acute respiratory infection. This most frequent class of infectious disease has an enormous impact ranging from lost work time for the common cold to widespread morbidity, mortality, and massive social burden for a possible pandemic high-pathogenicity Influenza A. More than twenty pathogens commonly infect people and cause similar early-stage presentations of acute respiratory symptoms, making it challenging for the physician to make appropriate decisions regarding cohorting, quarantining, and treatment in a timely fashion. Given the possible high transmissibility of respiratory pathogens and the wide spectrum of potential impact on the patient as well as on public health, the rapid differentiation between e.g. RSV infection and SARS Coronavirus is a crucial one. Respiratory infections thus provide a good example of where a molecular assay's sensitivity and speed in a multiplex context capable of detecting and differentiating between 15 - 20 possible pathogens would have enormous potential impact. Until recently however, the limitations to (RT)-PCR multiplexing as discussed theoretically above have made this impractical. The closest approach to this in general use has been Prodesse's 'Hexaplex' assay [14], which has been highly successful even with its limited six-target pathogen spectrum. Recent advances in (RT)-PCR technologies have however begun to address each of the limitations described in multiplex nucleic acid amplification methods. The remainder of this review will focus on describing one method of addressing each of the limitations and the successful clinical diagnostic application of these combined technologies in high multiplicity assays as applied to respiratory infection diagnostics, the Qiagen ('QIAplex') ResPlex I and ResPlex II.

Improving Sensitivity and Specificity

The QIAplex method used in these assays attempts to minimize the described problems in multiplex sensitivity and specificity through a single technical approach [15]. As with any conventional nucleic acid amplification approach, individual target-specific primers are designed using conventional criteria (GC content, amplicon length, secondary structure, Tm). Rather than a single forward and reverse primer for each target however, two of each are designed such that there are four potential slightly different products from each template (Forward Primer (FP)1 to Reverse Primer (RP)1; FP2 to RP1; FP1 to RP2; and FP2 to RP2). While this may at first seem counter-intuitive by making the reaction higher multiplicity, the rationale is that by choosing primer Tm values all as close to a single compromised optimum as possible one now has four chances to amplify a target at a fixed set of conditions and at least one of these is likely to perform acceptably.

A second aspect of the QIAplex method is that unlike conventional (RT)-PCR, each of the four target specific primers is present in the reaction at extremely low concentrations. This acts to simultaneously help limit sensitivity and specificity problems through simple kinetic considerations- at vanishingly small concentrations, primer-primer interactions are greatly minimized and primer - cross reacting target interactions are more readily outcompeted by proper primer-target interactions. The low primer concentration creates two novel problems however: primer depletion occurs rapidly when a complimentary target is present, and amplicon concentrations at that point are too low for easy detection by standard endpoint or real-time methods.

The solution employed for these two issues is to synthesize all target-specific forward primers in the multiplex assay with a common 5' 'tag' sequence, and similarly to tag all target specific reverse primers with a different 5' sequence. Ideally, these tag sequences should be designed to have similar Tms to each other, but somewhat below the Tm of the target-specific sequences. A single set of PCR 'SuperPrimers', complimentary to these flanking tags and at classical PCR concentrations, are then included in the reaction as well and a two-phase amplification process is used. In the first phase (consisting of hot start, enrichment, and tagging steps), a small number of thermocycles at a higher Tm, only the target-specific sequences can anneal and very limited amplification of target sequence(s) occurs. Because of the incorporation of flanking tags on forward and reverse target specific primers, a second phase of amplification (bulk amplification and final elongation steps) can then be done using a lower Tm at which the 'SuperPrimer' pair will proceed to amplify all of the products of the first amplification phase [15].

This approach thus addresses specificity through kinetic means (low target-specific primer concentrations) and by having four target specific primers for each target has an increased likelihood of successful first-phase amplification of all targets present at a single 'best compromise' thermocycling profile. In the second phase of amplification, as only the 'SuperPrimers' are active, the process is in effect no longer multiplex but a simplex PCR which can be run at its optimal conditions. Competition for reagents will still occur between different first-phase products however, with differences in length, sequence, and structure contributing to differential kinetics of second-phase amplification of each product. The fact that a single primer set amplifies all such targets however is likely to result in less kinetic difference between each target amplification rate than in the case of a traditional multiplex PCR, where these effects are compounded by primer-specific kinetics. In the case of a specimen positive for multiple targets, the final levels of each amplicon at reagent depletion will none the less be expected to be lower than in each of the equivalent simplex reactions individually. Sensitivity thus remains something of an issue with this approach, although to a lesser degree than in classical multiplex amplification strategies.

Improving Product Discrimination

This remaining sensitivity issue is addressed at the same time as the issue of product discrimination through the application of liquid-phase bead-coupled oligonucleotide array technology on the Luminex platform [16]. In this approach, up to 100 types of spectrally distinguishable polystyrene microspheres serve as carriers for affinity analytes [17]. Product to be analyzed is tagged with a fluorophore (generally phycoerythrin (PE)) and mixed with the conjugated bead mixture; analytes with affinity for their respective bead-attached partner bind, and the mixture is sampled via a flow-cytometer style approach wherein beads are sent single-file past a dual laser interrogation cell. The first laser allows a readout of the bead's dye signature, identifying it as one of the 100 possible bead classes and thus indirectly what the bead partner identity is. The second laser simultaneously examines the bead for surface-associated reporter (PE) presence, thereby detecting the presence or absence of the bead's target from the analysis mixture. In the case of the ResPlex I & II assays, target labeling is done through inclusion of a biotin tag on one of the assay SuperPrimers. Products of the second amplification phase are thus biotin tagged. These products are incubated with a set of reporter beads where one colour bead class is assigned to each potential amplification target. Each of these bead colour classes is in turn conjugated to an oligonucleotide complimentary to a sequence internal to each of the four possible first-phase amplification products for each target pathogen. These hybridization amplicon-capture sequences are chosen to all have similar Tm values such that all assay targets hybridize with similar efficiency. Once hybridized, the beads representing positive targets within a sample are thus attached to amplicon and finally to a PE reporter in a manner akin to a 'sandwich ELISA', while beads representing negative targets in the sample remain unconjugated. Luminex analysis of the bead set for each sample is then done, and PE signal associated with each bead colour class is measured and compared against a background fluorescence signal to determine positivity or negativity for each assay target.

Assay Performance Method and Related Considerations

I: Nucleic acid extraction. While any sample type amenable to nucleic acid extraction could serve as input for this method, we have in general only worked with 'cell free fluids' as a starting material. These have included serum and CSF but in the vast majority of cases have been nasopharyngeal wash (NPW) specimens as these are the primary respiratory diagnostic specimen obtained by our laboratory. In our lab, total nucleic acid isolation was generally performed with the Qiagen MagAttract M48 Virus kit on a BioRobot M48, although we have successfully processed smaller numbers of specimens by Qiagen manual spin column methods as well.

II: (RT)-PCR In their current commercially available format, the ResPlex assays consist of two separate multiplex reactions: ResPlex I is directed against DNA targets (bacterial pathogens plus particular Adenovirus serotypes and internal control; a total of 11 targets per reaction) while ResPlex II is directed against RNA targets (viral pathogens plus an internal control; a total of 13 targets per reaction). (Note that combinations of these two target types into a single higher-multiplicity panel is possible and will be discussed later in this review.) While single samples can be analyzed with these assays, the process lends itself best to mid-scale batch throughput based on the 96-well plate format accepted by the Luminex instrumentation. 10 ul aliquots of each sample nucleic extract are used as templates for PCR (ResPlex I) or RT-PCR (ResPlex II) using the supplied primer mixtures for each and either Qiagen HotStarTaq (PCR) or OneStep (RT-PCR) reagents. ResPlex I and II reactions are thermocycled in parallel by the manufacturer's recommended conditions. The reaction profile involves 5 steps for PCR (15 min, 95C; 15 cycles (94C, 30 sec; 52C, 1 min; 72C, 1 min); 6 cycles (94C, 15 sec; 70C, 90 sec); 30 cycles (94C, 15 sec; 52C, 15 sec; 72C, 15 sec); 72C, 3 min) and 6 for RT-PCR (identical to PCR with the addition of an initial 35 minute, 50C reverse transcription step). As the PCR reagents employed are a hot-start variety, we routinely run both reactions together on the RT-PCR program for simplicity; the PCR reactions will just idle during the RT step. Note that all required reagents are added prior to commencement of thermocycling no further addition of reagents or opening of reactions is required prior to analysis. This is of particular use in avoiding assay contamination which would be at an increased risk if e.g. the SuperPrimers had to be added partway through or if additional enzymes had to be added prior to assay analysis. *III: Hybridization.* Following thermocycling, 5 μl of each amplification product (ResPlex I and ResPlex II) is mixed with hybridization buffer and its respective oligonucleotide tagged bead array. Hybridization is carried out with gentle shaking at 52C for 10 minutes. As opposed to classical surface printed oligonucleotide arrays, the small size of the beads making up the suspension array has faster hybridization kinetics due to effective 3D diffusion. No wash steps are needed; asymmetry in the levels of the forward and reverse SuperPrimers leads to a preponderance of single stranded product of complimentary polarity to the oligonucleotide bead array. A further 5 minute incubation with streptavidin-phycoerythrin is then done and a stop buffer is added. Again, no wash steps are required as the instrumental analysis will only count PE fluorescence associated with a bead surface. *IV: Detection and scoring.* The hybridized oligonucleotide bead array is sampled by the Luminex instrumentation. A minimum number of beads to sample for each category can be set (generally 50 or 100) and for each bead category, corresponding to a single oligonucleotide capture probe, bead-associated PE mean fluorescence intensity (MFI) is measured. It should be noted that the observed MFI values can vary between versions of the Luminex instrument, as ones from some rebranding suppliers have different photomultiplier detector gain settings than others. In particular, a comparison of these assays as detected on BioPlex and LiquiChip instruments (from BioRad and Qiagen, respectively) conducted in the author's laboratory has shown MFI values to be 4.7-fold higher on the BioPlex than on the LiquiChip, which has a 'classical' instrument PMT gain setting. This higher MFI value was uniform across the entire dynamic measurement range and did not vary significantly between any particular bead class, allowing for simple comparison of data as obtained between the two versions of the instrument.

General Diagnostic Performance of the Assay

Early versions of the ResPlex I and II panels have been analyzed for analytical sensitivity by Amy Dean's lab [18] and others, notably at US CDC [19]. In general the assays show levels of analytical sensitivity expected from a nucleic acid amplification approach, with detection limits varying between targets but ranging from approximately 10 to 1000 target copies.

Following a preliminary analysis of clinical sensitivity and specificity of a previous version of the ResPlex panels in our own lab [20] we proceeded to a large-scale clinical validation study on the current versions. We examined a total of 1,742 clinical nasopharyngeal wash (NPW) specimens as submitted to our clinical diagnostic laboratory for diagnosis of acute respiratory symptoms. Results from the ResPlex I and II analysis of each specimen were compared with results of our standing diagnostic methods (Direct Fluorescence Assay (DFA) backed by viral cell culture for DFA-negative specimens)[21]. This approach allowed direct comparison of detection efficiencies for those pathogens detected by both methods: RSV (A&B), Influenza A, Influenza B, Parainfluenza virus (PIV) 1, PIV 2, PIV 3, and Adenovirus. Results of this study have been the subject of several conference presentations [22, 23, 24] and are described in more detail in an upcoming paper (manuscript under review at present). In general, our findings were that agreement between the methods was generally good, ranging from 100% (PIV 2) to ~85% (RSV). Adenovirus was a notable exception and showed very poor correlation between the two assays; this is likely attributable to a large extent to different strain specificities of the two methods (our DFA should detect 41 serotypes, while the ResPlex I targets only include serotypes 3, 4, 7, and 21). Not surprisingly, ResPlex reported slightly more positives for almost all shared targets than the comparison method; given the sensitivity of molecular methods as compared to DFA, this was expected. Rates of additional detection by ResPlex as compared to DFA ranged from an additional 18% (PIV 3; 1.5% vs 1.3% of all specimens, respectively) to 74% (Influenza B; 5.9% vs. 3.4% of all specimens, respectively). A smaller similar study using the same system has recently been published by Li and coworkers [25] and shows results similar to ours.

In general, signal to noise ratios as observed by this assay are on the range of 50 - 100 fold or higher; if background MFI values for a given target/instrument combination are for instance on the order of 5-15 MFI, a positive will yield a signal in the 1000 - 3000 range in most cases. While this makes identification of positive targets generally very easy, statistically rigorous positivity cutoff criteria should still be applied in order to detect weaker specimens and fully utilize the full analytical sensitivity of the test. While the author has found the manufacture's suggested cutoff of 250 MFI for all targets (as obtained on a classical Luminex platform; approximately 1250 on a BioPlex as calibrated to High RP1) works well in the majority of cases, there is some evidence in the author's laboratory to suggest that greater analytical sensitivity can be achieved by selecting target-specific cutoff criteria for each bead class such as the (average + N standard deviations) of a large set of blank analyses for each specific target. Target/ bead class specific comparisons of cutoff criteria as selected in this way demonstrate up to a 5-fold difference between one another, likely arising from differential hybridization efficiencies of each of the target-specific bead attached oligonucleotide sequences.

The author's lab has observed target MFI values against starting template copy number for a range of targets analyzed through this methodology to demonstrate a generally good log-linear relationship over a range of approximately 3 logs. While this may suggest the possibility of employing the method to extract quantitative data, the basis of the method on nucleic acid amplification combined with the well documented lack of correlation between starting template number and plateau-phase amplicon levels in PCR methods [26] would indicate that such an approach is unlikely to have real validity in making comparisons between different samples. While this could perhaps be improved upon by limiting the second-phase (SuperPrimer) amplification cycles such as to end within the linear PCR amplification phase, this would entail a loss of analytic sensitivity and ultimately limit utility of the assay. We observe that in the case of a sample positive for multiple targets, maximal positive signal intensities are very similar to those seen in singly positive specimens, suggesting that competition for reagents is not generally a limiting factor in detection of co-occurring targets.

Challenges, Future Directions, and Similar Methodologies

At least two other assay systems for respiratory pathogens, based on similar technologies to those described here, are in late-stage testing or early stages of clinical evaluation. One is from Tm Biosciences (now a division of Luminex) [27], and the other from EraGen [28, 29]. Both of these methods employ traditional multiplexing (as opposed to the QIAplex technology) followed by a similar hybridization / Luminex platform detection and classification approach as the ResPlex assays. While the theoretical considerations around traditional multiplexing described herein suggest these assays may be inherently susceptible to the shortcomings associated with that, data to date from both systems has been promising. This underlines the extreme power of the oligonucleotide bead-array approach to classify and detect targets so well as to at least partially overcome the sensitivity and specificity problems in its own right. Whether this will have the capacity to support higher levels of "traditional" multiplexing remains to be seen; to date, the QIAplex approach has been shown to be effective in assays with up to 26 separate targets in a single assay [15] while the other two methods have not presented data on anything over 19 targets (EraGen assay; [29]) or 20 targets (Luminex/Tm assay; [27]). Certainly, the prospect of having to re-design or add new amplification targets to one of these diagnostic panels in response to emerging or changing pathogen strains would seem less daunting for the QIAplex approach as opposed to classical multiplexing when issues of spurious primer interactions are considered.

A major issue for all assays of this general format is that of appropriate positive controls. With simplex assays, it is common practice to include a known positive specimen at least once in each assay batch. As assays become increasingly multiplex, this presents challenges. The inclusion of a single control positive for all targets in an assay is unlikely to work well, as the likelihood of squelching even with QIAplex technology is very high and a near-certainty with traditional multiplexing. At the opposite end of the spectrum, running individual positive controls for each of the assay targets with every sample batch is likely too costly and wasteful of resources to be feasible. The desirable goals of having any such controls also work to indicate efficient sample extraction, DNA and RNA integrity, effective reverse transcription, and lack of amplification inhibitors further complicate the issue. Of the three products described here, each has a different (and to the author's thinking, not completely adequate) approach. The ResPlex assays, both RP I and RP II, have an internal control in the form of making a human X-chromosome marker one of the assay targets. Successful amplification, classification, and detection of this target serves to indicate that intact DNA from patient cells has been recovered and is amplifiable; that PCR reagents (including the SuperPrimers shared by all targets) work; and that detection and classification steps are occurring appropriately. It does not however control for RNA integrity, failures in reverse transcription, or for the functionality of any single pathogen-specific reagents in the assay. Normally, a specimen is required to show an appropriate X-chromosome target signal to validate results on other targets; however the author has observed some instances where a positive pathogen signal is observed without this internal control being positive. Such cases likely occur in NPW samples with few host cells and large numbers of cell-free pathogen, and while interpretable as positive for the pathogen in question must remain indeterminate for all other assay targets. The Luminex/Tm assays include a spiked-in MS2 bacteriophage as a positive RNA control [30] and in its latest form also a lambda bacteriophage as a positive DNA control [27]: being exogenous, these can control for extraction, reverse transcription, PCR, presence of inhibitors, and downstream detection and classification steps. However, it falls short in not being able to identify samples where nucleic acid degradation has occurred prior to processing, and like the Qiagen assays it does not control for any target-specific reagents. The EraGen assay similarly employs two internally seeded controls (one RNA, one DNA) which offers the ability to independently control the reverse transcription and PCR phases of the assay but like the Luminex/Tm assay, fails to account for possible sample degradation prior to extraction. Each of these approaches has merit and the ideal approach will in the end likely be some combination of them. The author would propose both internal controls in the form of two host derived targets (one each DNA and RNA, to control for sample integrity, extraction, reverse transcription, amplification) as well as batch-included external controls for each target, run in parallel pools of ~5 targets each. It may be acceptable from a quality perspective to not run all positive controls for every target in every batch run, if there is an effective 'control rotation' schedule such that within a certain timeframe or number of tests that each test has been validated similar to the protocol suggested by Nolte et. al. [29]. In the QIAplex approach where only the first-round primers are target specific and the shared SuperPrimers are effectively validated on the internal control(s), such an external control rotation is particularly attractive from a quality assurance as well as economy of workflow perspective.

Aside from controls, there remain other areas where improvements to these assays are likely forthcoming. Additional targets, higher levels of multiplicity and in the case of ResPlex I and II, combination into a single clinical presentation-directed panel are all modifications the author would look for. The combined panel issue is somewhat of a contentious one; while a prototype single-reaction combination of ResPlex I and ResPlex II has been demonstrated to work well from a technical and implementation perspective in the author's hands, the clinical interpretation of simultaneous bacterial and viral detection is far from clear at present (discussed further, below).

A major difference between the ResPlex and current diagnostic methods as employed on this clinical specimen set was the frequency of multiple pathogen detection in specimens. As with most traditional methods, the DFA + culture system predominantly reports single pathogens as specimen workup stops with the first relevant pathogen detection. By contrast, ResPlex and similar approaches conduct all tests simultaneously; individual samples in our study were observed to contain as many as 5 simultaneous pathogens with an average of 1.53 pathogens per specimen. In the majority of cases these consisted of co-appearance of bacterial and viral targets; for the most part, multiple viruses were not observed. A detailed statistical analyses of these interactions (both positive and negative) is included in our upcoming manuscript and while some particular novel interactions are apparent, the general trends are in agreement with known respiratory infection pathology whereby primary viral infection can lead to secondary bacterial infection (or the reverse), while the majority of particularly RNA viruses tend to suppress or interfere with infection by other RNA viruses. What remains unclear at present is the true clinical significance of pathogen co-detections. The assays from Luminex and EraGen have chosen to sidestep this issue by only including viral targets in their current versions; it remains to be seen whether it is clinically wise to leave many likely pathogens within the appropriate presentation spectrum out of the assay for reasons of interpretive convenience. From a purely technical perspective, it seems to be counter to the entire rationale for developing a high multiplicity assay to intentionally leave out possible targets. For economic and workflow considerations, most historical diagnostic testing algorithms cease with the first relevant result and thus our understanding of the clinical course of respiratory infections is largely predicated on the assumption they were stand-alone pathogens. In reality, the situation may be much more complex with the range of clinical outcomes noted for a given pathogen being in part due to what, if any, co-infecting agents are present. One can certainly reasonably imagine cases where a clinical presentation results in a viral diagnosis and lack of antibiotic prescription when there is in fact a contributing bacterial component; the inappropriate treatment in this case may lead to increased severity or length of illness. This fascinating point will only be resolved by increased future application of multiplex methods such as ResPlex I and II with clinical correlation of the results.

CONCLUSIONS

Recent advances in oligonucleotide hybridization detection methodologies, primarily based around the Luminex platform, show great initial promise at overcoming some of the theoretical limitations on practical highly multiplex nucleic acid amplification assays. Particularly when combined with modifications such as the QIAplex method, implementations of these advances in currently available and in-development assays show good clinical sensitivity and specificity at target multiplicities well above those achievable by classical multiplex methods. These methods are already becoming well established within the research setting and are likely to become of greater importance as laboratories increase automation and application of molecular methods for diagnosis. Challenges to the widespread clinical implementation of these methods are primarily in the implementation of appropriate control strategies, in addressing regulatory issues (for example, it may be illegal in some settings to assay a patient sample for all targets within a multiplex panel if only specific targets were requested by the physician), and in generating a sufficient knowledge base to allow for evidence-based interpretation of the clinical significance of multiple pathogens within a single patient sample. None of these hurdles would appear extremely difficult, and we may well be approaching an era where an ordering physician will be able to request a single 'presentation-based' high multiplicity diagnostic panel capable of detecting or excluding a very large number of possible relevant targets within a six hour to one day turnaround time for any of a number of very common presentations. As costs for these approaches decrease with increased scale of implementation, the impact of these diagnostic methods in enabling appropriate and thus cost-effective medical management is increasingly likely to more than offset the cost of their deployment.

Acknowledgments

The author would like to thank Amy Lum for comments and critical reading of this chapter, and multiple scientists working on this topic (particularly Jian Han, Fredrick Nolte, and Leslie Krueger) for thought-provoking discussions on the technology and its applications.

References

[1] Chamberlain JS, Gibbs RA, Ranier JE, Nguyen PN, Caskey CT. Deletion screening of the Duchenne muscular dystrophy locus via multiplex DNA amplification. *Nucleic Acids Res* 1988; 16: 11141–11156.

[2] Edwards MC, Gibbs RA. Multiplex PCR: advantages, development, and applications. *PCR Methods Appl.* 1994 Feb;3(4):S65-75.

[3] Settanni L, Corsetti A. The use of multiplex PCR to detect and differentiate food- and beverage-associated microorganisms: a review. *J Microbiol Methods*. 2007 Apr;69(1): 1-22.

[4] Bej AK, Mahbubani MH, Miller R, DiCesare JL, Haff L, Atlas RM. Multiplex PCR amplification and immobilized capture probes for detection of bacterial pathogens and indicators in water. *Mol Cell Probes.* 1990 Oct;4(5):353-65.

[5] Vandenvelde C, Verstraete M, Van Beers D. Fast Multiplex polymerase chain reaction on boiled clinical samples for rapid viral diagnosis. *J Virol Methods.* 1990 Nov;30(2):215-27.

[6] Bourque SN, Valéro JR, Mercier J, Lavoie MC, Levesque RC. Multiplex polymerase chain reaction for detection and differentiation of the microbial insecticide Bacillus thuringiensis. *Appl Environ Microbiol.* 1993 Feb;59(2):523-7.

[7] Reue, K. mRNA Quantitation Techniques: Considerations for Experimental Design and Application. *J Nutr.* 1998 Nov;128(11):2038-44.

[8] Morrison TB, Weis JJ, Wittwer CT. Quantification of low-copy transcripts by continuous SYBR Green I monitoring during amplification. *Biotechniques* 1998; 24:954-8,960,962.

[9] Heid CA, Stevens J, Livak KJ, Williams PM. Real time quantitative PCR. *Genome Res.* 1996 Oct;6(10):986-94.

[10] Tyagi S, Kramer FR. Molecular beacons: probes that fluoresce upon hybridization. *Nat Biotechnol.* 1996 Mar;14(3):303-8.

[11] Thelwell N, Millington S, Solinas A, Booth J, Brown T. Mode of action and application of Scorpion primers to mutation detection. *Nucleic Acids Res.* 2000 Oct 1;28(19): 3752-61.

[12] Kilimann MW, Pizzuti A, Grompe M, Caskey CT. Point mutations and polymorphisms in the human dystrophin gene identified in genomic DNA sequences amplified by multiplex PCR. *Hum Genet.* 1992 May;89(3):253-8.

[13] http://www.corbettlifescience.com/control.cfm?page=6%2DColour%5FMultiplexing &langID=1 as of October 3, 2007

[14] Fan J, Henrickson KJ, Savatski LL. Rapid simultaneous diagnosis of infections with respiratory syncytial viruses A and B, influenza viruses A and B, and human parainfluenza virus types 1, 2, and 3 by multiplex quantitative reverse transcription-polymerase chain reaction-enzyme hybridization assay (Hexaplex). *Clin Infect Dis.* 1998 Jun;26(6):1397-402.

[15] Han J, Swan DC, Smith SJ, Lum SH, Sefers SE, Unger ER, Tang YW. Simultaneous amplification and identification of 25 human papillomavirus types with Templex technology. *J Clin Microbiol.* 2006 Nov;44(11):4157-62.

[16] Fulton RJ, McDade RL, Smith PL, Kienker LJ, Kettman JR Jr. Advanced multiplexed analysis with the FlowMetrix system. *Clin Chem.* 1997 Sep;43(9):1749-56.

[17] Dunbar SA. Applications of Luminex xMAP technology for rapid, high-throughput multiplexed nucleic acid detection. *Clin Chim Acta.* 2006 Jan;363(1-2):71-82.

[18] Amy B. Dean, Daryl M. Lamson, Jian Han, Kassi Cronin, Chiung-Mei Lu, Christy Teng-Renzelman, Shanjuan He, Lisa Varnavas, Amy Graham, Robert Vereen, Bill Spargo, Sara Griesemer, Seela M. Nattanmai, Meghan E. Fuschino, Norma Tavakoli, Kirsten St.George, Jill Taylor. Evaluation Of Genaco's Molecular Differential Diagnostic Technology For The Detection Of RNA Respiratory Viruses. Poster presentation at 21st annual Clinical Virology Symposium and Annual Meeting of the Pan American Society for Clinical Virology , Clearwater FL, July, 2005.

[19] Tondella, M.L.C, R. Benson, M.G. S. Carvalho, J.S. Sampson, D.F. Talkington, A. Whitney, E. Mothershed, L. McGee, G. Carlone, M. Moore, V. McClee, K. Cronin, J. Han., and B.S. Fields. Development and evaluation of a novel multiplex PCR technology for molecular differential diagnosis of bacterial respiratory diseases. Abstr. *Proceedings of the 106th General Meeting of American Society for Microbiology*, Orlando, FL, U.S.A. 2006.Abstr. C-269.

[20] Brunstein J, Thomas E. Direct screening of clinical specimens for multiple respiratory pathogens using the Genaco Respiratory Panels 1 and 2. *Diagn Mol Pathol.* 2006 Sep;15(3):169-73.

[21] Thomas E.et.al.VIRAP–A Quality Focused Rapid Respiratory Viral Testing Program. Presentation at the 22nd Annual Virology Symposium, Clearwater FL, 2005.

[22] Brunstein, J., Thomas E. Clinical Applications and Medical Economics of the Deployment of the QIAplex ResPlex I & II Molecular Diagnostics for Respiratory Pathogens. *Poster presentation M-6 at 23rd annual Clinical Virology Symposium and Annual Meeting of the Pan American Society for Clinical Virology* , Clearwater FL, April, 2007.

[23] Brunstein, J. McKinney S., Thomas, E. Patterns Of Respiratory Pathogen Co-Association As Observed By Multiplex Molecular Analysis Of 1742 Clinical Specimens. *Poster presentation at the AMMI – CACMID Annual conference, Halifax* NS, 2007

[24] Brunstein, J, Thomas, E. Rapid Pan-Respiratory Pathogen Screening of Nasopharyngeal Washes by Genaco Respiratory Infection Panels I & II. *Poster presentation at the AMMI/CACMID Annual conference*, Victoria BC, March 2006.

[25] Li H, McCormac MA, Estes RW, Sefers SE, Dare RK, Chappell JD, Erdman DD, Wright PF, Tang YW. Simultaneous detection and high-throughput identification of a panel of RNA viruses causing respiratory tract infections. *J Clin Microbiol.* 2007 Jul;45(7):2105-9.

[26] Morrison C, Gannon F. The impact of the PCR plateau phase on quantitative PCR. *Biochim Biophys Acta* 1994 1219: 493-8.

[27] Mahony J, Chong S, Merante F, Yaghoubian S, Sinha T, Lisle C, Janeczko R. Development of a respiratory virus panel test for detection of twenty human respiratory viruses by use of multiplex PCR and a fluid microbead-based assay. *J Clin Microbiol.* 2007 Sep;45(9):2965-70.

[28] Lee WM, Grindle K, Pappas T, Marshall DJ, Moser MJ, Beaty EL, Shult PA, Prudent JR, Gern JE. High-throughput, sensitive, and accurate multiplex PCR-microsphere flow cytometry system for large-scale comprehensive detection of respiratory viruses. *J Clin Microbiol.* 2007 Aug;45(8):2626-34.

[29] Nolte FS, Marshall DJ, Rasberry C, Schievelbein S, Banks GG, Storch GA, Arens MQ, Buller RS, Prudent JR. MultiCode-PLx system for multiplexed detection of seventeen respiratory viruses. *J Clin Microbiol.* 2007 Sep;45(9):2779-86.

[30] Do, MC., Forbes, BA.,Bender, C. Evaluation Of The Prodesse Proflu-1™ Real Time Assay And The Tm Bioscience Id-Tag Tm Respiratory Viral Panel For The Rapid Identification Of Respiratory Viruses. Poster presentation, Planet xMap conference, October 2007

In: Oligonucleotide Array Sequence Analysis
Editors: M.K. Moretti and L.J. Rizzo, pp. 457-467
ISBN: 978-1-60456-542-3

Chapter 18

ONCOLOGY APPLICATIONS OF MICROARRAY TECHNOLOGY DETECTING SMALL GENOMES VARIABILITY

Rossana Scavelli, Sergio Abrignani and Marco Crimi*

Functional Genomics Unit

National Institute of Molecular Genetics (INGM), Milan, Italy

ABSTRACT

Microarray is an exciting and promising technology for sequencing analysis, high-throughput mutation detection, SNP genotyping and gene expression profiling. Through the simultaneous analysis of thousands of genes, microarray technology is characterized by large-scale, highly-automatic and sensitive detection of genetic data at both qualitative and quantitative level.

The most recent clinical application of microarray technology is aimed at DNA genotyping [1]. A hot medical demand is the high-throughput scanning of cancer related DNA molecules, such as endogenous cytoplasmatic mitochondrial DNA (mtDNA) and exogenous DNA (i.e., viral genomes). Both mtDNA and viral genomes are characterized by a small size and a high exon/intron ratio, suggesting a perfect match with the microarray requirements.

Mitochondria are cytoplasmatic organelles that play a vital role in energy metabolism, producing the ATP required for cellular functions. Despite a poor genotype/phenotype correlation, complex human disorders are increasingly being linked to mtDNA mutations that effect mitochondrial function. Mutations in mtDNA can be considered as useful biomarkers associated with several diseases, such as cancer, aging and metabolic diseases [2]. Mutations throughout the mitochondrial genome are recurrent events in primary tumor tissues and in corresponding non-invasively collected body fluids. Thus, mtDNA mutation analysis may provide a molecular tool for the early detection and prognosis of cancer.

Other than mtDNA, exogenous DNAs have been reported to be correlated to cancer onset. Hepatocellular carcinoma is one of the most common cancers throughout the world. Several viral oncogenes, such as the Hepatitis C Virus (HCV) core protein, are related to hepatocarcinogenesis [3]. HCV is a serious, worldwide public health problem affecting an

* E-mail address: crimi@ingm.it. Tel. +39-02-00660211. Fax. +39-02-00660216. Correspondence to: Marco Crimi, PhD, INGM - Functional Genomics Unit, Via Francesco Sforza, 32, 20122 Milan, Italy

estimated 150–200 million individuals. HCV is a remarkably successful pathogen, establishing persistent infection in more than two-thirds of those who contract it. Its success is related to its abilities to blunt innate antiviral pathways and to evade adaptive immune responses. Up to 20% of the infected patients develop cirrhosis or liver cancer [4]. There is currently no effective vaccine. Although antiviral therapies have improved considerably in recent years, they are associated with significant toxicity and are effective in only about one-half of those treated [5]. Developing a specific microarray for HCV detection could have a direct effect on diagnostic applications.
Thus, the availability of such in-silico technology and the identification of potential cancer biomarkers in small genomes, such as mtDNA and HCV, could enable predictive cancer detection at early stages, improving patient management and treatment choices.

Introduction

Microarray is a powerful tool to enable sequencing on a scale that would be impossible using conventional analysis.

In the present chapter, we concentrate on the following main aspects:

- overview of microarray technology and oligonucleotide array sequencing
- sequencing array application on small genomes: mtDNA
- sequencing array application on small genomes: HCV

Overview of Microarray Technology and Oligonucleotide Array Sequencing

Microarray is a technology based on hybridization of nucleic acids on a solid matrix (such as glass, silicon or plastic) by a covalent attachment. The microarray (also known as DNA chip or Gene array) can be defined as an ordered collection of microspots, each spot containing a single defined specie of a nucleic acid. Taking advantage of the inherent nature of DNA-RNA or DNA-DNA interactions, the microarray surface will host the hybridization between two single-stranded nucleic acid molecules: the probe arrayed on the matrix and the target sample to be measured [6]. Since microarray development, this technology is really focusing new light on genomic measurements.

Microarray platforms are divided in two main categories: *spotted-* and *printed-*probes microarrays [7, 8]. In spotted microarrays the probes are DNA molecules obtained prior to their deposition on the array's surface by high-speed robots. The DNA molecules are more than 100nt (nucleotide) length of cDNAs or amplified PCR products [7]. Because of a quite low feature density (less than 10,000 spots per array) spotted microarrays are mainly used for gene expression applications [9].

The high-density printed arrays contain relatively short probes (usually oligonucleotides of about 25-40nt). Oligonucleotides are synthesized either directly on the array surface by photolithography and micro-mirror devices [10-12] or by ink-jet printing of oligonucleotide probe sequences [13]. Oligonucleotide arrays contain thousands of discrete features per chip, with each feature comprising several copies of a distinct probe sequence [14].

To date, the major application of the microarray technology implies the analyses of expression profiles of thousands of genes simultaneously. DNA microarray can also be applied to scan genomes in order to detect genetic variants such as mutations or Single Nucleotide Polymorphism (SNP) at specific locations [15].

The array sequencing, where probes complementary to target DNAs are arrayed on a matrix, was pioneered in the late 80's as alternative to gel-based approaches [16, 17]. Since the first manufactures, important efforts have been made to improve the performance of microarray technologies aimed at mutation detection, providing new tools for the molecular diagnostic of human disorders.

The Resequencing GeneChip® developed by Affymetrix (Santa Clara, CA, USA) was the first available array for high-throughput screening of thousands of sequences on a single platform. The GeneChip® is manufactured by photolithography technology and oligonucleotides are synthesized directly on the surface of quartz wafers.

Affymetrix technology for resequencing employs 8 different probes based on a reference sequence: four probes scanning the forward DNA strand and the other four for the reverse complementary. The features represent a series of overlapping oligonucleotide probes each of 25nt length. One feature accommodates just PM (Perfect Matching) nucleotides. PM features are completely complementary to the experimental sequence. Other three MM (Mis-Matching) features vary at the central (the 13th base) position to incorporate each alternative code letter (A, T, G or C) for the detection of both known and new SNPs. The other four features, customized as the PM and MM, hybridize to the reverse complement of the reference sequence. Following long range PCR pre-amplification, the biotin-labeled sample is hybridized against the array. If the target sequence matches with the reference sequence (and its complement) the PM features will generate the higher intensity compared with the MMs. Otherwise, if the sample contains a variant, MM features' fluorescence will reveal the mutation. Gene chip arrays allow the analysis of 600 kb of unidirectional sequence (or 300 kb of double stranded DNA) on a single experiment, with an estimated base calling accuracy higher than 99% [18].

In addition to the more popular Affymetrix arrays, several other oligonucleotide platforms have emerged in recent years. Illumina (San Diego, CA, USA) has created an alternative microarray technology (The BeadArray®) based on randomly arranged beads. The Illumina platform uses oligonucleotides attached to small beads which are randomly arrayed in small wells etched into wafers carrying the array [19]. A specific oligonucleotide sequence is assigned to each bead type, which is replicated about 30 times on the array. A serie of decoding hybridizations is used to identify every bead [20]. Multiple copies of each sequence-specific bead provide a precised and reliable measurement [19, 21].

Moreover, ASPER BIOTECH (Tartu, Estonia) uses Arrayed Primer EXtension (APEX) for rapid identification of mutations. APEX is a new resequencing method that combines the efficiency of a microarray-based assay with the Sanger sequencing. APEX technology is based on PCR amplified templates partially overlapping with 25-mer oligo immobilized probes. A template-dependent single nucleotide extention allows the nucleotide identification due to the presence of four unique fluorescently labeled terminator nucleotides [22].

Furthermore, several home made resequencing microarrays have been generated [23, 24]. However, detailed information about non-commercially available microarrays will not be provided in this chapter.

Since the unveiling of the human genome almost 7 years ago, the rapid improvements in sequencing technologies introduced us into the era of whole-genomes scanning. The complete sequencing of entire genomes is dramatically improving our understanding of various biological processes and disease conditions [25, 26]. In addition, the recent advances in microarray technologies have made the resequencing of entire genomes reliable in a single experiment, reducing cost, time and complexity compared to traditional methods (Table 1). Moreover, the flexibility makes oligonucleotide arrays ideal for detecting the broadest range of genomic features, such as missense or nonsense mutations, SNPs, or insertion-deletions. Genome-wide decoding with microarray technology can be applied in numerous approaches covering from detection of genetic variability to pathogens identification.

Table 1. Comparison of different strategies for mutation detection in small genomes

	Microarray	TaqMan	RFLP	Sequencing	HR
Cost	+	++	+++	++	++
Sensitivity	++	+++	+	+	+++
Automation	+++	++	+	++	++
Knowledge need	+++	++	+	++	++
Throughput	+++	++	+	+	++

(+++: excellent; ++: fine; +: satisfactory)
HR: Heteroduplex Resolution

Generally, small and compact genomes are of primary interest for microarray applications. Scientists have already designed custom arrays to assess variation in microbial genomes. In 1996, the entire sequence of the HIV-1 protease gene was screened with a microarray approach in 167 viral isolates [27]. Recently, in order to differentiate among multiple species causative the Severe Acute Respiratory Syndrome, SARS coronavirus was sequenced for all single nucleotide changes using an in-silico technology [28].

Moreover, researchers have demonstrated the value of the application of DNA microarray to other small-sized DNA molecules such as endogenous cytoplasmatic DNAs, as mtDNA, and exogenous viral DNAs such as HCV. Both genomes are characterized by a small size (about 16 kb and 9 kb respectively) and a high exon/intron ratio, suggesting to perfectly match a microarray analysis.

Sequencing Array Application on Small Genomes: mtDNA

Mitochondria are cytoplasmatic organelles that play a vital role in energy metabolism, producing the ATP required for cellular functions. The understanding of mitochondrial biology is a fundamental research goal in human genetics and medicine. Mitochondria generate much of the cellular energy by the process of OXidative PHOSphorilation (OXPHOS). As a by-product of energy production, the mitochondria generate most of the endogenous Reactive Oxygen Species (ROS) of the cell, which damage the mitochondria, mtDNAs, and the cell itself.

Mitochondria are derived from ancient bacterial symbionts and contain their own DNA, RNA, and protein synthesizing systems. Each human cell contains hundreds of mitochondria and thousands of mtDNAs. The mtDNA is maternally inherited and shows striking regional genetic variation. This regional variation was a major factor in permitting humans to adapt to the different global environments [2].

The mitochondrial genome is 16,569 bp and contains 37 genes that encode its own 22 tRNAs and two ribosomal RNAs required for mitochondrial protein synthesis, plus 13 polypeptides (subunits of the 'electron transport chain' involved in respiration) for the OXPHOS process [29, 30]. Mitochondria are assembled from both nuclear- and mitochondrial-encoded genes. So, there are 74 nuclear-encoded proteins that also contribute toward the electron transport chain [31]. All the mitochondrial genes are essential for proper organelle function.

The mtDNA has a very high mutation rate, presumably due to its chronic exposure to mitochondrial ROS. Mitochondrial ROS production, in turn, is modulated by regional mtDNA genetic variation and alterations in nuclear DNA expression of stress response genes [2].

When a new mtDNA mutation arises in a cell, a mixed intracellular population of mtDNAs is generated, a state known as heteroplasmy. As the heteroplasmic cell undergoes cytokinesis, the mutant and normal molecules are randomly distributed into the daughter cells. As a consequence of this replicative segregation, the proportion of mutant and normal mtDNAs can drift toward homoplasmic mutant or wild type [2].

Despite a poor genotype/phenotype correlation, complex human disorders are increasingly being linked to mtDNA mutations that affect mitochondrial function.

Hundreds of mtDNA point mutations and insertion/deletion polymorphisms have been reported, and current validation studies are addressing the sensitivity and specificity of potential mtDNA-based diagnostics [32-34].

Clinical and laboratory data have suggested that a wide variety of conditions are associated with changes in the mtDNA and/or proteins, including aging [35], cardiovascular disease [36], diabetes [37], macular degeneration [38], mental retardation [39], Parkinson's disease [40] Leigh Syndroms (LS) and MELAS (Mitochondrial Encephalomyopathy, Lactic Acidosis, and Stroke like episodes) syndrome [41]. In addition, mtDNA rearrangements have been found to be strikingly increased in a variety of age-related degenerative diseases, like Alzeimer's Disease (AD) [2].

Aging is also a risk factor for most solid tumors. Mitochondria were first suggested to contribute to carcinogenesis in 1973 when quantitative and qualitative electron microscopy analyses revealed structural differences in this organelle between cancer patients and normal controls [42]. Hence, the etiology of cancer might also be associated with mitochondrial dysfunction, damaged respiratory systems and ROS production [43, 44]. Because of the production of ROS toxic species, mtDNA is more likely to be damaged than nuclear DNA [45]. Moreover, mitochondria possess less efficient DNA repair mechanisms than the nucleus, therefore mtDNA mutations are more likely to persist in the nucleus and be representative of clonal expansion [46].

Since mtDNA lacks introns, it has been suggested that most mutations will occur in coding sequences and subsequent accumulation of mutations may lead to tumor formation [47]. The mtDNA is dependent upon the nuclear genome for transcription, translation, replication and repair, but precise mechanisms for how the two genomes interact and integrate with each other are poorly understood. However, in solid tumors elevated expression of OXPHOS mtDNA-encoded subunits may reflect mitochondrial adaptation to perturbations in cellular energy requirements [2].

Not only solid tumors but also hematological diseases including leukemias and lymphomas have been reported to contain mtDNA changes. The accumulation of mtDNA variants indicates a possible function as a molecular clock for organ degeneration [32, 47].

MtDNA mutations have also been observed in pre-malignant lesions, raising the possibility for early, pre-symptomatic disease detection [31]. Furthermore, mtDNA mutations have been detected in body fluids, indicative of the potential clinical utility of non-invasively collected specimens for early pre-symptomatic diagnosis.

Reductions in cancer can be expected if improved diagnostics and population screening technologies are implemented. Improvement in cancer prognosis can be primarily achieved via the development and validation of accurate early detection assays, which include strategies for noninvasively collecting samples coupled with automation and sensitive detection limits. Currently, no clinically applied DNA biomarkers linked to mtDNA exist for cancer detection. Thus, reliable, high-throughput assays are needed to determine the role of sequence variation in the mtDNA.

The first small genome sequencing array, was specifically aimed at variation analysis of the human mtDNA [48, 49].

In 2004 Affymetrix developed a human mitochondrial array-based sequencing platform for rapid and high-throughput analysis of mtDNA [50]. Oligonucleotide probes are synthesized in situ using the standard resequencing array tiling strategy above mentioned. This first generation of MitoChip (MitoChip v1.0) was able to sequence both strands of the entire mitochondrial coding region in a single assay.

However, the heteroplasmy detection sensitivity of the first version of the Mito-Chip was controversial. Mito-Chip 1.0 didn't allow a complete mutation analysis because of the poor range of the nucleotide call, just about 71%-74% [51]. A second limitation consisted of the lack of probes for sequencing the mitochondrial Displacement-loop (D-loop), the non coding region in the mtDNA molecule that controls both replication and transcription. As many of the somatic mtDNA mutations of primary tumors are found in the D-loop region [52, 53], additional conventional sequencing of the D-loop region was required to provide a complete mitochondrial genomic profiling. To create a complete high-throughput strategy for mtDNA analysis, a new second-generation version of the MitoChip (MitoChip v2.0) was developed [54]. The major difference is that the v2.0 chips are tilled with oligonucleotide probes for the entire 16.5 kb sequence of mtDNA. In addition, to increase sensitivity of mutation detection, the 500 most common haplotypes reported in the MitoMap public database (http://www.mitomap.org), are probed in the MitoChip v2.0.

The second generation Mito-Chip array shows an overall nucleotide call reproducibility of 97%-99,99% and harbor greater sensitivity in detecting mutations that are present only at low heteroplasmic level [55, 56].

Sequencing Array Application on Small Genomes: HCV

HCV is a remarkably successful pathogen, establishing persistent infection in more than two-thirds of infected people. Its success is related to its abilities to blunt innate antiviral pathways and to evade adaptive immune responses. HCV establishes persistent infection, despite the fact that it is recognized and targeted by innate, cellular, and humoral immune mechanisms [57].

HCV is an enveloped positive strand RNA virus in the genus Hepacivirus and the family Flaviviridae. HCV was identified as the infectious agent responsible for the majority of transfusion-associated non-A and non-B hepatitis, with the isolation of the first HCV genomic clones in 1989 [57]. The 9.6 kb RNA linear genome includes extensive RNA structures at its 5' and 3' ends, which are essential to translation and replication. Translation of the single, long open reading frame yields a polyprotein of approximately 3000 amino acids. The polyprotein is processed by host and viral proteases to release ten proteins that make up the viral particle. The HCV replication machinery is believed to take place in association with endoplasmic reticulum–derived membranous structures in the cytosol of infected cells [57].

Importantly, HCV's RNA polymerase lacks proofreading capability. The high rate of HCV replication together with its error-prone replication mechanism promote a rapid evolution under immune system and antiviral drug selection. There are six major genotypes and dozens of subtypes of HCV that differ in their geographic distribution and their responsiveness to antiviral therapy [58]. A number of studies have suggested that the severity of liver disease and the outcome of therapy might be genotype dependent [59]. It has been reported that about 20% of the infected may develop serious complications including cirrhosis, liver failure, or hepatocellular carcinoma [57]. We believe that the routin sequencing of the HCV genome in infected patients is important to better understand the viral life cycle and the molecular basis of the HCV infection, pathogenesis and treatment.

Although HCV antiviral therapies have improved considerably in recent years, they are still associated with significant toxicity. The treatment of HCV hepatitis with interferon/ribavirin is effective in only about one-half of patients and results in a varied response in terms of viral titer decrease and ultimate outcome [57]. Marked responders have a sharp decline in viral titer within a few days of treatment initiation, whereas in other patients there is no effect on the virus (poor responders) [60]. Different responsiveness could be due to two main factors: the variability of the HCV genome, that may confer resistance to the treatment, and SNPs or mRNAs level differences that are activated or silenced by the presence of the virus.

The development of a specific microarray for HCV detection could enable the characterization of genome subtypes directly effecting diagnostic applications, such as patient management, improvement of treatment choices and the limitation of infection's transmission. So far, tools for HCV whole-genome resequencing are missing. A preliminary cDNA microarray for HCV detection has been developed by spotting the HCV genes fragment on the surface of an amido-modified glass slide using a robotics system [61].

Alternatively, we predict the creation of an innovative sequencing chip for the effective detection of the genetic susceptibility of the host. Therefore, a standardized HCV microarray system would be essential for acquiring new diagnostic tools, although more studies are needed.

CONCLUSION

The development of early cancer detection tools is based on the availability of reliable biomarkers and miniaturized robotic assays. Microarray potentiality which enables early onset cancer detection is based on the identification of specific biomarkers linked to mtDNA and HCV genome.

The availability of in-silico technologies could strongly speed up the diagnosis and prognosis of cancer linked with mutations in small genomes. In addition, the end-out of mutations in mtDNA offers a diagnostic device for mtDNA-linked disorders. Therefore, the ultimate goal of the development of specific chips for small genomes, such as mtDNA and HCV, could be cancer detection at early stages, before cells become malignant.

ACKNOWLEDGMENTS

We gratefully acknowledge Francesco Marabita and Cristina Cheroni for their helpful suggestions and comments.

REFERENCES

[1] de Leon J, Susce MT, Murray-Carmichael E. The AmpliChip CYP450 genotyping test: Integrating a new clinical tool. *Mol Diagn Ther*. 2006;10(3):135-51.

[2] Wallace DC. A mitochondrial paradigm of metabolic and degenerative diseases, aging, and cancer: a dawn for evolutionary medicine. *Annu Rev Genet*. 2005;39:359-407.

[3] Rust C, Gores GJ. Apoptosis and liver disease. *Am J Med*. 2000 May;108(7):567-74.

[4] Afdhal NH. The natural history of hepatitis C. *Semin Liver Dis*. 2004;24 Suppl 2:3-8.

[5] Houghton M, Abrignani S. Prospects for a vaccine against the hepatitis C virus. *Nature*. 2005 Aug 18;436(7053):961-6.

[6] Southern E, Mir K, Shchepinov M. Molecular interactions on microarrays. *Nat Genet*. 1999 Jan;21(1 Suppl):5-9.

[7] Schena M, Shalon D, Davis RW, Brown PO. Quantitative monitoring of gene expression patterns with a complementary DNA microarray. *Science*. 1995 Oct 20;270(5235):467-70.

[8] Lockhart DJ, Dong H, Byrne MC, Follettie MT, Gallo MV, Chee MS, et al. Expression monitoring by hybridization to high-density oligonucleotide arrays. *Nat Biotechnol*. 1996 Dec;14(13):1675-80.

[9] Auburn RP, Kreil DP, Meadows LA, Fischer B, Matilla SS, Russell S. Robotic spotting of cDNA and oligonucleotide microarrays. *Trends Biotechnol*. 2005 Jul;23(7):374-9.

[10] McGall GHea. The efficiency of light-directed synthesis of DNA arrays on glass substrates. *J AmChemSoc*. 1997;119.

[11] Pease AC, Solas D, Sullivan EJ, Cronin MT, Holmes CP, Fodor SP. Light-generated oligonucleotide arrays for rapid DNA sequence analysis. *Proc Natl Acad Sci U S A*. 1994 May 24;91(11):5022-6.

[12] Fodor SP, Read JL, Pirrung MC, Stryer L, Lu AT, Solas D. Light-directed, spatially addressable parallel chemical synthesis. *Science*. 1991 Feb 15;251(4995):767-73.

[13] Wodicka L, Dong H, Mittmann M, Ho MH, Lockhart DJ. Genome-wide expression monitoring in Saccharomyces cerevisiae. *Nat Biotechnol*. 1997 Dec;15(13):1359-67.

[14] Mockler TC, Chan S, Sundaresan A, Chen H, Jacobsen SE, Ecker JR. Applications of DNA tiling arrays for whole-genome analysis. *Genomics*. 2005 Jan;85(1):1-15.

[15] Hacia JG. Resequencing and mutational analysis using oligonucleotide microarrays. *Nat Genet*. 1999 Jan;21(1 Suppl):42-7.

[16] Bains W, Smith GC. A novel method for nucleic acid sequence determination. *J Theor Biol.* 1988 Dec 7;135(3):303-7.

[17] Khrapko KR, Lysov Yu P, Khorlyn AA, Shick VV, Florentiev VL, Mirzabekov AD. An oligonucleotide hybridization approach to DNA sequencing. *FEBS Lett.* 1989 Oct 9;256(1-2):118-22.

[18] Cowell JK, Hawthorn L. The application of microarray technology to the analysis of the cancer genome. *Curr Mol Med.* 2007 Feb;7(1):103-20.

[19] Kuhn K, Baker SC, Chudin E, Lieu MH, Oeser S, Bennett H, et al. A novel, high-performance random array platform for quantitative gene expression profiling. *Genome Res.* 2004 Nov;14(11):2347-56.

[20] Gunderson KL, Kruglyak S, Graige MS, Garcia F, Kermani BG, Zhao C, et al. Decoding randomly ordered DNA arrays. *Genome Res.* 2004 May;14(5):870-7.

[21] Dunning MJ, Smith ML, Ritchie ME, Tavare S. beadarray: R classes and methods for Illumina bead-based data. Bioinformatics. 2007 Aug 15;23(16):2183-4.

[22] Jaakson K, Zernant J, Kulm M, Hutchinson A, Tonisson N, Glavac D, et al. Genotyping microarray (gene chip) for the ABCR (ABCA4) gene. *Hum Mutat.* 2003 Nov;22(5):395-403.

[23] Okou DT, Steinberg KM, Middle C, Cutler DJ, Albert TJ, Zwick ME. Microarray-based genomic selection for high-throughput resequencing. *Nat Methods.* 2007 Nov;4(11): 907-9.

[24] Denning L, Anderson JA, Davis R, Gregg JP, Kuzdenyi J, Maselli RA. High throughput genetic analysis of congenital myasthenic syndromes using resequencing microarrays. *PLoS ONE.* 2007;2(9):e918.

[25] Polak M, Dechaume A, Cave Dpharm H, Nimri R, Crosnier H, Sulmont V, et al. Heterozygous Missense Mutations in the Insulin Gene are linked to Permanent Diabetes appearing in the Neonatal Period or in Early-Infancy: A report from the French ND Study Group. *Diabetes.* 2008 Jan 2.

[26] Ferraris S, Clark S, Garelli E, Davidzon G, Moore SA, Kardon RH, et al. Progressive external ophthalmoplegia and vision and hearing loss in a patient with mutations in POLG2 and OPA1. *Arch Neurol.* 2008 Jan;65(1):125-31.

[27] Kozal MJ, Shah N, Shen N, Yang R, Fucini R, Merigan TC, et al. Extensive polymorphisms observed in HIV-1 clade B protease gene using high-density oligonucleotide arrays. *Nat Med.* 1996 Jul;2(7):753-9.

[28] Wong CW, Albert TJ, Vega VB, Norton JE, Cutler DJ, Richmond TA, et al. Tracking the evolution of the SARS coronavirus using high-throughput, high-density resequencing arrays. *Genome Res.* 2004 Mar;14(3):398-405.

[29] Toyokuni S, Okamoto K, Yodoi J, Hiai H. Persistent oxidative stress in cancer. *FEBS Lett.* 1995 Jan 16;358(1):1-3.

[30] Toescu EC, Myronova N, Verkhratsky A. Age-related structural and functional changes of brain mitochondria. *Cell Calcium.* 2000 Nov-Dec;28(5-6):329-38.

[31] Jakupciak JP, Dakubo GD, Maragh S, Parr RL. Analysis of potential cancer biomarkers in mitochondrial DNA. *Curr Opin Mol Ther.* 2006 Dec;8(6):500-6.

[32] Jakupciak JP, Wang W, Markowitz ME, Ally D, Coble M, Srivastava S, et al. Mitochondrial DNA as a cancer biomarker. *J Mol Diagn.* 2005 May;7(2):258-67.

[33] Parr RL, Dakubo GD, Thayer RE, McKenney K, Birch-Machin MA. Mitochondrial DNA as a potential tool for early cancer detection. Hum Genomics. 2006 Jan;2(4):252-7.

[34] Brandon MC, Lott MT, Nguyen KC, Spolim S, Navathe SB, Baldi P, et al. MITOMAP: a human mitochondrial genome database--2004 update. *Nucleic Acids Res.* 2005 Jan 1;33(Database issue):D611-3.

[35] Khaidakov M, Siegel ER, Shmookler Reis RJ. Direct repeats in mitochondrial DNA and mammalian lifespan. *Mech Ageing Dev.* 2006 Oct;127(10):808-12.

[36] Tsutsui H, Ide T, Kinugawa S. Mitochondrial oxidative stress, DNA damage, and heart failure. Antioxid Redox Signal. 2006 Sep-Oct;8(9-10):1737-44.

[37] Maassen JA, Janssen GM, t Hart LM. Molecular mechanisms of mitochondrial diabetes (MIDD). *Ann Med.* 2005;37(3):213-21.

[38] Jones M, Mitchell P, Wang JJ, Sue C. MELAS A3243G mitochondrial DNA mutation and age related maculopathy. *Am J Ophthalmol.* 2004 Dec;138(6):1051-3.

[39] Hattori Y, Takeoka M, Nakajima K, Ehara T, Koyama M. A heteroplasmic mitochondrial DNA 3310 mutation in the ND1 gene in a patient with type 2 diabetes, hypertrophic cardiomyopathy, and mental retardation. *Exp Clin Endocrinol Diabetes.* 2005 Jun;113(6):318-23.

[40] Bender A, Krishnan KJ, Morris CM, Taylor GA, Reeve AK, Perry RH, et al. High levels of mitochondrial DNA deletions in substantia nigra neurons in aging and Parkinson disease. *Nat Genet.* 2006 May;38(5):515-7.

[41] Crimi M, Galbiati S, Moroni I, Bordoni A, Perini MP, Lamantea E, et al. A missense mutation in the mitochondrial ND5 gene associated with a Leigh-MELAS overlap syndrome. *Neurology.* 2003 Jun 10;60(11):1857-61.

[42] Shumacher HR SI, Patel SB, Fisher DR. Mitochondrial, a clue to oncogenesis. *Lancet.* 1973;2.

[43] Chatterjee A, Mambo E, Sidransky D. Mitochondrial DNA mutations in human cancer. *Oncogene.* 2006 Aug 7;25(34):4663-74.

[44] Warburg O. On the origin of cancer cells. *Science.* 1956 Feb 24;123(3191):309-14.

[45] Yakes FM, Van Houten B. Mitochondrial DNA damage is more extensive and persists longer than nuclear DNA damage in human cells following oxidative stress. *Proc Natl Acad Sci U S A.* 1997 Jan 21;94(2):514-9.

[46] Fliss MS, Usadel H, Caballero OL, Wu L, Buta MR, Eleff SM, et al. Facile detection of mitochondrial DNA mutations in tumors and bodily fluids. *Science.* 2000 Mar 17;287(5460):2017-9.

[47] Penta JS, Johnson FM, Wachsman JT, Copeland WC. Mitochondrial DNA in human malignancy. *Mutat Res.* 2001 May;488(2):119-33.

[48] Chee M, Yang R, Hubbell E, Berno A, Huang XC, Stern D, et al. Accessing genetic information with high-density DNA arrays. *Science.* 1996 Oct 25;274(5287):610-4.

[49] Jones JB, Song JJ, Hempen PM, Parmigiani G, Hruban RH, Kern SE. Detection of mitochondrial DNA mutations in pancreatic cancer offers a "mass"-ive advantage over detection of nuclear DNA mutations. *Cancer Res.* 2001 Feb 15;61(4):1299-304.

[50] Maitra A, Cohen Y, Gillespie SE, Mambo E, Fukushima N, Hoque MO, et al. The Human MitoChip: a high-throughput sequencing microarray for mitochondrial mutation detection. *Genome Res.* 2004 May;14(5):812-9.

[51] Saladino FV, R. Cifola, I. Bordoni, A. Galbiati, S. Battaglia, C. Bresolin, N. Comi, G.P. Mitochondrial DNA microarray resequencing in Leber's hereditary optic neuropathy and other mitochondrial encephalomyopathies. *Journal of Neurology.* 2006 May 2006;253, suppl. 2, poster 200.

[52] Sanchez-Cespedes M, Parrella P, Nomoto S, Cohen D, Xiao Y, Esteller M, et al. Identification of a mononucleotide repeat as a major target for mitochondrial DNA alterations in human tumors. *Cancer Res*. 2001 Oct 1;61(19):7015-9.
[53] Tamori A, Nishiguchi S, Nishikawa M, Kubo S, Koh N, Hirohashi K, et al. Correlation between clinical characteristics and mitochondrial D-loop DNA mutations in hepatocellular carcinoma. *J Gastroenterol*. 2004 Nov;39(11):1063-8.
[54] Zhou S, Kassauei K, Cutler DJ, Kennedy GC, Sidransky D, Maitra A, et al. An oligonucleotide microarray for high-throughput sequencing of the mitochondrial genome. *J Mol Diagn*. 2006 Sep;8(4):476-82.
[55] *Halushka MKA, J.Bishe, B.R. Chakravarti, A. Cohn, R.D. Cutler, D.J. Johnson, N, Judge, D.P. Lu, J. Arking D.E.* Determination of mitochondrial sequence variants in familial heart disease using resequencing arrays [abstract 468]. 2006.
[56] Sui G, Zhou S, Wang J, Canto M, Lee EE, Eshleman JR, et al. Mitochondrial DNA mutations in preneoplastic lesions of the gastrointestinal tract: a biomarker for the early detection of cancer. *Mol Cancer*. 2006;5:73.
[57] Dustin LB, Rice CM. Flying under the radar: the immunobiology of hepatitis C. *Annu Rev Immunol*. 2007;25:71-99.
[58] Simmonds P, Alberti A, Alter HJ, Bonino F, Bradley DW, Brechot C, et al. A proposed system for the nomenclature of hepatitis C viral genotypes. *Hepatology*. 1994 May;19(5):1321-4.
[59] Bukh J ES, Purcell RH. Genetic heterogeneity of hepatitis C virus and related virus. Viral hepatitis and liver diseases:proceeding of IX triennal international symposium on viral hepatitis and liver disease. 1997.
[60] Brodsky LI, Wahed AS, Li J, Tavis JE, Tsukahara T, Taylor MW. A novel unsupervised method to identify genes important in the anti-viral response: application to interferon/ribavirin in hepatitis C patients. *PLoS ONE*. 2007;2(7):e584.
[61] Sun Z, Ma W, Wei M, Wang S, Zheng W. Identification of HCV-1b by low-density cDNA microarray-based assay. *Curr Microbiol*. 2007 Sep;55(3):211-6.

Rewied by: Rossana Mineri, PhD

Unit of Molecular Neurogenetic, Center for the Study of Mitochondrial Disorders, C. Besta Neurological Institute Foundation, Milan, Italy

In: Oligonucleotide Array Sequence Analysis
Editors: M.K. Moretti and L.J. Rizzo, pp. 469-490
ISBN: 978-1-60456-542-3

Chapter 19

OLIGONUCLEOTIDE ARRAYS IN FORENSIC DNA TESTING

Jason R.E. Shepard*
Department of Chemistry, University at Albany, SUNY
1400 Washington Ave., Albany, NY 12222

ABSTRACT

The past decade has seen the initiation, development and subsequent adoption of DNA analysis in forensic science. Universally accepted standard procedures and databases are in place to organize the analysis of DNA in a manageable fashion. The potential to improve and enhance these current resources exists, and as technology continues to develop, the boundaries of our current capabilities are constantly being challenged. Improvements and enhancements are needed; a backlog of forensic DNA samples is present throughout the United States, and efforts are underway to alleviate the strain associated with the delays in sample processing. In addition to the sample backlog, situations exist in which current DNA typing methodologies are sub-optimal, such as those examining extensively degraded material or those seeking better resolution of the Y-chromosome. Genetic information is continually being made available based on the sequencing and haplotyping of the human genome. This information, along with new technology, has identified other genetic polymorphisms with the potential to better address some of the current problems and solve the key issues involved in the field of DNA analysis in forensic science. Technologies employing higher throughput than the standard methods employed are better suited to capitalize on the limitless potential garnered from the sequence information and population data now available. Microarray analyses are one such format positioned to exploit this potential.

I. INTRODUCTION

Innovation is at the heart of new technology, which in turn, eventually allows for the incorporation of techniques, equipment and skills that can transform science. For technology to revolutionize a scientific discipline, a dramatic conversion is often required away from traditional methods. Technology must demonstrate its value and convenience in terms of

* E-mail address: jshepard@albany.edu

scientific utility, cost, and speed, among others, before widespread acceptance is met. An excellent example of this phenomenon is with the Polymerase Chain Reaction (PCR), which more than twenty years after the first PCR publication, is now a technique that is commonplace. The scientific efficacy of PCR is massive, and the cost for materials and instrumentation for such experiments are relatively minor as compared to the utility PCR provides. As technology development has continued to make PCR more affordable, it has now become a routine laboratory practice, and other disciplines and technologies have positioned themselves to better take advantage of PCR's potential.

In a similar manner to PCR, the biological sciences are undergoing another transformation related to manipulation of DNA. High throughput, parallel analysis of DNA is becoming commonplace; rapid analytical techniques to manipulate and interrogate DNA are becoming routine. Essentially, new technologies are enabling more rapid and accurate sequencing of entire genomes, which in turn, provides the genetic information for these techniques and instrumentation to be refined and applied to more complex problems. The Human Genome Project,[1] to sequence the entire human genome, was completed and the HapMap Project, to decipher the polymorphic differences in the human genome, is fully underway and draft versions have been released.[2] The knowledge garnered from these massive undertakings allows the focus of DNA manipulation to move away from discovery to application, from the entire genome to only those genetic polymorphisms of importance, such as cancer-related genes or forensic markers in crime laboratories. This focused approach to DNA analysis further refines and enhances the high-throughput nature of modern DNA interrogation.

DNA microarrays are one of many high-throughput formats that are following a similar trend to that experienced by PCR. A few decades ago, important advances in DNA analysis were achieved employing methods such as standard Southern blotting where a few gene targets were identified with agarose gels and radiolabeled probes. Now, hundreds to thousands, to tens of thousands of oligonucleotide sequences can be interrogated at one time. These oligonucleotide arrays are highly parallel formats with the ability to perform directed analysis of entire genomes. With the knowledge the scientific community has amassed related to genetic sequence analysis, a natural progression has led to advancements in specific applications where genomic differences can be identified, measured, and cataloged, most notably in the fields of medical diagnostics, pharmacology, and forensic science.

Forensic science is in the midst of this technological revolution that began with Sir Alec Jeffreys' advancement in genetic fingerprinting and Kerry Mullis's development of PCR and continues to blossom with the discovery of new genetic markers and novel technologies. The field of forensic sciences is broad, and encompasses everything from criminalistics, anthropology, toxicology, and biology, among many other subdivisions. Although DNA analysis in forensic biology got its start in the 1980s, almost a decade passed before DNA analysis made its way into routine forensic casework. Since then, technological advances in DNA sequencing and high throughput analysis has led to the common practice of examining DNA evidence in criminal prosecution cases and the formation of DNA databases. For modern criminal investigations, forensic scientists will routinely examine multiple biological samples related to crime scenes, so much that the general consensus is that a sample processing backlog exists. This backlog is a national problem adding unnecessary delays to criminal investigations and consuming valuable resources. To address this backlog and future demand, high-throughput methods employing parallel detection formats, such as oligonucleotide microarrays, are gaining acceptance for forensic-related analysis of DNA. This chapter will be concerned with the current state of oligonucleotide arrays with specific interest in their application to the field of forensic science.

II. Genomic Variation

The human genome is a complex collection of biological information primarily defined in terms of its hereditary characteristics. In its entirety, the human genome is approximately three billion nucleotides in length, the large majority of which does not differ between individuals and has no known function. Our genome is dynamic, containing numerous forms of variation, but because of its sheer size, the amount of variation across genomes is immense. Variations can be relatively uncomplicated in nature, represented as a single nucleotide base mismatch, or single nucleotide polymorphism (SNP). More complex differences also exist such as insertions, deletions, repeats, and rearrangements, the definition, of which are self-explanatory. Within the field of forensic science, the overwhelming preponderance of effort has been put into short tandem repeat (STR) marker analysis. Because of this fact, along with the sheer volume of SNPs and their recent emergence as useful markers, the focus of this chapter will keep within SNP and STR array methods.

Single Nucleotide Polymorphisms (SNPs)

SNPs are the most common variations in the human genome, representative of about 0.1% of the total sequence information. This percentage relates to more than 3 million SNPs, which have recently been cataloged by the International HapMap Consortium.[2] On average, with equal spacing and no other influencing factors, a SNP would expect to be positioned once every 1,000 bases along the genome. While the majority of SNPs have no known biological relevance or no recognized affect on inherited traits, many SNPs are known to be related to differential gene expression levels and/or have direct influence on protein function. In reality, many SNPs occur with some sort of connectivity, more specifically, with linkage disequilibrium. Linkage disequilibrium is the non-random association of alleles that occur in a population more or less frequently than would be predicted from a random event. Essentially, some SNPs can be considered as traveling together, such that not all SNPs should be thought of as having equal utility and the number of actual SNPs useful in forensic identification would be much less than their total number.

Short Tandem Repeats (STRs)

Within the definition of more complex variation characteristics are STRs. STRs are sections in the genome with consecutive sequence repeats. Thousands of repeat regions occur in the human genome, generally in the non-coding region. A tandem repeat can consist of a repeat unit of any length, ranging from a two nucleotide base repeat unit, up to ten or more nucleotide bases as the repeat sequence. STRs though, are categorized as those tandem repeats with a repeat unit between two and five nucleotide bases in length. When using a STR to compare genomes at a single locus, the length of the core repeat unit itself is generally constant between individuals, but the number of consecutive repeats in tandem will vary, and is determined from parental lineage. When generating STR profiles for forensic purposes, a number of different STR regions are compared to the database or a sample associated with a crime. The probability that two individuals have the same pattern of STR repeats across multiple loci decreases as the number of STRs compared increases. Across a dozen loci, a random match is extremely unlikely, as the level of discrimination of such a genetic profile generally has a resolution comparable to individualization.

In 1994, the DNA Identification Act was passed, authorizing the Federal Bureau of Investigation (FBI) to develop a national database of STR profiles. STRs were selected as the genetic markers of choice because they have a robust discriminatory power, can be rapidly separated based on size using PCR and capillary electrophoresis. For genetic profiling of individuals, the United States employs 13 consensus STR repeats, all with either tetra- or penta-nucleotide repeat units, plus the gender determinant amelogenin. The FBI coordinates a database system termed the Combined DNA Index System (CODIS) which is comprised of millions of forensic or criminal STR profiles and has aided tens of thousands of investigations.[3] While STRs have proven their utility in forensic identifications, related applications exist where further sample characterization may be necessary and other genetic markers can provide additional clarity. One set of markers with additional utility are those located on the Y-chromosome. Y-chromosome analysis is particularly valuable in sexual assault cases or issues involving paternal lineage. As the overwhelming majority of violent crimes are committed by males, further discrimination of genetic markers on this chromosome will provide higher confidence levels in prosecution of such crimes. A second set of markers directed at mitochondrial DNA (mtDNA) would also significantly contribute to forensic identifications. In particular, situations where nuclear DNA cannot be fully recovered would benefit, such as missing person cases examining skeletal remains or hair shafts, or any instance where DNA has been severely damaged or degraded. Mitochondrial DNA can be extracted from these types of materials. These genetic markers should be considered complementary to standard STR typing and have compelling rationale to be included in the analysis of DNA for forensic applications.

Probability

An important aspect of DNA analysis in forensic applications is the ability to statistically determine the certainty associated with a DNA profile. Each polymorphism explored has a natural level of occurrence, or allele frequency, which can be expressed as a probability, or more specifically, the likelihood that a match could occur randomly. Current STR methodology allows for an extremely high level of match certainty, with a full STR profile generally relating to a resolution probability at the level of a single person. The level of resolution is based on the number of genetic markers interrogated and their relative frequencies. The relative frequencies are estimates based on population sampling, to establish the extent to which a marker can occur in the overall population. In cases where individual determination is not possible, additional genetic markers are required. The overarching goal in forensic DNA analysis is to optimize the markers employed to provide the highest resolution possible, while minimizing the number of STRs interrogated. Recommendations for appropriate genetic markers are made by The Scientific Working Group on DNA Analysis

Methods, an organization of working scientists whose objective is to coordinate the work of the forensic DNA community. These genetic markers, in turn, provide a means to calculate the rarity of a genetic profile related to a criminal act. Ultimately, the trend has been to continually include greater numbers of genetic markers, as the inclusion of more STRs into a DNA profile would relate to a higher match probability. Because the inclusion of additional markers often comes at the sacrifice of time, expense, and throughput, novel technologies and methods are continually being developed and refined to deal with these issues. This chapter will detail the latest advances in forensic analysis of DNA in relation to microarray platforms. Microarrays are one such technology that addresses the need for rapid, cost-effective analysis in parallel, but have yet to cross into the mainstream of forensic DNA analysis.

Computation

A computational element exists that is relevant to DNA analysis in forensic science. The selection of appropriate genetic markers is a crucial factor in the probability of connecting a suspect to a crime. These markers are developed to provide the highest resolution possible, with a minimal number of markers. To aid in the identification and selection of relevant genetic markers, numerous computational methods are available. As these methods are of tremendous importance, a few fundamental efforts in this area will be briefly mentioned, but should not be considered the focus of this chapter.

Since the selection of appropriate genetic markers is the basis for profiling DNA specimens, any theoretical model to better elucidate their significance is of practical importance. The number of polymorphisms in the genome is massive, and the occurrence of linkage between genetic variations is known to occur. The sequence determination and the haplotyping of the human genome have now resulted in the compilation and annotation of libraries of genetic polymorphisms. In an effort to determine the individual and combined resolution of these polymorphisms, Krawczak et al. have proposed mathematical models as a measure to determine marker informativeness, while also accounting for spatial proximity associated with linkage disequilibrium.[4] Collins et al. have developed genome scanning methods that were capable of rapidly screening the entire draft sequence of the genome for all possible STRs.[5] Their rapid sequence comparison method not only identified SNPs and STRs, but provided a level of annotation to outline the average distances between markers to better gauge the potential linkage between them.

Every polymorphism has a potential to be used for discrimination purposes, but their individual value is determined by population studies. The knowledge garnered from computational analyses has obvious value in pre-screening polymorphisms to better understand their relevance in broader population analyses. Any high-throughput DNA analysis platform employed in forensics, and especially microarrays, has to initially undergo comprehensive analysis of the bioinformatic relevance not only between all genetic markers selected, but all sequences employed in the assay. Many techniques employ a level of multiplexing that involves dozens of sequences amplified and interrogated. Any cross amplification or interaction between loci sequences would severely complicate the ability to confidently profile a sample. With large numbers of probe sequences and PCR primers employed in a complete forensic DNA profile, this problem is not trivial.

III. Molecular Techniques

A common phenomenon in the fields of biodefense, medical diagnostics and forensic science, among others, is the adoption and acceptance of molecular techniques. Many of these fields are still anchored in traditional methods that have decades of expertise behind them and are considered definitive, or absolute. For example, in the detection of biowafare agents, culturing and organism growth are still considered the consensus method for organism identification. An enormous push has taken place to implement molecular diagnostic methods, such as real-time PCR, for rapid, early warning systems. While standard culturing can take days for some biothreat agents, real-time PCR can provide a positive result within a few hours. In many cases though, the real-time PCR positive result is only deemed a "presumptive positive" to provide first-responders initial caution and to begin basic institutional preparations, but until the positive result is confirmed with traditional methods, comprehensive actions in response to a pathogen is often delayed. This type of tiered response is born out of necessity. On one hand, to wait days to identify biothreat agent releases to the public could be considered illogical, impractical, and possibly immoral, as by then the organism can spread and any prophylactic treatment could be too late to be most effective. A counter issue would be the social and economical impact of multiple false warnings, such as repeated building or city evacuations, would have on the populace. A rapid, molecular identification of an organism allows an intermediate response to begin as soon as possible.

The field of forensic DNA analysis has also benefited enormously from molecular techniques not possible a decade ago. Even so, more rapid, more efficient technology is on the horizon. To take advantage of this technology, the forensic community must branch out of their comfort zone, which may mean building national, state, and local databases of SNPs alongside the STR databases that currently exist. Except for extreme instances, such as identification of remains at the World Trade Center after the terrorist attacks on September 11th, 2001, this has not yet occurred.

Technology

The sheer number of nuanced methodologies and formats available for performing molecular identification and DNA analysis is immense. Each of these techniques attempt to address some aspect related to the sensitivity, specificity, speed, throughput, cost and multiplex capabilities associated with the identification of polymorphic markers. The one commonality is the ability to multiplex, either in an amplification stage or an identification stage. To multiplex an assay, the specific method (amplification, identification, or both) is performed in parallel against the polymorphisms of interest. The parallel nature of these formats can be implemented in solution or on a surface, employing an array of specific sequences serving to interrogate the sample or act as primers for amplification. An excellent review by Sobrino et al. spells out the myriad of techniques and platforms that are being employed for SNP analysis.[6]

Oligonucleotide Microarrays

Microarrays are defined as a series of immobilized sequences used to address a sample in parallel. The traditional definition applies the target sample to be interrogated, to a series of immobilized probe oligonucleotides, resulting in a hybridization of the target molecules to the probe sequences on the surface. The first few applications to utilize microarrays measured RNA levels in gene expression and identified gene markers in disease diagnostics.[7-9] Before long, the high-throughput nature of microarray platforms allowed the technique to branch into a number of other applications, including sequencing, comparative genomic hybridization, and a plethora of other assays.[10-13] Multiple formats now exist that blur the traditional definition of a microarray by drawing on unique methods to multiplex the probe sequences, either employing them directly in solution or attaching them to microscale beads that are suspended in solution. Extensive reviews of microarray platforms are well documented and an in depth description is beyond the scope of this chapter.[14-18] A few of the major methods standard in microarray SNP analysis will be briefly discussed, allele specific hybridization (ASH), minisequencing, and allele specific primer extension (ASPE), as they are general in how arrays are utilized in forensic DNA analysis. ASH is simply the simultaneous, competitive hybridization of two DNA targets that differ at a single nucleotide

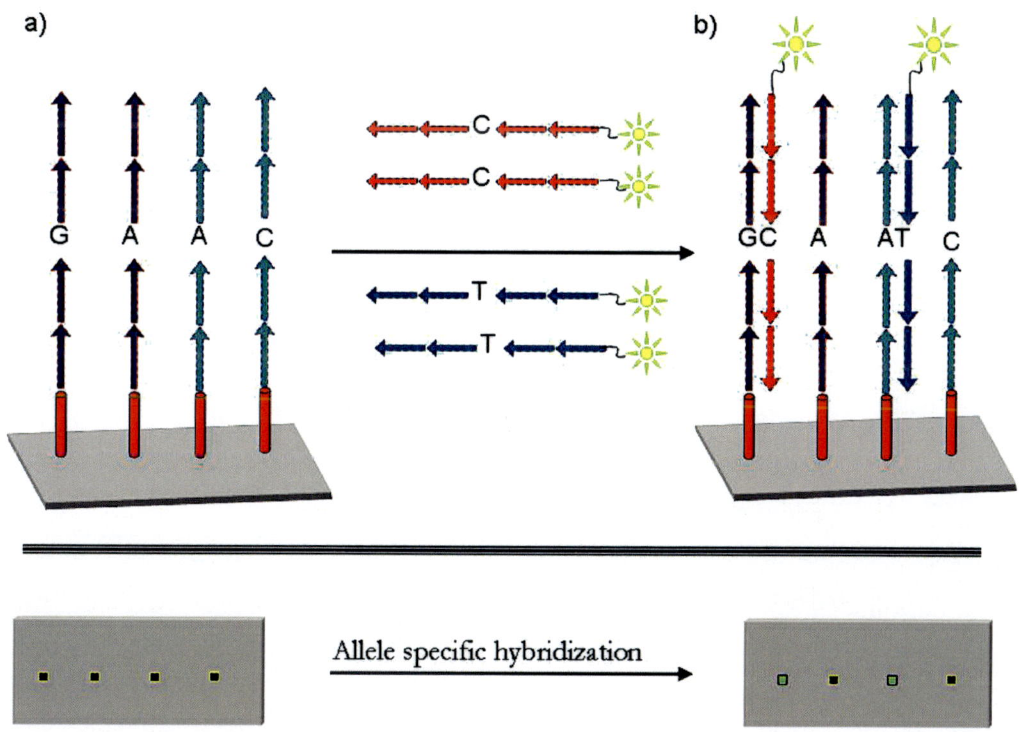

Figure 1. Allele specific hybridization. Competitive hybridization of target sequences (red and purple arrows, labeled with a green dye) to probe sequences immobilized onto a surface. a) The probe sequences are varied by a single SNP. b) Hybridization of labeled target indicates the correct SNP on the array.

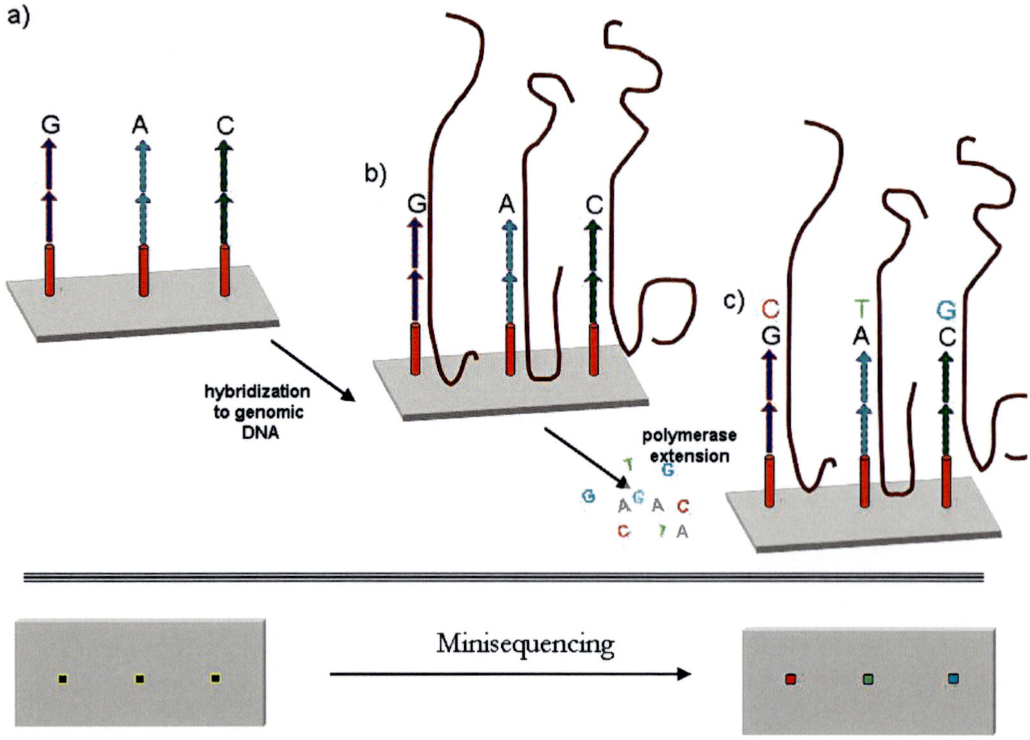

Figure 2. Minisequencing. a) Probe sequences immobilized on the surface are b) hybridized to genomic DNA (brown lines). The probe sequence serves as a primer for the genomic DNA template. c) Incubation with a polymerase and labeled dideoxynucleotides extends the primer sequence a single position, indicating the nucleotide complement to the genomic DNA at the SNP position.

position, such as a SNP position of interest (Figure 1). The polymorphic nucleotide is most commonly situated in the central region of the probe/target sequence and hybridization conditions allow only for the perfect match to remain bound to the probe sequence. Any probe-target hybrids with a mismatch would be less thermodynamically favored as compared to the perfect match. Another method commonly used in microarray analyses is minisequencing (Figure 2), which employs a polymerase to extend a primer/template duplex at the polymorphic position.[19] The primer binds the template immediately adjacent to the SNP position, and the polymerase incorporates the nucleotide complementary to the SNP. Many perturbations of the minisequencing method exist, but the simplest format would be the inclusion of a single dideoxynucleotide, labeled with a fluorophore, into the complementary SNP position. If four different indicators are linked to the four possible nucleotides, a nucleotide-specific response is determined at the SNP position. If an increased number of primers are simultaneously extended in this fashion, parallel determination of multiple SNPs is possible. The primers can be as an array format, immobilized directly on the surface with extension performed in the presence of hybridized target, or the entire assay can be performed in solution, and then followed by hybridization to an array. The ASPE method builds off the minisequencing concept (Figure 3). A primer extension occurs via a polymerase, as described previously, to determine the base pair complementary to the SNP. In this case though, multiple primers are employed, which terminate directly at the SNP position. The primer is extended beyond the SNP site, on only the primer where the DNA polymerase finds the perfect match. A second step is then performed, where the SNP is determined via hybridization to an array. These general microarray methods are all optimized for SNP determination. The adoption of microarrays into forensics has been limited due to the fact that STRs are less amenable to these microarray methods. STR variations are difficult to differentiate on arrays due to similar hybridization thermodynamics between repeat sequences of similar length.

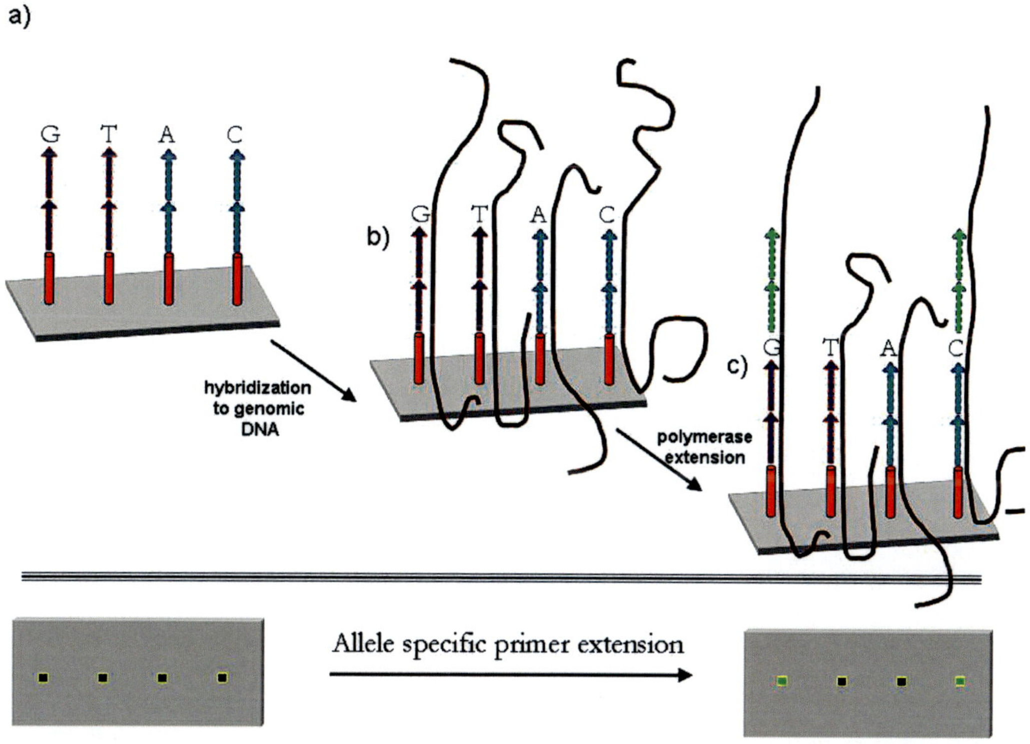

Figure 3. Allele specific primer extension. a) Probe sequences immobilized on the surface are b) hybridized to genomic DNA (brown lines). The probe sequence serves as a primer for the genomic DNA template. The terminal base on the primer is complementary to the SNP on the genomic DNA. c) Incubation with a polymerase and labeled labeled deoxynucleotides extends the primer sequence, indicating the nucleotide complement to the genomic DNA at the SNP position.

An important note is that a trend in microarray analysis is the continued miniaturization and the compatibility to microdevices, resulting in an overall decrease in sample requirement, while providing portability, higher throughput, increased speed, and better sensitivity. An issue with microarray formats though is that methods are generally considered destructive, and the sample material used in the assay is not recovered post-hybridization.

IV. Applications

Oligonucleotide microarray analyses in the field of forensic science have a number of applications. As each sub-field in forensics has different requirements and objectives, each also has different criteria that determine their utility. For example, to perform forensic analysis in a criminal case, substantial legal and ethical regulations exist to ensure the proper conditions are met. Where typing of a microbial pathogen or decomposed remains is performed, while still serious issues and requiring oversight, the requirements can often be much less rigorous as compared to criminal forensic applications. While criminal casework is very possibly the most common application in forensic DNA analysis, because of its specific requirements and objectives, criminalistics is possibly the least advanced in the incorporation of microarrays. This chapter will detail the current state of affairs of microarray analysis in forensic criminalistics and more specialized microarray applications such as microbial forensics and forensic anthropology, which are much more advanced in the adoption of microarrays.

A. Forensic Criminalistics

DNA analysis in forensic criminalistics, encompassing everything from the collection, examination, and identification of biological samples to crime laboratories, national databases, and the legal system, is based on STR analysis. So for oligonucleotide microarrays to provide full benefit, one of two things must occur, either adaptation of the technology to the field, or a blending of the field with the technology. The adaptation of microarray technology to forensic science has begun, where STR methodologies have already been incorporated into microarrays. The adjustment of microarrays to STR analysis, while problematic, is less complex because of the preference of the STR database to favor mid-level numbers of repeats (roughly 4-6 consecutive repeat units). Larger variations in the repeat unit lengths would provide a more difficult situation for incorporation into microarray analyses.

The opposite is beginning to happen as well, and the forensic community is also accommodating SNP analysis, and substantial forensic SNP work in currently being performed, albeit not centered on a crime laboratory database. STR analysis is viewed as the central tenet of forensic identification of DNA specimens, but not the complete answer, so assay development of specific Y-chromosome and SNP methods are gaining acceptance. While these types of methods are also not linked to the national crime lab databanks, they have demonstrated utility in unique situations, similar to how mtDNA analysis has evolved. For example, sexual assault and rape cases may involve the collection of a sample where the male (suspect) portion of the specimen is vastly underrepresented, as compared to the far greater level of female (victim) DNA. As the female genome is absent of Y-chromosome, focus on the recovery, identification, and match of a Y-chromosome profile to a suspect can better serve to link the culprit. The next two sections will describe the more recent advances in microarray analysis in forensic criminalistics for STR and SNP determination.

STR Capillary Electrophoresis Arrays

The standard processing of DNA specimens involves an isolation of the DNA, followed by a PCR amplification of particular STR loci. The STR amplifications are multiplexed, with up to 15 loci amplified in a single tube. Multiplexed amplification at this level is complex to perform in a single assay, and without the use of commercial kits, issues and possible cross amplification can occur. The amplifications are multiplexed with different reporter fluorophores and run on a capillary array electrophoresis instrument (CE). PCR fragments can be linked with up to five fluors and range in length from approximately 100 – 500 base pairs. The multiplex is then separated on the CE system for determination of the length of each STR. As this methodology is standard and has widespread usage, further detail will not be included herein, as other sufficient resources detail this method.[20-22]. While these types of CE separations are not technically oligonucleotide arrays in the classical sense, important advancements have been made in the field of CE in arrayed formats.

A number of multiplexed capillary electrophoresis systems have been developed. Yeung et al. employed a 96-channel CE system with microfabricated channels manufactured via photolithographic masks and chemical etching.[23] The microfabricated device, along with a high electrophoretic field enabled extremely fast separations and correct typing of STRs. Employing commercially available PCR multiplexing systems and a four-color labeling scheme, single base resolution was achieved in less than 30 minutes in all 96 wells simultaneously. The system compared favorably with standard CE instruments in terms of limits of detection and mixture analysis. Gao et al. have also employed a 96-capillary system for analysis of four STRs.[24] The results showed rapid, high resolution genotyping is possible in a high-throughput fashion. Dozens of runs were repeated with no loss in performance. Koumi et al. performed an evaluation of CE arrays as compared to slab-gel CE systems.[25] CE arrays were evaluated in terms of the precision, sensitivity, and genotyping capability. The array formats demonstrated increased speed and reduced hands on effort, while allowing an appropriate resolution for forensic casework. These CE array formats all increased the throughput by multiplexing the samples analyzed in parallel, but did not increase the number of polymorphisms addressed per sample.

STR Oligonucleotide Microarrays

The unnatural adaptation of technology to a specific application can be problematic. Many of the initial applications well suited for microarray analysis, such as gene expression studies and detection assays determining the presence/absence of specific gene targets, were developed because of the more amenable nature of the experiment to the technology. Further development has enabled microarray analysis for differentiation of specific polymorphisms, such as SNPs, by adjusting the stringency conditions of the assay. At first, a single base mismatch may seem to be the most problematic type of assay, as a single SNP represents the least amount of difference between two sequences. In actuality, situations like STRs, where multiple sequence length variations occur, along with several potential stable hybridization duplexes, can be even more complex. Microarray analysis of STRs are situations where an immobilized probe sequence, used to interrogate any number of repeats, would result in a myriad of similar hybridization efficiencies and a wide range of partial hybridization profiles. For an unknown, the potential exists for any number of target sequences to hybridize to the array, substantially complicating the determination of correct hybridization and correct repeat length. Although a difficult issue, a number of novel schemes have been developed to counter this inherent variation and allow for the determination of the repeat length of STRs via microarray analysis.

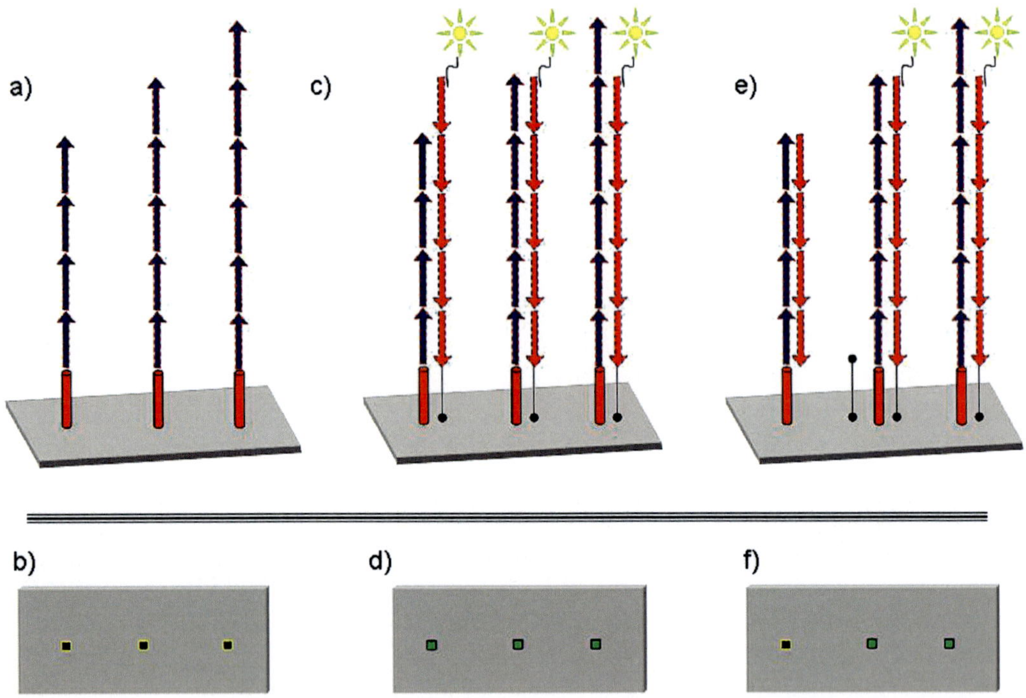

Figure 4. The Variable Length Probe Array concept, combining microarray hybridization of STRs and endonuclease treatment. a) Sequences complements to every possible repeat region are immobilized onto an array. Three complements, the tetra-, penta-, and hexa-repeats, are shown here for simplicity. A clamp sequence is included into the array design, shown as red cylinders. b) No hybridization signal is observed on the array. c) The labeled STR target is hybridized to the array, which non-specifically hybridizes to each position, d) creating a signal at every spot. e) After treatment with a single stranded nuclease, any target overhang is removed, f) creating signal on probe sequences with only the perfect STR complement length and above.

A format to perform STR microarray analysis, termed, variable length probe array (VLPA), was developed by Kemp et al., that combines a hybridization step with an enzymatic assay, post-hybridization.[26] The array is designed to have probe sequences complementary to every possible number of consecutive repeat, so would have a probe sequence complement to test for the di-repeat, tri-repeat, tetra-repeat, etc. on the array. To ensure proper target sequence alignment, a clamp sequence is included in the probe, adjacent to the repeat (Figure 4). For example, hybridization of a target STR sequence with five consecutive repeat units would result in three scenarios. The shorter probe sequences (less than penta-repeats) would result in a hybridization duplex with an overhang of the target penta-repeat sequence. The probe sequence complementary to the actual STR would align perfectly with the target repeat, and the probe sequences addressing longer repeats would have a probe overhang. Post hybridization, all probe sequences would then provide a signal based on the partial hybridization of the fluorescently-labeled STR target. The array is then subjected to an endonuclease to cleave any remaining single-stranded sequence overhang. The result of nuclease treatment is that any probe or target sequence not base-paired will be removed. With the example of the penta-repeat target sequence, post-nuclease treatment, the fluorescent signal will be measured at probe sequences addressing repeats of length five and above, as the fluorescent label on the target that is hybridized to shorter probe sequences will be cleaved by the nuclease. This VLPA ensures correct alignment of the STR with a clamp sequence and employs a novel signaling determination strategy, employing an enzymatic step. The work demonstrated the utility of this method for STR hybridizations to microarrays, but has not been applied outside of proof of principle, and did not address the difficulty involved in distinguishing the profile of a heterozygous profile, where two alleles would be present simultaneously.

Building off the VLPA clamp sequence, Pourmand et al. have also performed successful adaptation of STR analysis with microarrays.[27] In a similar fashion to Kemp et al., the clamp sequence was employed in the target hybridization of STRs, followed by two subsequent competitive hybridizations, rather than the enzymatic nuclease step described above. A schematic diagram of the competitive hybridization protocol is included in Figure 5. This microarray was again designed with probe sequences complementary to all possible repeat lengths. The clamp sequence ensures correct alignment of the target STR to the probe sequence during hybridization. Following target hybridization, the competitive hybridizations were performed to remove from the array, sequences with any unpaired single-stranded DNA, similar to the branch migration process of DNA recombination. The competitive hybridization process addresses any single-stranded sequence. Any target sequence extending over the probe sequence, meaning a longer target repeat length, and any probe sequence overhang, meaning a shorter target repeat length, was removed. Only target STRs of identical length to the probe sequence remained, signaling the proper STR length. This format was applied to the typing of two STRs from 20 samples, and can be readily scaled up to be directed at additional repeat sequences simultaneously. This assay type, performed on custom arrays, allowed the identification of STRs and represents the latest advancement in forensic analysis.

Another platform has successfully demonstrated the ability to interrogate STRs via and electronic microarray.[28] Radtkey et al. used high-stringency sandwich hybridizations on a system designed to facilitate rapid transport of target molecules to the array surface. This system employed an electric field to expedite delivery of the target sequences to the electrode/probe positions. The electric field not only concentrated the target at the probe, but also added a measure of fidelity to the hybridization, essentially allowing discrimination down to the level of single-base mismatch.[29] The platform was applied to hybridization of both tri-repeat and tetra-repeat sequences and performed with 100% concordance. This platform provided rapid analysis (<5 minutes) and alleviated the need to thermal or chemical treatment to ensure stringent hybridization.

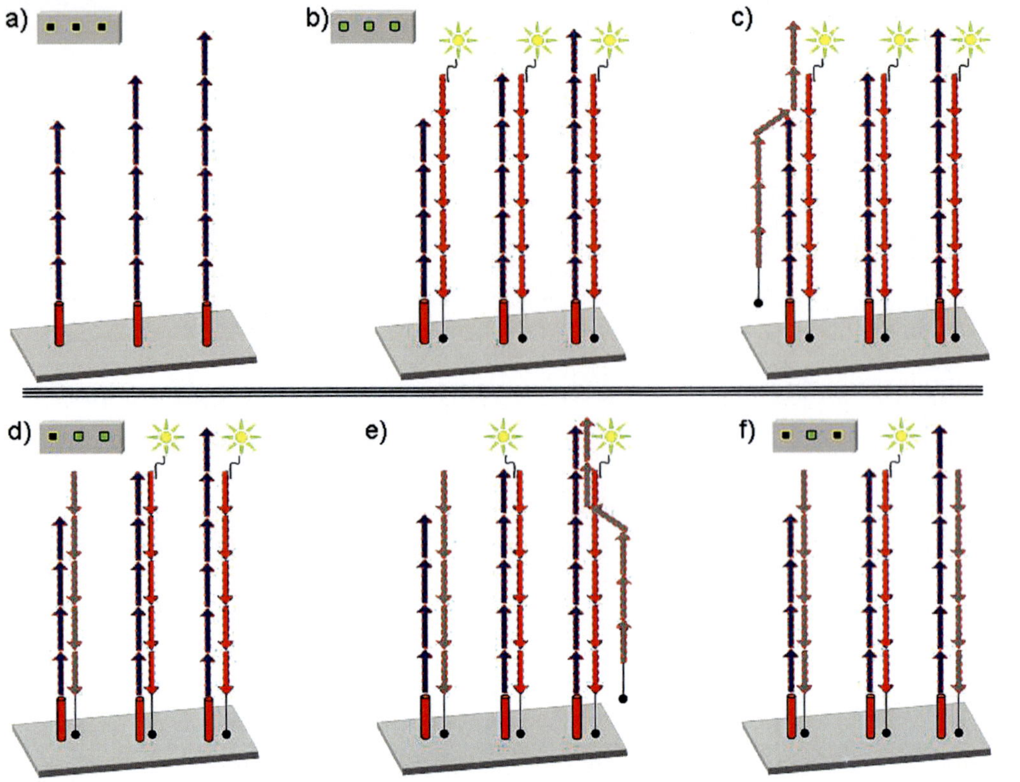

Figure 5. Microarray hybridization of STRs with competitive hybridization. a) Sequences complements to every possible repeat region are immobilized onto an array. Three complements, the tetra-, penta-, and hexa-repeats, are shown here for simplicity. A clamp sequence is included into the array design, shown as red cylinders. No hybridization signal is observed on the array. b) The labeled STR target is hybridized to the array, which non-specifically hybridizes to each position, creating a signal at every spot. c), d), and e) A series of competitive hybridizations are performed with a single stranded sequence (grey sequences), that remove any sequence but that of the perfect complement, f) leaving signal on probe sequences of only the perfect STR complement length.

SNP Oligonucleotide Microarrays

The utility of the microarray format has allowed the most progress in SNP determination. Forensic SNP analysis is not expected to replace or rival STR analysis in the near future, which is the genetic marker around which federal databases are built, but complement STR work and excel in areas where STR identification is problematic. STR analysis has limitations in further typing the Y-chromosome or with mtDNA applications. SNP analysis also has the advantage of low mutation rates and the ability to address short stretches of DNA, as low as 100 base pairs or less, which is important for analysis of degraded samples. SNP typing can be performed with a number of advanced molecular techniques, many of which have only been applied to forensics preliminarily. Advanced molecular techniques tend to be less labor intensive and more appropriate for rapid analysis, which is a benefit to a field already backlogged in sample processing. SNP markers are however, less informative than STR markers, resulting in the need to process more SNPs to provide the same level of resolution as a STR profile. An important point to note though is that the need to incorporate more SNP markers to achieve the same level of informativeness is countered in part, by the ability of SNP identification to be performed on many different high-throughput formats, in particular array-based platforms. Many of these platforms and techniques have recently demonstrated their ability to provide rapid, cost-effective analysis of multiple SNPs simultaneously. A number of excellent reviews exist on the topic of advanced techniques that may have further impact on forensic DNA analysis.[6, 30] Advances in array-based platforms used in SNP analysis of forensic samples will be discussed below.

Lareau and coworkers have published work employing the Affymetrix array system for SNP analysis of forensic samples.[31, 32] Directed at the Y-chromosome, two assays targeting 29 and 32 SNPs were developed. These low density arrays were fabricated by spotting allele specific probe sequences on slides. The target samples were prepared by multiple multiplex PCR preparations, which were then pooled for a final hybridization to the array. The methodology to interrogate the SNPs included both allele specific oligonucleotide hybridizations and minisequencing (single base extension) formats. Because a single color reporter was employed, a series of separate arrays was used to interrogate the presence of each base at the SNP region. This format essentially requires four different hybridizations per SNP, one for each possible nuclear base.

An array platform used to address 30 SNPs in the mtDNA region of the genome was performed by Nata and Hashiyada.[33] The array format employed was novel in that the probe sequences are not immobilized as traditionally performed with microarrays, but were affixed to microscale beads and used in solution. The beads, each coated with a specific probe DNA sequence, hybridized to the target sequences while free in solution and were interrogated post-hybridization via a flow cytometer. When a hybridized bead passed the detector, two signals were processed: one directly from the bead to determine the probe sequence that was being addressed, and a second color to determine the target SNP. Greater than 100 unrelated individuals were interrogated in this fashion, resulting in 50 different mtDNA genotypes in total, and 35 genotypes that were unique to the population interrogated. This solution-based array format, developed by Luminex Corporation, was designed to rapidly screen polymorphisms in a multiplexed fashion. Vallone and Butler have also employed the Luminex bead suspension array platform to examine 50 SNPs against 229 specimens.[34, 35] The array format was studied to determine the usefulness of SNPs on the Y-chromosome for forensic-related applications. Analysis of the specimens resulted in the identification of 18 unique haplogroups. A subset of samples was also used to directly compare the two methodologies commonly employed to detect DNA polymorphisms, allele specific hybridization and allele specific primer extension. The comparison resulted in 100% concordance across >2000 alleles between the assays, although a higher background signal noted with the ASH method could have potential issues in more complex or convoluted samples. In all, the assay consumed only 16 nanograms of DNA template, which is quite competitive to other pertinent typing formats.

A microarray for the analysis of SNP differences, directed at forensic applications, was performed by Divne et al.[36] A system based on solution-based minisequencing followed by hybridization to an array was performed to determine the polymorphisms associated with 12 nuclear SNPs and 21 mitochondrial SNPs in parallel. Primers annealed to the template sequence, at positions directly neighboring the SNP, and the minisequencing process was employed to extend the template sequence. The primers were extended only one position, at the SNP position, incorporating a single labeled dideoxynucleoside (ddNTP). The resulting single base extension product was hybridized to the array with a series of universal tags that were incorporated into the primer design. The universal tag system allowed the experimental design to be tailored to the polymorphisms without significant re-engineering of the array platform. The system addressed both coding regions and non-coding regions of the hypervariable region of the mitochondrial sequence, and was amenable to upscaling to statistically significant levels of SNP markers. This process did demonstrate the principle well on a smaller scale, but as the authors note, further development is required to increase the accuracy and reproducibility to levels acceptable with the field of forensics.

A relative tour-de-force of SNP genotyping on a microarray platform was produced by Sigurdsson et al.[37] The similar minisequencing format was used to interrogate 150 SNPs in the coding region of the mtDNA. A panel of 265 samples from populations throughout the world was used to differentiate almost 40,000 genotypes. The minisequencing reaction again employed the ddNTPs and a polymerase, in a four-color format, greatly reducing the overall size of the microarray required for analysis. The minisequencing primers all had a 5'universal tags which were then used to hybridize to the array. The array itself was divided into a series of separate arrays, created with 80 subarrays on a single surface (Figure 8 – array of arrays), enabling simultaneous hybridization of up to 80 samples, and each subarray was run against the panel of SNP targets. A measure of the genotypic diversity, or resolution power of the assay, was performed for the SNP panel as compared to that based on the entire sequence of the hypervariable region of the mtDNA, with comparable results. In addition, increased discrimination was possible for individuals with identical hypervariable regions. The genotyping results were used to predict the geographical origin of samples based on the maternal population of each specimen.

Quite possibly the most comprehensive method to address the polymorphisms in the mtDNA genome would be to sequence the entire region via a sequencing microarray. Maitra and coworkers employed an Affymetrix microarray product, the MitoChip, to sequence the entire ~16 kilobase mitochondrial sequence for cancer diagnostics.[38] While not used in forensic analysis, the MitoChip is essentially capable of sequencing >29 kb of DNA in a single experiment. The throughput of this format is such that ~84% of the complementary strand of the mtDNA can be analyzed in the same assay, based on 300 ng of genomic template. Vallone et al. evaluated the most recent iteration of the Affymetrix GeneChip Mitochondrial Resequencing Array.[39] The platform was found to have a high-throughput capacity and the capacity to detect many polymorphisms in a format suitable for numerous specific applications. However, the system did not provide full coverage of the mtDNA and had a relatively high requirement of starting material that would currently limit the platform's utility in more general forensic applications.

Tremendous growth in the science, technology, and instrumentation involved with forensic SNP analysis of DNA has occurred over the past few years. While numerous methods have been developed and applied to the problems associated with forensic analysis, no central methodology for SNP typing predominates. Each format has different strengths and limitations related to the sensitivity, assay speed, flexibility and reproducibility. A major focus of any new technique applied to DNA analysis must efficiently address the levels of multiplexing and throughput associated with forensic analysis to truly realize potential in the field.

B. Microbial Forensics

Microbial forensics encompasses the traditional ideology of microbiology, but with a focus of more modern molecular diagnostic techniques to provide a more detailed classification of pathogenic microorganisms and their origins.[40] By far, the incorporation of microarray technology has advanced furthest in the field of microbial forensics. Ironically, much of the advancements in identifying pathogens can be linked to the fact that no central oversight exists. While criminal forensics has created databases that have proved enormously beneficial in terms of apprehending repeat offenders and habitual criminals, a central database can also be restrictive if it is exclusive. This dichotomy stems from the fact that criminal databases are exclusively STR based, while microbial forensics is not tied directly to a central method or type of genetic marker, so can take full advantage of competing or complementary technologies. The increased usage of microarrays for pathogen identification is also due to the challenge of typing a microbe down to the strain level, while not trivial, can often be a more straightforward problem than individualization of a human sample.

A number of variables contribute to the ability to type microorganisms, including smaller genomes, more readily available sequence information, the current knowledge and development of specific genetic markers, and most importantly, the limited restrictions regarding the use of any advanced techniques. Because of these factors, conclusive identification can be determined with a few representative sequences. Detection of pathogenic microorganisms, including bacteria, viruses, protozoa and their representative toxins, is relevant to bioterrorism, contaminated food and water supplies, and public health in general. A need exists in each of these fields to identify and differentiate closely related strains with a high level of statistical confidence. Traditional microbial identification employs conventional techniques like organism growth/culturing and gel electrophoresis analysis as standard methods. But the difference between methods used to identify microbes and identify people is stark, and molecular techniques, such as real-time PCR and/or microarrays are much more common for pathogen identification. Molecular techniques provide faster, more definitive analysis, in a high-throughput fashion, testing multiple organisms simultaneously for multiple gene targets. Microarrays can incorporate enough markers to subtype multiple strains on a single platform, and novel strategies can be employed to take full advantage of a microarray's speed and utility.[41-43] As can be seen from the following examples, the performance of microbial arrays, far surpasses their inclusion in criminal forensics.

A microarray developed by Wilson et al. for microbial identification involved >50,000 oligonucleotide probes for the identification of 18 different pathogens.[44] Specific pathogens targeted in the array were *Bacillus anthracis*, *Clostridium botulinum* and *Yersinia pestis*, among others. This multi-pathogen identification microarray is unique in that it is directed at prokaryotic, eukaryotic and virus species on a single platform, covering 142 specific diagnostic regions of the pathogens. Each amplified diagnostic region targeted virulence genes or other markers associated with pathogenesis, as well as other previously uncharacterized regions. The assay encompassed close to 200 probe sequences per diagnostic region, and almost 1500 probes per organism. The microarray was capable of detection down to 500 femtograms of DNA, using spiked samples.

Wilse et al. employed a random 9-mer microarray with the intent to profile pathogens in an unbiased fashion, with no actual sequence specificity to microorganisms.[45] The unbiased approach was used to develop a protocol for statistical classification. The development of statistical protocols will be important in microbial forensics, as while many microarray fingerprints can be distinct, with emerging strains and closely related organisms, an incomplete profile could result, leading to incorrect classification. This current microarray protocol was directed at the differentiation of *Salmonella enterica* strains, having high sequence homology. The array was comprised of 192 different probe sequences and was used against 25 *Salmonella* isolates. The resulting fingerprint library was successful in differentiating many of the isolates, but a few distinct hybridization patterns resulted that were not representative of any of the initial classifications. This phenomenon of unrepresented subtypes would be expected for any organism that exists as a part of a substantially larger, closely-related pool.

A similar approach was performed by Chandler et al., who prepared custom microarrays of 9-mers to determine fingerprint profiles of *Bacillus* species down to the strain level.[46] The arrays consisted of 400 probe sequences generated randomly based on the *Escherichia coli* K12 genome. A dozen different organisms were hybridized to the arrays including *B. cereus*, *B. thuringiensis*, *B. anthracis*, as well as other non-*Bacillus* species. The protocol again used PCR as an initial step to selectively amplify the target sequences that would be hybridized to the array. Sub-typing down to strain level is an important issue, as numerous non-*anthracis*, *Bacillus* species exist, as well as non-pathogenic *B. anthracis* strains. The work outlined in this research demonstrated the utility of microarrays by differentiating organisms that would be problematic for gel analyses to distinguish.

Another microarray format was used to interrogate the genomic diversity across different strains of *E. coli* O157:H7. Jackson et al. employed a microarray to identify polymorphisms over 60,000 base pairs via a series of overlapping 29-mer probes.[47] A single chromosome, representing approximately 1% of the genome, was analyzed at 60 different regions, each 1,000 nuclear bases in length. The probe sequences were tiled, with each subsequent 29-mer overlapping the adjacent probe sequence by 24 nuclear bases. The intent was the identification of SNPs randomly distributed in the genome, without prior knowledge of their positioning. The microarray was prepared in a multi-well format, essentially allowing 12 simultaneous hybridizations, or 12 different strains, to be analyzed on a single photolithographic array, each with ~13,000 probe sequences. A total of 44 different isolates were tested and resulted in the identification of >150 SNP markers. The SNP pattern from this study showed unbalanced distribution, with some strains having no SNPs, while other strains exhibited hypervariable SNP profiles.

The major advantage of microarray analysis is the large degree of throughput that can be included in the assay design. The central goal of any forensic DNA application employing microarrays is to take advantage of the statistical, unequivocal certainty that can result from highly-parallel platforms. The large number of probe sequences can interrogate hundreds to thousands of sequences in parallel, directed down to the smallest of genomic polymorphisms. When functioning properly, the certainty provided by DNA analysis is irrefutable in the legal aspects of criminal prosecution. While the intent of this chapter is to describe and highlight the differences of how microarrays are incorporated into the different fields of forensic science, by no means is this a comprehensive description of such an enormous field of work, and other excellent research related to identification of pathogenic microorganisms exists.[48-54]

C. Forensic Anthropology

Numerous examples of forensic anthropology casework exist that employ microarrays. These situations deal with the origins, development, or biological characteristics associated with a population, or entail more specific situations, such as the identification of human remains. Human remains casework can involve missing individuals or situations related to mass graves and natural disasters, where multiple remains are discovered. The former situations are more global, requiring a broad population pool to screen for trends and variations based on geographical or physical influences, and are related to the social and cultural origins of a populace. Conversely, particularly with the identification of missing persons, the remains are generally compared with a known, such as a family member. For human remains or mass disasters, DNA analysis of limited recoverable biological material is often challenging because any DNA found has often been exposed to the harshest of conditions which severely limits exactly what tests can be performed. Microarray analyses have covered both ends of the spectrum related to forensic anthropology and have demonstrated their utility in the cases outlined below.

The SNP minisequencing methodology was applied to multiplexed typing of the Y-chromosome to evaluate genetic diversity in a Nordic population. Raitio et al. studied a panel of 25 SNPs and five microsatellites across five separate Finno-Ugric communities.[55] Greater than 10,000 genotypes and six separate haplotypes were determined in the study. The minisequencing format was performed in parallel, with the primers immobilized on a custom array made from a glass slide. A number of the SNPs analyzed were narrowly distributed and hinted at the presence of regional Nordic subpopulations. The genotype profile generated from these microarray hybridizations identified the most frequent shared haplotypes from distinct geographic regions and supported a hypothesis of two separate settlements to Finland.

A similar study was performed by Quintans et al., examining common Western European haplotypes based on 17 mtDNA SNPs.[56] Interestingly, the CE instrumental format was applied to SNP analysis, employing commercial multiplex reactions based on the minisequencing method. The dideoxynucleotide method was directed at the hypervariable regions of the mtDNA, applied to >250 samples, and analyzed with a four color assay and 16-lane CE. The capacity of this system, simultaneously interrogating 16 samples across 17 SNPs within 25 minutes, relates to close to 1,000 samples processed per day. The robustness of this method, along with the concept of applying STR technology to SNP analysis, is an excellent demonstration of how the future of DNA analysis in forensic science may evolve.

Forensic anthropological studies employing a microarray was performed for the identification of remains from mass graves in Croatia and Southern Bosnia.[57] Standard forensic DNA analysis employing STRs was performed where possible, but was successful in only 85% of the remains discovered. As STR analysis is performed with nuclear DNA, and nuclear DNA is often highly degraded in such situations, mtDNA analysis was performed on hair shafts and skeletal remains. This rapid haplotyping approach required PCR of the two hypervariable regions of the mtDNA. The PCR amplicons were hybridized to a microarray comprised of 27 probe sequences targeting eight regions of the mtDNA genome. The assay identified the remains of 105 missing individuals and produced 50 different haplotypes, where more than half were unique. The microarray analysis was also applied to the identification of mixed samples.

While the standard STR methods do not employ any type of array format, of particular importance to analysis of degraded samples, is the efforts to shorten the core STR amplicons. The practicality of STR analysis is diminished for analysis of damaged samples, as some STR regions interrogated have amplified regions consisting of ~500 base pairs. Degraded samples with fragmented template DNA will result in allelic dropout and partial STR profiles. The shortening of the STR amplicons would continue to allow nuclear DNA to be analyzed in such cases, avoiding the problems associated with mtDNA of the hypervariable regions, which have limited resolution power. Both Coble and Butler, and Opel et al. detail efforts to significantly reduce the size of the amplified STR regions to tolerate the highly degraded specimens.[58, 59] The original design of STR profiling included different amplicon lengths to take advantage of capillary electrophoresis separation based on size. The current improvements reduced the amplicons size, more beneficial to analyze degraded samples, but complicate the multiplexing levels previously achieved. The new, shorter primer sets could still align with the national STR database, and were applied to skeletal remains and directly compared to standard STR methods. A total of 31 specimens were tested with the shorter STR amplicons, called Miniplexes, generated a full profile 64% of the time, as compared to 16% success for the longer, standard STR system. The Miniplexes demonstrated their utility by resulting in complete profiles in 11 of 12 instances for all amplicons below 200 base pairs. The results presented in these works and others, illustrate that the current STR methods, while problematic in a few niche instances, can be improved upon or used in tandem with other methods to achieve results with a higher level of confidence and statistical significance. While STR analysis of the hypervariable regions of mtDNA is also ongoing, microarrays can move beyond these hypervariable regions for analysis, and target other polymorphisms that occur in mtDNA.

CONCLUSION

While DNA profiling for forensic applications has been in place for a decade, it still has not realized its full potential. The examination of DNA evidence is one of many important methods available to law enforcement and but commonly has principal influence in prosecution of violent crimes. Modern technology has been developed to aid in all areas of forensic DNA analysis, particularly in both the prosecution of criminals and the exoneration of innocent suspects. When used to its full capacity, technology can play a crucial role in advancing the goals of criminal justice, including prosecution and clearance rates. Because of forensic DNA analysis, enormous progress has been made in the certainty associated with convictions. But this progress has come with a cost. The definitive influence of DNA analysis also now makes it a relative necessity in determination of guilt or innocence, which has lead to enormous backlogs across the country on all levels, state, local, and federal. To aid jurisdictions, current technologies must better address the backlog and streamline throughput. The technology associated with forensic DNA analysis must address the existing reality and future potential requirements of criminal justice.

The forensic community has not yet taken advantage of the raw throughput power of microarrays. More advanced adoption of microarrays has occurred in microbial identification and subtyping, rather than other major veins of forensic science, mainly because there is no restriction on the technologies used. That is, because criminal forensics is married to the STR system and the CODIS databases, widespread adoption of high-throughput technologies have not been fully realized. While the current STR system is truly invaluable to the forensic community, it is also currently the central focus in a field that has important issues that need to be addressed, sooner rather than later. If methods to perform high-throughput forensic identification of DNA are available, such as SNP testing, they should soon be brought into context of inclusion into a national database as a complementary tool that can greatly benefit forensic science.

REFERENCES

[1] Venter, J.C., Adams, M.D., *et al.* 2001. *Science,* **291**: 1304-51.

[2] Consortium, T.I.H. 2007. *Nature,* **449**: 851-61.

[3] http://www.fbi.gov/hq/lab/codis/clickmap.htm, accessed 12/20/2007.

[4] Krawczak, M. 1999. *Electrophoresis,* **20**: 1676-81.

[5] Collins, J.R., Stephens, R.M., *et al.* 2003. *Genomics,* **82**: 10-9.

[6] Sobrino, B., Brion, M., *et al.* 2005. *Forensic Science International* 154: 181-94.

[7] DeRisi, J., Penland, L., *et al.* 1996. *Nat. Genet.* **14**: 457-60.

[8] Schena, M., Shalon, D., *et al.* 1996. *Proc. Natl. Acad. Sci. USA* 93: 10614-9.

[9] Schena, M., Shalon, D., *et al.* 1995. *Science,* **270**: 467-70.

[10] Hacia, J.G., Brody, L.C., *et al.* 1996. *Nat. Genet.* **14**: 441-7.

[11] Hacia, J.G. 1999. *Nat. Genet.* **21**: 42-7.

[12] Behr, M.A., Wilson, M.A., *et al.* 1999. *Science,* **284**: 1520-3.

[13] Perou, C.M., Jeffrey, S.S., *et al.* 1999. *Proc. Natl. Acad. Sci. USA* **96**(16): 9212-7.

[14] Epstein, J.R., Biran, I., *et al.* 2002. *Anal. Chim. Acta,* **469**: 3-36.

[15] Bally, M., Halter, M., *et al.* 2006. *Surface and Interface Analysis,* **38**: 1442-58.

[16] Park, H.G., Yang Song, J., *et al.* 2006. *Chemical Engineering Science,* **61**: 954-65.

[17] Wentzell, P.D. and Karakach, T.K. 2005. *Analyst,* **130**: 1331-6.

[18] Dufva, M. 2005. *Biomolecular Engineering* **22**: 173-84.

[19] Pastinen, T., Kurg, A., *et al.* 1997. *Genome Res.* **7**: 606-14.

[20] Butler, J.M., Buel, E., *et al.* 2004. *Electrophoresis,* 25: 1397-412.

[21] Butler, J.M. 2007. *BioTechniques, 43* **43**: Sii-Sv.

[22] Brettell, T.A., Butler, J.M., *et al.* 2007. *Anal. Chem.* **79**: 4365-84.

[23] Yeung, S.H.I., Greenspoon, S.A., *et al.* 2006. *Journal of Forensic Sciences,* **51**: 740-7.

[24] Gao, Q., Pang, H., *et al.* 1999. *Electrophoresis,* **20**: 1518-26.

[25] Koumi, P., Green, H.E., *et al.* 2004. *Electrophoresis,* **25**: 2227-41.

[26] Kemp, J.T., Davis, R.W., *et al.* 2005. *Journal of Forensic Science,* **50**(5): 1109-13.

[27] Pourmand, N., Caramuta, S., *et al.* 2007. *Proceedings of the National Academy of Sciences,* **104**: 6146-51.

[28] Radtkey, R., Feng, L., *et al.* 2000. *Nucl. Acids Res.* **28**: e17-.

[29] Huang, Y., Shirajian, J., *et al.* 2004. *Electrophoresis,* **25**: 3106-16.

[30] Butler, J.M. 2003. *Forensic Sci. Rev.* **15**(2): 91-111.

[31] Lareu, M., Sobrino, B., *et al.* 2003. *International Congress Series,* **1239**: 21-5.

[32] Sobrino, B., Lareu, M., *et al.* 2004. *International Congress Series,* **1261**: 331-3.

[33] Nata, M. and Hashiyada, M. 2004. *International Congress Series,* **1261**: 369-71.

[34] Vallone, P.M.B. and J.M. 2004. *Journal of Forensic Science,* **49**: 723-32.

[35] Vallone, P.M.B. and J. M. 2004. *International Congress Series,* **1261**: 85-7.
[36] Divne, A.M. and Allen, M. 2005. *Forensic Science International,* **154**: 111-21.
[37] Sigurdsson, S., Hedman, M., *et al.* 2006. *Genomics,* **87**: 534-42.
[38] Maitra, A., Cohen, Y., *et al.* 2004. *Genome Res.* **14**: 812-9.
[39] Vallone, P.M., Jakupciak, J.P., *et al.* 2007. *Forensic Science International: Genetics* **1**: 196-8.
[40] Pattnaik, P.,Jana, A.M. 2005. *Environmental Forensics* **6**: 197 - 204.
[41] Shepard, J.R.E., Danin-Poleg, Y., *et al.* 2005. *Anal. Chem.* **77**: 319-26.
[42] Shepard, J.R.E. 2006. *Anal. Chem.* **78**: 2478-86.
[43] Eom, H.S., Hwang, B.H., *et al.* 2007. *Biosens. Bioelect.* **22**(6): 845-53.
[44] Wilson, W.J., Strout, C.L., *et al.* 2002. *Molecular and Cellular Probes* **16**: 119-27.
[45] Willse, A., Straub, T.M., *et al.* 2004. *Nucl. Acids Res.* **32**: 1848-56.
[46] Chandler, D.P., Alferov, O., *et al.* 2006. *J. Clin. Microbiol.* **44**: 244-50.
[47] Jackson, S.A., Mammel, M.K., *et al.* 2007. *Forensic Science International* 168: 183-99.
[48] Call, D.R., Brockman, F.J., *et al.* 2001. *Int. J. Food Microbiol.* **67**: 71-80.
[49] Hoffmaster, A.R., Ravel, J., *et al.* 2004. *Proc. Natl. Acad. Sci. USA* **101**: 8449-54.
[50] Iqbal, S.S., Mayo, M.W., *et al.* 2000. *Biosens. Bioelect.* **15**: 549-78.
[51] Chizhikov, V., Rasooly, A., *et al.* 2001. *Appl. Environ. Microbiol.* **67**: 3258-63.
[52] Sergeev, N., Distler, M., *et al.* 2004. *Biosens. Bioelect.* **20**: 684-98.
[53] Gilbride, K.A., Lee, D.Y., *et al.* 2006. *Journal of Microbiological Methods* **66**: 1-20.
[54] Dorrell, N., Hinchliffe, S.J., *et al.* 2005. *Current Opinion in Microbiology* **8**: 620-6.
[55] Raitio, M., Lindroos, K., *et al.* 2001. *Genome Res.* **11**: 471-82.
[56] Quintans, B., Alvarez-Iglesias, V., *et al.* 2004. *Forensic Science International* **140**: 251-7.
[57] Gabriel, M.N., Calloway, C.D., *et al.* 2001. *Croatian Medical Journal* **42**(3): 328-35.
[58] Coble, M.D. and Butler, J.M. 2005. *Journal of Forensic Science* **50**(1): 43-53.
[59] Opel, K.L., Chung, D.T., *et al.* 2006. *Journal of Forensic Sciences* **51**: 351-6.

Index

B

C

D

E

F

G

H

I

J

K

L

M

P

Q

R

S

T

U

V

W

X

Y

Z